D0164361

Power System Analysis and Design

WITH PERSONAL COMPUTER APPLICATIONS

J. Duncan Glover / Mulukutla Sarma
Northeastern University

PWS-KENT
Publishing Company

20 Park Plaza
Boston, Massachusetts 02116

Copyright © 1987 by PWS Publishers.

All rights reserved. No part of this book may be reproduced, stored in a retrieval system, or transcribed, in any form or by any means—electronic, mechanical, photocopying, recording, or otherwise—without the prior written permission of PWS Publishers.

PWS-KENT Publishing Company is a division of Wadsworth, Inc.

Library of Congress Cataloging-in-Publication Data

Glover, J. Duncan.
Power system analysis and design.
1. Electric power systems—Data processing.
2. Microcomputers. I. Sarma, Mulukutla S.,
TK1005.G57 1986 621.31 86-25523
ISBN 0-534-07860-5

Printed in the United States of America

88 89 90 91 — 10 9 8 7 6 5 4 3

Sponsoring Editor: Robert Prior
Production Coordinator: Susan Graham
Production: Bernie Scheier/Miller • Scheier Associates
Manuscript Editor: Carol Dondrea
Interior Design: Susan Graham
Cover Design: Helen Walden
Cover Photo : © Bohdan Hrynewych/Stock Boston, Inc.
Interior Illustration: Carl Brown
Typesetting: Eyre & Spottiswoode
Printing and Binding: The Maple-Vail Book Manufacturing Group

621.31
G566p

To my wife Jaucione, to my children Tatiana and John Guedes, to my mother Eva, and to my dad in heaven, John Gladden Glover

PREFACE

The objective of this book is to present methods of power system analysis and design, particularly with the aid of a personal computer, in sufficient depth to give the student the basic theory at the undergraduate level. Our approach is designed to develop students' thinking process, enabling them to reach a sound understanding of a broad range of topics related to power system engineering, while motivating their interest in the electrical power industry. Because we believe that fundamental physical concepts underlie creative engineering and become the most valuable and permanent part of an engineering education, we have highlighted physical concepts while also giving due attention to mathematical techniques. Both theory and modeling are developed from simple beginnings so that they can be readily extended to new and complex situations.

This text is intended to be fully covered in a two-semester course offered to seniors and first-year graduate students. The organization of chapters and individual sections is flexible enough to give the instructor sufficient latitude in choosing topics to cover, especially in a one-semester course. The text is supported by an ample number of worked examples covering most of the theoretical points raised. There are numerous problems to be worked with a calculator as well as problems to be worked using a personal computer.

As background for this course, we assume that students have had courses in electric network theory (including transient analysis) and ordinary differential equations, and have been exposed to linear systems, matrix algebra, and computer programming. In addition, it would be helpful, but not necessary, to have had an electric machines course.

To help students orient themselves to the terminology, as well as serve as a brief review, we have included a chapter on Fundamentals. The chapter reviews phasor concepts and single-phase as well as three-phase circuits. Chapter 3 discusses Symmetrical Components and includes some applications. Chapters 4–6 examine Power Transformers, Transmission Line Parameters, and the Steady State Operation of Transmission Lines. Next are chapters on Power Flows, including the Newton-Raphson method, Symmetrical Faults, and Unsymmetrical Faults. The last chapters are on Power System Controls, including turbine-generator controls, load-frequency control, and economic dispatch; the Transient Operation of Transmission Lines, including surge protection; and Transient Stability, including the swing equation, the equal area criterion, and multimachine stability.

An innovative feature of this text is a personal computer software package that includes a user's manual and a diskette with computer programs for 9 of the chapters. The programs may be taught as an integral part of the course or may be omitted without loss of continuity. The programs are written in BASIC for the IBM and compatible personal computers. Operating instructions, flowcharts, and sample runs are given in the user's manual. Specific problems with a practical engineering design orientation to be solved with the help of a personal computer are included at the end of the chapters. The data-handling and number-crunching capabilities of the modern personal computer allow students to work on more difficult and realistic problems and make it an innovative tool in the learning process. The computer programs that accompany this text can also be used by students later in their careers and by professionals.

A second novel feature of this text is the up-front coverage of symmetrical components in Chapter 3, which enables modeling, on a three-phase basis, of power transformers in Chapter 4 and of transmission lines in Chapter 5. We have found that students are attracted to the matrix format and the resulting simplicity of symmetrical components, which enhances their interest in the use of the personal computer programs for power system design.

The third novel feature of this text is a self-contained chapter on transmission line transients. Just as methods for computing short circuit currents are important to circuit breaker selection and protective relaying, methods for computing transient overvoltages are a prerequisite to surge arrester selection and insulation coordination. After presenting the classical solution of transients, including traveling waves and the lattice diagram, we introduce discrete time models of transmission lines, which lead to the Transmission Line Transients computer program. The concepts of power system overvoltages, including lightning and switching surges as well as insulation coordination and surge protection, are also introduced.

Acknowledgments

The material in this text was gradually developed to meet the needs of the classes the authors have taught at universities in the United States and abroad over the past 20 years. The 12 chapters were written by the first author, J. D. Glover, with technical revisions provided by the second author, M. Sarma. The authors are indebted to many people who helped during the planning and writing of this book. The profound influence of earlier texts written on Power Systems, particularly by W. D. Stevenson, Jr., and the developments made by various outstanding engineers are gratefully acknowledged. Details of our sources can only be made through the references at the end of each chapter, as they are otherwise too numerous to mention. We are grateful to PWS electrical engineering editors Robert Prior and John Block for their broad knowledge and skills as well as diligence in publishing this text. The following reviewers have made many contributions to the final book: Frederick C. Brockhurst, Rose-Hulman Institute of Technology; Bell A. Cogbill, Northeastern University; Saul Goldberg, California Polytechnic State University; Mack Grady, University of Texas at Austin; Leonard F. Grigsby, Auburn University; Howard Hamilton, University of Pittsburgh; William F. Horton, California Polytechnic State University; W. H. Kersting, New Mexico State University; John Pavlat, Iowa State University; R. Ramakumar, Oklahoma State University; B. Don Russell, Texas A & M; Sheppard Salon, Rensselaer Polytechnic Institute; Stephen A. Sebo, Ohio State University; and Dennis O. Wiitanen, Michigan Technological University.

In conclusion, may we add that our objective in writing this text and the accompanying software package will have been fulfilled if the book is considered to be student-oriented, comprehensive, and up-to-date, with consistent notation and necessary detailed explanation at the level for which it is intended.

CONTENTS

LIST OF SYMBOLS AND UNITS

SYMBOL	DESCRIPTION
a	operator $1\underline{/120°}$
a_t	transformer turns ratio
A	area
A	transmission line parameter
$\boldsymbol{A}$	symmetrical components transformation matrix
B	loss coefficient
B	frequency bias constant
B	phasor magnetic flux density
B	transmission line parameter
C	capacitance
C	transmission line parameter
D	distance
D	transmission line parameter
E	phasor source voltage
E	phasor electric field strength
f	frequency
G	conductance
G	conductance matrix
H	normalized inertia constant
H	phasor magnetic field intensity
$i(t)$	instantaneous current
I	current magnitude (rms unless otherwise indicated)
I	phasor current
$\boldsymbol{I}$	vector of phasor currents
j	operator $1\underline{/90°}$
J	moment of inertia
l	length
ℓ	length
L	inductance
L	inductance matrix
N	number (of buses, lines, turns, etc.)
p.f.	power factor
$p(t)$	instantaneous power
P	real power
q	charge
Q	reactive power
r	radius
R	resistance

SYMBOL	DESCRIPTION
R	turbine-governor regulation constant
$\mathbf{R}$	resistance matrix
s	Laplace operator
S	apparent power
S	complex power
t	time
T	period
T	temperature
T	torque
$v(t)$	instantaneous voltage
V	voltage magnitude (rms unless otherwise indicated)
V	phasor voltage
$\boldsymbol{V}$	vector of phasor voltages
X	reactance
$\mathbf{X}$	reactance matrix
Y	phasor admittance
$\boldsymbol{Y}$	admittance matrix
Z	phasor impedance
$\boldsymbol{Z}$	impedance matrix

α	angular acceleration
α	transformer phase shift angle
β	current angle
β	area frequency response characteristic
δ	voltage angle
δ	torque angle
ϵ	permittivity
Γ	reflection or refraction coefficient
λ	magnetic flux linkage
λ	penalty factor
ϕ	magnetic flux
ρ	resistivity
τ	time in cycles
τ	transmission line transit time
θ	impedance angle
θ	angular position
μ	permeability
ν	velocity of propagation
ω	radian frequency

SI UNITS

A	ampere
C	coulomb
F	farad
H	henry
Hz	hertz
J	joule
kg	kilogram
m	meter
N	newton
rad	radian
s	second
S	siemen
VA	voltampere
var	voltampere reactive
W	watt
Wb	weber
Ω	ohm

ENGLISH UNITS

BTU	British thermal unit
cmil	circular mil
ft	foot
hp	horsepower
in	inch
mi	mile

CHAPTER 1

Introduction

Electrical engineers are concerned with every step in the process of generation, transmission, distribution, and utilization of electrical energy. The electric utility industry is probably the largest and most complex industry in the world. The electrical engineer who works in that industry will encounter challenging problems in designing future power systems to deliver increasing amounts of electrical energy in a safe, clean, and economical manner.

The objectives of this chapter are to review briefly the history of the electric utility industry, to discuss present and future trends in electric power systems, and to introduce a set of digital computer programs employed in power-system engineering that will be studied in later chapters.

Section 1.1

History of Electric Power Systems

In 1878 Thomas A. Edison began work on the electric light and formulated the concept of a centrally located power station with distributed lighting serving a surrounding area. He perfected his light by October 1879, and the opening of his historic Pearl Street Station in New York City on September 4, 1882, marked the beginning of the electric utility industry (see Figure 1.1). At Pearl Street, dc generators, then called dynamos, were driven by steam engines to supply an initial load of 30 kW for 110-V incandescent lighting to 59 customers in a 1-square-mile area. From this beginning in 1882 through 1972 the electric utility industry grew at a remarkable pace — a growth based on continuous reductions in the price of electricity due primarily to technological accomplishment and creative engineering.

The introduction of the practical dc motor by Sprague Electric, as well as the growth of incandescent lighting, promoted the expansion of Edison's dc systems. The development of three-wire 220-V dc systems allowed load to increase somewhat, but as transmission distances and loads continued to increase, voltage problems were encountered. These limitations of maximum distance and load were overcome in 1885 by William Stanley's development of a commercially practical transformer. Stanley installed an ac distribution system in Great Barrington, Massachusetts, to supply 150 lamps. With the transformer, the ability to transmit power at high voltage with corresponding lower current and lower line-voltage drops made ac more attractive than dc. The first single-phase ac line in the United States operated in 1889 in Oregon, between Oregon City and Portland — 21 km at 4 kV.

The growth of ac systems was further encouraged in 1888 when Nikola Tesla presented a paper at a meeting of the American Institute of Electrical Engineers describing two-phase induction and synchronous motors, which made evident the advantages of polyphase versus single-phase systems. The first three-phase line in Germany became operational in 1891, transmitting power 179 km at 12 kV. The first three-phase line in the United States (in California) became operational in 1893, transmitting power 12 km at 2.3 kV.

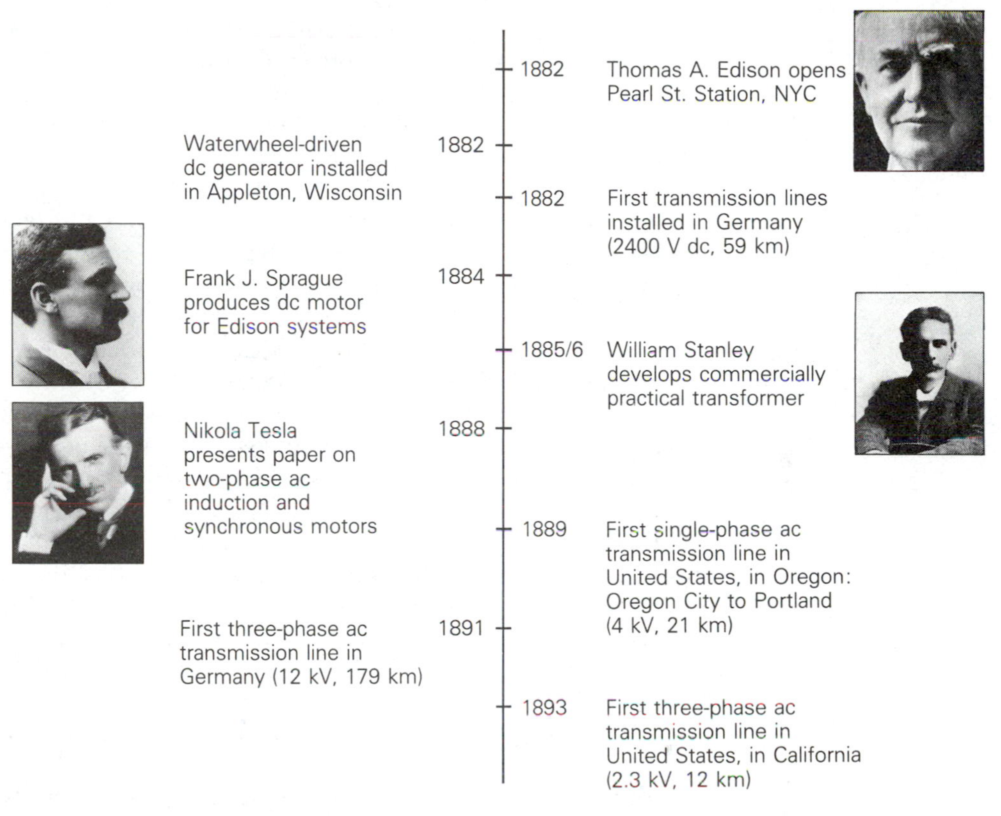

Figure 1.1 Milestones of the early electric utility industry [1]

The three-phase induction motor conceived by Tesla went on to become the workhorse of the industry.

In the same year that Edison's steam-driven generators were inaugurated, a waterwheel-driven generator was installed in Appleton, Wisconsin. Since then, most electric energy has been generated in steam-powered and in water-powered (called hydro) turbine plants. Today, steam turbines account for more than 85% of U.S. electric energy generation, whereas hydro turbines account for about 10%. Gas turbines are used in some cases during short periods to meet peak loads.

Steam plants are fueled primarily by coal, gas, oil, and uranium. Of these, coal is the most widely used fuel in the United States due to its abundance in the country. Although many of these coal-fueled power plants were converted to oil during the early 1970s, that trend has been reversed back to coal since the 1973/74 oil embargo, which caused an oil shortage and created a national desire to reduce dependency on foreign oil. In 1957 nuclear units with 90 MW steam-turbine capacity, fueled by uranium, were installed, and today nuclear units with 1280 MW steam-turbine capacity are in service. But the recent growth of nuclear capacity has been slowed by rising construction costs, licensing delays, and public opinion.

Other types of electric power generation are also being used, including wind-turbine generators; geothermal power plants, wherein energy in the form of steam or hot water is extracted from the earth's upper crust; solar cell arrays; and tidal power plants. These sources of energy cannot be ignored, but they are not expected to supply a large percentage of the world's future energy needs. On the other hand, nuclear fusion energy just may. Substantial research efforts now underway show nuclear fusion energy to be the most promising technology for producing safe, pollution-free, and economical electric energy in the next century and beyond. The fuel consumed in a nuclear fusion reaction is deuterium, of which a virtually inexhaustible supply is present in seawater.

The early ac systems operated at various frequencies including 25, 50, 60 and 133 Hz. In 1891, it was proposed that 60 Hz be the standard frequency in the United States. In 1893, 25-Hz systems were introduced with the synchronous converter. However, these systems were used primarily for railroad electrification (and many are now retired) since they had the disadvantage of causing incandescent lights to flicker. In California, the Los Angeles Department of Power and Water operated at 50 Hz, but converted to 60 Hz when power from the Hoover Dam became operational in 1937. In 1949 Southern California Edison also converted from 50 to 60 Hz. Today, the two standard frequencies for generation, transmission, and distribution of electric power in the world are 60 Hz (in the United States, Canada, Japan, Brazil) and 50 Hz (in Europe, the USSR, South America except Brazil, India, also Japan). The advantage of 60-Hz systems is that generators, motors, and transformers in these systems are generally smaller than 50-Hz equipment with the same ratings. The advantage of 50-Hz systems is that transmission lines and transformers have smaller reactances at 50 Hz than at 60 Hz.

As shown in Figure 1.2, the rate of growth of electric energy in the United

Figure 1.2

Growth of U.S. electric energy consumption [1,2,3]

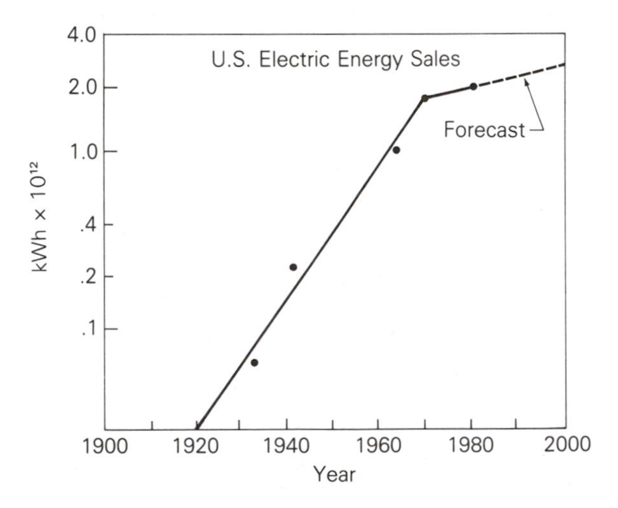

States was approximately 7% per year from 1902 to 1972. This corresponds to a doubling of electric energy consumption every 10 years over the 70-year period. In other words, every 10 years the industry installed a new electric

system equal in energy-producing capacity to the total of what it had built since the industry began. The annual growth rate slowed after the oil embargo of 1973/74, and from 1972 to 1982 kilowatt-hour consumption in the United States increased only 3% per year.

Along with increases in load growth, there have been continuing increases in the size of generating units (Table 1.1). The principal incentive to build larger units has been economy of scale — that is, a reduction in installed cost per kilowatt of capacity for larger units. However, there have also been steady improvements in generation efficiency. For example, in 1934 the average heat rate for steam generation in the U.S. electric industry was 17,950 BTU/kWh, which corresponds to 19% efficiency. By 1963 the average heat rate was 10,200 BTU/kWh, which corresponds to 33% efficiency. These improvements in thermal efficiency due to increases in unit size and in steam temperature and pressure, as well as to the use of steam reheat, have resulted in savings in fuel costs and overall operating costs.

Table 1.1

Growth of generator sizes in the United States [1]

HYDROELECTRIC GENERATORS		GENERATORS DRIVEN BY SINGLE-SHAFT, 3600 r/min FOSSIL-FUELED STEAM TURBINES	
SIZE	YEAR OF INSTALLATION	SIZE	YEAR OF INSTALLATION
MVA		MVA	
4	1895	5	1914
108	1941	50	1937
158	1966	216	1953
232	1973	506	1963
615	1975	907	1969
718	1978	1120	1974

There have been continuing increases, too, in transmission voltages (Table 1.2). From Edison's 220-V three-wire dc grid to 4-kV single-phase and 2.3-kV three-phase transmission, ac transmission voltages in the United States

Table 1.2

History of increases in three-phase transmission voltages in the United States [1]

VOLTAGE	YEAR OF INSTALLATION
kV	
2.3	1893
44	1897
150	1913
165	1922
230	1923
287	1935
345	1953
500	1965
765	1969

have risen progressively to 150, 230, 345, 500, and now 765 kV. And ultra-high voltages (UHV) above 1000 kV are now being studied. The incentives for increasing transmission voltages have been: (1) increases in transmission distance and transmission capacity, (2) smaller line-voltage drops, (3) reduced line losses, (4) reduced right-of-way requirements per MW transfer, and (5) lower capital and operating costs of transmission. Today, one 765-kV three-phase line can transmit thousands of megawatts over hundreds of kilometers.

The technological developments that have occurred in conjunction with ac transmission, including developments in insulation, protection, and control, are in themselves important. The following examples are noteworthy:

1. The suspension insulator
2. The high-speed relay system, currently capable of detecting short-circuit currents within one cycle (0.017 s)
3. High-speed, extra-high-voltage (EHV) circuit breakers, capable of interrupting up to 63-kA three-phase short-circuit currents within two cycles (0.033 s)
4. High-speed reclosure of EHV lines, which enables automatic return to service within a fraction of a second after a fault has been cleared
5. The EHV surge arrester, which provides protection against transient overvoltages due to lightning strikes and line-switching operations
6. Power-line carrier and microwave, as communication mechanisms for protecting, controlling, and metering transmission lines
7. The principle of insulation coordination applied to the design of an entire transmission system

In 1954 the first modern high-voltage dc (HVDC) transmission line was put into operation in Sweden between Vastervik and the island of Gotland in the Baltic sea; it operated at 100 kV for a distance of 100 km. The first HVDC line in the United States was the ±400-kV, 1360-km Pacific Intertie line installed between Oregon and California in 1970. As of 1983, three other HVDC lines up to ±400 kV had been installed in the United States, and a total of 24 HVDC lines up to ±533 kV had been installed worldwide.

For an HVDC line embedded in an ac system, electronic converters at both ends of the dc line operate as rectifiers and inverters. Since the cost of an HVDC transmission line is less than that of an ac line with the same capacity, the additional cost of converters for dc transmission is offset when the line is long enough. Studies have shown that overhead HVDC transmission is economical in the United States for transmission distances longer than about 600 km.

In the United States electric utilities grew first as isolated systems, with new ones continuously starting up throughout the country. Gradually, however, neighboring electric utilities began to interconnect, to operate in parallel. This improved both reliability and economy. Figure 1.3 shows major transmission in the United States in 1978: interconnected EHV lines as well as

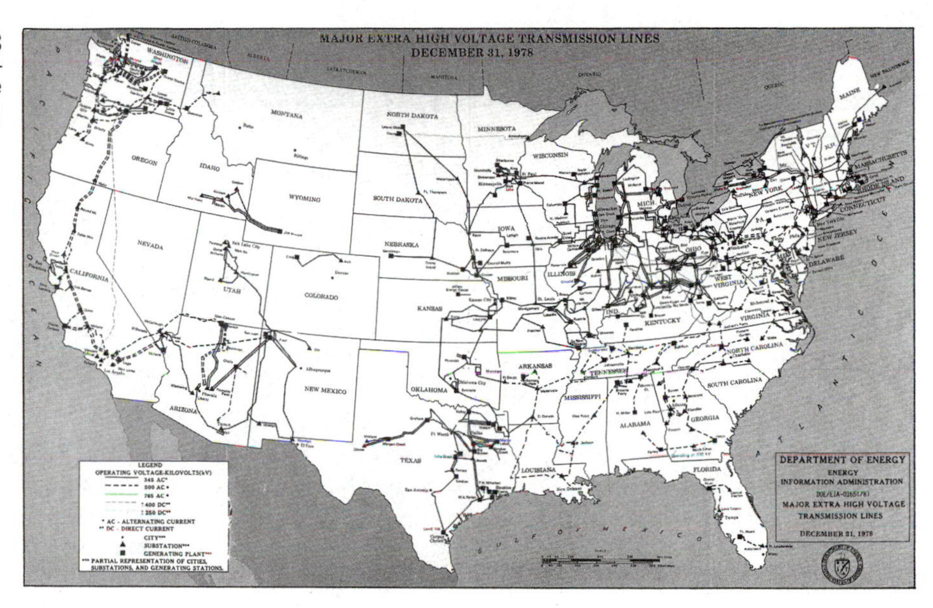

Figure 1.3

Major transmission in the United States — 1978 [4]

HVDC transmission lines. An interconnected system has many advantages. An interconnected company can draw upon another's rotating generator reserves during a time of need (such as a sudden generator outage or load increase), thereby maintaining continuity of service, increasing reliability, and reducing the total number of generators that need to be kept running under no-load conditions. Also, interconnected companies can schedule power transfers during normal periods to take advantage of energy-cost differences in respective areas, load diversity, time zone differences, and seasonal conditions. For example, utilities whose generation is primarily hydro can supply low-cost power during high-water periods in spring/summer, and can receive power from the interconnection during low-water periods in fall/winter. Interconnections also allow shared ownership of larger, more-efficient generating units.

While sharing the benefits of interconnected operation, each utility is obligated to help neighbors who are in trouble, to maintain scheduled intertie transfers during normal periods, and to participate in system frequency regulation.

In addition to the benefits/obligations of interconnected operation, there are disadvantages. Interconnections, for example, have increased fault currents that occur during short circuits, thus requiring the use of circuit breakers with higher interrupting capability. Furthermore, although overall system reliability and economy have improved dramatically through interconnection, there is a remote possibility that an initial disturbance may lead to a regional blackout, such as the one that occurred in 1966 in the northeastern United States.

Section 1.2

Present and Future Trends

Present trends indicate that the United States is becoming more electrified as it shifts away from a dependence on the direct use of fossil fuels. While U.S. consumption of all nonelectric forms of energy decreased by 15% from 1972 to 1982, electric energy consumption increased by 33%, which corresponds to an average growth rate of 3% per year during this period. As shown in Figure 1.2, the growth rate in the use of electric energy is projected to continue to increase at about 3% per year for the rest of this decade, and then gradually taper off by the year 2000 to about 2.5% per year. Although energy forecasts for the next 15 years are uncertain and based on economic and social factors that are subject to change, the predicted 3% annual growth rate over the next 10 years is considered necessary to generate the GNP anticipated over that period. The forecast that long-term annual energy growth will decline to 2.5% is based on predictions that the population growth rate will slacken, that conservation practices will continue, and that electrical energy will be used more efficiently. The cost of electricity, which is expected to rise somewhat higher than the inflation rate, is the major impediment to long-term growth.

Figure 1.4 shows the percentages of various fuels used to meet U.S.

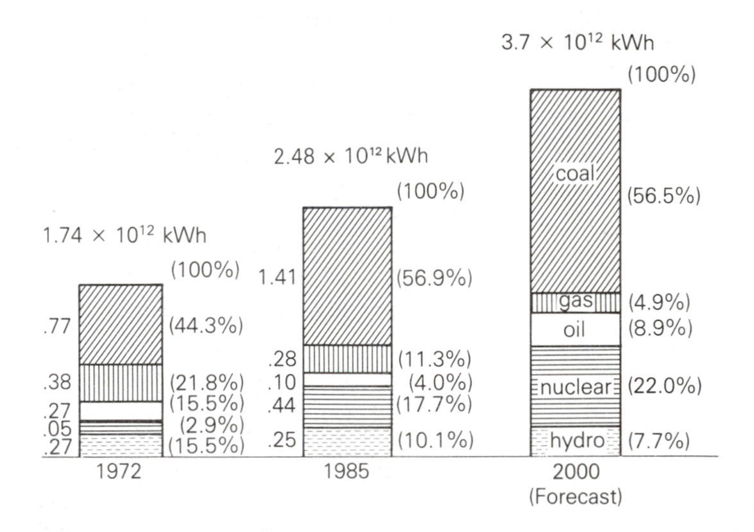

Figure 1.4

Electric energy generation in the United States, by principal fuel types [2, 3]

electric energy requirements over the recent past and those projected for the near future. Several trends are apparent in the chart. One is the continuing shift away from the use of gas and oil and toward the increasing use of coal. This trend is due primarily to the large amount of U.S. coal reserves, which, according to some estimates, is sufficient to meet U.S. energy needs for the next 500 years. An increase in nuclear fuel consumption is also evident; however, the increases seen in the figure are based only on the use of existing nuclear units and those presently under construction. Because of present financial, political, and social factors, no new orders for nuclear units are foreseen until after 1990. Another trend is the continuing percentage decline in projected hydroelectric energy consumption, which is based on the fact that major hydroelectric sites in the United States (except in Alaska) have been almost fully developed.

Figure 1.5

Installed generating capability in the United States, by principal fuel types [2,3]

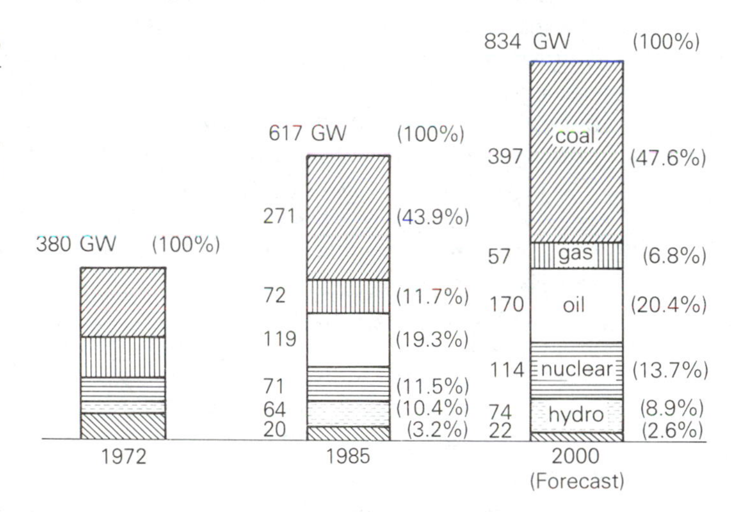

Figure 1.5 shows the recent past and projected generating capability in the United States, by principal fuel types. As shown, total generating capability is projected to reach 834 GW (1 GW = 1000 MW) by the year 2000. Due to presently high reserve margins, capacity problems will not be encountered through the 1980s. But adequate capacity throughout the 1990s is based on the forecast that a substantial number of large coal-fired units now in the planning stages will be brought on-line. Current lead times of about 10 years for licensing and construction of coal-fired units, however, may create insufficient reserve margins in some regions of the United States. Also, the projected total nuclear capacity of 114 GW by the year 2000 assumes that all units now under construction will be completed (by about 1991). Cancellation of any of these nuclear units would reduce this 114-GW figure.

Growth in transmission-line construction roughly correlates with generation additions. EHV transmission systems installed during the past 20 years are not yet loaded to peak transfer capacities, and should be capable of meeting anticipated load growth for the next 5 years. Thereafter, as new generation comes on-line and transmission loadings rise, transmission construction should increase.

Growth in distribution construction roughly correlates with growth in electric energy consumption. In absolute terms, distribution construction is projected to increase steadily during the next 15 years, but the annual growth rate should taper off by the year 2000, as the kWh consumption curve does in Figure 1.2.

Section 1.3

Computers in Power-System Engineering

As electric utilities have grown in size, and the number of interconnections has increased, planning for future expansion has become increasingly complex. The increasing cost of additions and modifications has made it imperative that utilities consider a range of design options, and perform detailed studies of the effects on the system of each option, based on a number of assumptions: normal and abnormal operating conditions, peak and off-peak loadings, and present and future years of operation. A large volume of network data must also be collected and accurately handled. To assist the engineer in this power-system planning, digital computers and highly developed computer programs are used. Such programs include power-flow, stability, short-circuits, and transients programs.

Power-flow programs compute the voltage magnitudes, phase angles, and transmission-line power flows for a network under steady-state operating conditions. Other results, including transformer tap settings and generator reactive power outputs, are also computed. Today's computers have sufficient storage and speed to compute in less than one minute power-flow solutions for networks with more than 2000 buses and 2500 transmission lines. High-speed printers then print out the complete solution in tabular form for analysis by the planning engineer. Also available are interactive power-flow programs, whereby power-flow results are displayed on CRTs in the form of single-line diagrams; the engineer uses these to modify the network with a light pen or from a keyboard and can readily visualize the results. The computer's large storage and high-speed capabilities allow the engineer to run the many different cases necessary to analyze and design transmission and generation-expansion options.

Stability programs are used to study power systems under disturbance conditions to determine whether synchronous generators and motors remain in synchronism. System disturbances can be caused by the sudden loss of a generator or transmission line, by sudden load increases or decreases, and by short circuits and switching operations. The stability program combines power-flow equations and machine-dynamic equations to compute the angular swings of machines during disturbances. The program also computes critical clearing times for network faults, and allows the engineer to investigate the effects of various machine parameters, network modifications, disturbance types, and control schemes.

Short-circuits programs are used to compute three-phase and line-to-ground faults in power-system networks in order to select circuit breakers for fault interruption, select relays that detect faults and control circuit breakers, and determine relay settings. Short-circuit currents are computed for each relay and circuit-breaker location, and for various system-operating conditions such as lines or generating units out of service, in order to determine minimum and maximum fault currents.

Transients programs compute the magnitudes and shapes of transient overvoltages and currents that result from lightning strikes and line-switching operations. The planning engineer uses the results of a transients program to determine insulation requirements for lines, transformers, and other equipment, and to select surge arresters that protect equipment against transient overvoltages.

Other computer programs for power-system planning include relay-coordination programs and distribution-circuits programs. Computer programs for generation-expansion planning include reliability analysis and loss-of-load-probability (LOLP) programs, production-cost programs, and investment-cost programs.

A computer software package, including a set of programs stored on floppy disk and a user's manual, accompanies this text. All the programs are written in BASIC to run on IBM and similar personal computers. Operating instructions, flow charts, and sample runs are given in the user's manual. The following sections in the text discuss the programs:

Section 2.5 Matrix Operations
Section 3.6 Symmetrical Components
Section 5.14 Line Constants
Section 6.8 Transmission Lines — Steady-State Operation
Section 7.8 Power Flow
Section 8.6 Symmetrical Short Circuits
Section 9.6 Short Circuits
Section 11.9 Transmission-Line Transients
Section 12.6 Transient Stability

These programs are intended to help users understand the process of power-system analysis and design. Specific problems with practical engineering design orientation that are to be solved with the programs are included in the text, along with conventional problems to be solved with only a calculator.

References

1. H. M. Rustebakke et al., *Electric Utility Systems Practice,* 4th ed. (New York: John Wiley and Sons, 1983).
2. North American Electric Reliability Council, *Electric Power Supply and Demand, 1985–1994* (Princeton, N.J., 1985).
3. "34th Annual Electric Utility Forecast," *Electrical World*, Vol. 197, No. 9 (September 1983): pp. 55–62.
4. U.S. Department of Energy, *Major Extra High Voltage Transmission Lines*, DOE/EIA-0165(78), December 31, 1978.

CHAPTER 2

Fundamentals

The objective of this chapter is to review basic concepts and establish terminology and notation. In particular, the following are reviewed: phasors, instantaneous power, complex power, and network equations. A set of matrix operations from the computer software package included with this text are described, and elementary aspects of balanced three-phase circuits are introduced. Students who have already had courses in electric network theory and basic electric machines should find this chapter primarily refresher material.

Section 2.1

Phasors

A sinusoidal voltage or current at constant frequency is characterized by two parameters: a maximum value and a phase angle. A voltage

$$v(t) = V_{max}\cos(\omega t + \delta) \tag{2.1.1}$$

has a maximum value V_{max} and a phase angle δ when referenced to $\cos(\omega t)$. The root-mean-square (rms) value, also called *effective value*, of the sinusoidal voltage is

$$V = \frac{V_{max}}{\sqrt{2}} \tag{2.1.2}$$

Euler's identity, $e^{j\phi} = \cos\phi + j\sin\phi$, can be used to express a sinusoid in terms of a phasor. For the above voltage,

$$\begin{aligned} v(t) &= \text{Re}[V_{max}e^{j(\omega t+\delta)}] \\ &= \text{Re}[\sqrt{2}(Ve^{j\delta})e^{j\omega t}] \end{aligned} \tag{2.1.3}$$

where $j = \sqrt{-1}$ and Re denotes "real part of." The rms phasor representation of the voltage is given in three forms — exponential, polar, and rectangular:

$$V = \underbrace{Ve^{j\delta}}_{\text{exponential}} = \underbrace{V\angle\delta}_{\text{polar}} = \underbrace{V\cos\delta + jV\sin\delta}_{\text{rectangular}} \tag{2.1.4}$$

A phasor can be easily converted from one form to another. Conversion from polar to rectangular is shown in the phasor diagram of Figure 2.1.

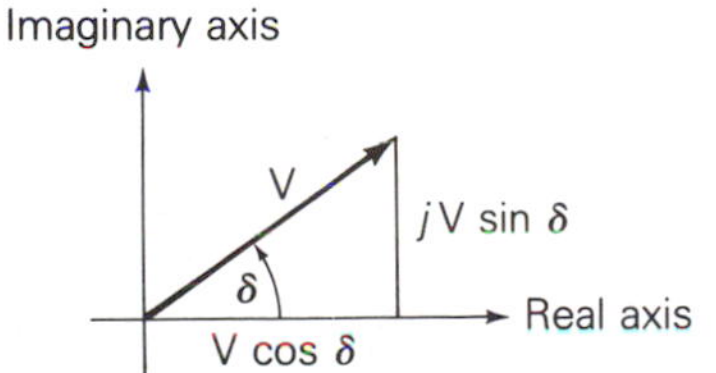

Figure 2.1 Phasor diagram for converting from polar to rectangular form

Euler's identity can be used to convert from exponential to rectangular form. As an example, the voltage

$$v(t) = 169.7\cos(\omega t + 60^\circ) \quad \text{volts} \tag{2.1.5}$$

has a maximum value $V_{max} = 169.7$ volts, a phase angle $\delta = 60^\circ$ when referenced to $\cos(\omega t)$, and an rms phasor representation in polar form of

$$V = 120\underline{/60^\circ} \quad \text{volts} \tag{2.1.6}$$

Also, the current

$$i(t) = 100\cos(\omega t + 45^\circ) \quad \text{A} \tag{2.1.7}$$

has a maximum value $I_{max} = 100$ A, an rms value $I = 100/\sqrt{2} = 70.7$ A, a phase angle of 45°, and a phasor representation

$$I = 70.7\underline{/45^\circ} = 70.7\,e^{j45^\circ} = 50 + j50 \quad \text{A} \tag{2.1.8}$$

The relationships between the voltage and current phasors for the three passive elements — resistor, inductor, and capacitor — are summarized in Figure 2.2, where sinusoidal-steady-state excitation and constant values of *R*, *L*, and *C* are assumed.

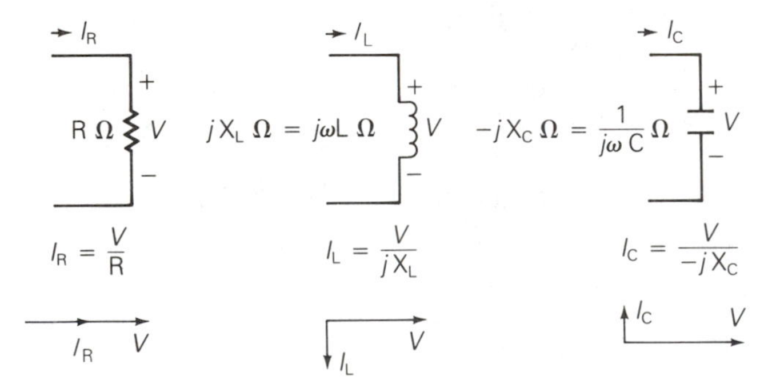

Figure 2.2 Summary of relationships between phasors *V* and *I* for constant R, L, and C elements with sinusoidal-steady-state excitation

When voltages and currents are discussed in this text, lowercase letters such as $v(t)$ and $i(t)$ indicate instantaneous values, uppercase letters such as V and I indicate rms values, and uppercase letters in italics such as *V* and *I* indicate rms phasors. When voltage or current values are specified, they shall be rms values unless otherwise indicated.

Section 2.2

Instantaneous Power in Single-Phase ac Circuits

Power is the rate of change of energy with respect to time. The unit of power is a watt, which is a joule per second. Instead of saying that a load absorbs energy at a rate given by the power, it is common practice to say that a load absorbs power. The instantaneous power in watts absorbed by an electrical load is the product of the instantaneous voltage across the load in volts and the instantaneous current into the load in amperes. Assume that the load voltage is

$$v(t) = V_{max} \cos(\omega t + \delta) \quad \text{volts} \tag{2.2.1}$$

We now investigate the instantaneous power absorbed by purely resistive, purely inductive, purely capacitive, and general RLC loads. We also introduce the concepts of real power, power factor, and reactive power. The physical significance of real and reactive power is also discussed.

Purely Resistive Load

For a purely resistive load, the current into the load is in phase with the load voltage, $I = V/R$, and the current into the resistive load is

$$i_R(t) = I_{Rmax} \cos(\omega t + \delta) \quad A \tag{2.2.2}$$

where $I_{Rmax} = V_{max}/R$. The instantaneous power absorbed by the resistor is

$$\begin{aligned} p_R(t) &= v(t) i_R(t) = V_{max} I_{Rmax} \cos^2(\omega t + \delta) \\ &= \tfrac{1}{2} V_{max} I_{Rmax} \{1 + \cos[2(\omega t + \delta)]\} \\ &= V I_R \{1 + \cos[2(\omega t + \delta)]\} \quad W \end{aligned} \tag{2.2.3}$$

As indicated by (2.2.3), the instantaneous power absorbed by the resistor has an average value

$$P_R = V I_R = \frac{V^2}{R} = I_R^2 R \quad W \tag{2.2.4}$$

plus a double-frequency term $V I_R \cos[2(\omega t + \delta)]$.

Purely Inductive Load

For a purely inductive load, the current lags the voltage by 90°, $I_L = V/(jX_L)$, and

$$i_L(t) = I_{Lmax} \cos(\omega t + \delta - 90°) \quad A \tag{2.2.5}$$

where $I_{Lmax} = V_{max}/X_L$, and $X_L = \omega L$ is the inductive reactance. The instantaneous power absorbed by the inductor is[1]

$$\begin{aligned} p_L(t) &= v(t)i_L(t) = V_{max}I_{Lmax}\cos(\omega t + \delta)\cos(\omega t + \delta - 90°) \\ &= \tfrac{1}{2}V_{max}I_{Lmax}\cos[2(\omega t + \delta) - 90°] \\ &= VI_L \sin[2(\omega t + \delta)] \quad \text{W} \end{aligned} \tag{2.2.6}$$

As indicated by (2.2.6), the instantaneous power absorbed by the inductor is a double-frequency sinusoid with *zero* average value.

Purely Capacitive Load

For a purely capacitive load, the current leads the voltage by 90°, $I_C = V/(-jX_C)$, and

$$i_C(t) = I_{Cmax}\cos(\omega t + \delta + 90°) \quad \text{A} \tag{2.2.7}$$

where $I_{Cmax} = V_{max}/X_C$, and $X_C = 1/(\omega C)$ is the capacitive reactance. The instantaneous power absorbed by the capacitor is

$$\begin{aligned} p_C(t) &= v(t)i_C(t) = V_{max}I_{Cmax}\cos(\omega t + \delta)\cos(\omega t + \delta + 90°) \\ &= \tfrac{1}{2}V_{max}I_{Cmax}\cos[2(\omega t + \delta) + 90°)] \\ &= -VI_C \sin[2(\omega t + \delta)] \quad \text{W} \end{aligned} \tag{2.2.8}$$

The instantaneous power absorbed by a capacitor is also a double-frequency sinusoid with *zero* average value.

General RLC Load

For a general load composed of RLC elements under sinusoidal-steady-state excitation, the load current is of the form

$$i(t) = I_{max}\cos(\omega t + \beta) \quad \text{A} \tag{2.2.9}$$

The instantaneous power absorbed by the load is then[1]

$$\begin{aligned} p(t) &= v(t)i(t) = V_{max}I_{max}\cos(\omega t + \delta)\cos(\omega t + \beta) \\ &= \tfrac{1}{2}V_{max}I_{max}\{\cos(\delta - \beta) + \cos[2(\omega t + \delta) - (\delta - \beta)]\} \\ &= VI\cos(\delta - \beta) + VI\cos(\delta - \beta)\cos[2(\omega t + \delta)] \\ &\qquad + VI\sin(\delta - \beta)\sin[2(\omega t + \delta)] \end{aligned}$$

$$p(t) = VI\cos(\delta - \beta)\{1 + \cos[2(\omega t + \delta)]\} + VI\sin(\delta - \beta)\sin[2(\omega t + \delta)]$$

Letting $I\cos(\delta - \beta) = I_R$ and $I\sin(\delta - \beta) = I_X$ gives

$$p(t) = \underbrace{VI_R\{1 + \cos[2(\omega t + \delta)]\}}_{p_R(t)} + \underbrace{VI_X\sin[2(\omega t + \delta)]}_{p_X(t)} \tag{2.2.10}$$

[1] Use the identity: $\cos A \cos B = \tfrac{1}{2}[\cos(A - B) + \cos(A + B)]$.

As indicated by (2.2.10), the instantaneous power absorbed by the load has two components: one can be associated with the power $p_R(t)$ absorbed by the resistive component of the load, and the other can be associated with the power $p_X(t)$ absorbed by the reactive (inductive or capacitive) component of the load. The first component $p_R(t)$ in (2.2.10) is identical to (2.2.3), where $I_R = I\cos(\delta - \beta)$ is the component of the load current in phase with the load voltage. The phase angle $(\delta - \beta)$ represents the angle between the voltage and current. The second component $p_X(t)$ in (2.2.10) is identical to (2.2.6) or (2.2.8), where $I_X = I\sin(\delta - \beta)$ is the component of load current 90° out of phase with the voltage.

Real Power

Equation (2.2.10) shows that the instantaneous power $p_R(t)$ absorbed by the resistive component of the load is a double-frequency sinusoid with average value P given by

$$P = VI_R = VI\cos(\delta - \beta) \quad W \tag{2.2.11}$$

The *average power* P is also called *real power* or *active power*. All three terms indicate the same quantity P given by (2.2.11).

Power Factor

The term $\cos(\delta - \beta)$ in (2.2.11) is called the *power factor*. The phase angle $(\delta - \beta)$, which is the angle between the voltage and current, is called the *power factor angle*. For dc circuits, the power absorbed by a load is the product of the dc load voltage and the dc load current; for ac circuits the average power absorbed by a load is the product of the rms load voltage V, rms load current I, and the power factor $\cos(\delta - \beta)$, as shown by (2.2.11). For inductive loads, the current lags the voltage, which means β is less than δ, and the power factor is said to be *lagging*. For capacitive loads, the current leads the voltage, which means β is greater than δ, and the power factor is said to be *leading*.

Reactive Power

The instantaneous power absorbed by the reactive part of the load, given by the component $p_X(t)$ in (2.2.10), is a double-frequency sinusoid with zero average value and with amplitude Q given by

$$Q = VI_X = VI\sin(\delta - \beta) \quad \text{var} \tag{2.2.12}$$

The term Q is given the name *reactive power*. Although it has the same units as real power, the usual practice is to define units of reactive power as volt-amperes reactive, or var.

Example 2.1 The voltage $v(t) = 141.4\cos(\omega t)$ is applied to a load consisting of a 10-Ω resistor in parallel with an inductive reactance $X_L = \omega L = 3.77\,\Omega$. Calculate the instantaneous power absorbed by the resistor and by the inductor. Also calculate the real and reactive power absorbed by the load, and the power factor.

Solution The circuit and phasor diagram are shown in Figure 2.3(a). The load voltage is

$$V = \frac{141.4}{\sqrt{2}}\underline{/0^\circ} = 100\underline{/0^\circ} \quad \text{volts}$$

The resistor current is

$$I_{\mathrm{R}} = \frac{V}{\mathrm{R}} = \frac{100}{10}\underline{/0^\circ} = 10\underline{/0^\circ} \quad \mathrm{A}$$

Figure 2.3

Circuit and phasor diagram for Example 2.1

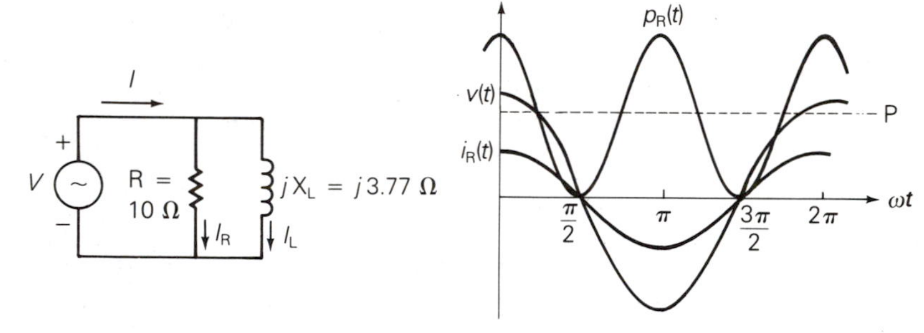

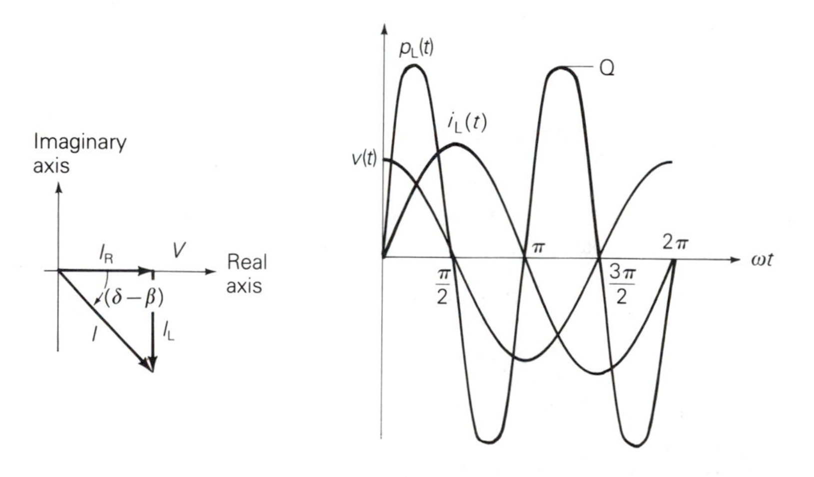

(a) Circuit and phasor diagram

(b) Waveforms

The inductor current is

$$I_L = \frac{V}{jX_L} = \frac{100}{(j3.77)}\underline{/0^\circ} = 26.53\underline{/-90^\circ} \quad \text{A}$$

$j = 1\angle 90$

The total load current is

$$I = I_R + I_L = 10 - j26.53 = 28.35\underline{/-69.34^\circ} \quad \text{A}$$

The instantaneous power absorbed by the resistor is, from (2.2.3),

$$p_R(t) = (100)(10)[1 + \cos(2\omega t)]$$
$$= 1000[1 + \cos(2\omega t)] \quad \text{W}$$

The instantaneous power absorbed by the inductor is, from (2.2.6),

$$p_L(t) = (100)(26.53)\sin(2\omega t)$$
$$= 2653\sin(2\omega t) \quad \text{W}$$

The real power absorbed by the load is, from (2.2.11),

$$P = VI\cos(\delta - \beta) = (100)(28.53)\cos(0^\circ + 69.34^\circ)$$
$$= 1000 \quad \text{W}$$

(*Note*: P is also equal to $VI_R = V^2/R$.)

The reactive power absorbed by the load is, from (2.2.12),

$$Q = VI\sin(\delta - \beta) = (100)(28.53)\sin(0^\circ + 69.34^\circ)$$
$$= 2653 \quad \text{var}$$

(*Note*: Q is also equal to $VI_L = V^2/X_L$.)

The power factor is

$$\text{p.f.} = \cos(\delta - \beta) = \cos(69.34^\circ) = 0.3528 \quad \text{lagging}$$

Voltage, current, and power waveforms are shown in Figure 2.3(b).

Note that $p_R(t)$ and $p_X(t)$, given by (2.2.10), are strictly valid only for a parallel R-X load. For a general RLC load, the voltages across the resistive and reactive components may not be in phase with the source voltage $v(t)$, resulting in additional phase shifts in $p_R(t)$ and $p_X(t)$ (see Problem 2.5). However, (2.2.11) and (2.2.12) for P and Q are valid for a general RLC load. ◁

Physical Significance of Real and Reactive Power

The physical significance of real power P is easily understood. The total energy absorbed by a load during a time interval T, consisting of one cycle of the sinusoidal voltage, is PT watt-seconds (Ws). During a time interval of n cycles, the energy absorbed is $P(nT)$ watt-seconds, all of which is absorbed by the resistive component of the load. A kilowatt-hour meter is designed to measure the energy absorbed by a load during a time interval $(t_2 - t_1)$, consisting of an integral number of cycles, by integrating the real power P over the time interval $(t_2 - t_1)$.

The physical significance of reactive power Q is not as easily understood. Q refers to the maximum value of the instantaneous power absorbed by the reactive component of the load. The instantaneous reactive power, given by the second term $p_X(t)$ in (2.2.10), is alternately positive and negative, and it expresses the reversible flow of energy to and from the reactive component of the load. Reactive power Q is a useful quantity when describing the operation of power systems (this will become evident in later chapters). As one example, shunt capacitors can be used in transmission systems to deliver reactive power and thereby increase voltage magnitudes during heavy load periods (see Chapter 6).

Section 2.3

Complex Power

For circuits operating in sinusoidal-steady-state, real and reactive power are conveniently calculated from complex power, defined below. Let the voltage across a circuit element be $V = \mathrm{V}\angle\delta$, and the current into the element be $I = \mathrm{I}\angle\beta$. Then the complex power S is the product of the voltage and the conjugate of the current:

$$S = VI^* = [\mathrm{V}\angle\delta][\mathrm{I}\angle\beta]^* = \mathrm{VI}\angle\delta - \beta$$

$$= \mathrm{VI}\cos(\delta - \beta) + j\mathrm{VI}\sin(\delta - \beta) \tag{2.3.1}$$

where $(\delta - \beta)$ is the angle between the voltage and current. Comparing (2.3.1) with (2.2.11) and (2.2.12), S is recognized as

$$S = \mathrm{P} + j\mathrm{Q} \tag{2.3.2}$$

The magnitude $\mathrm{S} = \mathrm{VI}$ of the complex power S is called the *apparent power*. Although it has the same units as P and Q, it is common practice to define the units of apparent power S as voltamperes or VA. The real power P is obtained by multiplying the apparent power $\mathrm{S} = \mathrm{VI}$ by the power factor p.f. $= \cos(\delta - \beta)$.

The procedure for determining whether a circuit element absorbs or delivers power is summarized in Figure 2.4 Figure 2.4(a) shows the *load convention*, where the current *enters* the positive terminal of the circuit element, and the complex power *absorbed* by the circuit element is calculated from (2.3.1). This equation shows that, depending on the value of $(\delta - \beta)$, P may have either a positive or negative value. If P is positive, then the circuit element absorbs positive real power. However, if P is negative, the circuit element absorbs negative real power, or alternatively, it delivers positive real power. Similarly, if Q is positive, the circuit element in Figure 2.4(a) absorbs positive reactive power. However, if Q is negative, the circuit element absorbs negative reactive power, or it delivers positive reactive power.

Figure 2.4

Load and generator conventions

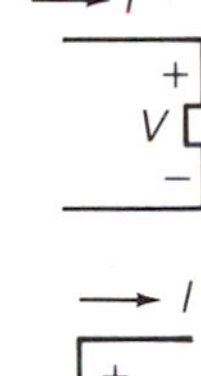

(a) *Load convention*. Current *enters* positive terminal of circuit element. If P is positive, then positive real power is *absorbed*. If Q is positive, then positive reactive power is *absorbed*. If P (Q) is negative, then positive real (reactive) power is *delivered*.

(b) *Generator convention*. Current *leaves* positive terminal of the circuit element. If P is positive, then positive real power is *delivered*. If Q is positive, then positive reactive power is *delivered*. If P (Q) is negative, then positive real (reactive) power is *absorbed*.

Figure 2.4(b) shows the *generator convention*, where the current *leaves* the positive terminal of the circuit element, and the complex power *delivered* is calculated from (2.3.1). When P is positive (negative) the circuit element *delivers* positive (negative) real power. Similarly, when Q is positive (negative), the circuit element *delivers* positive (negative) reactive power.

Example 2.2 A single-phase voltage source with $V = 100\angle 130°$ volts delivers a current $I = 10\angle 10°$ A, which leaves the positive terminal of the source. Calculate the source real and reactive power and state whether the source delivers or absorbs each of these.

Solution Since I leaves the positive terminal of the source, the generator convention is assumed, and the complex power delivered is, from (2.3.1),

$$S = VI^* = [100\angle 130°][10\angle 10°]^*$$

$$S = 1000\angle 120° = -500 + j866$$

$$\text{P} = \text{Re}[S] = -500 \quad \text{W}$$

$$\text{Q} = \text{Im}[S] = +866 \quad \text{var}$$

where Im denotes "imaginary part of." The source absorbs 500 W and delivers 866 var. Readers familiar with electric machines will recognize that one example of this source is a synchronous motor. When a synchronous motor operates at a leading power factor, it absorbs real power and delivers reactive power. ◁

The *load convention* is used for the RLC elements shown in Figure 2.2. Therefore, the complex power *absorbed* by any of these three elements can be calculated as follows. Assume a load voltage $V = \text{V}\angle\delta$. Then, from (2.3.1),

$$\text{resistor:} \quad S_R = VI_R^* = [\text{V}\angle\delta]\left[\frac{\text{V}}{\text{R}}\angle-\delta\right] = \frac{\text{V}^2}{\text{R}} \tag{2.3.3}$$

$$\text{inductor:} \quad S_L = VI_L^* = [\text{V}\angle\delta]\left[\frac{\text{V}}{-j\text{X}_L}\angle-\delta\right] = +j\frac{\text{V}^2}{\text{X}_L} \tag{2.3.4}$$

$$\text{capacitor:} \quad S_C = VI_C^* = [\text{V}\angle\delta]\left[\frac{\text{V}}{j\text{X}_C}\angle-\delta\right] = -j\frac{\text{V}^2}{\text{X}_C} \tag{2.3.5}$$

From these complex power expressions, the following can be stated:

A (positive-valued) resistor absorbs (positive) real power, $P_R = V^2/R$ W, and zero reactive power, $Q_R = 0$ var.

An inductor absorbs zero real power, $P_L = 0$ W, and positive reactive power, $Q_L = V^2/X_L$ var.

A capacitor *absorbs* zero real power, $P_C = 0$ W, and *negative* reactive power, $Q_C = -V^2/X_C$ var. Alternatively, a capacitor *delivers positive* reactive power, $+V^2/X_C$.

For a general load composed of RLC elements, complex power S is also calculated from (2.3.1). The real power $P = \text{Re}(S)$ absorbed by a passive load is always positive. The reactive power $Q = \text{Im}(S)$ absorbed by a load may be either positive or negative. When the load is inductive, the current lags the voltage, which means β is less than δ in (2.3.1), and the reactive power absorbed is positive. When the load is capacitive, the current leads the voltage, which means β is greater than δ, and the reactive power absorbed is negative; or, alternatively, the capacitive load delivers positive reactive power.

Complex power can be summarized graphically by use of the power triangle shown in Figure 2.5. As shown, the apparent power S, real power P, and reactive power Q form the three sides of the power triangle. The power factor angle $(\delta - \beta)$ is also shown, and the following expressions can be obtained:

$$S = \sqrt{P^2 + Q^2} \tag{2.3.6}$$

$$(\delta - \beta) = \tan^{-1}(Q/P) \tag{2.3.7}$$

$$Q = P\tan(\delta - \beta) \tag{2.3.8}$$

$$\text{p.f.} = \cos(\delta - \beta) = \frac{P}{S} = \frac{P}{\sqrt{P^2 + Q^2}} \tag{2.3.9}$$

Figure 2.5

Power triangle

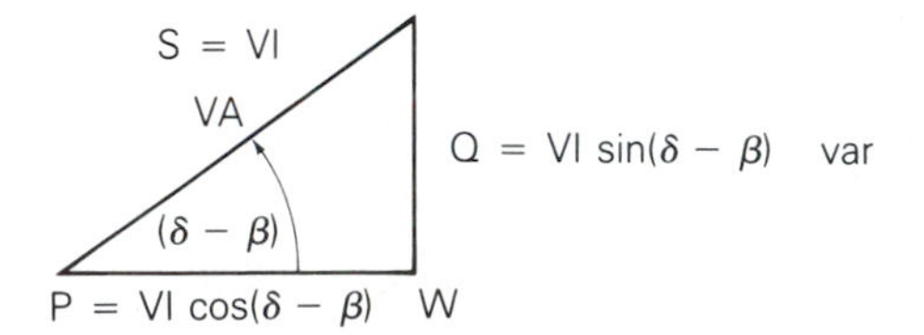

Example 2.3 A single-phase source delivers 100 kW to a load operating at a power factor of 0.8 lagging. Calculate the reactive power to be delivered by a capacitor connected in parallel with the load in order to raise the source power factor to 0.95 lagging. Also draw the power triangle for the source and load. Assume that the source voltage is constant, and neglect the line impedance between the source and load.

Solution The circuit and power triangle are shown in Figure 2.6. The real power

$P = P_S = P_R$ delivered by the source and absorbed by the load is not changed when the capacitor is connected in parallel with the load, since the capacitor delivers only reactive power Q_C. For the load, the power factor angle, reactive power absorbed, and apparent power are

$$\theta_L = (\delta - \beta_L) = \cos^{-1}(0.8) = 36.87°$$

$$Q_L = P \tan \theta_L = 100 \tan(36.87°) = 75 \quad \text{kvar}$$

$$S_L = \frac{P}{\cos \theta_L} = 125 \quad \text{kVA}$$

Figure 2.6

Circuit and power triangle for Example 2.3

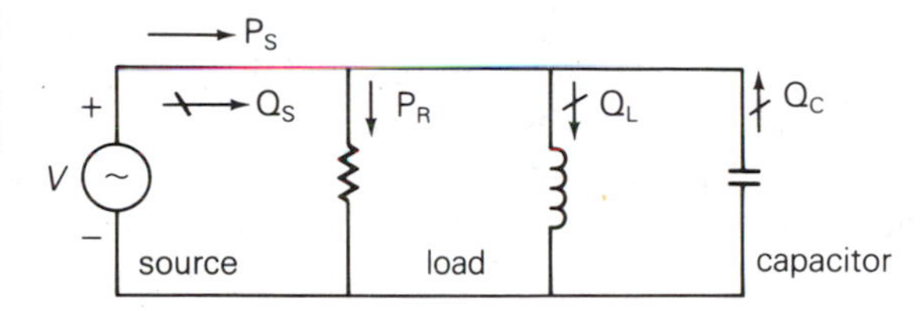

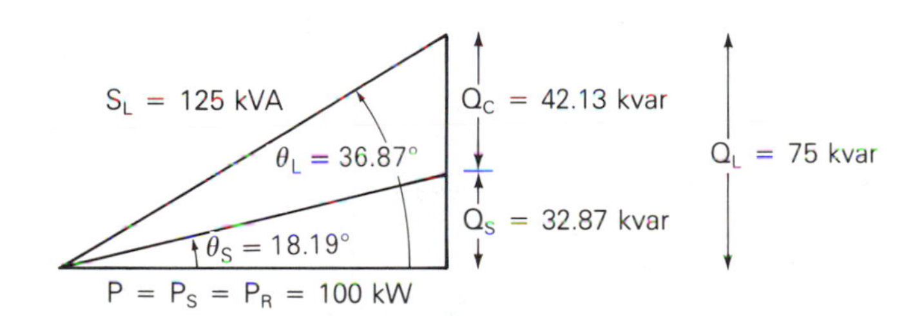

After the capacitor is connected, the power factor angle, reactive power delivered, and apparent power of the source are

$$\theta_S = (\delta - \beta_S) = \cos^{-1}(0.95) = 18.19°$$

$$Q_S = P \tan \theta_S = 100 \tan(18.19°) = 32.87 \quad \text{kvar}$$

$$S_S = \frac{P}{\cos \theta_S} = \frac{100}{0.95} = 105.3 \quad \text{kVA}$$

The capacitor delivers

$$Q_C = Q_L - Q_S = 75 - 32.87 = 42.13 \quad \text{kvar}$$

The method of connecting a capacitor in parallel with an inductive load is known as *power factor correction*. The effect of the capacitor is to increase the power factor of the source that delivers power to the load. Also, the source apparent power S_S decreases. As shown in Figure 2.6, the source apparent power for this example decreases from 125 kVA without the capacitor to 105.3 kVA with the capacitor. The source current $I_S = S_S/V$ also decreases.

When line impedance between the source and load is included, the decrease in source current results in lower line losses and lower line-voltage drops. The end result of power factor correction is improved efficiency and improved voltage regulation. ◁

Section 2.4

Network Equations

For circuits operating in sinusoidal-steady-state, Kirchhoff's current law (KCL) and voltage law (KVL) apply to phasor currents and voltages. Thus the sum of all phasor currents entering any node is zero and the sum of the phasor-voltage drops around any closed path is zero. Network analysis techniques based on Kirchhoff's laws, including nodal analysis, mesh or loop analysis, superposition, source transformations, and Thévenin's theorem or Norton's theorem, are useful for analyzing such circuits.

Various computer solutions of power-system problems are formulated from nodal equations, which can be systematically applied to circuits. The circuit shown in Figure 2.7, which is used here to review nodal analysis, is assumed to be operating in sinusoidal-steady-state; source voltages are represented by phasors E_{S1}, E_{S2}, and E_{S3}; circuit impedances are specified in ohms. Nodal equations are written in the following three steps:

Step 1 For a circuit with $(N + 1)$ nodes (also called buses), select one bus as the reference bus and define the voltages at the remaining buses with respect to the reference bus.

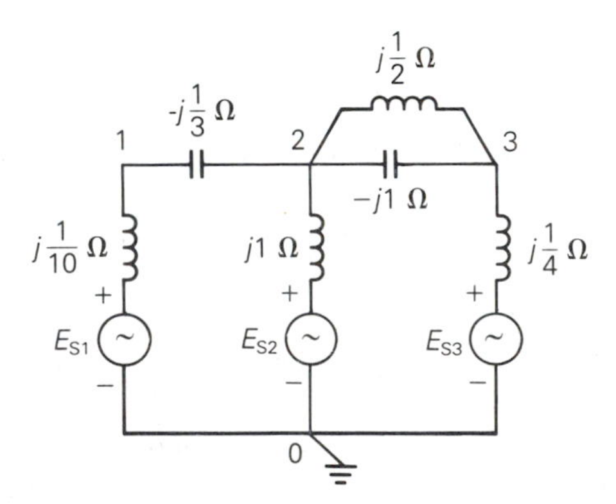

Figure 2.7 Circuit diagram for reviewing nodal analysis

The circuit in Figure 2.7 has four buses—that is, $N + 1 = 4$ or $N = 3$. Bus 0 is selected as the reference bus, and bus voltages V_{10}, V_{20}, and V_{30} are then defined with respect to bus 0.

Step 2 Transform each voltage source in series with an impedance to an equivalent current source in parallel with that impedance. Also, show admittance values instead of impedance values on the circuit diagram.

In Figure 2.8 equivalent current sources I_1, I_2, and I_3 are shown, and all impedances are converted to corresponding admittances.

Figure 2.8

Circuit of Figure 2.7 with equivalent current sources replacing voltage sources. Admittance values are also shown

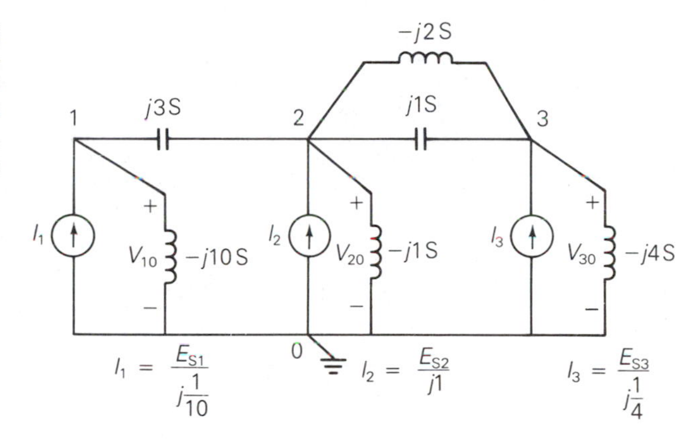

Step 3 Write nodal equations in matrix format as follows:

$$
\begin{bmatrix}
Y_{11} & Y_{12} & Y_{13} & \cdots & Y_{1N} \\
Y_{21} & Y_{22} & Y_{23} & \cdots & Y_{2N} \\
Y_{31} & Y_{32} & Y_{33} & \cdots & Y_{3N} \\
\vdots & \vdots & \vdots & & \vdots \\
Y_{N1} & Y_{N2} & Y_{N3} & \cdots & Y_{NN}
\end{bmatrix}
\begin{bmatrix}
V_{10} \\ V_{20} \\ V_{30} \\ \vdots \\ V_{NO}
\end{bmatrix}
=
\begin{bmatrix}
I_1 \\ I_2 \\ I_3 \\ \vdots \\ I_N
\end{bmatrix}
\tag{2.4.1}
$$

Using matrix notation, (2.4.1) becomes

$$\boldsymbol{YV} = \boldsymbol{I} \tag{2.4.2}$$

where $\boldsymbol{Y}$ is the $N \times N$ bus admittance matrix, $\boldsymbol{V}$ is the column vector of N bus voltages, and $\boldsymbol{I}$ is the column vector of N current sources. The elements Y_{kn} of the bus admittance matrix $\boldsymbol{Y}$ are formed as follows:

diagonal elements: Y_{kk} = sum of admittances connected to bus k $(k = 1, 2, \ldots, N)$ (2.4.3)

off-diagonal elements: Y_{kn} = − (sum of admittances connected between buses k and n) $(k \neq n)$ (2.4.4)

The diagonal element Y_{kk} is called the *self-admittance* or the *driving-point admittance* of bus k, and the off-diagonal element Y_{kn} for $k \neq n$ is called the *mutual admittance* or the *transfer admittance* between buses k and n. Since $Y_{kn} = Y_{nk}$, the matrix $\boldsymbol{Y}$ is symmetric.

For the circuit of Figure 2.8, (2.4.1) becomes

$$\begin{bmatrix} (j3-j10) & -(j3) & 0 \\ -(j3) & (j3-j1+j1-j2) & -(j1-j2) \\ 0 & -(j1-j2) & (j1-j2-j4) \end{bmatrix} \begin{bmatrix} V_{10} \\ V_{20} \\ V_{30} \end{bmatrix} = \begin{bmatrix} I_1 \\ I_2 \\ I_3 \end{bmatrix}$$

$$j \begin{bmatrix} -7 & -3 & 0 \\ -3 & 1 & 1 \\ 0 & 1 & -5 \end{bmatrix} \begin{bmatrix} V_{10} \\ V_{20} \\ V_{30} \end{bmatrix} = \begin{bmatrix} I_1 \\ I_2 \\ I_3 \end{bmatrix} \qquad (2.4.5)$$

The advantage of this method of writing nodal equations is that a digital computer can be used both to generate the admittance matrix $\boldsymbol{Y}$ and to solve (2.4.2) for the unknown bus voltage vector $\boldsymbol{V}$. Once a circuit is specified with the reference bus and other buses identified, the circuit admittances and their bus connections become computer input data for calculating the elements Y_{kn} via (2.4.3) and (2.4.4). After $\boldsymbol{Y}$ is calculated and the current source vector $\boldsymbol{I}$ is given as input, standard computer programs for solving simultaneous linear equations can then be used to determine the bus voltage vector $\boldsymbol{V}$. See Example 2.4 for an illustration.

When double subscripts are used to denote a voltage in this text, the voltage shall be that at the node identified by the first subscript with respect to the node identified by the second subscript. For example the voltage V_{10} in Figure 2.8 is the voltage at node 1 with respect to node 0. Also, a current I_{ab} shall indicate the current from node a to node b. Voltage polarity marks $(+/-)$ and current reference arrows $(\rightarrow)$ are not required when double subscript notation is employed. The polarity marks in Figure 2.8 for V_{10}, V_{20}, and V_{30}, although not required, are shown for clarity. The reference arrows for sources I_1, I_2, and I_3 in Figure 2.8 are required, however, since single subscripts are used for these currents. Matrices and vectors shall be indicated in this text by boldface type (for example, $\boldsymbol{Y}$ or $\boldsymbol{V}$).

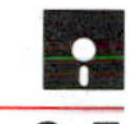

Section 2.5

Personal Computer Program: MATRIX OPERATIONS

The following matrix operation subroutines are included in the software package that accompanies this text:

1. RMA($\mathbf{A}$,$\mathbf{B}$,C,N,M) computes the matrix sum $\mathbf{C} = \mathbf{A} + \mathbf{B}$ of the two $N \times M$ real matrices $\mathbf{A}$ and $\mathbf{B}$.
2. CMA($\boldsymbol{A}$,$\boldsymbol{B}$,$\boldsymbol{C}$,N,M) computes the matrix sum $\boldsymbol{C} = \boldsymbol{A} + \boldsymbol{B}$ of the two $N \times M$ complex matrices $\boldsymbol{A}$ and $\boldsymbol{B}$.
3. RMM($\mathbf{A}$,$\mathbf{B}$,C,N,M,P) computes the matrix product $\mathbf{C} = \mathbf{AB}$ of the $N \times M$ real matrix $\mathbf{A}$ and the $M \times P$ real matrix $\mathbf{B}$.
4. CMM($\boldsymbol{A}$,$\boldsymbol{B}$,$\boldsymbol{C}$,N,M,P) computes the matrix product $\boldsymbol{C} = \boldsymbol{AB}$ of the $N \times M$ complex matrix $\boldsymbol{A}$ and the $M \times P$ complex matrix $\boldsymbol{B}$.
5. RMI($\mathbf{A}$,N) computes the matrix inverse $\mathbf{A}^{-1}$ of the $N \times N$ real matrix $\mathbf{A}$ whose determinant is assumed to be nonzero.
6. CMI($\boldsymbol{A}$,N) computes the matrix inverse $\boldsymbol{A}^{-1}$ of the $N \times N$ complex matrix $\boldsymbol{A}$, whose determinant is assumed to be nonzero.
7. RMT($\mathbf{A}$,$\mathbf{AT}$,N,M) computes the matrix transpose $\mathbf{AT} = \mathbf{A}^{\mathrm{T}}$ of the $N \times M$ matrix $\mathbf{A}$ by interchanging its rows and columns.
8. CMT($\boldsymbol{A}$,$\boldsymbol{AT}$,N,M) computes the matrix transpose $\boldsymbol{AT} = \boldsymbol{A}^{\mathrm{T}}$ of the $N \times M$ complex matrix $\boldsymbol{A}$.
9. CMC($\boldsymbol{A}$,$\boldsymbol{AC}$,N,M) computes the complex conjugate $\boldsymbol{AC} = \boldsymbol{A}^*$ of the $N \times M$ complex matrix $\boldsymbol{A}$.

In this text, the matrix inverse, transpose, and complex conjugate operations shall be denoted by the superscripts -1, T, and *, respectively.

Example 2.4 The voltage sources shown in the circuit of Figure 2.7 are $E_{S1} = 1\underline{/0^\circ}$, $E_{S2} = 1.5\underline{/30^\circ}$, and $E_{S3} = 2.0\underline{/-30^\circ}$ volts. Using these sources as well as the bus connections and impedances shown in Figure 2.7 as input data, write a computer program to:

1. Convert the voltage sources to current sources and the impedances to corresponding admittances.
2. Compute the 3×3 admittance matrix $\boldsymbol{Y}$.
3. Compute $\boldsymbol{Y}^{-1}$ using the CMI subroutine.
4. Compute the bus voltage vector $\boldsymbol{V}$ from (2.4.2) using the CMM subroutine.

Solution Table 2.1 shows the solution, including the program and input data and output data listings.

Table 2.1(a) Example 2.4 computer program listing

```
1    '***** EXAMPLE 2.4 *****
5    '***** BUS DATA *****
10   DATA 1,1.0,0,0,0.1
20   DATA 2,1.29904,0.75,0,1.0
30   DATA 3,1.73206,-1.0,0,0.25
40   DIM NBUS(3), ESR(3), ESI(3), RSRC(3),
         XSRC(3)
50   FOR I = 1 TO 3
60   READ NBUS(I), ESR(I), ESI(I), RSRC(I),
         XSRC(I)
70   NEXT I
100  '***** LINE DATA *****
110  DATA 1,2,0,-0.3333333333333
120  DATA 2,3,0,-1.0
130  DATA 2,3,0,0.5
140  DIM NBUS1(3), NBUS2(3), RLINE(3),
         XLINE(3)
150  FOR L = 1 TO 3
160  READ NBUS1(L), NBUS2(L), RLINE(L),
         XLINE(L)
170  NEXT L
200  '***** STEP 1 *****
210  'CONVERT VOLTAGE SOURCES TO
         CURRENT SOURCES
220  DIM ISR(3,1), ISI(3,1)
230  FOR I = 1 TO 3
235  ZSQ = RSRC(I)^2 + XSRC(I)^2
240  ISR(I,1) = (ESR(I)*RSRC(I)
         + ESI(I)*XSRC(I))/ZSQ
250  ISI(I,1) = (-ESR(I)*XSRC(I)
         + ESI(I)*RSRC(I))/ZSQ
260  NEXT I
300  '**STEP 2: COMPUTE Y = YR + jYI**
350  '***** OFF-DIAGONAL ELEMENTS *****
360  DIM YR(3,3), YI(3,3)
370  FOR I = 1 TO 3
380  FOR J = 1 TO 3
385  YR(I,J) = 0
390  YI(I,J) = 0
400  NEXT J
410  NEXT I
420  FOR I = 1 TO 3
430  FOR J = 1 TO 3
435  IF I = J THEN GOTO 480
440  FOR L = 1 TO 3
450  IF (NBUS1(L) = I) AND (NBUS2(L) = J)
         THEN 455 ELSE 470
455  ZLSQ = RLINE(L)^2 + XLINE(L)^2
457  YR(I,J) = YR(I,J) - RLINE(L)/ZLSQ
460  YI(I,J) = YI(I,J) + XLINE(L)/ZLSQ
462  YR(J,I) = YR(I,J)
465  YI(J,I) = YI(I,J)
470  NEXT L
480  NEXT J
490  NEXT I
500  '***** DIAGONAL ELEMENTS *****
540  FOR I = 1 TO 3
550  FOR J = 1 TO 3
560  IF J = I THEN 580
565  YR(I,I) = YR(I,I) - YR(I,J)
570  YI(I,I) = YI(I,I) - YI(I,J)
580  NEXT J
590  NEXT I
600  FOR I = 1 TO 3
605  ZSQ = RSRC(I)^2 + XSRC(I)^2
608  YR(I,I) = YR(I,I) + RSRC(I)/ZSQ
610  YI(I,I) = YI(I,I) - XSRC(1)/ZSQ
615  NEXT I
700  '*** STEP 3: INVERT Y ***
720  N = 3
730  DIM AR(3,3), AI(3,3), YINVR(3,3), YINVI(3,3)
740  FOR I = 1 TO 3
750  FOR J = 1 TO 3
755  AR(I,J) = YR(I,J)
760  AI(I,J) = YI(I,J)
770  NEXT J
780  NEXT I
790  GOSUB 6000
800  FOR I = 1 TO 3
810  FOR J = 1 to 3
820  YINVR(I,J) = AR(I,J)
830  YINVI(I,J) = AI(I,J)
840  NEXT J
850  NEXT I
900  '** STEP 4: COMPUTE COMPLEX VOLTAGE
         VECTOR**
920  DIM VR(3,1), VI(3,1), BR(3,3)
925  DIM BI(3,3), CR(3,1), CI(3,1)
990  FOR I = 1 TO 3
1000 BR(I,1) = ISR(I,1)
1010 BI(I,1) = ISI(I,1)
1020 NEXT I
1030 N = 3
1040 M = 3
1050 P = 1
1055 GOSUB 4000
```

```
1060 FOR I = 1 TO 3
1070 VR(I,1) = CR(I,1)
1080 VI(I,1) = CI(I,1)
1090 NEXT I
2000 '***** PRINT INPUT/OUTPUT DATA *****
2010 PRINT"*****  BUS DATA *****"
2020 PRINT
2030 PRINT TAB(2); "BUS#"; TAB(17); "SOURCE
        VOLTAGE";
2035 PRINT TAB(40); "SOURCE IMPEDANCE"
2040 PRINT TAB(14); "REAL PART"; TAB(24);
        "IMAG PART";
2042 PRINT TAB(37); "REAL PART"; TAB(48);
        "IMAG PART"
2045 PRINT TAB(14);"  VOLTS  ";
2047 PRINT TAB(24);" VOLTS"; TAB(45); "OHMS"
2050 PRINT "——————————————————"
2065 FOR I = 1 TO 3
2070 PRINT TAB(2); NBUS(I); TAB(14); ESR(I);
2072 PRINT TAB(27);"j"; ESI(I); TAB(40); RSRC(I)
2074 PRINT TAB(52); "j"; XSRC(I)
2075 NEXT I
2090 PRINT
2095 PRINT
2110 PRINT "*****  "LINE DATA" *****"
2130 PRINT TAB(2); "FROM"; TAB(12); " TO ";
2135 PRINT TAB(25); "LINE IMPEDANCE"
2140 PRINT TAB(3); "BUS"; TAB(11); "BUS";
2142 PRINT TAB(22); "REAL PART"; TAB(34);
        "IMAG PART"
2144 PRINT TAB(30); "OHMS"
2150 PRINT "——————————————————"
2160 PRINT
2165 FOR L = 1 TO 3
2170 PRINT TAB(2); NBUS1(L); TAB(12);
        NBUS2(L);
2172 PRINT TAB(25); RLINE(L); TAB(37); "j";
        XLINE(L)
2175 NEXT L
2210 PRINT "*****  BUS VOLTAGES *****"
2220 PRINT
2230 PRINT TAB(2); "BUS#"; TAB(22); "BUS
        VOLTAGE"
2240 PRINT TAB(14); "REAL PART"; TAB(30);
        "IMAG PART"
2245 PRINT TAB(14); "  VOLTS  "; TAB(30);
        "  VOLTS  "
2250 PRINT "——————————————————"
2265 FOR I = 1 TO 3
2270 PRINT TAB(2); NBUS(I); TAB(10); VR(I, 1);
2272 PRINT TAB(27); "j"; VI(I,1)
2275 NEXT I
2297 PRINT "***************************************"
3000 END
```

Table 2.1(b)

Example 2.4 input/output data files

BUS DATA				
BUS#	SOURCE VOLTAGE		SOURCE IMPEDANCE	
	REAL PART	IMAG PART	REAL PART	IMAG PART
	VOLTS	VOLTS	OHMS	OHMS
1	1	*j* 0	0	*j* .1
2	1.29904	*j* .75	0	*j* 1
3	1.73206	*j* − 1	0	*j* .25

LINE DATA			
FROM BUS	TO BUS	LINE IMPEDANCE	
		REAL PART	IMAG PART
		OHMS	OHMS
1	2	0	*j* − .3333333333333
2	3	0	*j* − 1
2	3	0	*j* .5

BUS VOLTAGES		
BUS#	BUS VOLTAGE	
	REAL PART	IMAG PART
	VOLTS	VOLTS
1	1.1525324137931	j − .00862068965517
2	.64409103448291	j .02011494252872
3	1.5144662068966	j − .79597701149423

◁

Section 2.6

Balanced Three-phase Circuits

In this section we introduce the following topics for balanced three-phase circuits: Y connections, line-to-neutral voltages, line-to-line voltages, line currents, Δ loads, Δ–Y conversions, and equivalent line-to-neutral diagrams.

Balanced-Y Connections

Figure 2.9 shows a three-phase Y-connected (or "wye-connected") voltage source feeding a balanced-Y-connected load. For a Y connection, the neutrals of each phase are connected. In Figure 2.9 the source neutral connection is labeled bus n and the load neutral connection is labeled bus N. The three-phase source is assumed to be ideal since source impedances are neglected. Also neglected are the line impedances between the source and load terminals and the neutral impedance between buses n and N. The three-phase load is *balanced*, which means the load impedances in all three phases are identical.

Figure 2.9

Circuit diagram of a three-phase Y-connected source feeding a balanced-Y load

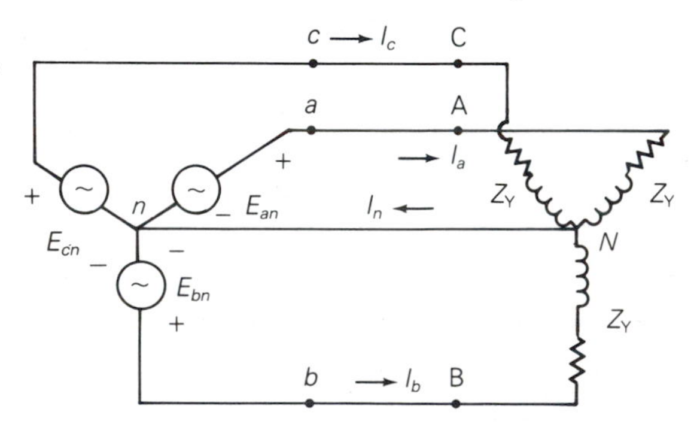

Balanced Line-to-Neutral Voltages

In Figure 2.9, the terminal buses of the three-phase source are labeled a, b, and c, and the source line-to-neutral voltages are labeled E_{an}, E_{bn}, and E_{cn}. The source is *balanced* when these voltages have equal magnitudes and an equal

120°-phase difference between any two phases. An example of balanced three-phase line-to-neutral voltages is

$$
\begin{aligned}
E_{an} &= 10\underline{/0^\circ} \\
E_{bn} &= 10\underline{/-120^\circ} = 10\underline{/+240^\circ} \\
E_{cn} &= 10\underline{/+120^\circ} = 10\underline{/-240^\circ} \quad \text{volts}
\end{aligned}
\tag{2.6.1}
$$

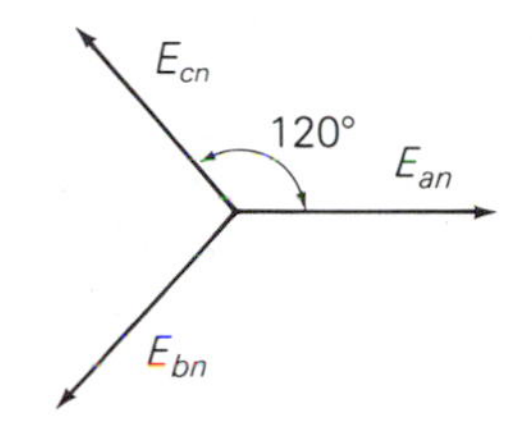

Figure 2.10

Phasor diagram of balanced positive-sequence line-to-neutral voltages with E_{an} as the reference

where the line-to-neutral voltage magnitude is 10 volts and E_{an} is the reference phasor. The phase sequence is called *positive sequence* or *abc* sequence when E_{an} leads E_{bn} by 120° and E_{bn} leads E_{cn} by 120°. The phase sequence is called *negative sequence* or *acb* sequence when E_{an} leads E_{cn} by 120° and E_{cn} leads E_{bn} by 120°. The voltages in (2.6.1) are positive-sequence voltages, since E_{an} leads E_{bn} by 120°. The corresponding phasor diagram is shown in Figure 2.10.

Balanced Line-to-Line Voltages

The voltages E_{ab}, E_{bc}, and E_{ca} between phases are called line-to-line voltages. Writing a KVL equation for a closed path around buses *a*, *b*, and *n* in Figure 2.9,

$$E_{ab} = E_{an} - E_{bn} \tag{2.6.2}$$

For the line-to-neutral voltages of (2.6.1),

$$E_{ab} = 10\underline{/0^\circ} - 10\underline{/-120^\circ} = 10 - 10\left[\frac{-1 - j\sqrt{3}}{2}\right]$$

$$E_{ab} = \sqrt{3}(10)\left(\frac{\sqrt{3} + j1}{2}\right) = \sqrt{3}(10\underline{/30^\circ}) \quad \text{volts} \tag{2.6.3}$$

Similarly, the line-to-line voltages E_{bc} and E_{ca} are

$$
\begin{aligned}
E_{bc} &= E_{bn} - E_{cn} = 10\underline{/-120^\circ} - 10\underline{/+120^\circ} \\
&= \sqrt{3}(10\underline{/-90^\circ}) \quad \text{volts}
\end{aligned}
\tag{2.6.4}
$$

$$
\begin{aligned}
E_{ca} &= E_{cn} - E_{an} = 10\underline{/+120^\circ} - 10\underline{/0^\circ} \\
&= \sqrt{3}(10\underline{/150^\circ}) \quad \text{volts}
\end{aligned}
\tag{2.6.5}
$$

The line-to-line voltages of (2.6.3) – (2.6.5) are also balanced, since they have equal magnitudes of $\sqrt{3}(10)$ volts and 120° displacement between any two phases. Comparison of these line-to-line voltages with the line-to-neutral voltages of (2.6.1) leads to the following conclusion:

In a balanced three-phase Y-connected system with positive-sequence sources, the line-to-line voltages are $\sqrt{3}$ times the line-to-neutral voltages and lead by 30°. That is,

$$
\begin{aligned}
E_{ab} &= \sqrt{3}\, E_{an}\underline{/+30^\circ} \\
E_{bc} &= \sqrt{3}\, E_{bn}\underline{/+30^\circ} \\
E_{ca} &= \sqrt{3}\, E_{cn}\underline{/+30^\circ}
\end{aligned}
\tag{2.6.6}
$$

This very important result is summarized in Figure 2.11. In Figure 2.11(a) each phasor begins at the origin of the phasor diagram. In Figure 2.11(b) the line-to-line voltages form an equilateral triangle with vertices labeled a, b, c corresponding to buses a, b, and c of the system; the line-to-neutral voltages begin at the vertices and end at the center of the triangle, which is labeled n for neutral bus n. Also, the clockwise sequence of the vertices abc in Figure 2.11(b) indicates positive-sequence voltages. In both diagrams, E_{an} is the reference. However, the diagrams could be rotated to align with any other reference.

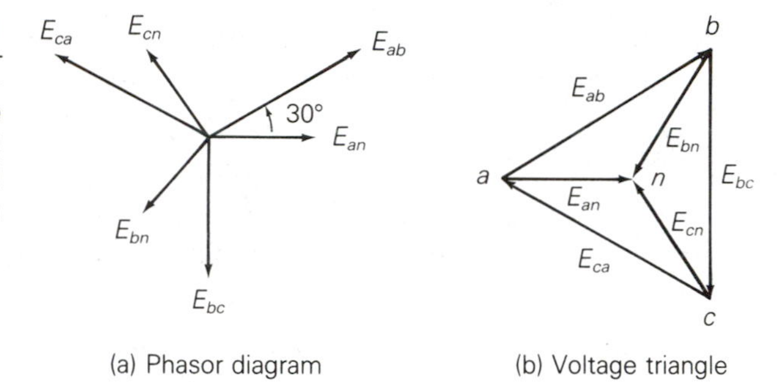

Figure 2.11 Positive-sequence line-to-neutral and line-to-line voltages in a balanced three-phase Y-connected system

Since the balanced line-to-line voltages form a closed triangle in Figure 2.11, their sum is zero. In fact, the sum of line-to-line voltages $(E_{ab} + E_{bc} + E_{ca})$ is *always* zero, even if the system is unbalanced, since these voltages form a closed path around buses a, b, and c. Also, in a balanced system the sum of the line-to-neutral voltages $(E_{an} + E_{bn} + E_{cn})$ equals zero.

Balanced Line Currents

Since the impedance between the source and load neutrals in Figure 2.9 is neglected, buses n and N are at the same potential, $E_{nN} = 0$. Accordingly, a separate KVL equation can be written for each phase, and the line currents can be written by inspection:

$$\begin{aligned} I_a &= E_{an}/Z_Y \\ I_b &= E_{bn}/Z_Y \\ I_c &= E_{cn}/Z_Y \end{aligned} \tag{2.6.7}$$

For example, if each phase of the Y-connected load has an impedance $Z_Y = 2\underline{/30°}\ \Omega$, then

$$
\begin{aligned}
I_a &= \frac{10\underline{/0°}}{2\underline{/30°}} = 5\underline{/-30°} \quad \text{A} \\
I_b &= \frac{10\underline{/-120°}}{2\underline{/30°}} = 5\underline{/-150°} \quad \text{A} \\
I_c &= \frac{10\underline{/+120°}}{2\underline{/30°}} = 5\underline{/90°} \quad \text{A}
\end{aligned}
\tag{2.6.8}
$$

The line currents are also balanced, since they have equal magnitudes of 5 A and 120° displacement between any two phases. The neutral current I_n is determined by writing a KCL equation at bus N in Figure 2.9.

$$I_n = I_a + I_b + I_c \tag{2.6.9}$$

Using the line currents of (2.6.8),

$$
\begin{aligned}
I_n &= 5\underline{/-30°} + 5\underline{/-150°} + 5\underline{/90°} \\
I_n &= 5\left(\frac{\sqrt{3} - j1}{2}\right) + 5\left(\frac{-\sqrt{3} - j1}{2}\right) + j5 = 0
\end{aligned}
\tag{2.6.10}
$$

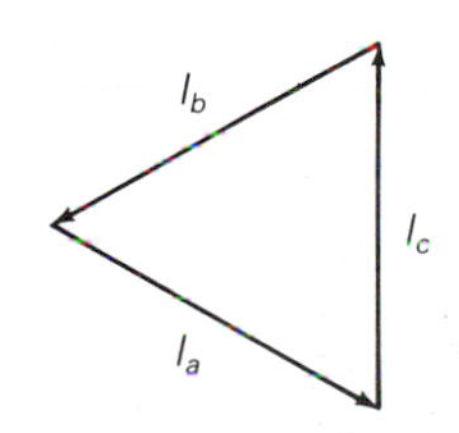

Figure 2.12

Phasor diagram of line currents in a balanced three-phase system

The phasor diagram of the line currents is shown in Figure 2.12. Since these line currents form a closed triangle, their sum, which is the neutral current I_n, is zero. In general, the sum of any balanced three-phase set of phasors is zero, since balanced phasors form a closed triangle. Thus, although the impedance between neutrals n and N in Figure 2.9 is assumed to be zero, the neutral current will be zero for *any* neutral impedance ranging from short circuit (0 Ω) to open circuit (∞ Ω), as long as the system is balanced. If the system is not balanced — which could occur if the source voltages, load impedances, or line impedances were unbalanced — then the line currents will not be balanced and a neutral current I_n may flow between buses n and N.

Balanced Δ Loads

Figure 2.13 shows a three-phase Y-connected source feeding a balanced-Δ-connected (or "delta-connected") load. For a balanced-Δ connection, equal

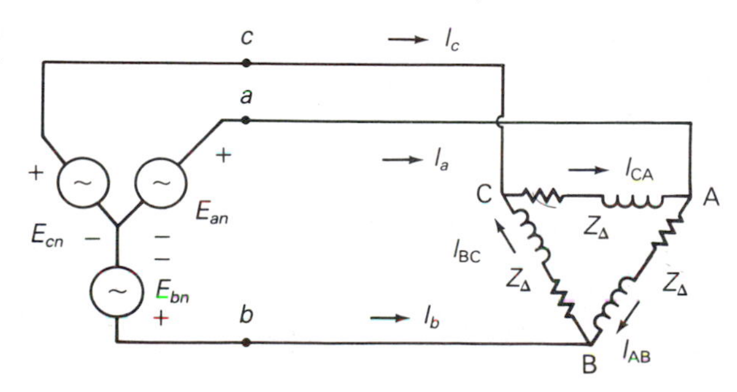

Figure 2.13

Circuit diagram of a Y-connected source feeding a balanced-Δ load

load impedances Z_Δ are connected in a triangle whose vertices form the buses, labeled A, B, and C in Figure 2.13. The Δ connection does not have a neutral bus.

Since the line impedances are neglected in Figure 2.13, the source line-to-line voltages are equal to the load line-to-line voltages, and the Δ-load currents I_{AB}, I_{BC}, and I_{CA} are

$$\begin{aligned} I_{AB} &= E_{ab}/Z_\Delta \\ I_{BC} &= E_{bc}/Z_\Delta \\ I_{CA} &= E_{ca}/Z_\Delta \end{aligned} \tag{2.6.11}$$

For example, if the line-to-line voltages are given by (2.6.3) – (2.6.5) and if $Z_\Delta = 5\underline{/30^\circ}\ \Omega$, then the Δ-load currents are

$$\begin{aligned} I_{AB} &= \sqrt{3}\left(\frac{10\underline{/30^\circ}}{5\underline{/30^\circ}}\right) = 3.464\underline{/0^\circ} \quad \text{A} \\ I_{BC} &= \sqrt{3}\left(\frac{10\underline{/-90^\circ}}{5\underline{/30^\circ}}\right) = 3.464\underline{/-120^\circ} \quad \text{A} \\ I_{CA} &= \sqrt{3}\left(\frac{10\underline{/150^\circ}}{5\underline{/30^\circ}}\right) = 3.464\underline{/+120^\circ} \quad \text{A} \end{aligned} \tag{2.6.12}$$

Also, the line currents can be determined by writing a KCL equation at each bus of the Δ load, as follows:

$$\begin{aligned} I_a &= I_{AB} - I_{CA} = 3.464\underline{/0^\circ} - 3.464\underline{/120^\circ} = \sqrt{3}(3.464\underline{/-30^\circ}) \\ I_b &= I_{BC} - I_{AB} = 3.464\underline{/-120^\circ} - 3.464\underline{/0^\circ} = \sqrt{3}(3.464\underline{/-150^\circ}) \\ I_c &= I_{CA} - I_{BC} = 3.464\underline{/120^\circ} - 3.464\underline{/-120^\circ} = \sqrt{3}(3.464\underline{/+90^\circ}) \end{aligned} \tag{2.6.13}$$

Both the Δ-load currents given by (2.6.12) and the line currents given by (2.6.13) are balanced. Thus the sum of balanced Δ-load currents $(I_{AB} + I_{BC} + I_{CA})$ equals zero. The sum of line currents $(I_a + I_b + I_c)$ is always zero for a Δ-connected load even if the system is unbalanced, since there is no neutral wire. Comparison of (2.6.12) and (2.6.13) leads to the following conclusion:

> For a balanced-Δ load supplied by a balanced positive-sequence source, the line currents into the load are $\sqrt{3}$ times the Δ-load currents and lag by 30°. That is,
>
> $$\begin{aligned} I_a &= \sqrt{3}\, I_{AB}\underline{/-30^\circ} \\ I_b &= \sqrt{3}\, I_{BC}\underline{/-30^\circ} \\ I_c &= \sqrt{3}\, I_{CA}\underline{/-30^\circ} \end{aligned} \tag{2.6.14}$$

This result is summarized in Figure 2.14.

Figure 2.14

Phasor diagram of line currents and load currents for a balanced-Δ load

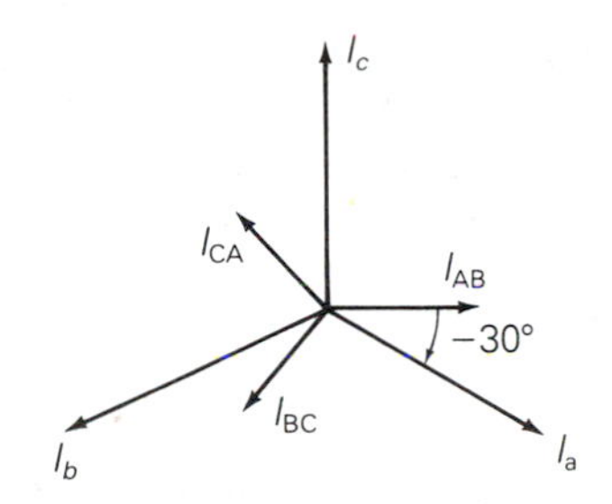

Δ–Y Conversion for Balanced Loads

Figure 2.15 shows the conversion of a balanced-Δ load to a balanced-Y load. If balanced voltages are applied, then these loads will be equivalent as viewed from their terminal buses A, B, and C when the line currents into the Δ load are the same as the line currents into the Y load. For the Δ load,

$$I_A = \sqrt{3} I_{AB} \underline{/-30^\circ} = \frac{\sqrt{3} E_{AB} \underline{/-30^\circ}}{Z_\Delta} \tag{2.6.15}$$

and for the Y load,

$$I_A = \frac{E_{AN}}{Z_Y} = \frac{E_{AB} \underline{/-30^\circ}}{\sqrt{3} Z_Y} \tag{2.6.16}$$

Figure 2.15

Δ–Y conversion for balanced loads

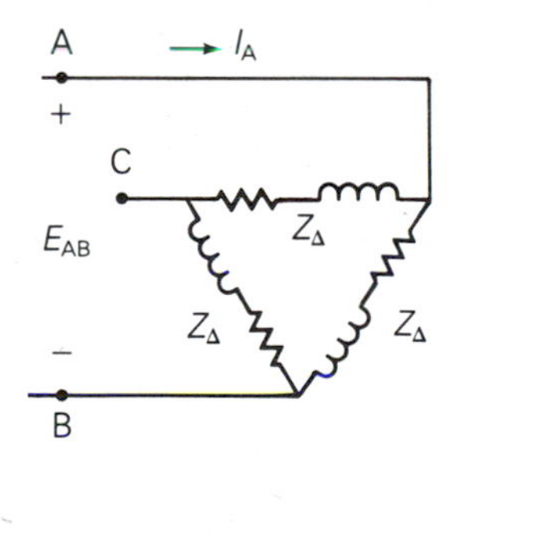

(a) Balanced-Δ load

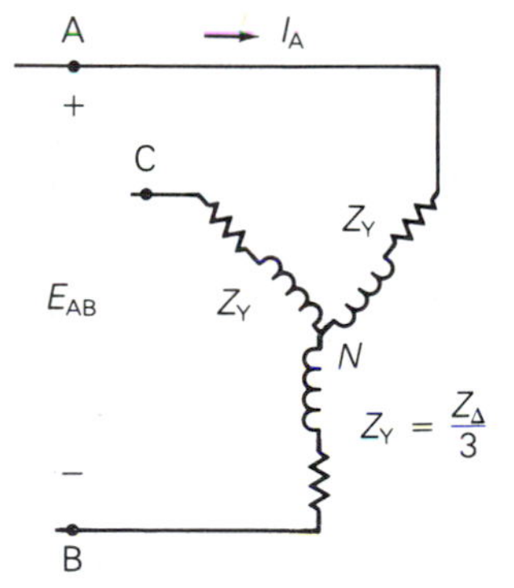

(b) Equivalent balanced-Y load

Comparison of (2.6.15) and (2.6.16) indicates that I_A will be the same for both the Δ and Y loads when

$$Z_Y = \frac{Z_\Delta}{3} \tag{2.6.17}$$

Also, the other line currents I_B and I_C into the Y load will equal those into the Δ load when $Z_Y = Z_\Delta/3$, since these loads are balanced. Thus a balanced-Δ load can be converted to an equivalent balanced-Y load by dividing the Δ-load impedance by 3. The angles of these Δ- and equivalent Y-load impedances are the same. Similarly, a balanced-Y load can be converted to an equivalent balanced-Δ load using $Z_\Delta = 3Z_Y$.

Example 2.5 A balanced, positive-sequence, Y-connected voltage source with $E_{ab} = 480\underline{/0^\circ}$ volts is applied to a balanced-Δ load with $Z_\Delta = 30\underline{/40^\circ}\ \Omega$. The line impedance between the source and load is $Z_L = 1\underline{/85^\circ}\ \Omega$ for each phase. Calculate the line currents, the Δ-load currents, and the voltages at the load terminals.

Solution The solution is most easily obtained as follows. First convert the Δ load to an equivalent Y. Then connect the source and Y-load neutrals with a zero-ohm neutral wire. The connection of the neutral wire has no affect on the circuit, since the neutral current $I_n = 0$ in a balanced system. The resulting circuit is shown in Figure 2.16. The line currents are

$$I_a = \frac{E_{an}}{Z_L + Z_Y} = \frac{\dfrac{480}{\sqrt{3}}\underline{/-30^\circ}}{1\underline{/85^\circ} + \dfrac{30}{3}\underline{/40^\circ}}$$

$$= \frac{277.1\underline{/-30^\circ}}{(0.0876 + j0.9962) + (7.660 + j6.428)}$$

$$= \frac{277.1\underline{/-30^\circ}}{(7.747 + j7.424)} = \frac{277.1\underline{/-30^\circ}}{10.73\underline{/43.78^\circ}} = 25.83\underline{/-73.78^\circ} \quad \text{A}$$

$$I_b = 25.83\underline{/166.22^\circ} \quad \text{A} \tag{2.6.18}$$

$$I_c = 25.83\underline{/46.22^\circ} \quad \text{A}$$

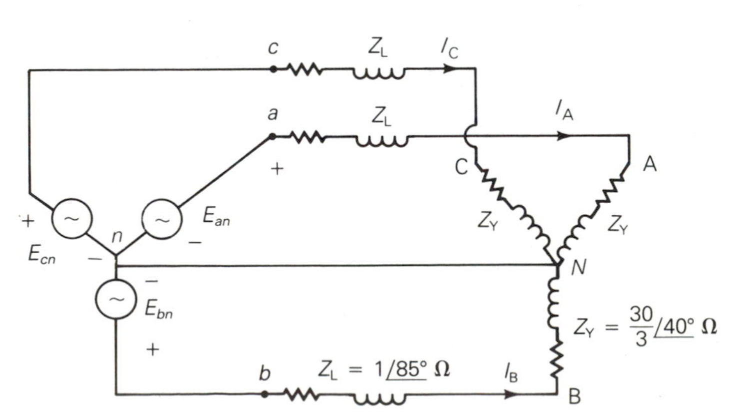

Figure 2.16 Circuit diagram for Example 2.5

The Δ-load currents are, from (2.6.14),

$$I_{AB} = \frac{I_a}{\sqrt{3}}\underline{/+30^\circ} = \frac{25.83}{\sqrt{3}}\underline{/-73.78^\circ + 30^\circ} = 14.91\underline{/-43.78^\circ} \quad \text{A}$$

$$I_{BC} = 14.91\underline{/-163.78^\circ} \quad \text{A} \tag{2.6.19}$$

$$I_{CA} = 14.91\underline{/+76.22^\circ} \quad \text{A}$$

The voltages at the load terminals are

$$E_{AB} = Z_\Delta I_{AB} = (30\underline{/40^\circ})(14.91\underline{/-43.78^\circ}) = 447.3\underline{/-3.780^\circ}$$

$$E_{BC} = 447.3\underline{/-123.78^\circ} \tag{2.6.20}$$

$$E_{CA} = 447.3\underline{/116.22^\circ} \quad \text{volts}$$

◁

Equivalent Line-to-Neutral Diagrams

When working with balanced three-phase circuits, only one phase need be analyzed. Δ loads can be converted to Y loads, and all source and load neutrals can be connected with a zero-ohm neutral wire without changing the solution. Then one phase of the circuit can be solved. The voltages and currents in the other two phases are equal in magnitude to and $\pm 120^\circ$ out of phase with those of the solved phase. Figure 2.17 shows an equivalent line-to-neutral diagram for one phase of the circuit in Example 2.5.

Figure 2.17

Equivalent line-to-neutral diagram for the circuit of Example 2.5

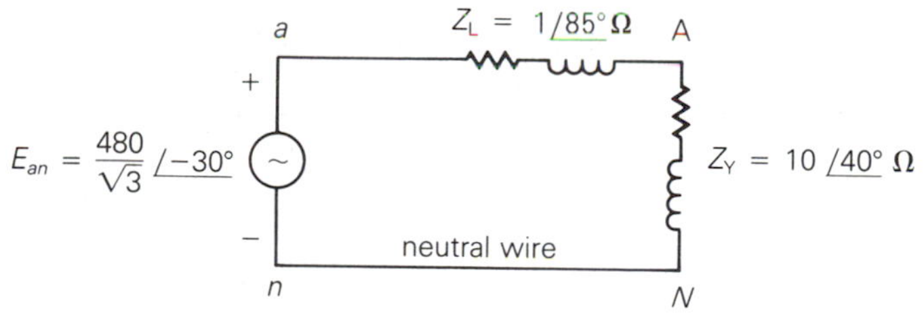

a $Z_L = 1\underline{/85^\circ}\,\Omega$ A

+

$E_{an} = \frac{480}{\sqrt{3}}\underline{/-30^\circ}$

$Z_Y = 10\underline{/40^\circ}\,\Omega$

−

neutral wire

n N

When discussing three-phase systems in this text, voltages shall be rms line-to-line voltages unless otherwise indicated. This is standard industry practice.

Section 2.7

Power in Balanced Three-phase Circuits

In this section we discuss instantaneous power and complex power for balanced three-phase generators and motors and for balanced-Y and Δ-impedance loads.

Instantaneous Power: Balanced Three-phase Generators

Figure 2.18 shows a Y-connected generator represented by three voltage sources with their neutrals connected at bus n and by three identical generator impedances Z_g. Assume that the generator is operating under balanced steady-state conditions with the instantaneous generator terminal voltage given by

$$v_{an}(t) = \sqrt{2}V_{LN}\cos(\omega t + \delta) \quad \text{volts} \tag{2.7.1}$$

and with the instantaneous current leaving the positive terminal of phase a given by

$$i_a(t) = \sqrt{2}I_L\cos(\omega t + \beta) \quad \text{A} \tag{2.7.2}$$

where V_{LN} is the rms line-to-neutral voltage and I_L is the rms line current. The instantaneous power $p_a(t)$ delivered by phase a of the generator is

$$\begin{aligned} p_a(t) &= v_{an}(t)i_a(t) \\ &= 2V_{LN}I_L\cos(\omega t + \delta)\cos(\omega t + \beta) \\ &= V_{LN}I_L\cos(\delta - \beta) + V_{LN}I_L\cos(2\omega t + \delta + \beta) \quad \text{W} \end{aligned} \tag{2.7.3}$$

Figure 2.18

Y-connected generator

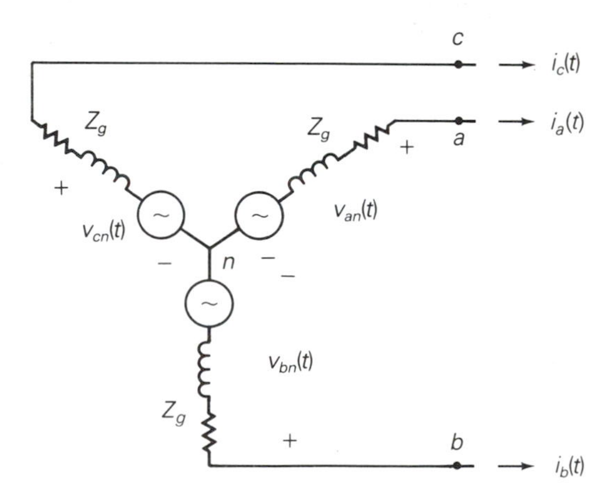

Assuming balanced operating conditions, the voltages and currents of phases b and c have the same magnitudes as those of phase a and are $\pm 120°$

out of phase with phase a. Therefore the instantaneous power delivered by phase b is

$$p_b(t) = 2V_{LN}I_L \cos(\omega t + \delta - 120°)\cos(\omega t + \beta - 120°)$$

$$= V_{LN}I_L \cos(\delta - \beta) + V_{LN}I_L \cos(2\omega t + \delta + \beta - 240°) \quad \text{W} \quad (2.7.4)$$

and by phase c,

$$p_c(t) = 2V_{LN}I_L \cos(\omega t + \delta + 120°)\cos(\omega t + \beta + 120°)$$

$$= V_{LN}I_L \cos(\delta - \beta) + V_{LN}I_L \cos(2\omega t + \delta + \beta + 240°) \quad \text{W} \quad (2.7.5)$$

The total instantaneous power $p_{3\theta}(t)$ delivered by the three-phase generator is the sum of the instantaneous powers delivered by each phase. Using (2.7.3)–(2.7.5):

$$p_{3\theta}(t) = p_a(t) + p_b(t) + p_c(t)$$

$$= 3V_{LN}I_L \cos(\delta - \beta) + V_{LN}I_L[\cos(2\omega t + \delta + \beta) + \cos(2\omega t + \delta + \beta - 240°) + \cos(2\omega t + \delta + \beta + 240°)] \quad \text{W} \quad (2.7.6)$$

The three cosine terms within the brackets of (2.7.6) can be represented by a *balanced* set of three phasors. Therefore the sum of these three terms is zero for any value of δ, for any value of β, and for all values of t. Equation (2.7.6) then reduces to

$$p_{3\theta}(t) = P_{3\theta} = 3V_{LN}I_L \cos(\delta - \beta) \quad \text{W} \quad (2.7.7)$$

Equation (2.7.7) can be written in terms of the line-to-line voltage V_{LL} instead of the line-to-neutral voltage V_{LN}. Under balanced operating conditions, $V_{LN} = V_{LL}/\sqrt{3}$ and

$$P_{3\theta} = \sqrt{3}V_{LL}I_L \cos(\delta - \beta) \quad \text{W} \quad (2.7.8)$$

Inspection of (2.7.8) leads to the following conclusion:

> The total instantaneous power delivered by a three-phase generator under balanced operating conditions is not a function of time, but a *constant*, $p_{3\theta}(t) = P_{3\theta}$.

Instantaneous Power: Balanced Three-phase Motors and Impedance Loads

The total instantaneous power absorbed by a three-phase motor under balanced steady-state conditions is also a constant. Figure 2.18 can be used to represent a three-phase motor by reversing the line currents to enter rather than leave the positive terminals. Then (2.7.1)–(2.7.8), valid for power *delivered* by a generator, are also valid for power *absorbed* by a motor. These equations are also valid for the instantaneous power absorbed by a balanced three-phase impedance load.

Complex Power: Balanced Three-phase Generators

The phasor representations of the voltage and current in (2.7.1) and (2.7.2) are

$$V_{an} = \mathrm{V_{LN}}\underline{/\delta} \quad \text{volts} \tag{2.7.9}$$

$$I_a = \mathrm{I_L}\underline{/\beta} \quad \mathrm{A} \tag{2.7.10}$$

where I_a leaves positive terminal "a" of the generator. The complex power S_a delivered by phase a of the generator is

$$\begin{aligned} S_a &= V_{an} I_a^* = \mathrm{V_{LN} I_L}\underline{/(\delta - \beta)} \\ &= \mathrm{V_{LN} I_L} \cos(\delta - \beta) + j\mathrm{V_{LN} I_L} \sin(\delta - \beta) \end{aligned} \tag{2.7.11}$$

Under balanced operating conditions, the complex powers delivered by phases b and c are identical to S_a, and the total complex power $S_{3\theta}$ delivered by the generator is

$$\begin{aligned} S_{3\theta} &= S_a + S_b + S_c = 3S_a \\ &= 3\mathrm{V_{LN} I_L}\underline{/(\delta - \beta)} \\ &= 3\mathrm{V_{LN} I_L} \cos(\delta - \beta) + j3\mathrm{V_{LN} I_L} \sin(\delta - \beta) \end{aligned} \tag{2.7.12}$$

In terms of the total real and reactive powers,

$$S_{3\theta} = \mathrm{P}_{3\theta} + j\mathrm{Q}_{3\theta} \tag{2.7.13}$$

where

$$\begin{aligned} \mathrm{P}_{3\theta} &= \mathrm{Re}(S_{3\theta}) = 3\mathrm{V_{LN} I_L} \cos(\delta - \beta) \\ &= \sqrt{3}\mathrm{V_{LL} I_L} \cos(\delta - \beta) \quad \mathrm{W} \end{aligned} \tag{2.7.14}$$

and

$$\begin{aligned} \mathrm{Q}_{3\theta} &= \mathrm{Im}(S_{3\theta}) = 3\mathrm{V_{LN} I_L} \sin(\delta - \beta) \\ &= \sqrt{3}\mathrm{V_{LL} I_L} \sin(\delta - \beta) \quad \mathrm{var} \end{aligned} \tag{2.7.15}$$

Also, the total apparent power is

$$\mathrm{S}_{3\theta} = |S_{3\theta}| = 3\mathrm{V_{LN} I_L} = \sqrt{3}\mathrm{V_{LL} I_L} \quad \mathrm{VA} \tag{2.7.16}$$

Complex Power: Balanced Three-phase Motors

The preceding expressions for complex, real, reactive, and apparent power *delivered* by a three-phase generator are also valid for the complex, real, reactive, and apparent power *absorbed* by a three-phase motor.

Complex Power: Balanced-Y and -Δ Impedance Loads

Equations (2.7.13)–(2.7.16) are also valid for balanced-Y and -Δ impedance loads. For a balanced-Y load, the line-to-neutral voltage across the phase a load impedance and the current entering the positive terminal of that load impedance can be represented by (2.7.9) and (2.7.10). Then (2.7.11)–(2.7.16) are valid for the power absorbed by the balanced-Y load.

For a balanced-Δ load, the line-to-line voltage across the phase a–b load impedance and the current into the positive terminal of that load impedance can be represented by

$$V_{ab} = \text{V}_{\text{LL}}\underline{/\delta} \quad \text{volts} \tag{2.7.17}$$

$$I_{ab} = \text{I}_{\Delta}\underline{/\beta} \quad \text{A} \tag{2.7.18}$$

where V_{LL} is the rms line-to-line voltage and I_{Δ} is the rms Δ-load current. The complex power S_{ab} absorbed by the phase a–b load impedance is then

$$S_{ab} = V_{ab} I_{ab}^{*} = \text{V}_{\text{LL}}\,\text{I}_{\Delta}\underline{/(\delta - \beta)} \tag{2.7.19}$$

The total complex power absorbed by the Δ load is

$$\begin{aligned} S_{3\theta} &= S_{ab} + S_{bc} + S_{ca} = 3S_{ab} \\ &= 3\text{V}_{\text{LL}}\text{I}_{\Delta}\underline{/(\delta - \beta)} \\ &= 3\text{V}_{\text{LL}}\text{I}_{\Delta}\cos(\delta - \beta) + j3\text{V}_{\text{LL}}\text{I}_{\Delta}\sin(\delta - \beta) \end{aligned} \tag{2.7.20}$$

Rewriting (2.7.19) in terms of the total real and reactive power,

$$S_{3\theta} = \text{P}_{3\theta} + j\text{Q}_{3\theta} \tag{2.7.21}$$

$$\begin{aligned} \text{P}_{3\theta} &= \text{Re}(S_{3\theta}) = 3\text{V}_{\text{LL}}\text{I}_{\Delta}\cos(\delta - \beta) \\ &= \sqrt{3}\text{V}_{\text{LL}}\text{I}_{\text{L}}\cos(\delta - \beta) \quad \text{W} \end{aligned} \tag{2.7.22}$$

$$\begin{aligned} \text{Q}_{3\theta} &= \text{Im}(S_{3\theta}) = 3\text{V}_{\text{LL}}\text{I}_{\Delta}\sin(\delta - \beta) \\ &= \sqrt{3}\text{V}_{\text{LL}}\text{I}_{\text{L}}\sin(\delta - \beta) \quad \text{var} \end{aligned} \tag{2.7.23}$$

where the Δ-load current I_{Δ} is expressed in terms of the line current $\text{I}_{\text{L}} = \sqrt{3}\text{I}_{\Delta}$ in (2.7.22) and (2.7.23). Also, the total apparent power is

$$\text{S}_{3\theta} = |S_{3\theta}| = 3\text{V}_{\text{LL}}\text{I}_{\Delta} = \sqrt{3}\text{V}_{\text{LL}}\text{I}_{\text{L}} \quad \text{VA} \tag{2.7.24}$$

Equations (2.7.21)–(2.7.24) developed for the balanced-Δ load are identical to (2.7.13)–(2.7.16).

Section 2.8

Advantages of Balanced Three-phase versus Single-phase Systems

Figure 2.19 shows three separate single-phase systems. Each single-phase system consists of the following identical components: (1) a generator represented by a voltage source and a generator impedance Z_g; (2) a forward and return conductor represented by two series line impedances Z_{L}; (3) a load represented by an impedance Z_Y. The three single-phase systems, although completely separated, are drawn in a Y configuration in the figure to illustrate two advantages of three-phase systems.

Figure 2.19

Three single-phase systems

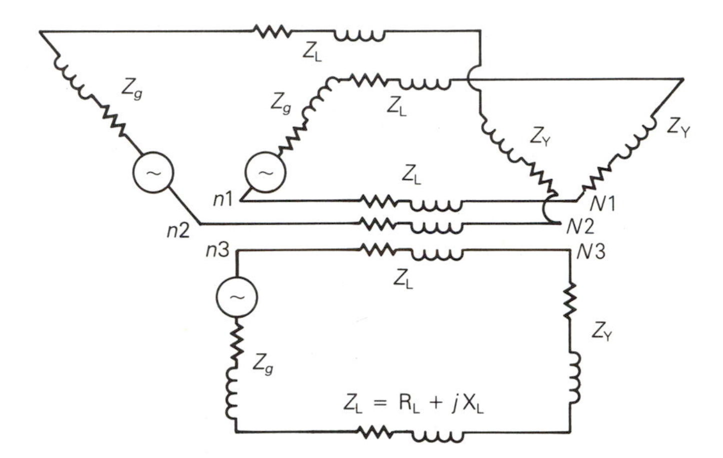

Each separate single-phase system requires that *both* the forward and return conductors have a current capacity (or *ampacity*) equal to or greater than the load current. However, if the source and load neutrals in Figure 2.19 are connected to form a three-phase system, and if the source voltages are balanced with equal magnitudes and with 120° displacement between phases, then the neutral current will be zero [see (2.6.10)] and the three neutral conductors can be removed. Thus the balanced three-phase system, while delivering the same power to the three load impedances Z_Y, requires only half the number of conductors needed for the three separate single-phase systems. Also, the total I^2R line losses in the three-phase system are only half those of the three separate single-phase systems, and the line-voltage drop between the source and load in the three-phase system is half that of each single-phase system. Therefore one advantage of balanced three-phase systems over separate single-phase systems is reduced capital and operating costs of transmission and distribution, as well as better voltage regulation.

Some three-phase systems such as Δ-connected systems and three-wire Y-connected systems do not have any neutral conductor. However, the majority of three-phase systems are four-wire Y-connected systems, where a grounded neutral conductor is used. Neutral conductors are used to reduce transient overvoltages, which can be caused by lightning strikes and by line-switching operations, and to carry unbalanced currents, which can occur during unsymmetrical short-circuit conditions. Neutral conductors for transmission lines are typically smaller in size and ampacity than the phase conductors because the neutral current is nearly zero under normal operating conditions. Thus the cost of a neutral conductor is substantially less than that of a phase conductor. The capital and operating costs of three-phase transmission and distribution systems with or without neutral conductors are substantially less than those of separate single-phase systems.

A second advantage of three-phase systems is that the total instantaneous electric power delivered by a three-phase generator under balanced steady-state conditions is (nearly) constant, as shown in Section 2.7. A three-phase generator (constructed with its field winding on one shaft and with its three-phase windings equally displaced by 120° on the stator core) will also have a nearly constant mechanical input power under balanced steady-state conditions, since the mechanical input power equals the electrical output power plus the small generator losses. Furthermore, the mechanical shaft torque, which equals mechanical input power divided by mechanical radian frequency ($T_{mech} = P_{mech}/\omega_m$) is nearly constant.

On the other hand, the equation for the instantaneous electric power delivered by a single-phase generator under balanced steady-state conditions is the same as the instantaneous power delivered by one phase of a three-phase generator, given by $p_a(t)$ in (2.7.3). As shown in that equation, $p_a(t)$ has two components: a constant and a double-frequency sinusoid. Both the mechanical input power and the mechanical shaft torque of the single-phase generator will have corresponding double-frequency components that create shaft vibration and noise, which could cause shaft failure in large machines. Accordingly, most electric generators and motors rated 5 kVA and higher are constructed as three-phase machines in order to produce nearly constant torque and thereby minimize shaft vibration and noise.

Problems

Section 2.1

2.1 Given the complex numbers $A_1 = 5\underline{/30^\circ}$ and $A_2 = -3 + j4$, (a) convert A_1 to rectangular form; (b) convert A_2 to polar and exponential form; (c) calculate $A_3 = (A_1 + A_2)$, giving your answer in polar form; (d) calculate $A_4 = A_1 A_2$, giving your answer in rectangular form; (e) calculate $A_5 = A_1/(A_2^*)$, giving your answer in exponential form.

2.2 The instantaneous voltage across a circuit element is $v(t) = 359.3\cos(\omega t + 15^\circ)$ volts, and the instantaneous current entering the positive terminal of the circuit element is $i(t) = 100\sin(\omega t + 5^\circ)$ A. For both the current and voltage, determine: (a) the maximum value; (b) the rms value; (c) the phasor expression, using $\cos(\omega t)$ as the reference.

Section 2.2

2.3 For the circuit element of Problem 2.2, calculate: (a) the instantaneous power absorbed; (b) the real power (state whether it is delivered or absorbed); (c) the reactive power (state whether delivered or absorbed); (d) the power factor (state whether lagging or leading).

[*Note*: By convention the power factor $\cos(\delta - \beta)$ is positive. If $|\delta - \beta|$ is greater than 90°, then the reference direction for current may be reversed, resulting in a positive value of $\cos(\delta - \beta)$].

2.4 The voltage $v(t) = 359.3\cos(\omega t)$ volts is applied to a load consisting of a 10-Ω resistor in parallel with a capacitive reactance $X_C = 25\,\Omega$. Calculate: (a) the instantaneous power absorbed by the resistor; (b) the instantaneous power absorbed by the capacitor; (c) the real power absorbed by the resistor; (d) the reactive power delivered by the capacitor; (e) the load power factor.

2.5 Repeat Problem 2.4 if the resistor and capacitor are connected in series.

Section 2.3

2.6 Consider a single-phase load with an applied voltage $v(t) = 150\cos(\omega t + 10°)$ volts and load current $i(t) = 5\cos(\omega t - 50°)$ A. (a) Determine the power triangle. (b) Find the power factor and specify whether it is lagging or leading. (c) Calculate the reactive power supplied by capacitors in parallel with the load that correct the power factor to 0.9 lagging.

2.7 A circuit consists of two impedances, $Z_1 = 20\underline{/30°}\,\Omega$ and $Z_2 = 25\underline{/60°}\,\Omega$, in parallel, supplied by a source voltage $V = 100\underline{/60°}$ volts. Determine the power triangle for each of the impedances and for the source.

2.8 An industrial plant consisting primarily of induction motor loads absorbs 500 kW at 0.6 power factor lagging. (a) Compute the required kVA rating of a shunt capacitor to improve the power factor to 0.9 lagging. (b) If a synchronous motor rated 500 hp with 90% efficiency operating at rated load and at unity power factor is added to the plant instead of the capacitor, calculate the resulting power factor. Assume constant voltage. (1 hp = 0.747 kW)

2.14

2.9 The real power delivered by a source to two impedances, $Z_1 = 3 + j4\,\Omega$ and $Z_2 = 10\,\Omega$, connected in parallel, is 1100 W. Determine (a) the real power absorbed by each of the impedances and (b) the source current.

2.15

2.10 A single-phase source has a terminal voltage $V = 120\underline{/0°}$ volts and a current $I = 10\underline{/30°}$ A, which leaves the positive terminal of the source. Determine the real and reactive power, and state whether the source is delivering or absorbing each.

Section 2.4

2.11 For the circuit shown in Figure 2.20, convert the voltage sources to equivalent current sources and write nodal equations in matrix format using bus 0 as the reference bus. Do not solve the equations.

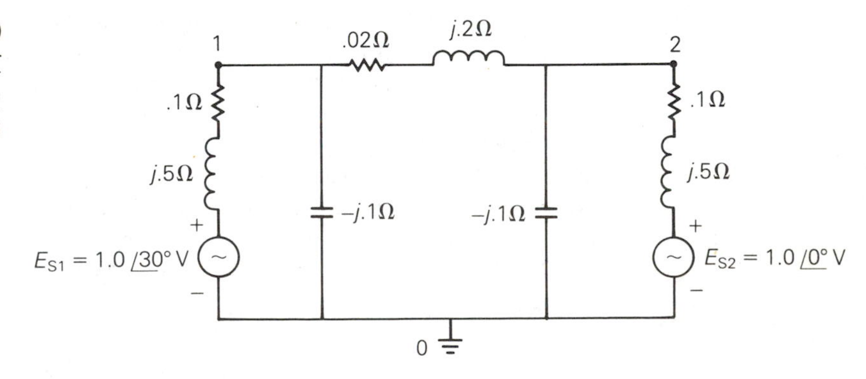

Figure 2.20 Circuit diagram for problems 2.11, 2.12 and 2.13

2.12 For the circuit shown in Figure 2.20, write a computer program that uses the sources, impedances, and bus connections as input data to: (a) compute the 2 × 2 bus admittance matrix Y; (b) convert the voltage sources to current sources and compute the vector of source currents into buses 1 and 2.

Section 2.5

2.13 For the circuit shown in Figure 2.20, use the CMI and CMM subroutines, as well as the results of Problem 2.12, to compute the bus voltages V_{10} and V_{20}.

2.14 Use the CMI subroutine to invert the 3×3 matrix A given by (3.1.8) to verify the resulting inverse matrix A^{-1} given by (3.1.10). Note that in (3.1.8), $a = 1\underline{/120^\circ}$.

Sections 2.6 and 2.7

2.15 A balanced three-phase 208-V source supplies a balanced three-phase load. If the line current I_A is measured to be 10 A, and is in phase with the line-to-line voltage V_{BC}, find the per-phase load impedance if the load is (a) Y-connected, (b) Δ-connected.

2.16 A three-phase 25-kVA, 480-V, 60-Hz alternator, operating under balanced steady-state conditions, supplies a line current of 20 A per phase at a 0.8 lagging power factor and at rated voltage. Determine the power triangle for this operating condition.

2.17 A balanced Δ-connected impedance load with $(12 + j9)\,\Omega$ per phase is supplied by a balanced three-phase 60-Hz, 208-V source. (a) Calculate the line current, the total real and reactive power absorbed by the load, the load power factor, and the apparent load power. (b) Sketch a phasor diagram showing the line currents, the line-to-line source voltages, and the Δ-load currents. Assume positive sequence and use V_{ab} as the reference.

2.18 Two balanced Y-connected loads, one drawing 10 kW at 0.8 p.f. lagging and the other 15 kW at 0.9 p.f. leading, are connected in parallel and supplied by a balanced three-phase Y-connected, 480-V source. (a) Determine the source current. (b) If the load neutrals are connected to the source neutral by a zero-ohm neutral wire through an ammeter, what will the ammeter read?

2.19 Three identical impedances $Z_\Delta = 30\underline{/30^\circ}\,\Omega$ are connected in Δ to a balanced three-phase 208-V source by three identical line conductors with impedance $Z_L = (0.8 + j0.6)\,\Omega$ per line. (a) Calculate the line-to-line voltage at the load terminals. (b) Repeat part (a) when a Δ-connected capacitor bank with reactance $(-j60)\,\Omega$ per phase is connected in parallel with the load.

2.20 Two three-phase generators supply a three-phase load through separate three-phase lines. The load absorbs 30 kW at 0.8 p.f. lagging. The line impedance is $(1.4 + j1.6)\,\Omega$ per phase between generator G1 and the load, and $(0.8 + j1)\,\Omega$ per phase between generator G2 and the load. If generator G1 supplies 15 kW at 0.8 p.f. lagging, with a terminal voltage of 460 V line-to-line, determine: (a) the voltage at the load terminals; (b) the voltage at the terminals of generator G2; and (c) the real and reactive power supplied by generator G2. Assume balanced operation.

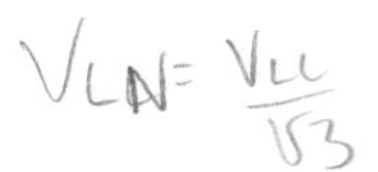

2.21 Figure 2.21 gives the general Δ–Y transformation. (a) Show that the general transformation reduces to that given in Figure 2.15 for a balanced three-phase load. (b) Determine the impedances of the equivalent Y for the following Δ impedances: $Z_{AB} = j10$, $Z_{BC} = j20$, and $Z_{CA} = -j25\,\Omega$.

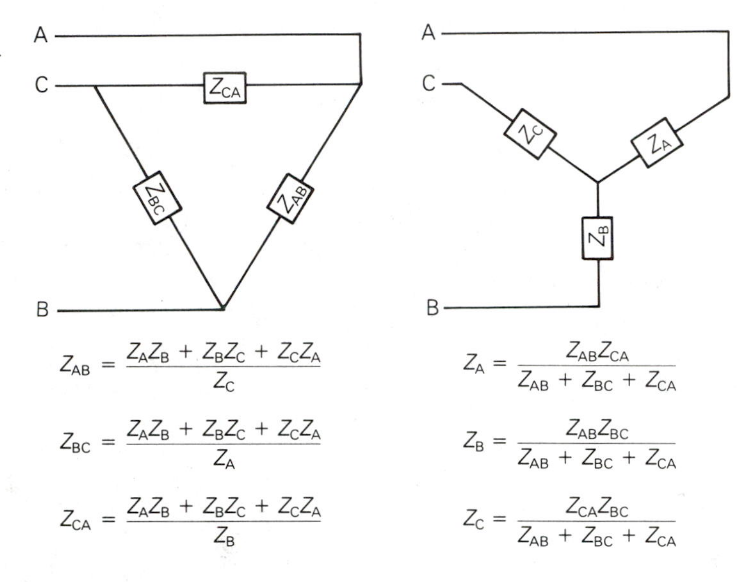

Figure 2.21
General Δ–Y transformation

References

1. W. H. Hayt, Jr., and J. E. Kemmerly, *Engineering Circuit Analysis,* 3d ed. New York: McGraw-Hill, 1978.
2. W. A. Blackwell, and L. L. Grigsby, *Introductory Network Theory* (Boston: PWS Publishers, 1985).
3. A. E. Fitzgerald, D. E. Higginbotham, and A. Grabel, *Basic Electrical Engineering* (New York: McGraw-Hill, 1981).
4. W. D. Stevenson, Jr., *Elements of Power System Analysis,* 4th ed. (New York: McGraw-Hill, 1982).

CHAPTER 3

Symmetrical Components

The method of symmetrical components, first developed by C. L. Fortescue in 1918, is a powerful technique for analyzing unbalanced three-phase systems. Fortescue defined a linear transformation from phase components to a new set of components called *symmetrical components*. The advantage of this transformation is that for balanced three-phase networks the equivalent circuits obtained for the symmetrical components, called *sequence networks*, are separated into three uncoupled networks. Furthermore, for unbalanced three-phase systems, the three sequence networks are connected only at points of unbalance. As a result, sequence networks for many cases of unbalanced three-phase systems are relatively easy to analyze.

The symmetrical component method is basically a modeling technique that permits systematic analysis and design of three-phase systems. Decoupling a detailed three-phase network into three simpler sequence networks reveals complicated phenomena in more simplistic terms. Sequence network results can then be superposed to obtain three-phase network results. As an example, the application of symmetrical components to unsymmetrical short-circuit studies (see Chapter 9) is indispensable.

The objective of this chapter is to introduce the concept of symmetrical components in order to lay a foundation and provide a framework for later chapters covering both equipment models as well as power-system analysis and design methods. In Section 3.1, symmetrical components are defined. In Sections 3.2, 3.3, and 3.4 sequence networks of loads, series impedances, and rotating machines are presented. Complex power in sequence networks is discussed in Section 3.5, and symmetrical components computer programs are presented in Section 3.6. Although Fortescue's original work is valid for polyphase systems with n phases, only three-phase systems will be considered here.

Section 3.1

Definition of Symmetrical Components

Assume that a set of three-phase voltages designated V_a, V_b, and V_c is given. In accordance with Fortescue, these phase voltages are resolved into the following three sets of sequence components:

1. *Zero-sequence* components, consisting of three phasors with equal magnitudes and with zero phase displacement, as shown in Figure 3.1(a).
2. *Positive-sequence* components, consisting of three phasors with equal magnitudes, $\pm 120°$ phase displacement, and positive sequence, as in Figure 3.1(b).
3. *Negative-sequence* components, consisting of three phasors with equal magnitudes, $\pm 120°$ phase displacement, and negative sequence, as in Figure 3.1(c).

Figure 3.1

Resolving phase voltages into three sets of sequence components

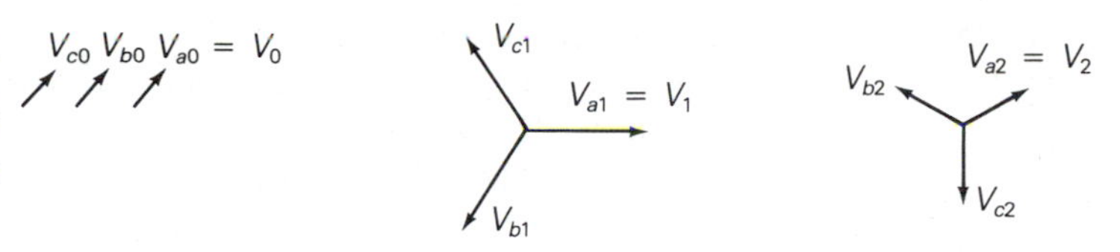

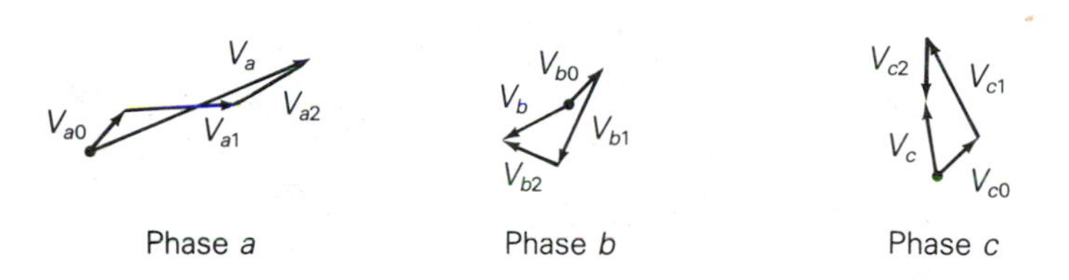

In this text we will work only with the zero-, positive-, and negative-sequence components of phase a, which are V_{a0}, V_{a1}, and V_{a2}, respectively. For simplicity, we drop the subscript a and denote these sequence components as V_0, V_1, and V_2. They are defined by the following transformation:

$$\begin{bmatrix} V_a \\ V_b \\ V_c \end{bmatrix} = \begin{bmatrix} 1 & 1 & 1 \\ 1 & a^2 & a \\ 1 & a & a^2 \end{bmatrix} \begin{bmatrix} V_0 \\ V_1 \\ V_2 \end{bmatrix} \tag{3.1.1}$$

$$\text{where } a = 1\underline{/120°} = \frac{-1}{2} + j\frac{\sqrt{3}}{2} \tag{3.1.2}$$

Writing (3.1.1) as three separate equations:

$$V_a = V_0 + V_1 + V_2 \tag{3.1.3}$$

$$V_b = V_0 + a^2V_1 + aV_2 \tag{3.1.4}$$

$$V_c = V_0 + aV_1 + a^2V_2 \tag{3.1.5}$$

In (3.1.2) a is a complex number with unit magnitude and a 120° phase angle. When any phasor is multiplied by a, that phasor rotates by 120° (counterclockwise). Similarly, when any phasor is multiplied by $a^2 = (1\underline{/120°})(1\underline{/120°}) = 1\underline{/240°}$, the phasor rotates by 240°. Table 3.1 lists some common identities involving a.

Table 3.1

Common identities involving $a = 1\underline{/120°}$

$a^4 = a = 1\underline{/120°}$
$a^2 = 1\underline{/240°}$
$a^3 = 1\underline{/0°}$
$1 + a + a^2 = 0$
$1 - a = \sqrt{3}\underline{/-30°}$
$1 - a^2 = \sqrt{3}\underline{/+30°}$
$a^2 - a = \sqrt{3}\underline{/270°}$
$ja = 1\underline{/210°}$
$1 + a = -a^2 = 1\underline{/60°}$
$1 + a^2 = -a = 1\underline{/-60°}$
$a + a^2 = -1 = 1\underline{/180°}$

The complex number a is similar to the well-known complex number $j = \sqrt{-1} = 1\underline{/90°}$. Thus the only difference between j and a is that the angle of j is 90°, and that of a is 120°.

Equation (3.1.1) can be rewritten more compactly using matrix notation. We define the following vectors $\boldsymbol{V}_p$ and $\boldsymbol{V}_s$, and matrix $\boldsymbol{A}$:

$$\boldsymbol{V}_p = \begin{bmatrix} V_a \\ V_b \\ V_c \end{bmatrix} \tag{3.1.6}$$

$$\boldsymbol{V}_s = \begin{bmatrix} V_0 \\ V_1 \\ V_2 \end{bmatrix} \tag{3.1.7}$$

$$\boldsymbol{A} = \begin{bmatrix} 1 & 1 & 1 \\ 1 & a^2 & a \\ 1 & a & a^2 \end{bmatrix} \tag{3.1.8}$$

$\boldsymbol{V}_p$ is the column vector of phase voltages, $\boldsymbol{V}_s$ is the column vector of sequence voltages, and $\boldsymbol{A}$ is a 3×3 transformation matrix. Using these definitions, (3.1.1) becomes

$$\boldsymbol{V}_p = \boldsymbol{A}\boldsymbol{V}_s \tag{3.1.9}$$

The inverse of the $\boldsymbol{A}$ matrix is

$$\boldsymbol{A}^{-1} = \tfrac{1}{3}\begin{bmatrix} 1 & 1 & 1 \\ 1 & a & a^2 \\ 1 & a^2 & a \end{bmatrix} \tag{3.1.10}$$

Equation (3.1.10) can be verified by showing that the product $\boldsymbol{A}\boldsymbol{A}^{-1}$ is the unit matrix. Also, premultiplying (3.1.9) by $\boldsymbol{A}^{-1}$ gives

$$\boldsymbol{V}_s = \boldsymbol{A}^{-1}\,\boldsymbol{V}_p \tag{3.1.11}$$

Using (3.1.6), (3.1.7), and (3.1.10), then (3.1.11) becomes

$$\begin{bmatrix} V_0 \\ V_1 \\ V_2 \end{bmatrix} = \frac{1}{3}\begin{bmatrix} 1 & 1 & 1 \\ 1 & a & a^2 \\ 1 & a^2 & a \end{bmatrix}\begin{bmatrix} V_a \\ V_b \\ V_c \end{bmatrix} \tag{3.1.12}$$

Writing (3.1.12) as three separate equations,

$$V_0 = \tfrac{1}{3}(V_a + V_b + V_c) \tag{3.1.13}$$

$$V_1 = \tfrac{1}{3}(V_a + aV_b + a^2V_c) \tag{3.1.14}$$

$$V_2 = \tfrac{1}{3}(V_a + a^2V_b + aV_c) \tag{3.1.15}$$

Equation (3.1.13) shows that there is no zero-sequence voltage in a *balanced* three-phase system because the sum of three balanced phasors is zero. In an unbalanced three-phase system, line-to-neutral voltages may have a zero-sequence component. But line-to-line voltages never have a zero-sequence component, since by KVL their sum is always zero.

The symmetrical component transformation can also be applied to currents, as follows. Let

$$\boldsymbol{I}_p = \boldsymbol{A}\boldsymbol{I}_s \tag{3.1.16}$$

where $\boldsymbol{I}_p$ is a vector of phase currents,

$$\boldsymbol{I}_p = \begin{bmatrix} I_a \\ I_b \\ I_c \end{bmatrix} \tag{3.1.17}$$

and I_s is a vector of sequence currents,

$$I_s = \begin{bmatrix} I_0 \\ I_1 \\ I_2 \end{bmatrix} \tag{3.1.18}$$

Also,

$$I_s = A^{-1} I_p \tag{3.1.19}$$

Equations (3.1.16) and (3.1.19) can be written as separate equations as follows. The phase currents are

$$I_a = I_0 + I_1 + I_2 \tag{3.1.20}$$

$$I_b = I_0 + a^2 I_1 + a I_2 \tag{3.1.21}$$

$$I_c = I_0 + a I_1 + a^2 I_2 \tag{3.1.22}$$

and the sequence currents are

$$I_0 = \tfrac{1}{3}(I_a + I_b + I_c) \tag{3.1.23}$$

$$I_1 = \tfrac{1}{3}(I_a + a I_b + a^2 I_c) \tag{3.1.24}$$

$$I_2 = \tfrac{1}{3}(I_a + a^2 I_b + a I_c) \tag{3.1.25}$$

In a three-phase Y-connected system, the neutral current I_n is the sum of the line currents:

$$I_n = I_a + I_b + I_c \tag{3.1.26}$$

Comparing (3.1.26) and (3.1.23),

$$I_n = 3 I_0 \tag{3.1.27}$$

The neutral current equals three times the zero-sequence current. In a balanced Y-connected system, line currents have no zero-sequence component, since the neutral current is zero. Also, in any three-phase system with no neutral path, such as a Δ-connected system or a three-wire Y-connected system with an ungrounded neutral, line currents have no zero-sequence component.

The following three examples further illustrate symmetrical components.

Example 3.1 Calculate the sequence components of the following balanced line-to-neutral voltages with *abc* sequence:

$$V_p = \begin{bmatrix} V_{an} \\ V_{bn} \\ V_{cn} \end{bmatrix} = \begin{bmatrix} 277\underline{/0^\circ} \\ 277\underline{/-120^\circ} \\ 277\underline{/+120^\circ} \end{bmatrix} \quad \text{volts}$$

Solution Using (3.1.13)–(3.1.15):

$$V_0 = \tfrac{1}{3}[277\underline{/0^0} + 277\underline{/-120^\circ} + 277\underline{/+120^\circ}] = 0$$

$$V_1 = \tfrac{1}{3}[277\underline{/0^\circ} + 277\underline{/(-120^\circ + 120^\circ)} + 277\underline{/(120^\circ + 240^\circ)}]$$
$$= 277\underline{/0^\circ} \quad \text{volts} = V_{an}$$

$$V_2 = \tfrac{1}{3}[277\underline{/0^\circ} + 277\underline{/(-120^\circ + 240^\circ)} + 277\underline{/(120^\circ + 120^\circ)}]$$
$$= \tfrac{1}{3}[277\underline{/0^\circ} + 277\underline{/120^\circ} + 277\underline{/240^\circ}] = 0$$

This example illustrates the fact that balanced three-phase systems with *abc* sequence (or positive sequence) have no zero-sequence or negative-sequence components. For this example, the positive-sequence voltage V_1 equals V_{an}, and the zero-sequence and negative-sequence voltages are both zero. ◁

Example 3.2 A Y-connected load has balanced currents with *acb* sequence given by

$$I_p = \begin{bmatrix} I_a \\ I_b \\ I_c \end{bmatrix} = \begin{bmatrix} 10\underline{/0^\circ} \\ 10\underline{/+120^\circ} \\ 10\underline{/-120^\circ} \end{bmatrix} \quad \text{A}$$

Calculate the sequence currents.

Solution Using (3.1.23)–(3.1.25):

$$I_0 = \tfrac{1}{3}[10\underline{/0^\circ} + 10\underline{/120^\circ} + 10\underline{/-120^\circ}] = 0$$

$$I_1 = \tfrac{1}{3}[10\underline{/0^\circ} + 10\underline{/(120^\circ + 120^\circ)} + 10\underline{/(-120^\circ + 240^\circ)}]$$

$$= \tfrac{1}{3}[10\underline{/0^\circ} + 10\underline{/240^\circ} + 10\underline{/120^\circ}] = 0$$

$$I_2 = \tfrac{1}{3}[10\underline{/0^\circ} + 10\underline{/(120^\circ + 240^\circ)} + 10\underline{/(-120^\circ + 120^\circ)}]$$

$$= 10\underline{/0^\circ}\ \text{A} = I_a$$

This example illustrates the fact that balanced three-phase systems with *acb* sequence (or negative sequence) have no zero-sequence or positive-sequence components. For this example the negative-sequence current I_2 equals I_a, and the zero-sequence and positive-sequence currents are both zero. ◁

Example 3.3 A three-phase line feeding a balanced-Y load has one of its phases (phase b) open. The load neutral is grounded, and the unbalanced line currents are

$$I_p = \begin{bmatrix} I_a \\ I_b \\ I_c \end{bmatrix} = \begin{bmatrix} 10\underline{/0^\circ} \\ 0 \\ 10\underline{/120^\circ} \end{bmatrix} \quad \text{A}$$

Calculate the sequence currents and the neutral current.

Solution The circuit is shown in Figure 3.2. Using (3.1.23)–(3.1.25):

$$I_0 = \tfrac{1}{3}[10\underline{/0^\circ} + 0 + 10\underline{/120^\circ}]$$

$$= 3.333\underline{/60^\circ} \quad \text{A}$$

$$I_1 = \tfrac{1}{3}[10\underline{/0^\circ} + 0 + 10\underline{/(120^\circ + 240^\circ)}] = 6.667\underline{/0^\circ} \quad \text{A}$$

$$I_2 = \tfrac{1}{3}[10\underline{/0^\circ} + 0 + 10\underline{/(120^\circ + 120^\circ)}]$$

$$= 3.333\underline{/-60^\circ} \quad \text{A}$$

Using (3.1.26) the neutral current is

$$I_n = (10\underline{/0^\circ} + 0 + 10\underline{/120^\circ})$$

$$= 10\underline{/60^\circ} \text{ A} = 3I_0$$

This example illustrates the fact that *unbalanced* three-phase systems may have nonzero values for all sequence components. Also, the neutral current equals three times the zero-sequence current, as given by (3.1.27).

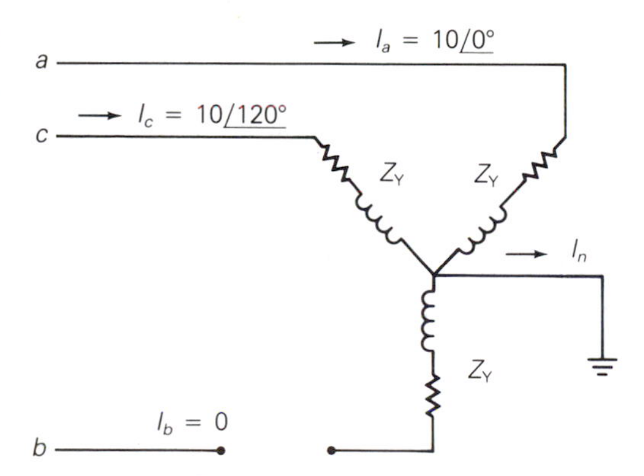

Figure 3.2 Circuit for Example 3.3

◁

Section 3.2

Sequence Networks of Impedance Loads

Figure 3.3 shows a balanced-Y impedance load. The impedance of each phase is designated Z_Y, and a neutral impedance Z_n is connected between the load neutral and ground. Note from Figure 3.3 that the line-to-ground voltage V_{ag} is

$$\begin{aligned} V_{ag} &= Z_Y I_a + Z_n I_n \\ &= Z_Y I_a + Z_n(I_a + I_b + I_c) \\ &= (Z_Y + Z_n) I_a + Z_n I_b + Z_n I_c \end{aligned} \tag{3.2.1}$$

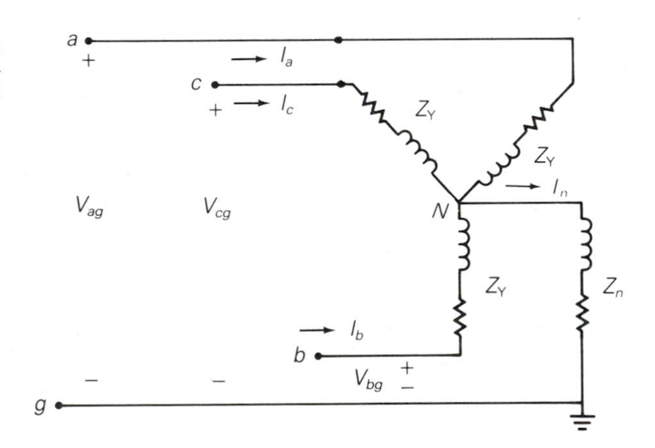

Figure 3.3

Balanced-Y impedance load

Similar equations can be written for V_{bg} and V_{cg}:

$$V_{bg} = Z_n I_a + (Z_Y + Z_n) I_b + Z_n I_c \tag{3.2.2}$$

$$V_{cg} = Z_n I_a + Z_n I_b + (Z_Y + Z_n) I_c \tag{3.2.3}$$

Equations (3.2.1)–(3.2.3) can be rewritten in matrix format:

$$\begin{bmatrix} V_{ag} \\ V_{bg} \\ V_{cg} \end{bmatrix} = \begin{bmatrix} (Z_Y + Z_n) & Z_n & Z_n \\ Z_n & (Z_Y + Z_n) & Z_n \\ Z_n & Z_n & (Z_Y + Z_n) \end{bmatrix} \begin{bmatrix} I_a \\ I_b \\ I_c \end{bmatrix} \tag{3.2.4}$$

Equation (3.2.4) is written more compactly as

$$\boldsymbol{V}_p = \boldsymbol{Z}_p \boldsymbol{I}_p \tag{3.2.5}$$

where $\boldsymbol{V}_p$ is the vector of line-to-ground voltages (or phase voltages), $\boldsymbol{I}_p$ is the vector of line currents (or phase currents), and $\boldsymbol{Z}_p$ is the 3×3 phase impedance matrix shown in (3.2.4). Equations (3.1.9) and (3.1.16) can now be used in (3.2.5) to determine the relationship between the sequence voltages and currents, as follows:

$$\boldsymbol{A}\boldsymbol{V}_s = \boldsymbol{Z}_p \boldsymbol{A}\boldsymbol{I}_s \tag{3.2.6}$$

Premultiplying both sides of (3.2.6) by $\boldsymbol{A}^{-1}$ gives

$$\boldsymbol{V}_s = (\boldsymbol{A}^{-1}\boldsymbol{Z}_p\boldsymbol{A})\boldsymbol{I}_s \tag{3.2.7}$$

or

$$\boldsymbol{V}_s = \boldsymbol{Z}_s \boldsymbol{I}_s \tag{3.2.8}$$

where

$$\boldsymbol{Z}_s = \boldsymbol{A}^{-1}\boldsymbol{Z}_p\boldsymbol{A} \tag{3.2.9}$$

The impedance matrix $\boldsymbol{Z}_s$ defined by (3.2.9) is called the *sequence impedance matrix*. Using the definition of $\boldsymbol{A}$, its inverse $\boldsymbol{A}^{-1}$, and $\boldsymbol{Z}_p$ given by (3.1.8), (3.1.10), and (3.2.4), the sequence impedance matrix $\boldsymbol{Z}_s$ for the balanced-Y load is

$$\boldsymbol{Z}_s = \frac{1}{3}\begin{bmatrix} 1 & 1 & 1 \\ 1 & a & a^2 \\ 1 & a^2 & a \end{bmatrix} \begin{bmatrix} (Z_Y + Z_n) & Z_n & Z_n \\ Z_n & (Z_Y + Z_n) & Z_n \\ Z_n & Z_n & (Z_Y + Z_n) \end{bmatrix} \begin{bmatrix} 1 & 1 & 1 \\ 1 & a^2 & a \\ 1 & a & a^2 \end{bmatrix} \tag{3.2.10}$$

Performing the indicated matrix multiplications in (3.2.10), and using the identity $(1 + a + a^2) = 0$,

$$\boldsymbol{Z}_s = \tfrac{1}{3}\begin{bmatrix} 1 & 1 & 1 \\ 1 & a & a^2 \\ 1 & a^2 & a \end{bmatrix}\begin{bmatrix} (Z_Y + 3Z_n) & Z_Y & Z_Y \\ (Z_Y + 3Z_n) & a^2 Z_Y & aZ_Y \\ (Z_Y + 3Z_n) & aZ_Y & a^2 Z_Y \end{bmatrix}$$

$$= \begin{bmatrix} (Z_Y + 3Z_n) & 0 & 0 \\ 0 & Z_Y & 0 \\ 0 & 0 & Z_Y \end{bmatrix} \tag{3.2.11}$$

As shown in (3.2.11), the sequence impedance matrix $\boldsymbol{Z}_s$ for the balanced-Y load of Figure 3.3 is a diagonal matrix. Since $\boldsymbol{Z}_s$ is diagonal, (3.2.8) can be written as three *uncoupled* equations. Using (3.1.7), (3.1.18), and (3.2.11) in (3.2.8),

$$\begin{bmatrix} V_0 \\ V_1 \\ V_2 \end{bmatrix} = \begin{bmatrix} (Z_Y + 3Z_n) & 0 & 0 \\ 0 & Z_Y & 0 \\ 0 & 0 & Z_Y \end{bmatrix}\begin{bmatrix} I_0 \\ I_1 \\ I_2 \end{bmatrix} \tag{3.2.12}$$

Rewriting (3.2.12) as three separate equations,

$$V_0 = (Z_Y + 3Z_n)I_0 = Z_0 I_0 \tag{3.2.13}$$

$$V_1 = Z_Y I_1 = Z_1 I_1 \tag{3.2.14}$$

$$V_2 = Z_Y I_2 = Z_2 I_2 \tag{3.2.15}$$

As shown in (3.2.13), the zero-sequence voltage V_0 depends only on the zero-sequence current I_0 and the impedance $(Z_Y + 3Z_n)$. This impedance is called the *zero-sequence impedance* and is designated Z_0. Also, the positive-sequence voltage V_1 depends only on the positive-sequence current I_1 and an impedance $Z_1 = Z_Y$ called the *positive-sequence impedance*. Similarly, V_2 depends only on I_2 and the *negative-sequence impedance* $Z_2 = Z_Y$.

Equations (3.2.13)–(3.2.15) can be represented by the three networks shown in Figure 3.4. These networks are called the *zero-sequence, positive-sequence,* and *negative-sequence networks.* As shown, each sequence network is separate, uncoupled from the other two. The separation of these sequence networks is a consequence of the fact that $\boldsymbol{Z}_s$ is a diagonal matrix for a balanced-Y load. This separation underlies the advantage of symmetrical components.

Figure 3.4

Sequence networks of a balanced-Y load

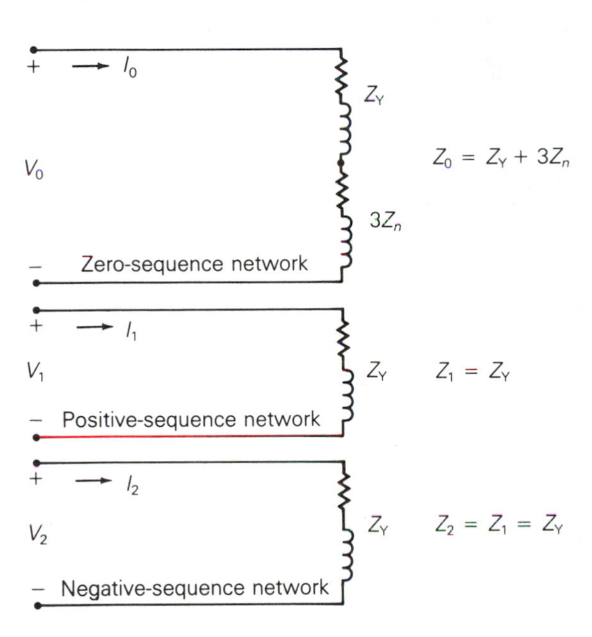

Note that the neutral impedance does not appear in the positive- and negative-sequence networks of Figure 3.4. This illustrates the fact that positive- and negative-sequence currents do not flow in neutral impedances. However, the neutral impedance is multiplied by 3 and placed in the zero-sequence network of the figure. The voltage $I_0(3Z_n)$ across the impedance $3Z_n$ is the voltage drop (I_nZ_n) across the neutral impedance Z_n in Figure 3.3, since $I_n = 3I_0$.

When the neutral of the Y load in Figure 3.3 has no return path, then the neutral impedance Z_n is infinite and the term $3Z_n$ in the zero-sequence network of Figure 3.4 becomes an open circuit. Under this condition of an open neutral, no zero-sequence current exists. However, when the neutral of the Y load is solidly grounded with a zero-ohm conductor, then the neutral impedance is zero and the term $3Z_n$ in the zero-sequence network becomes a short circuit. Under this condition of a solidly grounded neutral, zero-sequence current I_0 can exist when there is a zero-sequence voltage caused by unbalanced voltages applied to the load.

Figure 2.15 shows a balanced-Δ load and its equivalent balanced-Y load. Since the Δ load has no neutral connection, the equivalent Y load in Figure 2.15 has an open neutral. The sequence networks of the equivalent Y load corresponding to a balanced-Δ load are shown in Figure 3.5. As shown, the equivalent Y impedance $Z_Y = Z_\Delta/3$ appears in each of the sequence networks. Also, the zero-sequence network has an open circuit, since $Z_n = \infty$ corresponds to an open neutral. No zero-sequence current occurs in the equivalent Y load.

Figure 3.5 Sequence networks for an equivalent Y representation of a balanced-Δ load

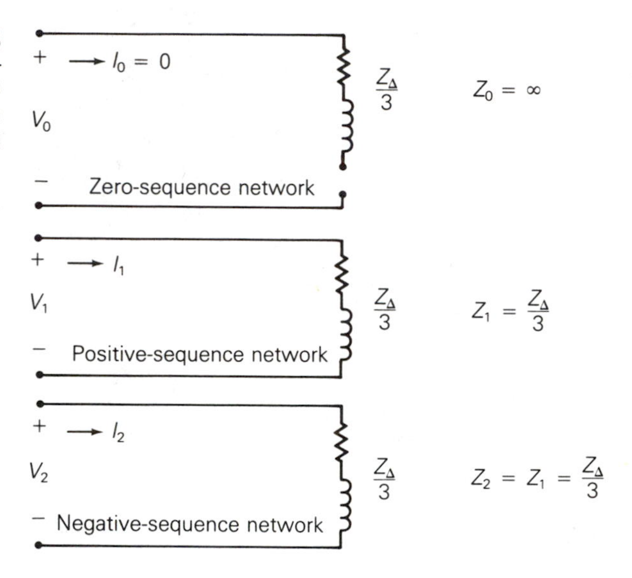

The sequence networks of Figure 3.5 represent the balanced-Δ load as viewed from its terminals, but they do not represent the internal load characteristics. The currents I_0, I_1, and I_2 in Figure 3.5 are the sequence components of the line currents feeding the Δ load, not the load currents within the Δ. The Δ load currents, which are related to the line currents by (2.6.14), are not shown in Figure 3.5.

Example 3.4 A balanced-Y load is in parallel with a balanced-Δ-connected capacitor bank. The Y load has an impedance $Z_Y = (3 + j4)\,\Omega$ per phase, and its neutral is grounded through an inductive reactance $X_n = 2\,\Omega$. The capacitor bank has a reactance $X_c = 30\,\Omega$ per phase. Draw the sequence networks for this load and calculate the load-sequence impedances.

Solution The sequence networks are shown in Figure 3.6. As shown, the Y-load impedance in the zero-sequence network is in series with three times the neutral impedance. Also, the Δ-load branch in the zero-sequence network is open, since no zero-sequence current flows into the Δ load. In the positive- and negative-sequence circuits, the Δ-load impedance is divided by 3 and placed in parallel with the Y-load impedance. The equivalent sequence impedances are

$$Z_0 = Z_Y + 3Z_n = 3 + j4 + 3(j2) = 3 + j10 \quad \Omega$$

$$Z_1 = Z_Y//(Z_\Delta/3) = \frac{(3 + j4)(-j30/3)}{3 + j4 - j(30/3)}$$

$$= \frac{(5\underline{/53.13^\circ})(10\underline{/-90^\circ})}{6.708\underline{/-63.44^\circ}} = 7.454\underline{/26.57^\circ} \quad \Omega$$

$$Z_2 = Z_1 = 7.454\underline{/26.57^\circ} \quad \Omega$$

◁

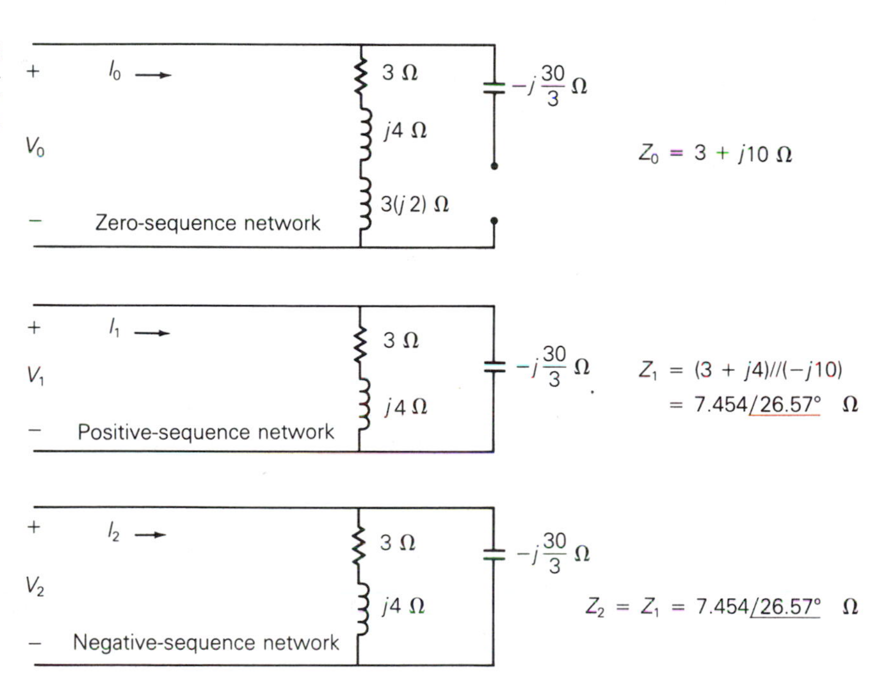

Figure 3.6
Sequence networks for Example 3.4

Figure 3.7 shows a general three-phase linear impedance load. The load could represent a balanced load such as the balanced-Y or -Δ load, or an unbalanced impedance load. The general relationship between the line-to-ground voltages and line currents for this load can be written as

$$\begin{bmatrix} V_{ag} \\ V_{bg} \\ V_{cg} \end{bmatrix} = \begin{bmatrix} Z_{aa} & Z_{ab} & Z_{ac} \\ Z_{ab} & Z_{bb} & Z_{bc} \\ Z_{ac} & Z_{bc} & Z_{cc} \end{bmatrix} \begin{bmatrix} I_a \\ I_b \\ I_c \end{bmatrix} \tag{3.2.16}$$

$$\text{or } \boldsymbol{V}_p = \boldsymbol{Z}_p \boldsymbol{I}_p \tag{3.2.17}$$

where $\boldsymbol{V}_p$ is the vector of line-to-neutral (or phase) voltages, $\boldsymbol{I}_p$ is the vector of line (or phase) currents, and $\boldsymbol{Z}_p$ is a 3×3 phase impedance matrix. It is assumed here that the load is nonrotating, and that $\boldsymbol{Z}_p$ is a symmetric matrix, which corresponds to a bilateral network.

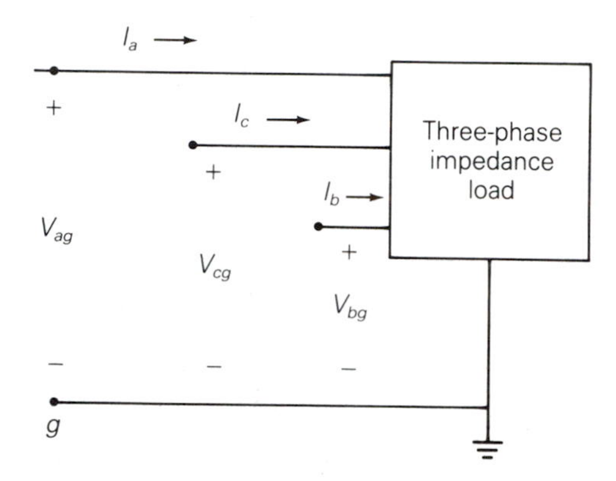

Figure 3.7

General three-phase impedance load (linear, bilateral network, non-rotating equipment)

Since (3.2.17) has the same form as (3.2.5), the relationship between the sequence voltages and currents for the general three-phase load of Figure 3.6 is the same as that of (3.2.8) and (3.2.9), which are rewritten here:

$$\boldsymbol{V}_s = \boldsymbol{Z}_s \boldsymbol{I}_s \tag{3.2.18}$$

$$\boldsymbol{Z}_s = \boldsymbol{A}^{-1} \boldsymbol{Z}_p \boldsymbol{A} \tag{3.2.19}$$

The sequence impedance matrix $\boldsymbol{Z}_s$ given by (3.2.19) is a 3×3 matrix with nine sequence impedances, defined as follows:

$$\boldsymbol{Z}_s = \begin{bmatrix} Z_0 & Z_{01} & Z_{02} \\ Z_{10} & Z_1 & Z_{12} \\ Z_{20} & Z_{21} & Z_2 \end{bmatrix} \tag{3.2.20}$$

The diagonal impedances Z_0, Z_1, and Z_2 in this matrix are the self-impedances of the zero-, positive-, and negative-sequence networks. The off-diagonal impedances are the mutual impedances between sequence networks. Using the definitions of $\boldsymbol{A}$, $\boldsymbol{A}^{-1}$, $\boldsymbol{Z}_p$, and $\boldsymbol{Z}_s$, (3.2.19) is

$$\begin{bmatrix} Z_0 & Z_{01} & Z_{02} \\ Z_{10} & Z_1 & Z_{12} \\ Z_{20} & Z_{21} & Z_2 \end{bmatrix} = \frac{1}{3} \begin{bmatrix} 1 & 1 & 1 \\ 1 & a & a^2 \\ 1 & a^2 & a \end{bmatrix} \begin{bmatrix} Z_{aa} & Z_{ab} & Z_{ac} \\ Z_{ab} & Z_{bb} & Z_{bc} \\ Z_{ac} & Z_{bc} & Z_{cc} \end{bmatrix} \begin{bmatrix} 1 & 1 & 1 \\ 1 & a^2 & a \\ 1 & a & a^2 \end{bmatrix} \tag{3.2.21}$$

Performing the indicated multiplications in (3.2.21), and using the identity $(1 + a + a^2) = 0$, the following separate equations can be obtained (see Problem 3.11):

Diagonal sequence impedances

$$Z_0 = \tfrac{1}{3}(Z_{aa} + Z_{bb} + Z_{cc} + 2Z_{ab} + 2Z_{ac} + 2Z_{bc}) \tag{3.2.22}$$

$$Z_1 = Z_2 = \tfrac{1}{3}(Z_{aa} + Z_{bb} + Z_{cc} - Z_{ab} - Z_{ac} - Z_{bc}) \tag{3.2.23}$$

Off-diagonal sequence impedances

$$Z_{01} = Z_{20} = \tfrac{1}{3}(Z_{aa} + a^2 Z_{bb} + aZ_{cc} - aZ_{ab} - a^2 Z_{ac} - Z_{bc}) \tag{3.2.24}$$

$$Z_{02} = Z_{10} = \tfrac{1}{3}(Z_{aa} + aZ_{bb} + a^2 Z_{cc} - a^2 Z_{ab} - aZ_{ac} - Z_{bc}) \tag{3.2.25}$$

$$Z_{12} = \tfrac{1}{3}(Z_{aa} + a^2 Z_{bb} + aZ_{cc} + 2aZ_{ab} + 2a^2 Z_{ac} + 2Z_{bc}) \tag{3.2.26}$$

$$Z_{21} = \tfrac{1}{3}(Z_{aa} + aZ_{bb} + a^2 Z_{cc} + 2a^2 Z_{ab} + 2aZ_{ac} + 2Z_{bc}) \tag{3.2.27}$$

A *symmetrical load* is defined as a load whose sequence impedance matrix is diagonal; that is, all the mutual impedances in (3.2.24)–(3.2.27) are zero. Equating these mutual impedances to zero and solving, the following conditions for a symmetrical load are determined. When both

$$Z_{aa} = Z_{bb} = Z_{cc} \tag{3.2.28}$$

and

$$Z_{ab} = Z_{ac} = Z_{bc} \tag{3.2.29}$$

(conditions for a symmetrical load)

then

$$Z_{01} = Z_{10} = Z_{02} = Z_{20} = Z_{12} = Z_{21} = 0 \tag{3.2.30}$$

$$Z_0 = Z_{aa} + 2Z_{ab} \tag{3.2.31}$$

$$Z_1 = Z_2 = Z_{aa} - Z_{ab} \tag{3.2.32}$$

The conditions for a symmetrical load are that the diagonal phase impedances be equal and that the off-diagonal phase impedances be equal. These conditions can be verified by using (3.2.28) and (3.2.29) with the identity $(1 + a + a^2) = 0$ in (3.2.24)–(3.2.27) to show that all the mutual sequence impedances are zero. Note that the positive- and negative-sequence impedances are equal for a symmetrical load, as shown by (3.2.32), and for a nonsymmetrical load, as shown by (3.2.23). This is always true for linear, symmetric impedances that represent nonrotating equipment such as transformers and transmission lines. However, the positive- and negative-sequence impedances of rotating equipment such as generators and motors are generally not equal. Note also that the zero-sequence impedance Z_0 is not equal to the positive- and negative-sequence impedances of a symmetrical load unless the mutual phase impedances $Z_{ab} = Z_{ac} = Z_{bc}$ are zero.

The sequence networks of a symmetrical impedance load are shown in Figure 3.8. Since the sequence impedance matrix $\boldsymbol{Z}_s$ is diagonal for a symmetrical load, the sequence networks are separate or uncoupled.

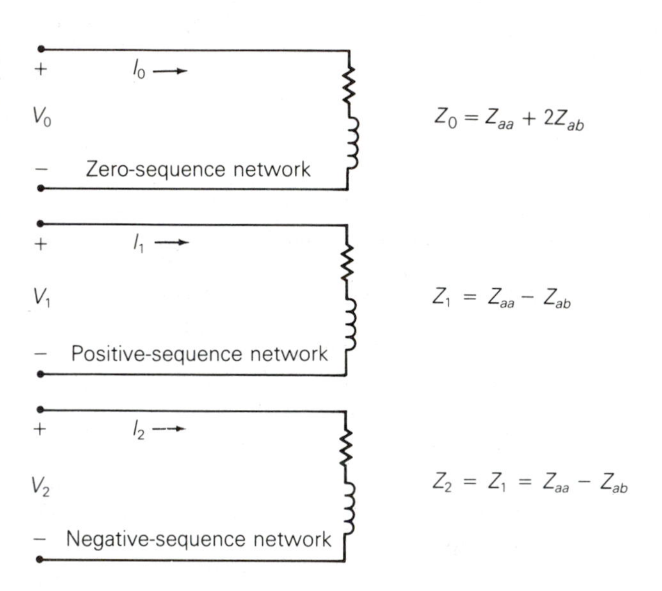

$Z_0 = Z_{aa} + 2Z_{ab}$

$Z_1 = Z_{aa} - Z_{ab}$

$Z_2 = Z_1 = Z_{aa} - Z_{ab}$

Figure 3.8 Sequence networks of a three-phase symmetrical impedance load (linear, bilateral network, nonrotating equipment)

Section 3.3

Sequence Networks of Series Impedances

Figure 3.9 shows series impedances connected between two three-phase buses denoted *abc* and *a′b′c′*. Self-impedances of each phase are denoted Z_{aa}, Z_{bb}, and

Z_{cc}. In general, the series network may also have mutual impedances between phases. The voltage drops across the series-phase impedances are given by

$$\begin{bmatrix} V_{an} - V_{a'n} \\ V_{bn} - V_{b'n} \\ V_{cn} - V_{c'n} \end{bmatrix} = \begin{bmatrix} V_{aa'} \\ V_{bb'} \\ V_{cc'} \end{bmatrix} = \begin{bmatrix} Z_{aa} & Z_{ab} & Z_{ac} \\ Z_{ab} & Z_{bb} & Z_{bc} \\ Z_{ac} & Z_{bc} & Z_{cc} \end{bmatrix} \begin{bmatrix} I_a \\ I_b \\ I_c \end{bmatrix} \tag{3.3.1}$$

Figure 3.9

Three-phase series impedances (linear, bilateral network, nonrotating equipment)

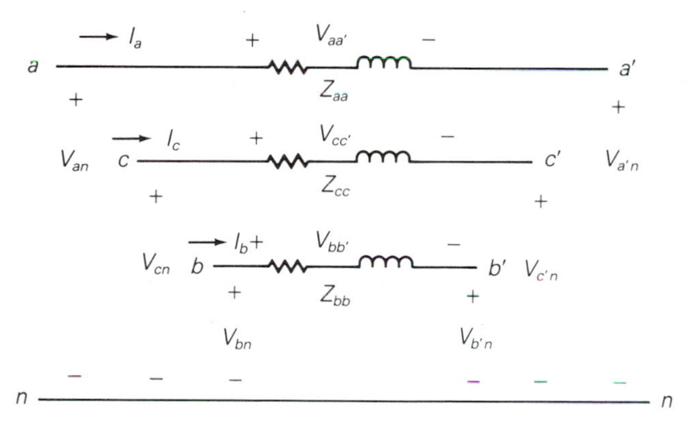

Both self-impedances and mutual impedances are included in (3.3.1). It is assumed that the impedance matrix is symmetric, which corresponds to a bilateral network. It is also assumed that these impedances represent non-rotating equipment. Typical examples are series impedances of transmission lines and of transformers. Equation (3.3.1) has the following form:

$$\boldsymbol{V}_p - \boldsymbol{V}_{p'} = \boldsymbol{Z}_p \boldsymbol{I}_p \tag{3.3.2}$$

where $\boldsymbol{V}_p$ is the vector of line-to-neutral voltages at bus abc, $\boldsymbol{V}_{p'}$ is the vector of line-to-neutral voltages at bus $a'b'c'$, $\boldsymbol{I}_p$ is the vector of line currents, and $\boldsymbol{Z}_p$ is the 3×3 phase impedance matrix for the series network. Equation (3.3.2) is now transformed to the sequence domain in the same manner that the load-phase impedances were transformed in Section 3.2. Thus,

$$\boldsymbol{V}_s - \boldsymbol{V}_{s'} = \boldsymbol{Z}_s \boldsymbol{I}_s \tag{3.3.3}$$

where

$$\boldsymbol{Z}_s = \boldsymbol{A}^{-1} \boldsymbol{Z}_p \boldsymbol{A} \tag{3.3.4}$$

From the results of Section 3.2, this sequence impedance $\boldsymbol{Z}_s$ matrix is diagonal under the following conditions:

$$\left.\begin{aligned} Z_{aa} &= Z_{bb} = Z_{cc} \\ \text{and} \\ Z_{ab} &= Z_{ac} = Z_{bc} \end{aligned}\right\} \text{conditions for symmetrical series impedances} \tag{3.3.5}$$

When the phase impedance matrix $\boldsymbol{Z}_p$ of (3.3.1) has both equal self-impedances and equal mutual impedances, then (3.3.4) becomes

$$Z_s = \begin{bmatrix} Z_0 & 0 & 0 \\ 0 & Z_1 & 0 \\ 0 & 0 & Z_2 \end{bmatrix} \tag{3.3.6}$$

where

$$Z_0 = Z_{aa} + 2Z_{ab} \tag{3.3.7}$$

and

$$Z_1 = Z_2 = Z_{aa} - Z_{ab} \tag{3.3.8}$$

and (3.3.3) becomes three uncoupled equations, written as follows:

$$V_0 - V_{0'} = Z_0 I_0 \tag{3.3.9}$$

$$V_1 - V_{1'} = Z_1 I_1 \tag{3.3.10}$$

$$V_2 - V_{2'} = Z_2 I_2 \tag{3.3.11}$$

Equations (3.3.9)–(3.3.11) are represented by the three uncoupled sequence networks shown in Figure 3.10. From the figure it is apparent that for symmetrical series impedances, positive-sequence currents produce only positive-sequence voltage drops. Similarly, negative-sequence currents produce only negative-sequence voltage drops, and zero-sequence currents produce only zero-sequence voltage drops. However, if the series impedances are not symmetrical, then $\boldsymbol{Z}_s$ is not diagonal, the sequence networks are coupled, and the voltage drop across any one sequence network depends on all three sequence currents.

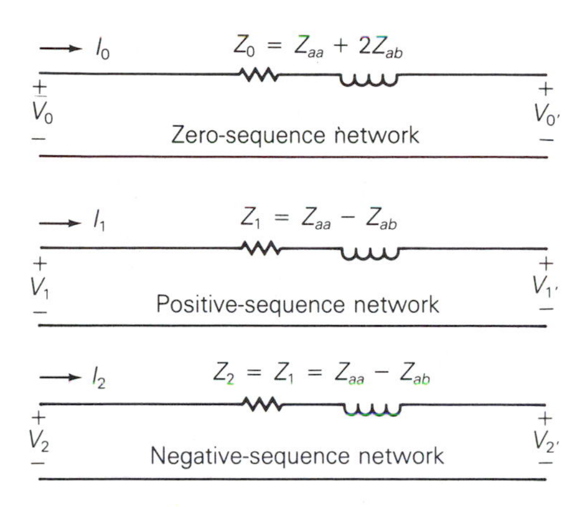

Figure 3.10

Sequence networks of three-phase symmetrical series impedances (linear, bilateral network, nonrotating equipment)

Section 3.4

Sequence Networks of Rotating Machines

A Y-connected synchronous generator grounded through a neutral impedance Z_n is shown in Figure 3.11. The internal generator voltages are designated E_a, E_b, and E_c, and the generator line currents are designated I_a, I_b, and I_c.

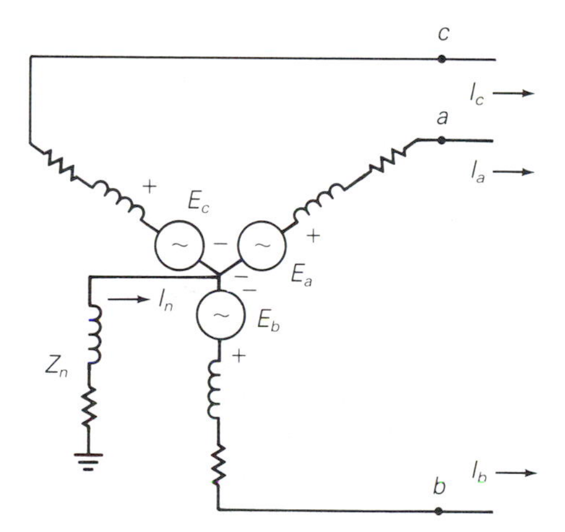

Figure 3.11

Y-connected synchronous generator

The sequence networks of the generator are shown in Figure 3.12. Since a three-phase synchronous generator is designed to produce balanced internal phase voltages E_a, E_b, E_c with only a positive-sequence component, a source voltage E_{g1} is included only in the positive-sequence network. The sequence components of the line-to-ground voltages at the generator terminals are denoted V_0, V_1, and V_2 in Figure 3.12.

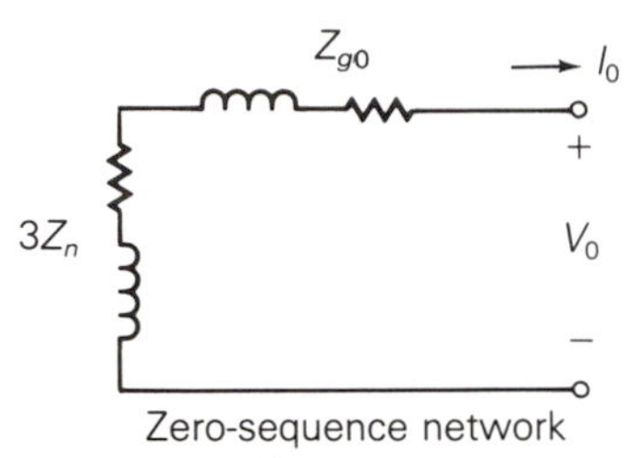

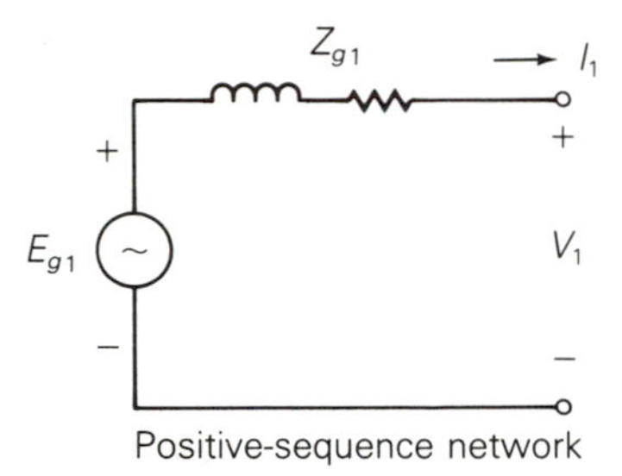

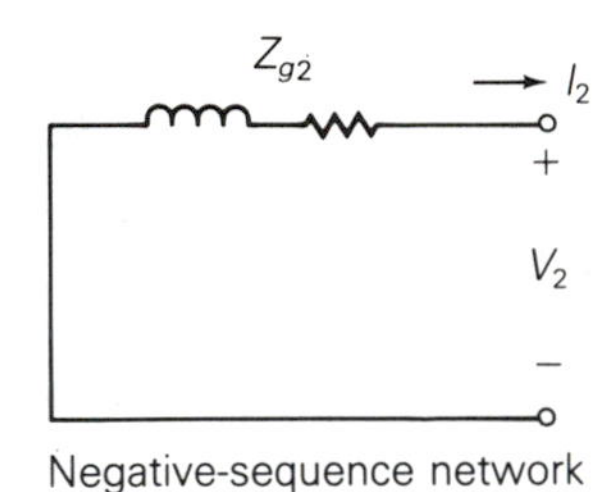

Figure 3.12

Sequence networks of a Y-connected synchronous generator

The voltage drop in the generator neutral impedance is Z_nI_n, which can be written as $(3Z_n)I_0$, since, from (3.1.27), the neutral current is three times the zero-sequence current. Since this voltage drop is due only to zero-sequence current, an impedance $(3Z_n)$ is placed in the zero-sequence network of Figure 3.12 in series with the generator zero-sequence impedance Z_{g0}.

The sequence impedances of rotating machines are generally not equal. A detailed analysis of machine-sequence impedances is given in machine theory texts.[1] Only a brief explanation is given here.

When a synchronous generator stator has balanced three-phase positive-sequence currents under steady-state conditions, the net mmf produced by these positive-sequence currents rotates at the synchronous rotor speed in the same direction as that of the rotor. Under this condition, a high value of magnetic flux penetrates the rotor, and the positive-sequence impedance Z_{g1} has a high value. Under steady-state conditions, the positive-sequence generator impedance is called the *synchronous impedance*.

When a synchronous generator stator has balanced three-phase negative-sequence currents, the net mmf produced by these currents rotates at synchronous speed in the direction opposite to that of the rotor. With respect to the rotor, the net mmf is not stationary but rotates at twice synchronous speed. Under this condition, currents are induced in the rotor windings that prevent the magnetic flux from penetrating the rotor. As such, the negative-sequence impedance Z_{g2} is less than the positive-sequence synchronous impedance.

When a synchronous generator has only zero-sequence currents, which are line (or phase) currents with equal magnitude and phase, then the net mmf produced by these currents is theoretically zero. The generator zero-sequence impedance Z_{g0} is the smallest sequence impedance and is due to leakage flux, end turns, and harmonic flux from windings that do not produce a perfectly sinusoidal mmf.

Typical values of machine-sequence impedances are listed in Table A.1. The positive-sequence machine impedance is synchronous, transient, or subtransient. *Synchronous* impedances are used for steady-state conditions, such as in power-flow studies, which are described in Chapter 7. *Transient* impedances are used for stability studies, which are described in Chapter 12, and *subtransient* impedances are used for short-circuit studies, which are described in Chapters 8 and 9. Unlike the positive-sequence impedances, a machine has only one negative-sequence impedance and only one zero-sequence impedance.

The sequence networks for three-phase synchronous motors and for three-phase induction motors are shown in Figure 3.13. Synchronous motors have the same sequence networks as synchronous generators, except that the sequence currents for synchronous motors are referenced *into* rather than out of the sequence networks. Also, induction motors have the same sequence networks as synchronous motors, except that the positive-sequence voltage source E_{m1} is removed. Induction motors do not have a dc source of magnetic flux in their rotor circuits, and therefore E_{m1} is zero (or a short circuit).

The sequence networks shown in Figures 3.12 and 3.13 are simplified networks for rotating machines. The networks do not take into account such phenomena as machine saliency, saturation effects, and more complicated transient effects. These simplified networks, however, are in many cases accurate enough for power-system studies.

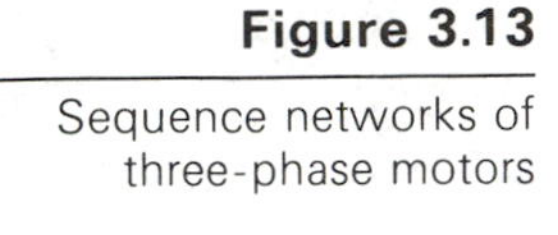

Figure 3.13 Sequence networks of three-phase motors

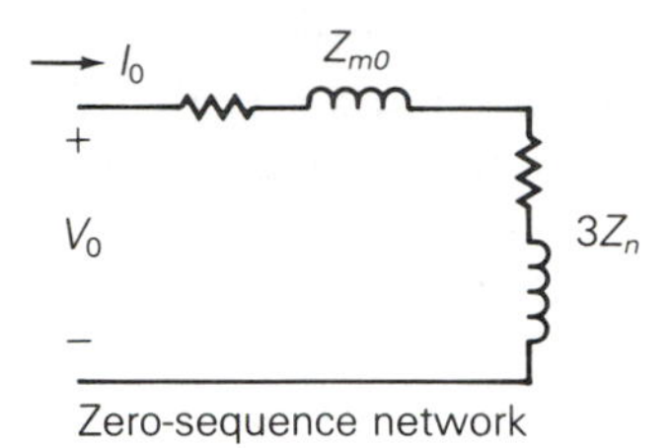

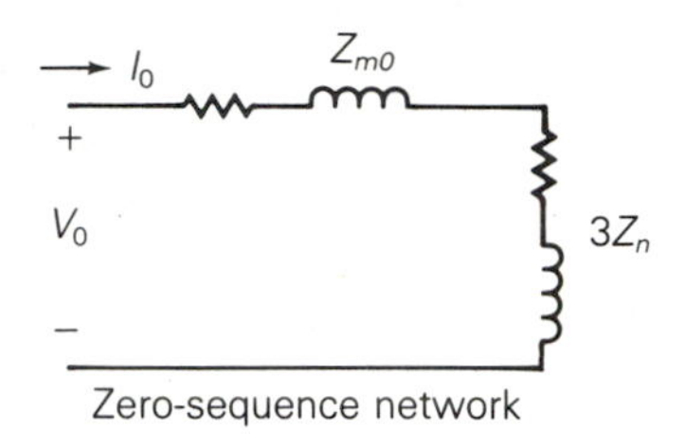

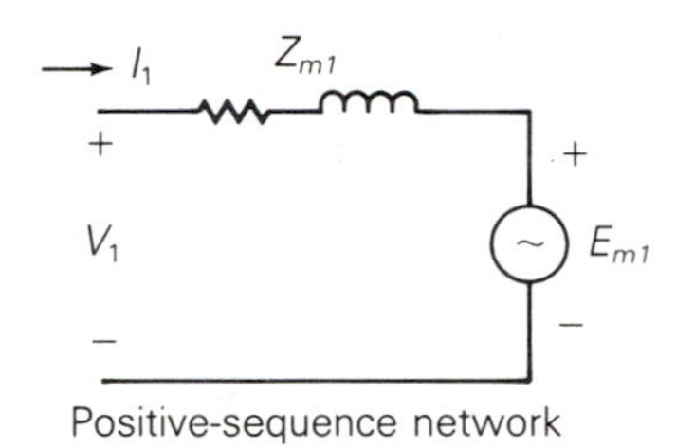

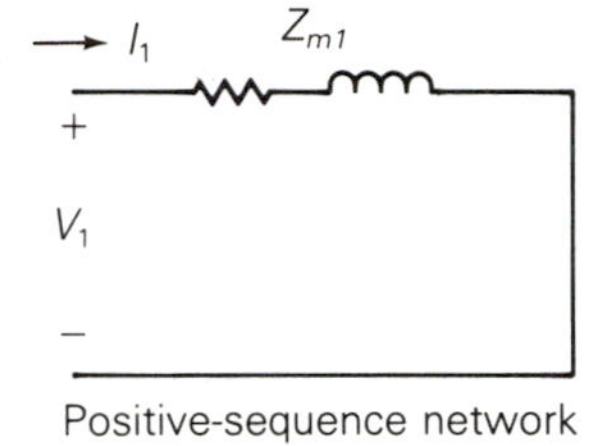

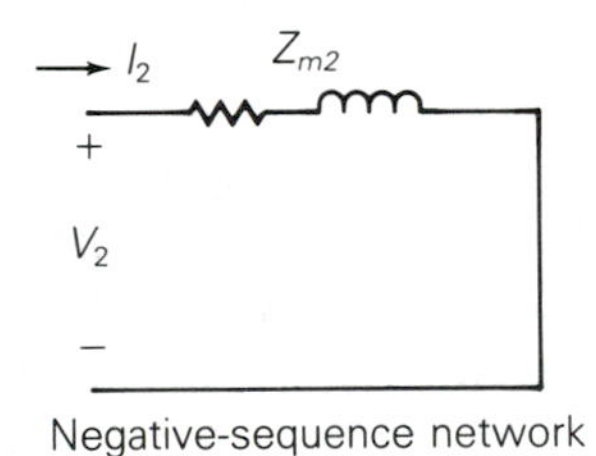

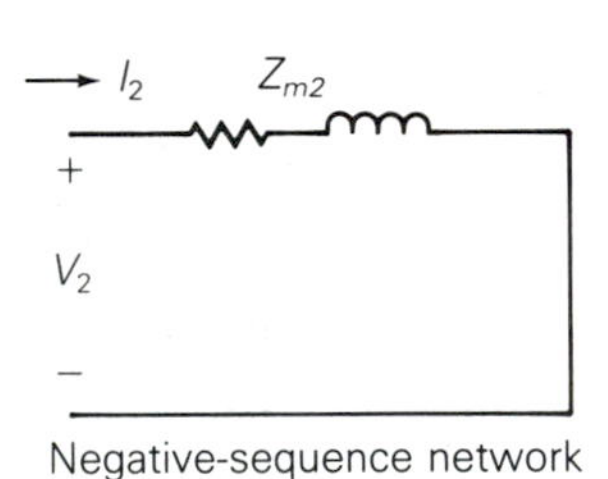

(a) Synchronous motor

(b) Induction motor

Example 3.5 Draw the sequence networks for the circuit of Example 2.5 and calculate the sequence components of the line current. Assume that the generator neutral is grounded through an impedance $Z_n = j10\,\Omega$, and that the generator sequence impedances are $Z_{g0} = j1\,\Omega$, $Z_{g1} = j15\,\Omega$, and $Z_{g2} = j3\,\Omega$.

Solution The sequence networks are shown in Figure 3.14. They are obtained by interconnecting the sequence networks for a balanced-Δ load, for series-line impedances, and for a synchronous generator, which are given in Figures 3.5, 3.10, and 3.12.

It is clear from Figure 3.14 that $I_0 = I_2 = 0$ since there are no sources in the zero- and negative-sequence networks. Also, the positive-sequence generator terminal voltage V_1 equals the generator line-to-neutral terminal

Figure 3.14

Sequence networks for Example 3.5

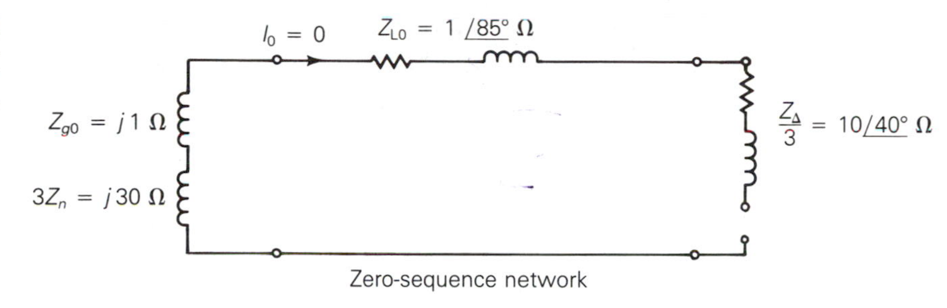

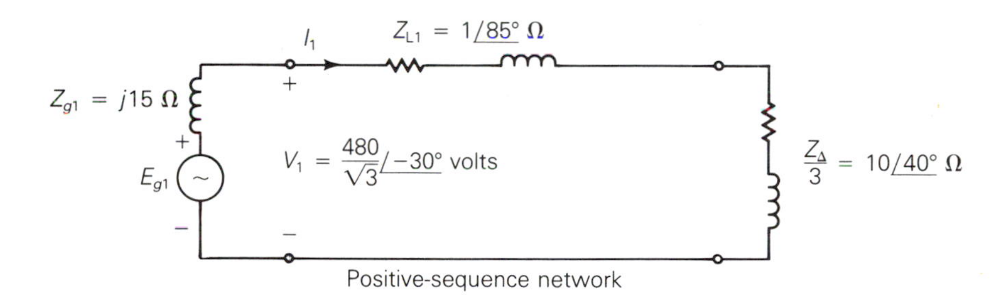

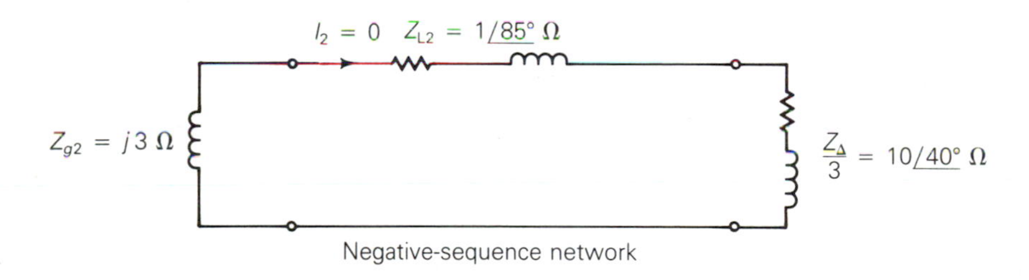

voltage. Therefore, from the positive-sequence network shown in the figure and from the results of Example 2.5,

$$I_1 = \frac{V_1}{(Z_{L1} + \frac{1}{3}Z_\Delta)} = 25.83\underline{/-73.78°}\text{ A} = I_a$$

Note that from (3.1.20), I_1 equals the line current I_a since $I_0 = I_2 = 0$. ◁

The following example illustrates the superiority of using symmetrical components for analyzing unbalanced systems.

Example 3.6 A Y-connected voltage source with the following unbalanced voltage is applied to the balanced line and load of Example 2.5.

$$\begin{bmatrix} V_{ag} \\ V_{bg} \\ V_{cg} \end{bmatrix} = \begin{bmatrix} 277\underline{/0^\circ} \\ 260\underline{/-120^\circ} \\ 295\underline{/+115^\circ} \end{bmatrix} \quad \text{volts}$$

The source neutral is solidly grounded. Using the method of symmetrical components, calculate the source currents I_a, I_b, and I_c.

Solution Using (3.1.13)–(3.1.15), the sequence components of the source voltages are:

$$V_0 = \tfrac{1}{3}(277\underline{/0^\circ} + 260\underline{/-120^\circ} + 295\underline{/115^\circ})$$
$$= 7.4425 + j14.065 = 15.912\underline{/62.11^\circ} \quad \text{volts}$$
$$V_1 = \tfrac{1}{3}(277\underline{/0^\circ} + 260\underline{/-120^\circ + 120^\circ} + 295\underline{/115^\circ + 240^\circ})$$
$$= \tfrac{1}{3}(277\underline{/0^\circ} + 260\underline{/0^\circ} + 295\underline{/-5^\circ})$$
$$= 276.96 - j8.5703 = 277.1\underline{/-1.772^\circ} \quad \text{volts}$$
$$V_2 = \tfrac{1}{3}(277\underline{/0^\circ} + 260\underline{/-120^\circ + 240^\circ} + 295\underline{/115^\circ + 120^\circ})$$
$$= \tfrac{1}{3}(277\underline{/0^\circ} + 260\underline{/120^\circ} + 295\underline{/235^\circ})$$
$$= -7.4017 - j5.4944 = 9.218\underline{/216.59^\circ} \quad \text{volts}$$

These sequence voltages are applied to the sequence networks of the line and load, as shown in Figure 3.15. The sequence networks of this figure are uncoupled, and the sequence components of the source currents are easily calculated as follows:

$$I_0 = 0$$

$$I_1 = \frac{V_1}{Z_{L1} + \dfrac{Z_\Delta}{3}} = \frac{277.1\underline{/-1.772^\circ}}{10.73\underline{/43.78^\circ}} = 25.82\underline{/-45.55^\circ} \quad \text{A}$$

$$I_2 = \frac{V_2}{Z_{L2} + \dfrac{Z_\Delta}{3}} = \frac{9.218\underline{/216.59^\circ}}{10.73\underline{/43.78^\circ}} = 0.8591\underline{/172.81^\circ} \quad \text{A}$$

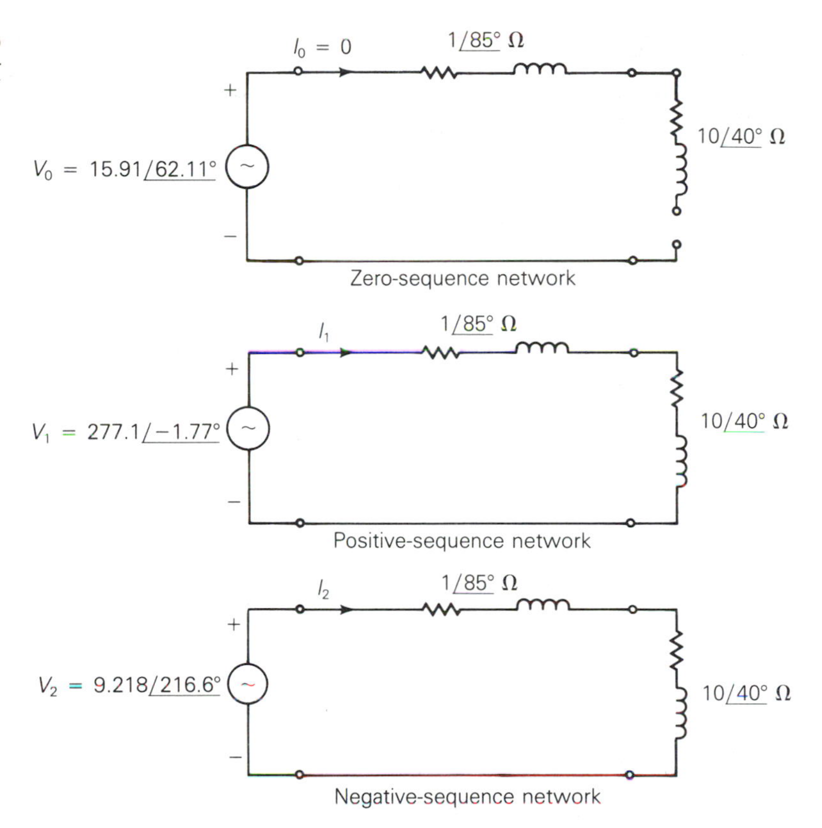

Figure 3.15 Sequence networks for Example 3.6

Using (3.1.20)–(3.1.22), the source currents are:

$$I_a = (0 + 25.82\underline{/-45.55^\circ} + 0.8591\underline{/172.81^\circ})$$

$$= 17.23 - j18.32 = 25.15\underline{/-46.76^\circ} \quad \text{A}$$

$$I_b = (0 + 25.82\underline{/-45.55^\circ + 240^\circ} + 0.8591\underline{/172.81^\circ + 120^\circ})$$

$$= (25.82\underline{/194.45^\circ} + 0.8591\underline{/292.81^\circ})$$

$$= -24.67 - j7.235 = 25.71\underline{/196.34^\circ} \quad \text{A}$$

$$I_c = (0 + 25.82\underline{/-45.55^\circ + 120^\circ} + 0.8591\underline{/172.81^\circ + 240^\circ})$$

$$= (25.82\underline{/74.45^\circ} + 0.8591\underline{/52.81^\circ})$$

$$= 7.441 + j25.56 = 26.62\underline{/73.77^\circ} \quad \text{A}$$

The reader should calculate the line currents for this example without using symmetrical components, in order to verify this result and to compare the two solution methods (see Problem 3.15). Without symmetrical components, coupled KVL equations must be solved. With symmetrical components, the conversion from phase to sequence components decouples the networks as well as the resulting KVL equations, as shown above. ◁

Section 3.5

Power in Sequence Networks

The power delivered to a three-phase network can be determined from the power delivered to the sequence networks. Let S_p denote the total complex power delivered to the three-phase load of Figure 3.7, which can be calculated from

$$S_p = V_{ag}I_a^* + V_{bg}I_b^* + V_{cg}I_c^* \tag{3.5.1}$$

Equation (3.5.1) is also valid for the total complex power delivered by the three-phase generator of Figure 3.11, or for the complex power delivered to any three-phase bus. Rewriting (3.5.1) in matrix format,

$$S_p = [V_{ag}\ V_{bg}\ V_{cg}] \begin{bmatrix} I_a^* \\ I_b^* \\ I_c^* \end{bmatrix}$$

$$= \boldsymbol{V}_p^{\mathrm{T}}\boldsymbol{I}_p^* \tag{3.5.2}$$

where T denotes transpose and * denotes complex conjugate. Now, using (3.1.9) and (3.1.16),

$$S_p = (\boldsymbol{A}\boldsymbol{V}_s)^{\mathrm{T}}(\boldsymbol{A}\boldsymbol{I}_s)^*$$

$$= \boldsymbol{V}_s^{\mathrm{T}}[\boldsymbol{A}^{\mathrm{T}}\boldsymbol{A}^*]\boldsymbol{I}_s^* \tag{3.5.3}$$

Using the definition of $\boldsymbol{A}$, which is (3.1.8), to calculate the term within the brackets of (3.5.3), and noting that a and a^2 are conjugates,

$$A^{T}A^{*} = \begin{bmatrix} 1 & 1 & 1 \\ 1 & a^2 & a \\ 1 & a & a^2 \end{bmatrix}^{T} \begin{bmatrix} 1 & 1 & 1 \\ 1 & a^2 & a \\ 1 & a & a^2 \end{bmatrix}^{*}$$

$$= \begin{bmatrix} 1 & 1 & 1 \\ 1 & a^2 & a \\ 1 & a & a^2 \end{bmatrix} \begin{bmatrix} 1 & 1 & 1 \\ 1 & a & a^2 \\ 1 & a^2 & a \end{bmatrix}$$

$$= \begin{bmatrix} 3 & 0 & 0 \\ 0 & 3 & 0 \\ 0 & 0 & 3 \end{bmatrix} = 3U \tag{3.5.4}$$

Equation (3.5.4) can now be used in (3.5.3) to obtain

$$S_p = 3V^{T}{}_{s}I_s^{*}$$

$$= 3[V_0 + V_1 + V_2] \begin{bmatrix} I_0^* \\ I_1^* \\ I_2^* \end{bmatrix} \tag{3.5.5}$$

$$S_p = 3(V_0 I_0^* + V_1 I_1^* + V_2 I_2^*)$$

$$= 3S_s \tag{3.5.6}$$

Thus, the total complex power S_p delivered to a three-phase network equals *three* times the total complex power S_s delivered to the sequence networks.

The reason the factor of 3 occurs in (3.5.6) is that $A^{T}A^{*} = 3U$, as shown by (3.5.4). It is possible to eliminate this factor of 3 by defining a new transformation matrix $A_1 = (1/\sqrt{3})\,A$ such that $A_1^{T}A_1^{*} = U$, which means that A_1 is a *unitary* matrix. Using A_1 instead of A, the total complex power delivered to a three-phase network would equal the total complex power delivered to the sequence networks. However, standard industry practice for symmetrical components is to use $\boldsymbol{A}$, defined by (3.1.8).

Example 3.7 Calculate S_p and S_s delivered by the three-phase source in Example 3.6. Verify that $S_p = 3S_s$.

Solution Using (3.5.1),

$$\begin{aligned} S_p &= (277\underline{/0°})(25.15\underline{/+46.76°}) + (260\underline{/-120°})(25.71\underline{/-196.34°}) \\ &\quad + (295\underline{/115°})(26.62\underline{/-73.77°}) \\ &= 6967\underline{/46.76°} + 6685\underline{/43.66°} + 7853\underline{/41.23°} \\ &= 15{,}520 + j14{,}870 = 21{,}490\underline{/43.78°} \quad \text{VA} \end{aligned}$$

In the sequence domain,

$$\begin{aligned} S_s &= V_0 I_0^* + V_1 I_1^* + V_2 I_2^* \\ &= 0 + (277.1\underline{/-1.772°})(25.82\underline{/45.55°}) + (9.218\underline{/216.59°}) \\ &\quad (0.8591\underline{/-172.81°}) \\ &= 7155\underline{/43.78°} + 7.919\underline{/43.78°} \\ &= 5172 + j4958 = 7163\underline{/43.78°} \quad \text{VA} \end{aligned}$$

Also,

$$3S_s = 3(7163\underline{/43.78°}) = 21{,}490\underline{/43.78°} = S_p$$

◁

Section 3.6

Personal Computer Program: SYMMETRICAL COMPONENTS

The following symmetrical component subroutines are included in the personal computer software package that accompanies this text.

1. SEQVEC($\boldsymbol{V}_p$, $\boldsymbol{V}_s$) computes the complex sequence vector $\boldsymbol{V}_s = \boldsymbol{A}^{-1}\boldsymbol{V}_p$ for any three-phase complex vector $\boldsymbol{V}_p$, where

$$\boldsymbol{V}_p = \begin{bmatrix} V_a \\ V_b \\ V_c \end{bmatrix} \qquad \boldsymbol{V}_s = \begin{bmatrix} V_0 \\ V_1 \\ V_2 \end{bmatrix}$$

The complex vector $\boldsymbol{V}_p$ may be a three-phase voltage vector, a three-phase current vector, or any complex vector with three components.

2. PHAVEC($\boldsymbol{V}_s$, $\boldsymbol{V}_p$) computes the complex phase vector $\boldsymbol{V}_p = \boldsymbol{A}\boldsymbol{V}_s$ for any complex sequence vector $\boldsymbol{V}_s$.
3. SEQIMP($\boldsymbol{Z}_p$, $\boldsymbol{Z}_s$) computes the 3 × 3 complex sequence impedance matrix $\boldsymbol{Z}_s = \boldsymbol{A}^{-1}\boldsymbol{Z}_p\boldsymbol{A}$ for any 3 × 3 complex phase impedance matrix $\boldsymbol{Z}_p$.

Example 3.8 A three-phase impedance load, as shown in Figure 3.7, is represented by the following phase impedance matrix:

$$\boldsymbol{Z}_p = \begin{bmatrix} (10+j30) & (5+j20) & (5+j20) \\ (5+j20) & (10+j30) & (5+j20) \\ (5+j20) & (5+j20) & (10+j30) \end{bmatrix} \quad \Omega$$

A Y-connected source with the following unbalanced line-to-ground voltages is applied to the load:

$$\boldsymbol{V}_p = \begin{bmatrix} 277\underline{/0^\circ} \\ 260\underline{/-120^\circ} \\ 295\underline{/+115^\circ} \end{bmatrix} \quad \text{volts}$$

The source neutral is solidly grounded. Use the personal computer software package to calculate the following:

a. Load-sequence impedance matrix $\boldsymbol{Z}_s$
b. Load-sequence voltage vector $\boldsymbol{V}_s$
c. Load-sequence current vector $\boldsymbol{I}_s$
d. Load-phase current vector $\boldsymbol{I}_p$
e. Total complex power delivered to the sequence networks, $S_s = \boldsymbol{V}_s^{\mathrm{T}}\boldsymbol{I}_s^*$
f. Total complex power delivered to the three-phase load, $S_p = \boldsymbol{V}_p^{\mathrm{T}}\boldsymbol{I}_p^*$

Solution Table 3.2 shows the solution to Example 3.8, including the input data and output data listings. Note that the sequence impedance matrix is diagonal, since the conditions for a symmetrical load are satisfied; that is, $Z_{aa} = Z_{bb} = Z_{cc} = (10 + j30)\,\Omega$ and $Z_{ab} = Z_{ac} = Z_{bc} = (5 + j20)\,\Omega$. Note also that in addition to the subroutines listed in this section, the subroutines CMI,

CMM, CMT, and CMC described in Section 2.5 are employed to compute $\boldsymbol{I}_s$, S_s, and S_p. As shown in Table 3.2, the relationship $S_p = 3S_s$ is verified for this example.

◁

a.

$$\boldsymbol{Z}_p = \begin{bmatrix} +1.000D+01 \;\; j+3.000D+01 & +5.000D+00 \;\; j+2.000D+01 & +5.000D+00 \;\; j+2.000D+01 \\ +5.000D+00 \;\; j+2.000D+01 & +1.000D+01 \;\; j+3.000D+01 & +5.000D+00 \;\; j+2.000D+01 \\ +5.000D+00 \;\; j+2.000D+01 & +5.000D+00 \;\; j+2.000D+01 & +1.000D+01 \;\; j+3.000D+01 \end{bmatrix}$$

$$\boldsymbol{Z}_s = \begin{bmatrix} +2.000D+01 \;\; j+7.000D+01 & +1.000D-12 \;\; j+0.000D+00 & -1.000D-12 \;\; j+0.000D+00 \\ -2.000D-13 \;\; j-7.000D-13 & +5.000D+00 \;\; j+1.000D+01 & -3.000D-13 \;\; j-3.000D-13 \\ -2.000D-13 \;\; j-7.000D-13 & -1.600D-13 \;\; j-1.000D-13 & +5.000D+00 \;\; j+1.000D+01 \end{bmatrix}$$

b. $$\boldsymbol{V}_p = \begin{bmatrix} +2.770D+02 & j+0.000D+00 \\ -1.300D+02 & j-2.252D+02 \\ -1.247D+02 & j+2.674D+02 \end{bmatrix} \qquad \boldsymbol{V}_s = \begin{bmatrix} +7.443D+00 & j+1.406D+01 \\ +2.770D+02 & j-8.570D+00 \\ -7.402D+00 & j-5.494D+00 \end{bmatrix}$$

c. $$\boldsymbol{I}_s = \begin{bmatrix} +2.138D-01 & j-4.526D-02 \\ +1.039D+01 & j-2.250D+01 \\ -7.356D-01 & j+3.724D-01 \end{bmatrix}$$

d. $$\boldsymbol{I}_p = \begin{bmatrix} +9.868D+00 & j-2.217D+01 \\ -2.442D+01 & j+1.383D+00 \\ +1.519D+01 & j+2.065D+01 \end{bmatrix}$$

e. $S_s = +3.075D+03 \quad j+6.154D+03$

f. $S_p = +9.224D+03 \quad j+1.846D+04$

Table 3.2 Solution to Example 3.8

Problems

Section 3.1

3.1 Using the operator $a = 1\underline{/120^\circ}$, evaluate the following in polar form: (a) $(a-1)/(1+a-a^2)$ (b) $(a^2+a+j)/(ja+a^2)$ (c) $(1+a)(1+a^2)$ (d) $(a-a^2)(a^2-1)$

3.2 Determine the symmetrical components of the following line currents: (a) $I_a = 5\underline{/90^\circ}$, $I_b = 5\underline{/320^\circ}$, $I_c = 5\underline{/220^\circ}$ A (b) $I_a = j50$, $I_b = 50$, $I_c = 0$ A

3.3 Find the phase voltages V_{an}, V_{bn}, and V_{cn} whose sequence components are: $V_0 = 50\underline{/80^\circ}$, $V_1 = 100\underline{/0^\circ}$, $V_2 = 50\underline{/90^\circ}$.

3.4 One line of a three-phase generator is open-circuited, while the other two are short-circuited to ground. The line currents are $I_a = 0$, $I_b = 1000\underline{/150^\circ}$, and $I_c = 1000\underline{/30^\circ}$ A. Find the symmetrical components of these currents. Also find the current into the ground.

3.5 Given the line-to-ground voltages $V_{ag} = 280\underline{/0^\circ}$, $V_{bg} = 250\underline{/-110^\circ}$, and $V_{cg} = 290\underline{/130^\circ}$ volts, calculate (a) the sequence components of the line-to-ground voltages, denoted V_{Lg0}, V_{Lg1}, and V_{Lg2}; (b) line-to-line voltages V_{ab}, V_{bc}, and V_{ca}; and (c) sequence components of the line-to-line voltages V_{LL0}, V_{LL1}, and V_{LL2}. Also, verify the following general relation: $V_{LL0} = 0$, $V_{LL1} = \sqrt{3}V_{Lg1}\underline{/+30^\circ}$, and $V_{LL2} = \sqrt{3}V_{Lg2}\underline{/-30^\circ}$ volts.

Section 3.2

3.6 The currents in a Δ load are $I_{ab} = 10\underline{/0^\circ}$, $I_{bc} = 15\underline{/-90^\circ}$, and $I_{ca} = 20\underline{/90^\circ}$ A. Calculate (a) the sequence components of the Δ-load currents, denoted $I_{\Delta 0}$, $I_{\Delta 1}$, $I_{\Delta 2}$; (b) the line currents I_a, I_b, and I_c, which feed the Δ load; and (c) sequence components of the line currents I_{L0}, I_{L1}, and I_{L2}. Also, verify the following general relation: $I_{L0} = 0$, $I_{L1} = \sqrt{3}I_{\Delta 1}\underline{/-30^\circ}$, and $I_{L2} = \sqrt{3}I_{\Delta 2}\underline{/+30^\circ}$ A.

3.7 The voltages given in Problem 3.5 are applied to a balanced-Y load consisting of $(12 + j16)$ ohms per phase. The load neutral is solidly grounded. Draw the sequence networks and calculate I_0, I_1, and I_2, the sequence components of the line currents. Then calculate the line currents I_a, I_b, and I_c.

3.8 Repeat Problem 3.7 with the load neutral open.

3.9 Repeat Problem 3.7 for a balanced-Δ load consisting of $(12 + j16)$ ohms per phase.

3.10 Repeat Problem 3.7 for the load shown in Example 3.4 (Figure 3.6).

3.11 Perform the indicated matrix multiplications in (3.2.21) and verify the sequence impedances given by (3.2.22)–(3.2.27).

Section 3.3

3.12 Repeat Problem 3.7 but include balanced three-phase line impedances of $(3 + j4)$ ohms per phase between the source and load.

Section 3.4

3.13 A balanced-Y-connected generator with terminal voltage $V_{bc} = 200\underline{/0^\circ}$ volts is connected to a balanced-Δ load whose impedance is $10\underline{/40^\circ}$ ohms per phase. The line impedance between the source and load is $0.5\underline{/80^\circ}$ ohm for each phase. The generator neutral is grounded through an impedance of $j5$ ohms. The generator sequence impedances are given by $Z_{g0} = j7$, $Z_{g1} = j15$, and $Z_{g2} = j10$ ohms. Draw the sequence networks for this system and determine the sequence components of the line currents.

3.14 In a three-phase system, a synchronous generator supplies power to a 200-volt synchronous motor through a line having an impedance of $0.5\underline{/80^\circ}$ ohm per phase. The motor draws 5 kW at 0.8 p.f. leading and at rated voltage. The neutrals of both the generator and motor are grounded through impedances of $j5$ ohms. The sequence impedances of both machines are $Z_0 = j5$, $Z_1 = j15$, and $Z_2 = j10$ ohms. Draw the sequence networks for this system and find the line-to-line voltage at the generator terminals. Assume balanced three-phase operation.

3.15 Calculate the source currents in Example 3.6 without using symmetrical components. Compare your solution method with that of Example 3.6. Which method is easier?

Section 3.5

3.16 For Problem 3.7, calculate the real and reactive power delivered to the three-phase load.

Section 3.6

3.17 Solve Problem 3.2 using the personal computer subroutine SEQVEC.

3.18 Solve Problem 3.3 using the personal computer subroutine PHAVEC.

3.19 Repeat Example 3.8, but include balanced three-phase line impedances of $(1 + j5)$ ohms per phase between the source and load.

References

1. Westinghouse Electric Corporation, *Applied Protective Relaying* (Newark, N.J.: 1976).
2. P. M. Anderson, *Analysis of Faulted Power Systems* (Ames: Iowa State University Press, 1973).
3. W. D. Stevenson, Jr., *Elements of Power System Analysis*, 4th ed. (New York: McGraw-Hill, 1982).

CHAPTER 4

Power Transformers

The power transformer is a major power-system component that permits economical power transmission with high efficiency and low series-voltage drops. Since electric power is proportional to the product of voltage and current, low current levels (and therefore low I^2R losses and low IZ voltage drops) can be maintained for given power levels via high voltages. Power transformers transform ac voltage and current to optimum levels for generation, transmission, distribution, and utilization of electric power.

The development in 1885 by William Stanley of a commercially practical transformer was what made ac power systems more attractive than dc power systems. The ac system with a transformer overcame voltage problems encountered in dc systems as load levels and transmission distances increased. Today's modern power transformers have nearly 100% efficiency, with ratings up to and beyond 1300 MVA.

In this chapter, basic transformer theory is reviewed, and equivalent circuits for practical transformers operating under sinusoidal-steady-state conditions are developed. Models of single-phase two-winding, three-phase two-winding, and three-phase three-winding transformers, as well as autotransformers and regulating transformers are given. Also, the per-unit system, which simplifies power-system analysis by eliminating the ideal transformer winding in transformer equivalent circuits, is introduced in this chapter and used throughout the remainder of the text.

Section 4.1

The Ideal Transformer

Figure 4.1 shows a basic single-phase two-winding transformer, where the two windings are wrapped around a magnetic core [1, 2, 3]. It is assumed here that the transformer is operating under sinusoidal-steady-state excitation. Shown in the figure are the phasor voltages E_1 and E_2 across the windings, and the phasor currents I_1 entering winding 1, which has N_1 turns, and I_2 leaving winding 2, which has N_2 turns. A phasor flux Φ_c set up in the core and a magnetic field intensity phasor H_c are also shown. The core has a cross-sectional area denoted A_c, a mean length of the magnetic circuit l_c, and a magnetic permeability μ_c, assumed constant.

Figure 4.1

Basic single-phase two-winding transformer

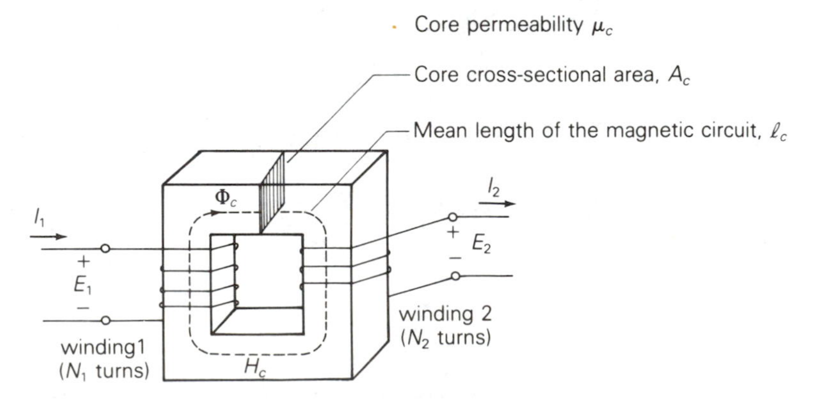

For an ideal transformer, the following are assumed:

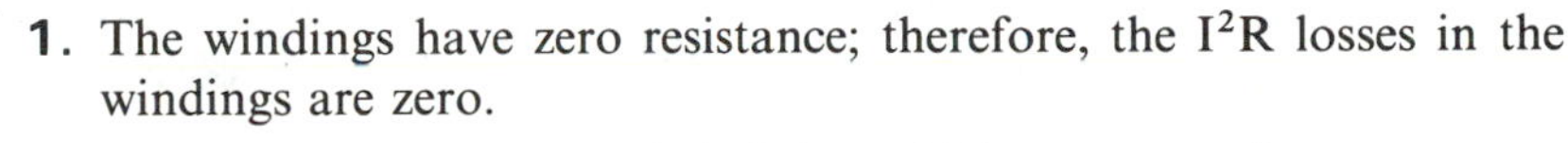

1. The windings have zero resistance; therefore, the I^2R losses in the windings are zero.
2. The core permeability μ_c is infinite, which corresponds to zero core reluctance.
3. There is no leakage flux; that is, the entire flux Φ_c is confined to the core and links both windings.
4. There are no core losses.

A schematic representation of a two-winding transformer is shown in Figure 4.2. Ampere's and Faraday's laws can be used along with the preceding assumptions to derive the ideal transformer relationships. Ampere's law states that the tangential component of the magnetic field intensity vector integrated along a closed path equals the net current enclosed by that path; that is,

$$\oint H_{\tan} dl = I_{\text{enclosed}} \tag{4.1.1}$$

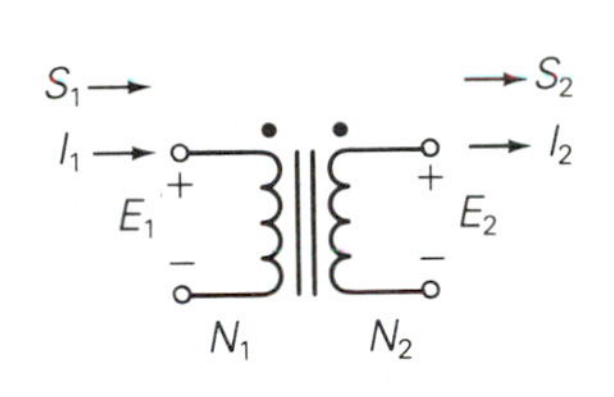

Figure 4.2

Schematic representation of a single-phase two-winding transformer

If the core center line shown in Figure 4.1 is selected as the closed path, and if H_c is constant along the path as well as tangent to the path, then (4.1.1) becomes

$$H_c l_c = N_1 I_1 - N_2 I_2 \tag{4.1.2}$$

Note that the current I_1 is enclosed N_1 times and I_2 is enclosed N_2 times, one time for each turn of the coils. Also, using the right-hand rule*, current I_1 contributes to clockwise flux but current I_2 contributes to counterclockwise flux. Thus, in (4.1.2) the net current enclosed is $N_1 I_1 - N_2 I_2$. For constant core permeability μ_c, the magnetic flux density B_c within the core, also constant, is

$$B_c = \mu_c H_c \quad \text{Wb/m}^2 \tag{4.1.3}$$

and the core flux Φ_c is

$$\Phi_c = B_c \text{A}_c \quad \text{Wb} \tag{4.1.4}$$

Using (4.1.3) and (4.1.4) in (4.1.2) yields

$$N_1 I_1 - N_2 I_2 = l_c B_c / \mu_c = \left(\frac{l_c}{\mu_c \text{A}_c}\right) \Phi_c \tag{4.1.5}$$

We define core reluctance R_c as

$$\text{R}_c = \frac{l_c}{\mu_c \text{A}_c} \tag{4.1.6}$$

Then (4.1.5) becomes

$$N_1 I_1 - N_2 I_2 = \text{R}_c \Phi_c \tag{4.1.7}$$

*The right-hand rule for a coil is as follows: Wrap the fingers of your right hand around the coil in the direction of the current. Your right thumb then points in the direction of the flux.

Equation (4.1.7) can be called "Ohm's law" for the magnetic circuit, wherein the net magnetomotive force mmf $= N_1I_1 - N_2I_2$ equals the product of the core reluctance R_c and the core flux Φ_c. Reluctance R_c, which impedes the establishment of flux in a magnetic circuit, is analogous to resistance in an electric circuit. For an ideal transformer, μ_c is assumed infinite, which, from (4.1.6), means that R_c is zero, and (4.1.7) becomes

$$N_1I_1 = N_2I_2 \tag{4.1.8}$$

In practice, power transformer windings and cores are contained within enclosures, and the winding directions are not visible. One way of conveying winding information is to place a dot at one end of each winding such that when current enters a winding at the dot, it produces an mmf acting in the *same* direction. This dot convention is shown in the schematic of Figure 4.2. The dots are conventionally called *polarity marks*.

Equation (4.1.8) is written for current I_1 *entering* its dotted terminal and current I_2 *leaving* its dotted terminal. As such, I_1 and I_2 are *in phase*, since $I_1 = (N_2/N_1)I_2$. If the direction chosen for I_2 were reversed, such that both currents entered their dotted terminals, then I_1 would be 180° *out of phase* with I_2.

Faraday's law states that the voltage $e(t)$ induced across an N-turn winding by a time-varying flux $\phi(t)$ linking the winding is

$$e(t) = N\frac{d\phi(t)}{dt} \tag{4.1.9}$$

Assuming a sinusoidal-steady-state flux with constant frequency ω, and representing $e(t)$ and $\phi(t)$ by their phasors E and Φ, (4.1.9) becomes

$$E = N(j\omega)\Phi \tag{4.1.10}$$

For an ideal transformer, the entire flux is assumed to be confined to the core, linking both windings. From Faraday's law, the induced voltages across the windings of Figure 4.1 are

$$E_1 = N_1(j\omega)\Phi_c \tag{4.1.11}$$

$$E_2 = N_2(j\omega)\Phi_c \tag{4.1.12}$$

Dividing (4.1.11) by (4.1.12) yields

$$\frac{E_1}{E_2} = \frac{N_1}{N_2} \tag{4.1.13}$$

or

$$\frac{E_1}{N_1} = \frac{E_2}{N_2} \tag{4.1.14}$$

The dots shown in Figure 4.2 indicate that the voltages E_1 and E_2, both of which have their + polarities at the dotted terminals, are in phase. If the polarity chosen for one of the voltages in Figure 4.1 were reversed, then E_1 would be 180° out of phase with E_2.

The turns ratio a_t is defined as follows:

$$a_t = \frac{N_1}{N_2} \tag{4.1.15}$$

Using a_t in (4.1.8) and (4.1.14), the basic relations for an ideal single-phase two-winding transformer are

$$E_1 = \left(\frac{N_1}{N_2}\right) E_2 = a_t E_2 \tag{4.1.16}$$

$$I_1 = \left(\frac{N_2}{N_1}\right) I_2 = \frac{I_2}{a_t} \tag{4.1.17}$$

Two additional relations concerning complex power and impedance can be derived from (4.1.16) and (4.1.17) as follows. The complex power entering winding 1 in Figure 4.2 is

$$S_1 = E_1 I_1^* \tag{4.1.18}$$

Using (4.1.16) and (4.1.17),

$$S_1 = E_1 I_1^* = (a_t E_2)\left(\frac{I_2}{a_t}\right)^* = E_2 I_2^* = S_2 \tag{4.1.19}$$

As shown by (4.1.19), the complex power S_1 entering winding 1 equals the complex power S_2 leaving winding 2. That is, an ideal transformer has no real or reactive power loss.

If an impedance Z_2 is connected across winding 2 of the ideal transformer in Figure 4.2, then

$$Z_2 = \frac{E_2}{I_2} \tag{4.1.20}$$

This impedance, when measured from winding 1, is

$$Z_2' = \frac{E_1}{I_1} = \frac{a_t E_2}{I_2/a_t} = a_t^2 Z_2 = \left(\frac{N_1}{N_2}\right)^2 Z_2 \tag{4.1.21}$$

Thus, the impedance Z_2 connected to winding 2 is referred to winding 1 by multiplying Z_2 by a_t^2, the square of the turns ratio.

Example 4.1

A single-phase two-winding transformer is rated 20 kVA, 480/120 V, 60 Hz. A source connected to the 480-V winding supplies an impedance load connected to the 120-V winding. The load absorbs 15 kVA at 0.8 p.f. lagging when the load voltage is 118 V. Assume that the transformer is ideal and calculate the following:

a. The voltage across the 480-V winding

b. The load impedance

c. The load impedance referred to the 480-V winding

d. The real and reactive power supplied to the 480-V winding

Solution

a. The circuit is shown in Figure 4.3, where winding 1 denotes the 480-V winding and winding 2 denotes the 120-V winding. Selecting the load voltage E_2 as the reference,

$$E_2 = 118\underline{/0^\circ} \quad \text{V}$$

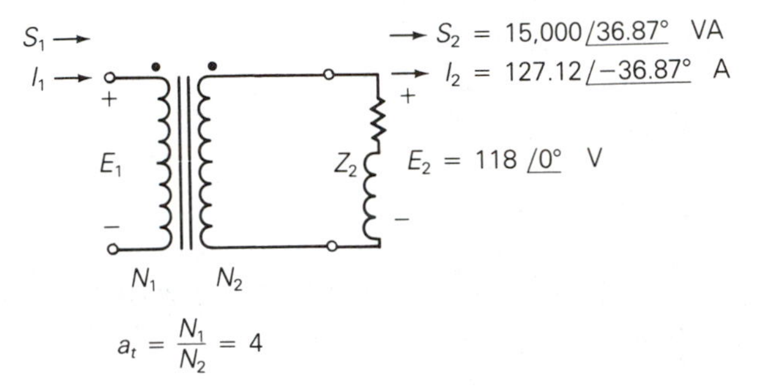

Figure 4.3 Circuit for Example 4.1

The turns ratio is, from (4.1.13),

$$a_t = \frac{N_1}{N_2} = \frac{\text{E}_{1\text{rated}}}{\text{E}_{2\text{rated}}} = \frac{480}{120} = 4$$

and the voltage across winding 1 is

$$E_1 = a_t E_2 = 4(118\underline{/0^\circ}) = 472\underline{/0^\circ} \quad \text{V}$$

b. The complex power S_2 absorbed by the load is

$$S_2 = E_2 I_2^* = 118 I_2^* = 15{,}000\underline{/\cos^{-1}(0.8)} = 15{,}000\underline{/36.87^\circ} \quad \text{VA}$$

Solving, the load current I_2 is

$$I_2 = 127.12\underline{/-36.87^\circ} \quad \text{A}$$

The load impedance Z_2 is

$$Z_2 = \frac{E_2}{I_2} = \frac{118\underline{/0^\circ}}{127.12\underline{/-36.87^\circ}} = 0.9283\underline{/36.87^\circ} \quad \Omega$$

c. From (4.1.21), the load impedance referred to the 480-V winding is

$$Z_2' = a_t^2 Z_2 = (4)^2(0.9283\underline{/36.87^\circ}) = 14.85\underline{/36.87^\circ} \quad \Omega$$

d. From (4.1.19)

$$S_1 = S_2 = 15{,}000\underline{/36.87^\circ} = 12{,}000 + j9000$$

Thus, the real and reactive powers supplied to the 480-V winding are

$$\text{P}_1 = \text{Re}\, S_1 = 12{,}000 \text{ W} = 12 \text{ kW}$$

$$\text{Q}_1 = \text{Im}\, S_1 = 9000 \text{ var} = 9 \text{ kvar}$$

◁

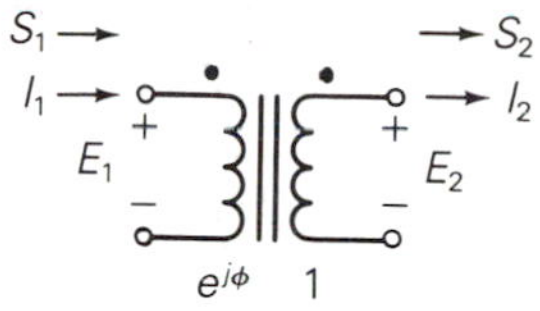

Figure 4.4 shows a schematic of a conceptual single-phase, phase-shifting transformer. This transformer is not an idealization of an actual transformer since it is physically impossible to obtain a complex turns ratio. It will be used later in this chapter as a mathematical model for representing phase shift of three-phase transformers. As shown in Figure 4.4, the complex turns ratio a_t is defined for the phase-shifting transformer as

$$a_t = \frac{e^{j\phi}}{1} = e^{j\phi} \tag{4.1.22}$$

where ϕ is the phase-shift angle. The transformer relations are then

$$E_1 = a_t E_2 = e^{j\phi} E_2 \tag{4.1.23}$$

$$I_1 = \frac{I_2}{a_t^*} = e^{j\phi} I_2 \tag{4.1.24}$$

Note that the phase angle of E_1 leads the phase angle of E_2 by ϕ. Similarly, I_1 leads I_2 by the angle ϕ. But the magnitudes are unchanged. That is, $|E_1| = |E_2|$ and $|I_1| = |I_2|$.

From these two relations, the following two additional relations are derived:

$$S_1 = E_1 I_1^* = \left(a_t E_2\right)\left(\frac{I_2}{a_t^*}\right)^* = E_2 I_2^* = S_2 \tag{4.1.25}$$

$$Z_2' = \frac{E_1}{I_1} = \frac{a_t E_2}{\frac{1}{a_t^*} I_2} = |a_t|^2 Z_2 = Z_2 \tag{4.1.26}$$

$S_1 \rightarrow$ $\rightarrow S_2$, $I_1 \rightarrow$ $\rightarrow I_2$, E_1, E_2, $e^{j\phi}$ 1

$a_t = e^{j\phi}$

$E_1 = a_t E_2 = e^{j\phi} E_2$

$I_1 = \frac{I_2}{a_t^*} = e^{j\phi} I_2$

$S_1 = S_2$

$Z_2' = Z_2$

Figure 4.4 Schematic representation of a conceptual single-phase, phase-shifting transformer

Thus, per-unit impedance is unchanged when it is referred from one side of an ideal phase-shifting transformer to the other. Also, the ideal phase-shifting transformer has no real or reactive power losses since $S_1 = S_2$.

Note that (4.1.23) and (4.1.24) for the phase-shifting transformer are the same as (4.1.16) and (4.1.17) for the ideal physical transformer except for the complex conjugate (*) in (4.1.24). The complex conjugate for the phase-shifting transformer is required to make $S_1 = S_2$ (complex power into winding 1 equals complex power out of winding 2), as shown in (4.1.25).

Section 4.2

Equivalent Circuits for Practical Transformers

Figure 4.5 shows an equivalent circuit for a practical single-phase two-winding transformer, which differs from the ideal transformer as follows:

1. The windings have resistance.
2. The core permeability μ_c is finite.
3. The magnetic flux is not entirely confined to the core.
4. There are real and reactive power losses in the core.

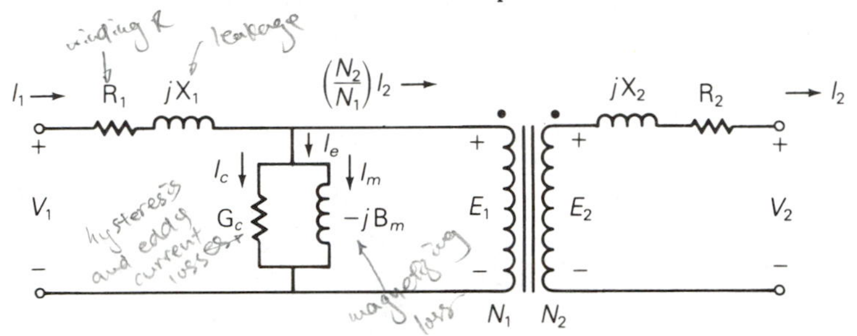

Figure 4.5

Equivalent circuit of a practical single-phase two-winding transformer

The resistance R_1 is included in series with winding 1 of the figure to account for I^2R losses in this winding. A reactance X_1, called the leakage reactance of winding 1, is also included in series with winding 1 to account for the leakage flux of winding 1. This leakage flux is the component of the flux that links winding 1 but does not link winding 2; it causes a voltage drop $I_1(jX_1)$, which is proportional to I_1 and leads I_1 by 90°. There is also a reactive power loss $I_1^2X_1$ associated with this leakage reactance. Similarly, there is a resistance R_2 and a leakage reactance X_2 in series with winding 2.

Equation (4.1.7) shows that for finite core permeability μ_c, the total mmf is not zero. Dividing (4.1.7) by N_1 and using (4.1.11), we get

$$I_1 - \left(\frac{N_2}{N_1}\right)I_2 = \frac{R_c}{N_1}\Phi_c = \frac{R_c}{N_1}\left(\frac{E_1}{j\omega N_1}\right) = -j\left(\frac{R_c}{\omega N_1^2}\right)E_1 \tag{4.2.1}$$

Defining the term on the right-hand side of (4.2.1) to be I_m, called *magnetizing* current, it is evident that I_m lags E_1 by 90°, and can be represented by a shunt inductor with susceptance $B_m = \left(\frac{R_c}{\omega N_1^2}\right)$ mhos.* However, in reality there is an additional shunt branch, represented by a resistor with conductance G_c mhos, which carries a current I_c, called the *core loss* current. I_c is in phase with E_1. When the core loss current I_c is included, (4.2.1) becomes

$$I_1 - \left(\frac{N_2}{N_1}\right)I_2 = I_c + I_m = (G_c - jB_m)E_1 \tag{4.2.2}$$

*The units of admittance, conductance, and susceptance, which in the SI system are siemens (with symbol S), are also called mhos (with symbol ℧) or ohms^{-1} (with symbol Ω^{-1}).

The equivalent circuit of Figure 4.5, which includes the shunt branch with admittance $(G_c - jB_m)$ mhos, satisfies the KCL equation (4.2.2). Note that when winding 2 is open ($I_2 = 0$) and when a sinusoidal voltage V_1 is applied to winding 1, then (4.2.2) indicates that the current I_1 will have two components: the core loss current I_c and the magnetizing current I_m. Associated with I_c is a real power loss $I_c^2/G_c = E_1^2 G_c$ W. This real power loss accounts for both hysteresis and eddy current losses within the core. Hysteresis loss occurs because a cyclic variation of flux within the core requires energy dissipated as heat. As such, hysteresis loss can be reduced by the use of special high grades of alloy steel as core material. Eddy current loss occurs because induced currents called eddy currents flow within the magnetic core perpendicular to the flux. As such, eddy current loss can be reduced by constructing the core with laminated sheets of alloy steel. Associated with I_m is a reactive power loss $I_m^2/B_m = E_1^2 B_m$ var. This reactive power is required to magnetize the core. The phasor sum $(I_c + I_m)$ is called the *exciting* current I_e.

Figure 4.6

Equivalent circuits for a practical single-phase two-winding transformer

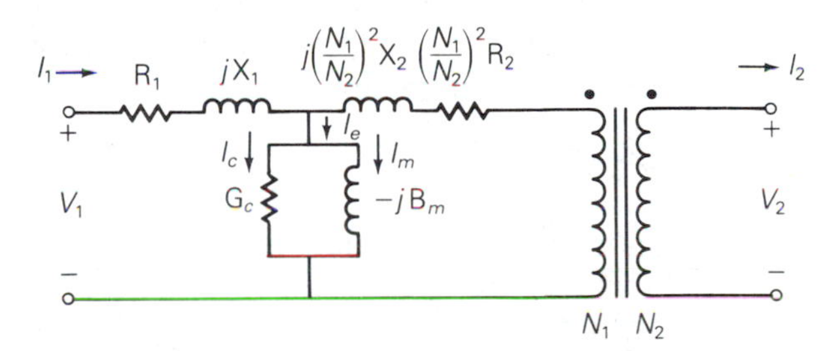

(a) R_2 and X_2 are referred to winding 1

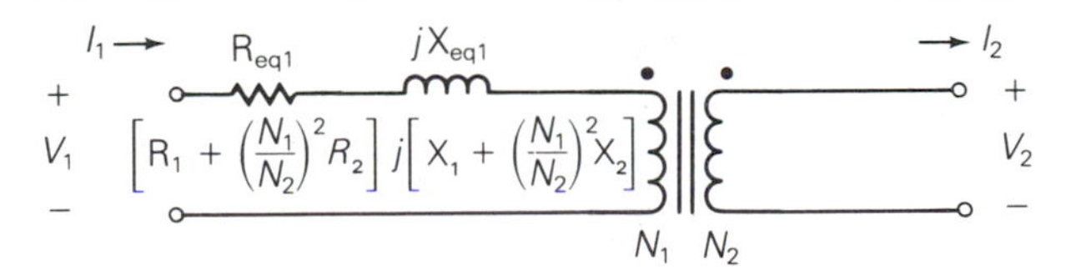

(b) Neglecting exciting current

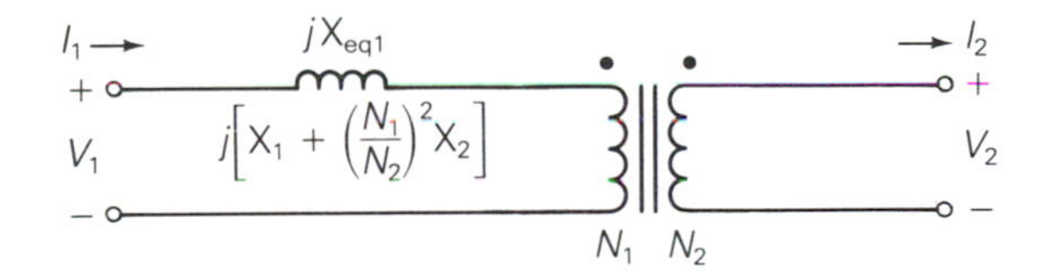

(c) Neglecting exciting current and I^2R winding loss

Figure 4.6 shows three alternative equivalent circuits for a practical single-phase two-winding transformer. In Figure 4.6(a) the resistance R_2 and leakage reactance X_2 of winding 2 are referred to winding 1 via (4.1.21). In Figure 4.6(b) the shunt branch is omitted, which corresponds to neglecting the exciting current. Since the exciting current is typically about 5% of rated current, neglecting it in power-system studies is often valid unless transformer efficiency or exciting current phenomena are of particular concern. For large power transformers rated more than 500 kVA, the winding resistances, which are small compared to the leakage reactances, can often be neglected, as shown in Figure 4.6(c).

Thus a practical transformer operating in sinusoidal steady state is equivalent to an ideal transformer with external impedance and admittance branches, as shown in Figure 4.6. The external branches can be evaluated from short-circuit and open-circuit tests, as illustrated by the following example.

Example 4.2 A single-phase two-winding transformer is rated 20 kVA, 480/120 volts, 60 Hz. During a short-circuit test, where rated current at rated frequency is applied to the 480-volt winding (denoted winding 1), with the 120-volt winding (winding 2) shorted, the following readings are obtained: $V_1 = 35$ volts, $P_1 = 300$ W. During an open-circuit test, where rated voltage is applied to winding 2, with winding 1 open, the following readings are obtained: $I_2 = 12$ A, $P_2 = 200$ W.

a. From the short-circuit test, determine the equivalent series impedance $Z_{eq1} = R_{eq1} + jX_{eq1}$ referred to winding 1. Neglect the shunt admittance.

b. From the open-circuit test, determine the shunt admittance $Y_m = G_c - jB_m$ referred to winding 1. Neglect the series impedance.

Solution

a. The equivalent circuit for the short-circuit test is shown in Figure 4.7(a), where the shunt admittance branch is neglected. Rated current for winding 1 is

$$I_{1\text{rated}} = \frac{S_{\text{rated}}}{V_{1\text{rated}}} = \frac{20 \times 10^3}{480} = 41.667 \quad \text{A}$$

R_{eq1}, Z_{eq1}, and X_{eq1} are then determined as follows:

$$R_{eq1} = \frac{P_1}{I_{1\text{rated}}^2} = \frac{300}{(41.667)^2} = 0.1728 \quad \Omega$$

$$|Z_{eq1}| = \frac{V_1}{I_{1\text{rated}}} = \frac{35}{41.667} = 0.8400 \quad \Omega$$

$$X_{eq1} = \sqrt{Z_{eq1}^2 - R_{eq1}^2} = 0.8220 \quad \Omega$$

$$Z_{eq1} = R_{eq1} + jX_{eq1} = 0.1728 + j0.8220 = 0.8400\underline{/78.13^\circ} \quad \Omega$$

b. The equivalent circuit for the open-circuit test is shown in Figure 4.7(b), where the series impedance is neglected. From (4.1.16),

$$V_1 = E_1 = a_t E_2 = \frac{N_1}{N_2} V_{2\text{rated}} = \frac{480}{120}(120) = 480 \text{ volts}$$

Figure 4.7

Circuits for Example 4.2

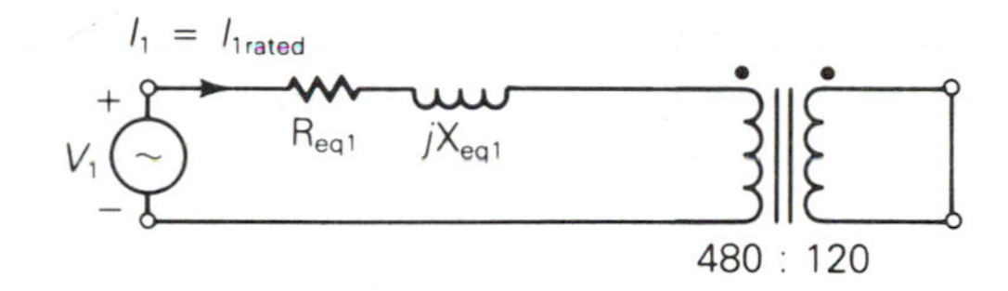

(a) Short-circuit test (neglecting shunt admittance)

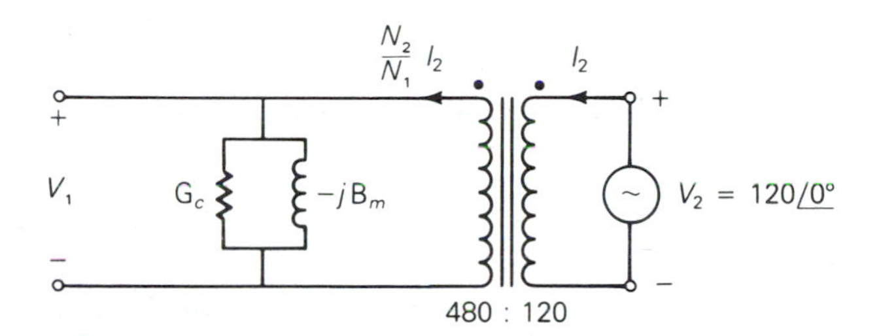

(b) Open-circuit test (neglecting series impedance)

G_c, Y_m, and B_m are then determined as follows:

$$G_c = \frac{P_2}{V_1^2} = \frac{200}{(480)^2} = 0.000868 \quad S$$

$$|Y_m| = \frac{\left(\frac{N_2}{N_1}\right) I_2}{V_1} = \frac{\left(\frac{120}{480}\right)(12)}{480} = 0.00625 \quad S$$

$$B_m = \sqrt{Y_m^2 - G_c^2} = \sqrt{(0.00625)^2 - (0.000868)^2} = 0.00619 \quad S$$

$$Y_m = G_c - jB_m = 0.000868 - j0.00619 = 0.00625\underline{/-82.02°} \quad S$$

Note that the equivalent series impedance is usually evaluated at rated current from a short-circuit test, and the shunt admittance is evaluated at rated voltage from an open-circuit test. For small variations in transformer operation near rated conditions, the impedance and admittance values are often assumed constant. ◁

The following are not represented by the equivalent circuit of Figure 4.5:

1. Saturation
2. Inrush current
3. Nonsinusoidal exciting current
4. Surge phenomena

They are briefly discussed in the following sections.

Saturation In deriving the equivalent circuit of the ideal and practical transformers, we have assumed constant core permeability μ_c and the linear relationship $B_c = \mu_c H_c$ of (4.1.3). However, the relationship between B and H for ferromagnetic materials used for transformer cores is nonlinear and multivalued. Figure 4.8 shows a set of B–H curves for a grain-oriented electrical steel typically used in transformers. As shown, each curve is multivalued, which is caused by hysteresis. For many engineering applications, the B–H curves can be adequately described by the dashed line drawn through the curves in Figure 4.8. Note that as H increases, the core becomes saturated; that is, the curves flatten out as B increases above 1 Wb/m^2. If the magnitude of the voltage applied to a transformer is too large, the core will saturate and a high magnetizing current will flow. In a well-designed transformer, the applied peak voltage causes the peak flux density in steady state to occur at the knee of the B–H curve, with a corresponding low value of magnetizing current.

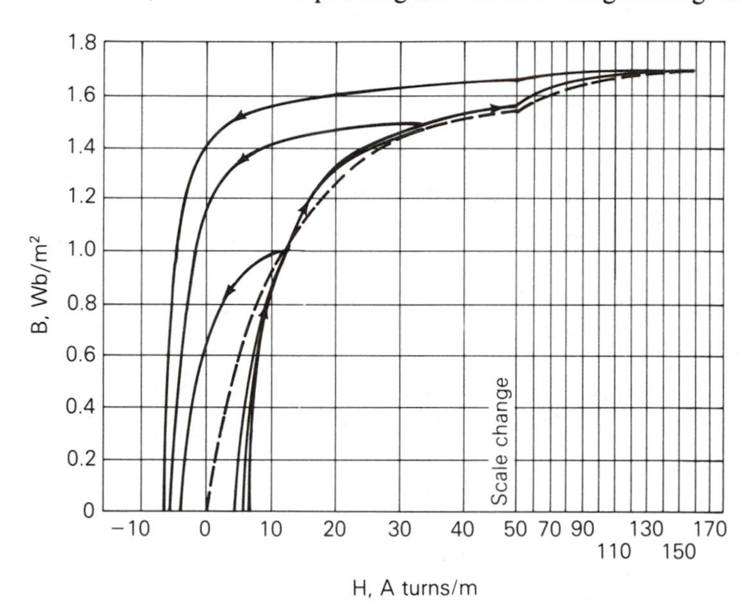

Figure 4.8

B–H curves for M-5 grain-oriented electrical steel 0.012 in. thick (Armco Inc.)

Inrush Current When a transformer is first energized, a transient current much larger than rated transformer current can flow for several cycles. This current, called *inrush current*, is nonsinusoidal and has a large dc component. To understand the cause of inrush, assume that before energization, the transformer core is magnetized with a residual flux density $B(0) = +1.5\,Wb/m^2$ (near the knee of the dotted curve in Figure 4.8). If the transformer is then energized when the source voltage is positive and increasing, Faraday's law, (4.1.9), will cause the flux density $B(t)$ to increase further, since

$$B(t) = \frac{\phi(t)}{A} = \frac{1}{NA}\int_0^t e(t)\,dt + B(0)$$

As $B(t)$ moves into the saturation region of the B–H curve, large values of $H(t)$ will occur, and, from Ampere's law, (4.1.1), corresponding large values of current $i(t)$ will flow for several cycles until it has dissipated. Since normal inrush currents can be as large as abnormal short-circuit currents in transformers, transformer protection schemes must be able to distinguish between these two types of currents.

Nonsinusoidal Exciting Current When a sinusoidal voltage is applied to one winding of a transformer with the other winding open, the flux $\phi(t)$ and flux density $B(t)$ will, from Faraday's law, (4.1.9), be very nearly sinusoidal in steady state. But the magnetic field intensity $H(t)$ and the resulting exciting current will not be sinusoidal in steady state, due to the nonlinear B–H curve. If the exciting current is measured and analyzed by Fourier analysis techniques, one finds that it has a fundamental component and a set of odd harmonics. The principal harmonic is the third, whose rms value is typically about 40% of the total rms exciting current. However, the nonsinusoidal nature of exciting current is usually neglected unless harmonic effects are of direct concern, because the exciting current itself is only about 5% of rated current for power transformers.

Surge Phenomena When power transformers are subjected to transient overvoltages caused by lightning or switching surges, the capacitances of the transformer windings have important effects on transient response. Transformer winding capacitances and response to surges are discussed in Chapter 11.

Section 4.3

The Per-Unit System

Power-system quantities such as voltage, current, power, and impedance are often expressed in per-unit or percent of specified base values. For example, if a base voltage of 20 kV is specified, then the voltage 18 kV is $(18/20) = 0.9$ per unit or 90%. Calculations can then be made with per-unit quantities rather than with the actual quantities.

One advantage of the per-unit system is that by properly specifying base quantities, the transformer equivalent circuit can be simplified. The ideal transformer winding can be eliminated, such that voltages, currents, and external impedances and admittances expressed in per-unit do not change when they are referred from one side of a transformer to the other. This can be a significant advantage even in a power system of moderate size, where hundreds of transformers may be encountered. The per-unit system allows us to avoid the possibility of making serious calculation errors when referring quantities from one side of a transformer to the other. Another advantage of the per-unit system is that the per-unit impedances of electrical equipment of similar type usually lie within a narrow numerical range when the equipment ratings are used as base values. Because of this, per-unit impedance data can be checked rapidly for gross errors by someone familiar with per-unit quantities. In addition, manufacturers usually specify the impedances of machines and transformers in per-unit or percent of nameplate rating.

Per-unit quantities are calculated as follows:

$$\text{per-unit quantity} = \frac{\text{actual quantity}}{\text{base value of quantity}} \tag{4.3.1}$$

where *actual quantity* is the value of the quantity in the actual units. The base value has the same units as the actual quantity, thus making the per-unit quantity dimensionless. Also, the base value is always a real number. Therefore, the angle of the per-unit quantity is the same as the angle of the actual quantity.

Two independent base values can be arbitrarily selected at one point in a power system. Usually the base voltage V_{baseLN} and base complex power $S_{base1\phi}$ are selected for either a single-phase circuit or for one phase of a three-phase circuit. Then, in order for electrical laws to be valid in the per-unit system, the following relations must be used for other base values:

$$P_{base1\phi} = Q_{base1\phi} = S_{base1\phi} \tag{4.3.2}$$

$$I_{base} = \frac{S_{base1\phi}}{V_{baseLN}} \tag{4.3.3}$$

$$Z_{base} = R_{base} = X_{base} = \frac{V_{baseLN}}{I_{base}} = \frac{V^2_{baseLN}}{S_{base1\phi}} \tag{4.3.4}$$

$$Y_{base} = \frac{1}{Z_{base}} \tag{4.3.5}$$

In (4.3.2)–(4.3.5) the subscripts LN and 1ϕ denote "line-to-neutral" and "per-phase," respectively, for three-phase circuits. These equations are also valid for single-phase circuits, where subscripts can be omitted.

By convention, we adopt the following two rules for base quantities:

1. The value of $S_{base1\phi}$ is the same for the entire power system of concern.
2. The ratio of the voltage bases on either side of a transformer is selected to be the same as the ratio of the transformer voltage ratings.

With these two rules, a per-unit impedance remains unchanged when referred from one side of a transformer to the other.

Example 4.3 A single-phase two-winding transformer is rated 20 kVA, 480/120 volts, 60 Hz. The equivalent leakage impedance of the transformer referred to the 120-volt winding, denoted winding 2, is $Z_{eq2} = 0.0525\underline{/78.13^\circ}\ \Omega$. Using the transformer ratings as base values, determine the per-unit leakage impedance referred to winding 2 and referred to winding 1.

Solution The values of S_{base}, V_{base1}, and V_{base2} are, from the transformer ratings,

$$S_{base} = 20\ \text{kVA}, \qquad V_{base1} = 480\ \text{volts}, \qquad V_{base2} = 120\ \text{volts}$$

Using (4.3.4), the base impedance on the 120-volt side of the transformer is

$$Z_{\text{base}2} = \frac{V^2_{\text{base}2}}{S_{\text{base}}} = \frac{(120)^2}{20{,}000} = 0.72 \quad \Omega$$

Then, using (4.3.1), the per-unit leakage impedance referred to winding 2 is

$$Z_{\text{eq}2\text{p.u.}} = \frac{Z_{\text{eq}2}}{Z_{\text{base}2}} = \frac{0.0525\underline{/78.13^\circ}}{0.72} = 0.0729\underline{/78.13^\circ} \quad \text{per unit}$$

If $Z_{\text{eq}2}$ is referred to winding 1,

$$Z_{\text{eq}1} = a_t^2 Z_{\text{eq}2} = \left(\frac{N_1}{N_2}\right)^2 Z_{\text{eq}2} = \left(\frac{480}{120}\right)^2 (0.0525\underline{/78.13^\circ})$$
$$= 0.84\underline{/78.13^\circ} \quad \Omega$$

The base impedance on the 480-volt side of the transformer is

$$Z_{\text{base}1} = \frac{V^2_{\text{base}1}}{S_{\text{base}}} = \frac{(480)^2}{20{,}000} = 11.52 \quad \Omega$$

and the per-unit leakage reactance referred to winding 1 is

$$Z_{\text{eq}1\text{p.u.}} = \frac{Z_{\text{eq}1}}{Z_{\text{base}1}} = \frac{0.84\underline{/78.13^\circ}}{11.52} = 0.0729\underline{/78.13^\circ} \text{ per unit} = Z_{\text{eq}2\text{p.u.}}$$

Thus, the *per-unit* leakage impedance remains unchanged when referred from winding 2 to winding 1. This has been achieved by specifying

$$\frac{V_{\text{base}1}}{V_{\text{base}2}} = \frac{V_{\text{rated}1}}{V_{\text{rated}2}} = \left(\frac{480}{120}\right)$$

◁

Figure 4.9 shows three per-unit circuits of a single-phase two-winding transformer. The ideal transformer, shown in Figure 4.9(a), satisfies the per-unit relations $E_{1\text{p.u.}} = E_{2\text{p.u.}}$, and $I_{1\text{p.u.}} = I_{2\text{p.u.}}$, which can be derived as follows. First divide (4.1.16) by $V_{\text{base}1}$:

$$E_{1\text{p.u.}} = \frac{E_1}{V_{\text{base}1}} = \frac{N_1}{N_2} \times \frac{E_2}{V_{\text{base}1}} \tag{4.3.6}$$

Then, using $V_{\text{base}1}/V_{\text{base}2} = V_{\text{rated}1}/V_{\text{rated}2} = N_1/N_2$,

$$E_{1\text{p.u.}} = \frac{N_1}{N_2} \frac{E_2}{\left(\frac{N_1}{N_2}\right) V_{\text{base}2}} = \frac{E_2}{V_{\text{base}2}} = E_{2\text{p.u.}} \tag{4.3.7}$$

Similarly, divide (4.1.17) by $I_{\text{base}1}$:

$$I_{1\text{p.u.}} = \frac{I_1}{I_{\text{base}1}} = \frac{N_2}{N_1} \frac{I_2}{I_{\text{base}1}} \tag{4.3.8}$$

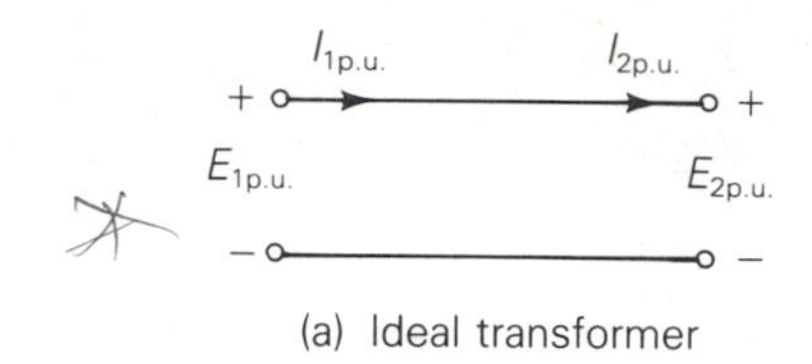

(a) Ideal transformer

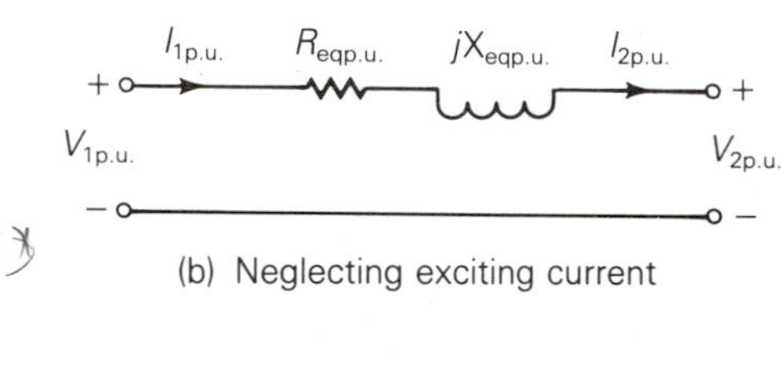

(b) Neglecting exciting current

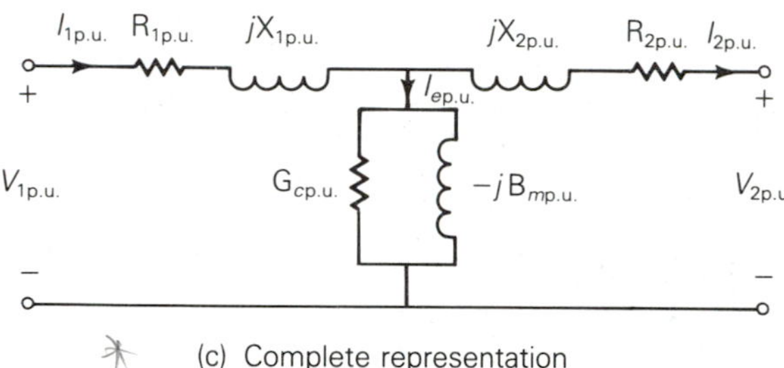

(c) Complete representation

Figure 4.9 Per-unit equivalent circuits of a single-phase two-winding transformer

Then, using $I_{base1} = S_{base}/V_{base1} = S_{base}/[(N_1/N_2)V_{base2}] = (N_2/N_1)\,I_{base2}$,

$$I_{1p.u.} = \frac{N_2}{N_1}\frac{I_2}{\left(\dfrac{N_2}{N_1}\right)I_{base2}} = \frac{I_2}{I_{base2}} = I_{2p.u.} \tag{4.3.9}$$

Thus, the ideal transformer winding in Figure 4.2 is eliminated from the per-unit circuit in Figure 4.9(a). The per-unit leakage impedance is included in Figure 4.9(b), and the per-unit shunt admittance branch is added in Figure 4.9(c) to obtain the complete representation.

When only one component, such as a transformer, is considered, the nameplate ratings of that component are usually selected as base values. When several components are involved, however, the system base values may be different from the nameplate ratings of any particular device. It is then necessary to convert the per-unit impedance of a device from its nameplate ratings to the system base values. To convert a per-unit impedance from "old" to "new" base values, use

$$Z_{p.u.new} = \frac{Z_{actual}}{Z_{basenew}} = \frac{Z_{p.u.old}Z_{baseold}}{Z_{basenew}} \tag{4.3.10}$$

or, from (4.3.4),

$$Z_{p.u.new} = Z_{p.u.old}\left(\frac{V_{baseold}}{V_{basenew}}\right)^2\left(\frac{S_{basenew}}{S_{baseold}}\right) \tag{4.3.11}$$

Example 4.4 Three zones of a single-phase circuit are identified in Figure 4.10(a). The zones are connected by transformers T_1 and T_2, whose ratings are also shown. Using base values of 30 kVA and 240 volts in zone 1, draw the per-unit circuit and determine the per-unit impedances and the per-unit source voltage. Then calculate the load current both in per-unit and in amperes. Transformer winding resistances and shunt admittance branches are neglected.

Figure 4.10

Circuits for Example 4.4

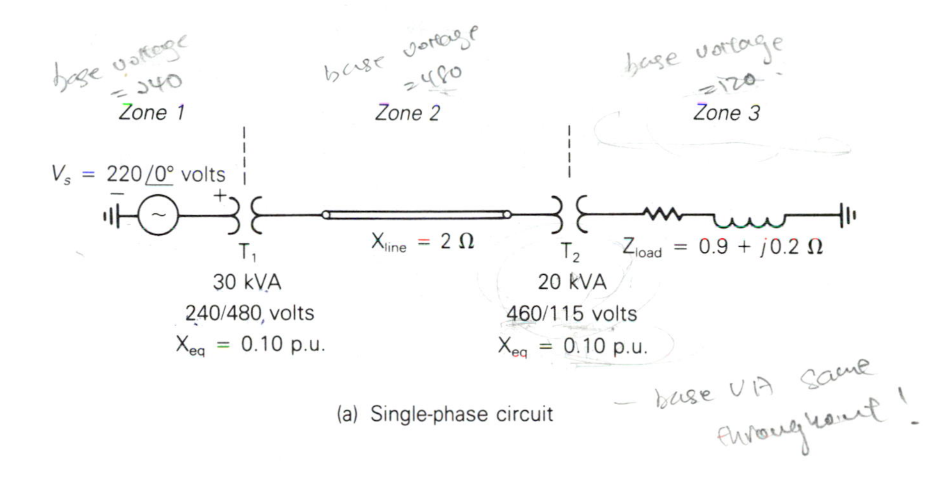

(a) Single-phase circuit

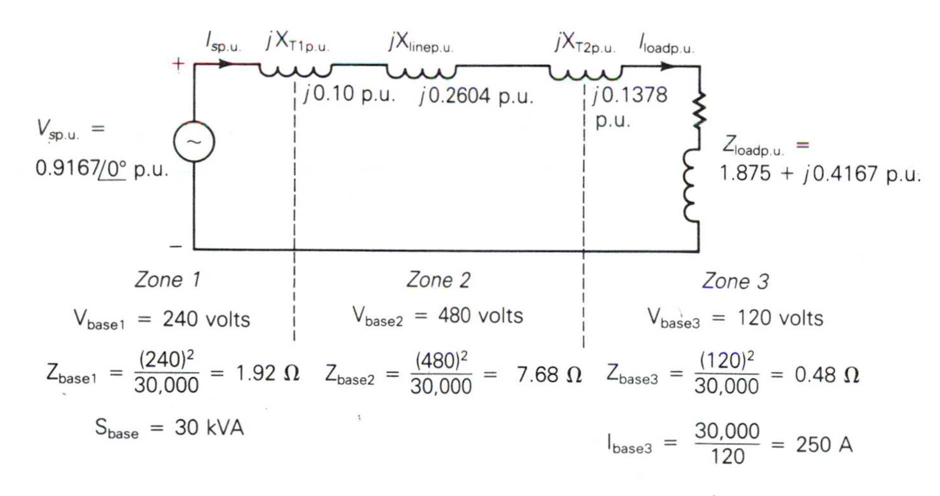

(b) Per-unit circuit

Solution First the base values in each zone are determined. $S_{base} = 30$ kVA is the same for the entire network. Also, $V_{base1} = 240$ volts, as specified for zone 1. When moving across a transformer, the voltage base is changed in proportion to the transformer voltage ratings. Thus,

$$V_{\text{base}\,2} = \left(\frac{480}{240}\right)\left(240\right) = 480 \quad \text{volts}$$

and

$$V_{\text{base}\,3} = \left(\frac{115}{460}\right)\left(480\right) = 120 \quad \text{volts}$$

The base impedances in zones 2 and 3 are

$$Z_{\text{base}\,2} = \frac{V_{\text{base}\,2}^2}{S_{\text{base}}} = \frac{480^2}{30{,}000} = 7.68 \quad \Omega$$

and

$$Z_{\text{base}\,3} = \frac{V_{\text{base}\,3}^2}{S_{\text{base}}} = \frac{120^2}{30{,}000} = 0.48 \quad \Omega$$

and the base current in zone 3 is

$$I_{\text{base}\,3} = \frac{S_{\text{base}}}{V_{\text{base}\,3}} = \frac{30{,}000}{120} = 250 \quad \text{A}$$

Next, the per-unit circuit impedances are calculated using the system base values. Since $S_{\text{base}} = 30\,\text{kVA}$ is the same as the kVA rating of transformer T_1, and $V_{\text{base}\,1} = 240$ volts is the same as the voltage rating of the zone 1 side of transformer T_1, the per-unit leakage reactance of T_1 is the same as its nameplate value, $X_{T1\text{p.u.}} = 0.1$ per unit. But the per-unit leakage reactance of transformer T_2 must be converted from its nameplate rating to the system base. Using (4.3.11) and $V_{\text{base}\,2} = 480$ volts,

$$X_{T2\text{p.u.}} = \left(0.10\right)\left(\frac{460}{480}\right)^2\left(\frac{30{,}000}{20{,}000}\right) = 0.1378 \quad \text{per unit}$$

Alternatively, using $V_{\text{base}\,3} = 120$ volts,

$$X_{T2\text{p.u.}} = \left(0.10\right)\left(\frac{115}{120}\right)^2\left(\frac{30{,}000}{20{,}000}\right) = 0.1378 \quad \text{per unit}$$

which gives the same result. The line, which is located in zone 2, has a per-unit reactance

$$X_{\text{linep.u.}} = \frac{X_{\text{line}}}{Z_{\text{base}\,2}} = \frac{2}{7.68} = 0.2604 \quad \text{per unit}$$

and the load, which is located in zone 3, has a per-unit impedance

$$Z_{\text{loadp.u.}} = \frac{Z_{\text{load}}}{Z_{\text{base}\,3}} = \frac{0.9 + j0.2}{0.48} = 1.875 + j0.4167 \quad \text{per unit}$$

The per-unit circuit is shown in Figure 4.10(b), where the base values for each zone, per-unit impedances, and the per-unit source voltage are shown.

The per-unit load current is then easily calculated from Figure 4.10(b) as follows:

$$I_{\text{loadp.u.}} = I_{\text{sp.u.}} = \frac{V_{\text{sp.u.}}}{j(X_{\text{T1p.u.}} + X_{\text{linep.u.}} + X_{\text{T2p.u.}}) + Z_{\text{loadp.u.}}}$$

$$= \frac{0.9167\underline{/0^\circ}}{j(0.10 + 0.2604 + 0.1378) + (1.875 + j0.4167)}$$

$$= \frac{0.9167\underline{/0^\circ}}{1.875 + j0.9149} = \frac{0.9167\underline{/0^\circ}}{2.086\underline{/26.01^\circ}}$$

$$= 0.4395\underline{/-26.01^\circ} \quad \text{per unit}$$

The actual load current is

$$I_{\text{load}} = (I_{\text{loadp.u.}})I_{\text{base3}} = (0.4395\underline{/-26.01^\circ})(250) = 109.9\underline{/-26.01^\circ} \quad \text{A}$$

Note that the per-unit equivalent circuit of Figure 4.10(b) is relatively easy to analyze, since ideal transformer windings have been eliminated by proper selection of base values. ◁

Balanced three-phase circuits can be solved in per-unit on a per-phase basis after converting Δ-load impedances to equivalent Y impedances. Base values can be selected either on a per-phase basis or on a three-phase basis. Equations (4.3.1)–(4.3.5) remain valid for three-phase circuits on a per-phase basis. Usually $S_{\text{base}3\phi}$ and V_{baseLL} are selected, where the subscripts 3ϕ and LL denote "three-phase" and "line-to-line," respectively. Then the following relations must be used for other base values:

$$S_{\text{base}1\phi} = \frac{S_{\text{base}3\phi}}{3} \tag{4.3.12}$$

$$V_{\text{baseLN}} = \frac{V_{\text{baseLL}}}{\sqrt{3}} \tag{4.3.13}$$

$$S_{\text{base}3\phi} = P_{\text{base}3\phi} = Q_{\text{base}3\phi} \tag{4.3.14}$$

$$I_{\text{base}} = \frac{S_{\text{base}1\phi}}{V_{\text{baseLN}}} = \frac{S_{\text{base}3\phi}}{\sqrt{3}V_{\text{baseLL}}} \tag{4.3.15}$$

$$Z_{\text{base}} = \frac{V_{\text{baseLN}}}{I_{\text{base}}} = \frac{V_{\text{baseLN}}^2}{S_{\text{base}1\phi}} = \frac{V_{\text{baseLL}}^2}{S_{\text{base}3\phi}} \tag{4.3.16}$$

$$R_{\text{base}} = X_{\text{base}} = Z_{\text{base}} = \frac{1}{Y_{\text{base}}} \tag{4.3.17}$$

Example 4.5 As in Example 2.5, a balanced-Y-connected voltage source with $E_{ab} = 480\underline{/0^\circ}$ volts is applied to a balanced-Δ load with $Z_\Delta = 30\underline{/40^\circ}$ Ω. The line impedance between the source and load is $Z_L = 1\underline{/85^\circ}$ Ω for each phase. Calculate the per-unit and actual current in phase a of the line using $S_{base3\phi} = 10\,kVA$ and $V_{baseLL} = 480$ volts.

Solution First convert Z_Δ to an equivalent Z_Y; the equivalent line-to-neutral diagram is shown in Figure 2.17. The base impedance is, from (4.3.16),

$$Z_{base} = \frac{V^2_{baseLL}}{S_{base3\phi}} = \frac{(480)^2}{10{,}000} = 23.04 \quad \Omega$$

The per-unit line and load impedances are

$$Z_{Lp.u.} = \frac{Z_L}{Z_{base}} = \frac{1\underline{/85^\circ}}{23.04} = 0.04340\underline{/85^\circ} \quad \text{per unit}$$

and

$$Z_{Yp.u.} = \frac{Z_Y}{Z_{base}} = \frac{10\underline{/40^\circ}}{23.04} = 0.4340\underline{/40^\circ} \quad \text{per unit}$$

Also,

$$V_{baseLN} = \frac{V_{baseLL}}{\sqrt{3}} = \frac{480}{\sqrt{3}} = 277 \quad \text{volts}$$

and

$$E_{anp.u.} = \frac{E_{an}}{V_{baseLN}} = \frac{277\underline{/-30^\circ}}{277} = 1.0\underline{/-30^\circ} \quad \text{per unit}$$

The per-unit equivalent circuit is shown in Figure 4.11. The per-unit line current in phase a is then

$$I_{ap.u.} = \frac{E_{anp.u.}}{Z_{Lp.u.} + Z_{Yp.u.}} = \frac{1.0\underline{/-30^\circ}}{0.04340\underline{/85^\circ} + 0.4340\underline{/40^\circ}}$$

$$= \frac{1.0\underline{/-30^\circ}}{(0.00378 + j0.04323) + (0.3325 + j0.2790)}$$

$$= \frac{1.0\underline{/-30^\circ}}{0.3362 + j0.3222} = \frac{1.0\underline{/-30^\circ}}{0.4657\underline{/43.78^\circ}}$$

$$= 2.147\underline{/-73.78^\circ} \quad \text{per unit}$$

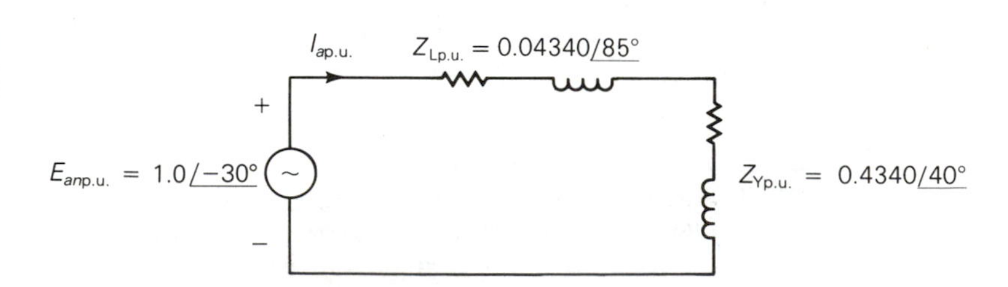

Figure 4.11
Circuit for Example 4.5

The base current is

$$I_{\text{base}} = \frac{S_{\text{base}3\phi}}{\sqrt{3}V_{\text{baseLL}}} = \frac{10{,}000}{\sqrt{3}(480)} = 12.03 \quad \text{A}$$

and the actual phase a line current is

$$I_a = (2.147\underline{/-73.78^\circ})(12.03) = 25.83\underline{/-73.78^\circ} \quad \text{A}$$

◁

Section 4.4

Three-phase Transformer Connections and Phase Shift

Three identical single-phase two-winding transformers may be connected to form a three-phase bank. Four ways to connect the windings are Y–Y, Y–Δ,

Figure 4.12

Three-phase two-winding Y–Y transformer bank

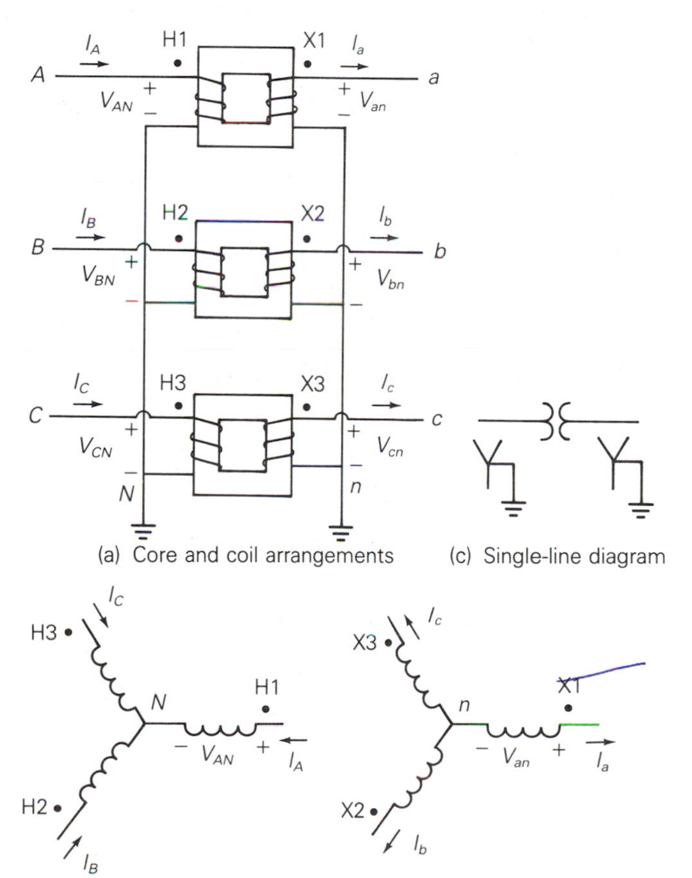

Δ–Y, and Δ–Δ. For example, Figure 4.12 shows a three-phase Y–Y bank. Figure 4.12(a) shows the core and coil arrangements. The American standard for marking three-phase transformers substitutes H1, H2, and H3 on the high-voltage terminals and X1, X2, and X3 on the low-voltage terminals in place of the polarity dots. Also, in this text, we will use uppercase letters *ABC* to identify phases on the high-voltage side of the transformer and lowercase letters *abc* to identify phases on the low-voltage side of the transformer. In Figure 4.12(a) the transformer high-voltage terminals H1, H2, and H3 are connected to phases A, B, and C, and the low-voltage terminals X1, X2, and X3 are connected to phases a, b, and c, respectively.

Figure 4.12(b) shows a schematic representation of the three-phase Y–Y transformer. Windings on the same core are drawn in parallel, and the phasor relationship for balanced positive-sequence operation is shown. For example, high-voltage winding H1–N is on the same magnetic core as low-voltage winding X1–n in Figure 4.12(b). Also, V_{AN} is in phase with V_{an}. Figure 4.12(c) shows a single-line diagram of a Y–Y transformer. A single-line diagram shows one phase of a three-phase network with the neutral wire omitted and with components represented by symbols rather than equivalent circuits.

The phases of a Y–Y or a Δ–Δ transformer can be labeled so there is no phase shift between corresponding quantities on the low- and high-voltage windings. But for Y–Δ and Δ–Y transformers, there is always a phase shift. Figure 4.13 shows a Y–Δ transformer. The labeling of the windings and the schematic representation are in accordance with the American standard, which is as follows:

> In either a Y–Δ or Δ–Y transformer, positive-sequence quantities on the high-voltage side shall lead their corresponding quantities on the low-voltage side by 30°.

As shown in Figure 4.13(b), V_{AN} leads V_{an} by 30°.

The positive-sequence phasor diagram shown in Figure 4.13(b) can be constructed via the following five steps, which are also indicated in Figure 4.13:

Step 1 Assume that positive-sequence voltages are applied to the Y winding. Draw the positive-sequence phasor diagram for these voltages.

Step 2 Move phasor A–N next to terminals A–N in Figure 4.13(a). Identify the ends of this line in the same manner as in the phasor diagram. Similarly, move phasors B–N and C–N next to terminals B–N and C–N in Figure 4.13(a).

Step 3 For each single-phase transformer, the voltage across the low-voltage winding must be in phase with the voltage across the high-voltage winding, assuming an ideal transformer. Therefore, draw a line next to each low-voltage winding parallel to the corresponding line already drawn next to the high-voltage winding.

Figure 4.13

Three-phase two-winding Y–Δ transformer bank

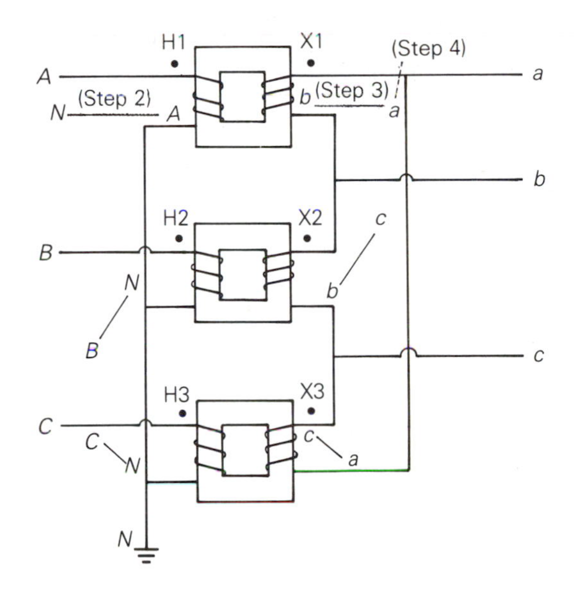

(a) Core and coil arrangement

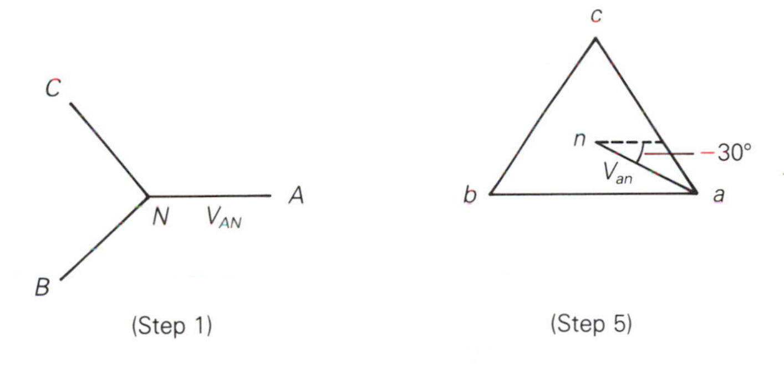

(b) Positive-sequence phasor diagram

Step 4 Label the ends of the lines drawn in Step 3 by inspecting the polarity marks. For example, phase A is connected to dotted terminal H1, and A appears on the *right* side of line A–N. Therefore, phase a, which is connected to dotted terminal X1, must be on the *right* side, and b on the left side of line a–b. Similarly, phase B is connected to dotted terminal H2, and B is *down* on line B–N. Therefore, phase b, connected to dotted terminal X2, must be *down* on line b–c. Similarly, c is *up* on line c–a.

Step 5 Bring the three lines labeled in Step 4 together to complete the phasor diagram for the low-voltage Δ winding. Note that V_{AN} leads V_{an} by 30° in accordance with the American standard.

Example 4.6 Construct the negative-sequence diagram for the Y–Δ transformer shown in Figure 4.13. Determine the negative-sequence phase shift of this transformer.

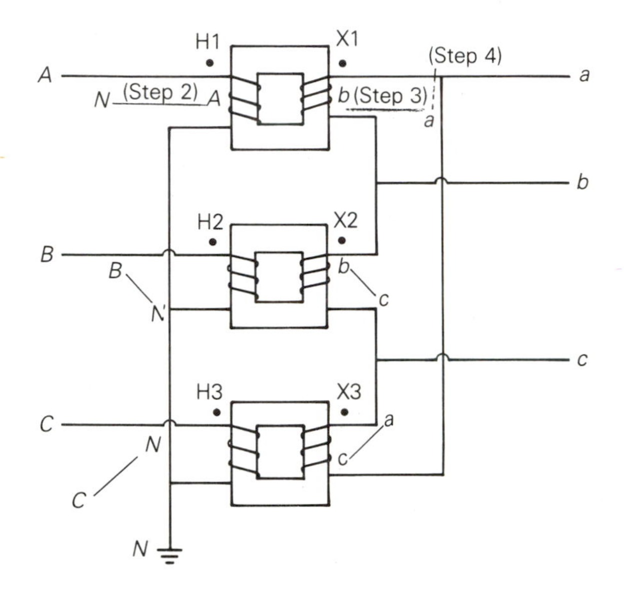

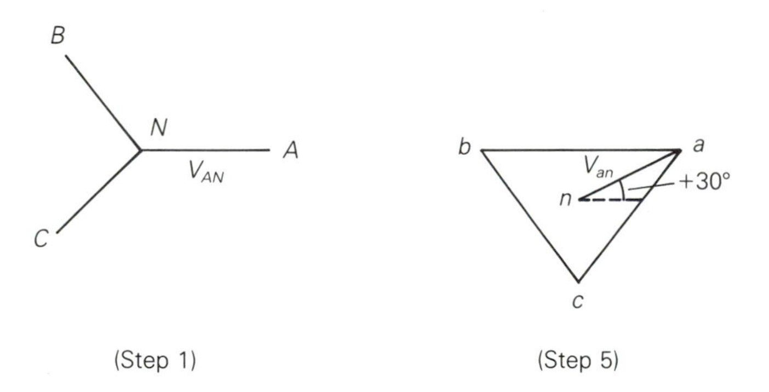

Figure 4.14 Example 4.6–Construction of negative-sequence phasor diagram for Y–Δ transformer bank

Solution The negative-sequence diagram, shown in Figure 4.14, is constructed from the following five steps, as outlined above:

Step 1 Draw the phasor diagram of negative-sequence voltages, which are assumed to be applied to the Y winding.

Step 2 Move the phasors *A–N*, *B–N*, and *C–N* next to the high-voltage Y windings.

Step 3 For each single-phase transformer, draw a line next to the low-voltage winding that is parallel to the line drawn in Step 2 next to the high-voltage winding.

Step 4 Label the lines drawn in Step 3. For example, phase *B*, which is connected to dotted terminal H2, is shown *up* on line *B–N*;

therefore, phase b, which is connected to dotted terminal X2, must be *up* on line b–c.

Step 5 Bring the lines drawn in Step 4 together to form the negative-sequence phasor diagram for the low-voltage Δ winding.

As shown in Figure 4.14, the high-voltage phasors *lag* the low-voltage phasors by 30°. Thus the negative-sequence phase shift is the reverse of the positive-sequence phase shift. ◁

The Δ–Y transformer is commonly used as a generator step-up transformer, where the Δ winding is connected to the generator terminals and the Y winding is connected to a transmission line. One advantage of a high-voltage Y winding is that a neutral point N is provided for grounding on the high-voltage side. With a permanently grounded neutral, the insulation requirements for the high-voltage transformer windings are reduced. The high-voltage insulation can be graded or tapered from maximum insulation at terminals ABC to minimum insulation at grounded terminal N. One advantage of the Δ winding is that the undesirable third harmonic magnetizing current, caused by the nonlinear core B–H characteristic, remains trapped inside the Δ winding. Third harmonic currents are (triple-frequency) zero-sequence currents, which cannot enter or leave a Δ connection, but can flow within the Δ. The Y–Y transformer is seldom used because of difficulties with third harmonic exciting current.

The Δ–Δ transformer has the advantage that one phase can be removed for repair or maintenance while the remaining phases continue to operate as a three-phase bank. This *open*-Δ connection permits balanced three-phase operation with the kVA rating reduced to 58% of the original bank (see Problem 4.14).

Figure 4.15

Transformer core configurations

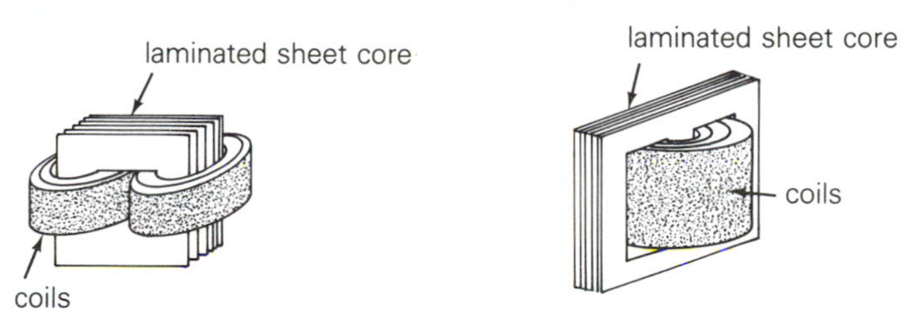

(a) Single-phase core type

(b) Single-phase shell type

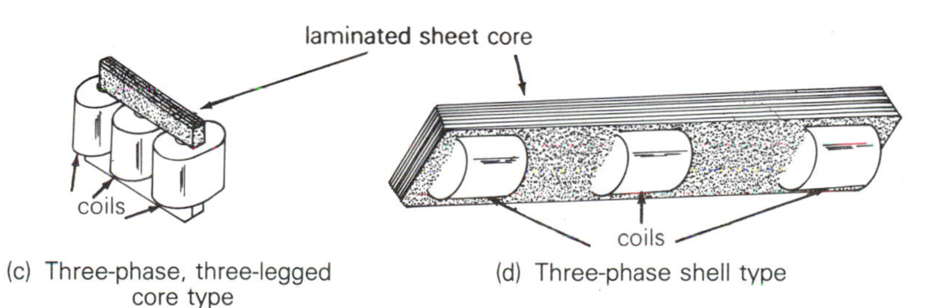

(c) Three-phase, three-legged core type

(d) Three-phase shell type

Instead of a bank of three single-phase transformers, all six windings may be placed on a common three-phase core to form a three-phase transformer, as shown in Figure 4.15. The three-phase core contains less iron than the three single-phase units; therefore, it costs less, weighs less, requires less floor space, and has a slightly higher efficiency. However, a winding failure would require replacement of an entire three-phase transformer, compared to replacement of only one phase of a three-phase bank. In either case, the equivalent circuits developed here and subsequent analyses are the same.*

Section 4.5

Per-Unit Sequence Models of Three-phase Two-winding Transformers

Figure 4.16(a) is a schematic representation of an ideal Y–Y transformer grounded through neutral impedances Z_N and Z_n. Figure 4.16(b–d) show the per-unit sequence networks of this ideal transformer. Throughout the remainder of this text per-unit quantities will be used unless otherwise indicated. Also, the subscript "p.u.," used to indicate a per-unit quantity, will be omitted in most cases.

By convention, we adopt the following two rules for selecting base quantities:

1. A common S_{base} is selected for both the H and X terminals.
2. The ratio of the voltage bases V_{baseH}/V_{baseX} is selected to be equal to the ratio of the rated line-to-line voltages $V_{ratedHLL}/V_{VratedXLL}$.

When balanced positive-sequence currents or balanced negative-sequence currents (as in Example 3.2) are applied to the transformer, the neutral currents are zero and there are no voltage drops across the neutral impedances. Therefore, the per-unit positive- and negative-sequence networks of the ideal Y–Y transformer, Figure 4.16(b) and (c), are the same as the per-unit single-phase ideal transformer, Figure 4.9(a).

Zero-sequence currents have equal magnitudes and equal phase angles. When per-unit sequence currents $I_{A0} = I_{B0} = I_{C0} = I_0$ are applied to the high-voltage windings of an ideal Y–Y transformer, the neutral current $I_N = 3I_0$ flows through the neutral impedance Z_N, with a voltage drop $(3Z_N)I_0$. Also, per-unit zero-sequence current I_0 flows in each low-voltage winding [from (4.3.9)], and therefore $3I_0$ flows through neutral impedance Z_n, with a voltage drop $(3I_0)Z_n$. The per-unit zero-sequence network, which includes the impedances $(3Z_N)$ and $(3Z_n)$, is shown in Figure 4.16(b).

*We note that the zero-sequence circuit of a three-phase shell-type transformer is not the same as the zero-sequence circuit of a three-phase core-type transformer [4].

Figure 4.16

Ideal Y–Y transformer

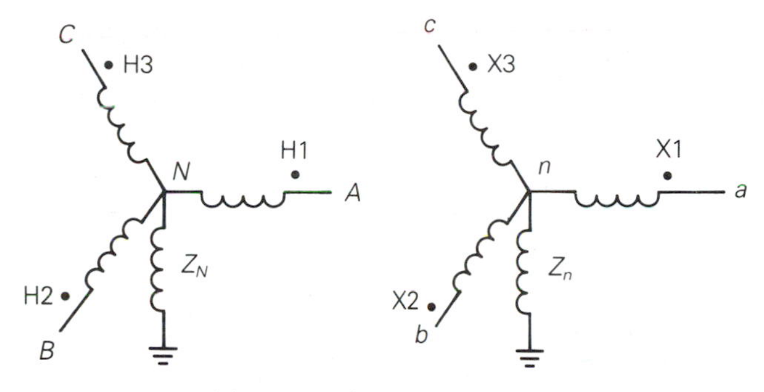

(a) Schematic representation

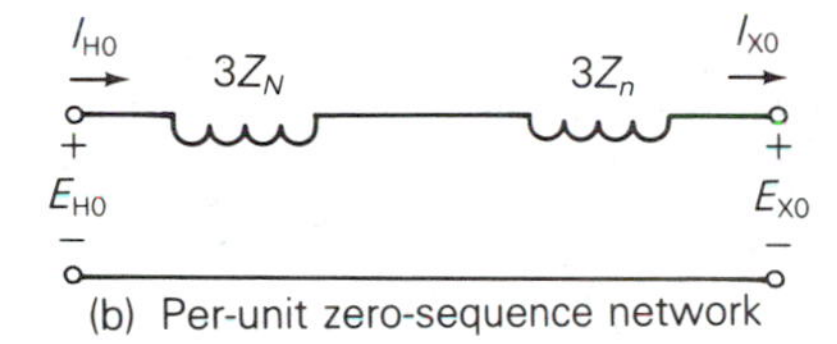

(b) Per-unit zero-sequence network

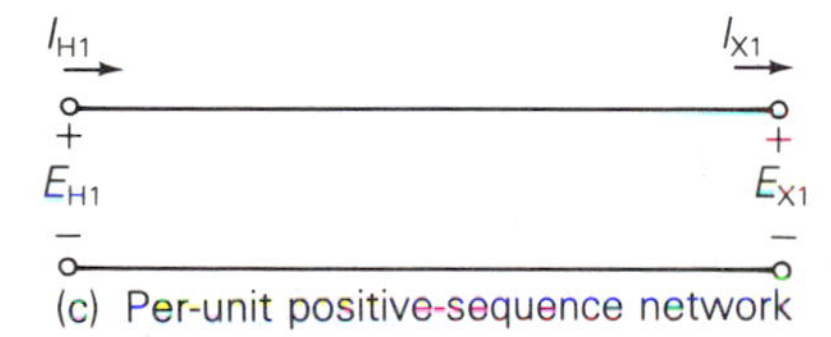

(c) Per-unit positive-sequence network

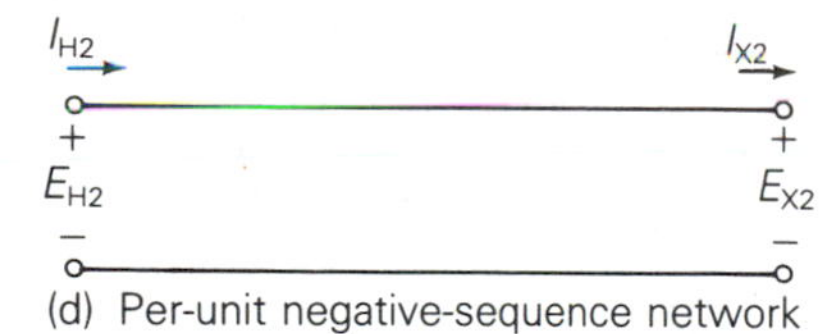

(d) Per-unit negative-sequence network

Note that if either one of the neutrals of an ideal transformer is ungrounded, then no zero sequence can flow in either the high- or low-voltage windings. For example, if the high-voltage winding has an open neutral, then $I_N = 3I_0 = 0$, which in turn forces $I_0 = 0$ on the low-voltage side. This can be shown in the zero-sequence network of Figure 4.16(b) by making $Z_N = \infty$, which corresponds to an open circuit.

The per-unit sequence networks of a practical Y–Y transformer are shown in Figure 4.17(a). These networks are obtained by adding external impedances to the sequence networks of the ideal transformer, as follows. The leakage impedances of the high-voltage windings are series impedances like the series impedances shown in Figure 3.9, with no coupling between phases ($Z_{ab} = 0$). If the phase a, b, and c windings have equal leakage impedances $Z_H = R_H + jX_H$, then the series impedances are *symmetrical* with sequence

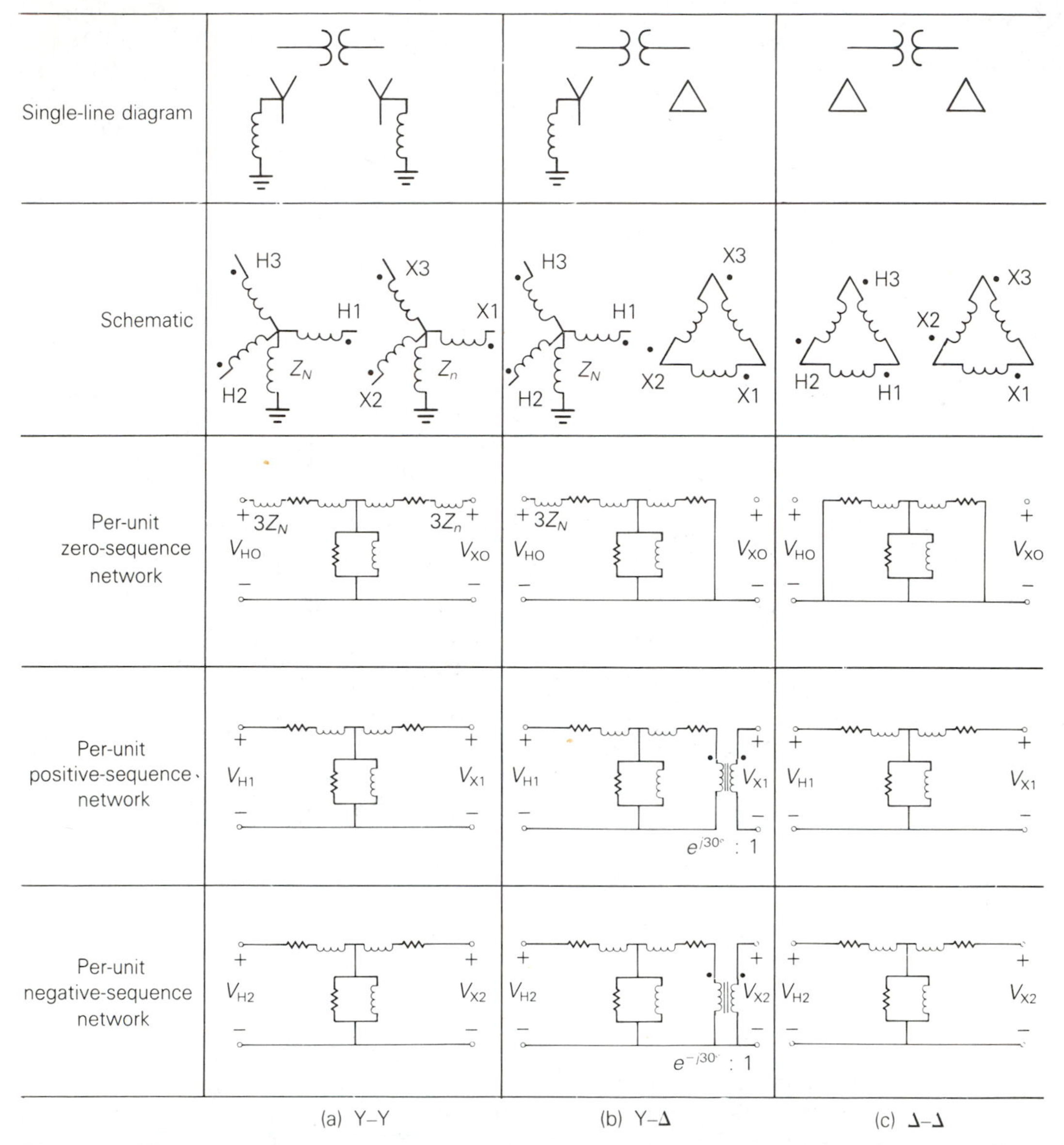

Figure 4.17 Per-unit sequence networks of practical Y–Y, Y–Δ, and Δ–Δ transformers

networks, as shown in Figure 3.10, where $Z_{H0} = Z_{H1} = Z_{H2} = Z_H$. Similarly, the leakage impedances of the low-voltage windings are symmetrical series impedances with $Z_{X0} = Z_{X1} = Z_{X2} = Z_X$. These series leakage impedances are shown in per-unit in the sequence networks of Figure 4.17(a).

The shunt branches of the practical Y–Y transformer, which represent exciting current, are equivalent to the Y load of Figure 3.3. Each phase in Figure 3.3 represents a core loss resistor in parallel with a magnetizing inductance. Assuming these are the same for each phase, then the Y load is *symmetrical*, and the sequence networks are shown in Figure 3.4. These shunt branches are also shown in Figure 4.17(a). Note that $(3Z_N)$ and $(3Z_n)$ have already been included in the zero-sequence network.

The per-unit positive- and negative-sequence transformer impedances of the practical Y–Y transformer in Figure 4.17(a) are identical, which is always true for nonrotating equipment. The per-unit zero-sequence network, however, depends on the neutral impedances Z_N and Z_n.

The per-unit sequence networks of the Y–Δ transformer, shown in Figure 4.17(b), have the following features:

1. The per-unit impedances do not depend on the winding connections. That is, the per-unit impedances of a transformer that is connected Y–Y, Y–Δ, Δ–Y, or Δ–Δ are the same. However, the base voltages do depend on the winding connections.
2. A phase shift is included in the per-unit positive- and negative-sequence networks. For the American standard, the positive-sequence voltages and currents on the high-voltage side of the Y–Δ transformer lead the corresponding quantities on the low-voltage side by 30°. For negative sequence, the high-voltage quantities lag by 30°
3. Zero-sequence currents can flow in the Y winding if there is a neutral connection, and corresponding zero-sequence currents flow within the Δ winding. But no zero-sequence current enters or leaves the Δ winding.

The phase shifts in the positive- and negative-sequence networks of Figure 4.17(b) are represented by the phase-shifting transformer of Figure 4.4. Also, the zero-sequence network of Figure 4.17(b) provides a path on the Y side for zero-sequence current to flow, but no zero-sequence current can enter or leave the Δ side.

The per-unit sequence networks of the Δ–Δ transformer, shown in Figure 4.17(c), have the following features:

1. The positive- and negative-sequence networks, which are identical, are the same as those for the Y–Y transformer. It is assumed that the windings are labeled so there is no phase shift. Also, the per-unit impedances do not depend on the winding connections, but the base voltages do.
2. Zero-sequence currents *cannot* enter or leave either Δ winding, although they can circulate within the Δ windings.

Example 4.7 Three single-phase two-winding transformers, each rated 400 MVA, 13.8/199.2 kV, with leakage reactance $X_{eq} = 0.10$ per unit, are connected to form a three-phase bank. Winding resistances and exciting current are neglected. The high-voltage windings are connected in Y. A three-phase load operating under balanced positive-sequence conditions on the high-voltage side absorbs 1000 MVA at 0.90 p.f. lagging, with $V_{AN} = 199.2\underline{/0°}$ kV. Determine the voltage V_{an} at the low-voltage bus if the low-voltage windings are connected (a) in Y, (b) in Δ.

Solution Since balanced operation occurs for this example, zero- and negative-sequence currents and voltages are zero. The positive-sequence network is shown in Figure 4.18. Using the transformer bank ratings as base quantities, $S_{base3\phi} = 1200$ MVA, $V_{baseHLL} = 345$ kV, and $I_{baseH} = 1200/(345\sqrt{3}) = 2.008$ kA. The per-unit load voltage and load current are then

$$V_{H1} = V_{AN} = 1.0\underline{/0°} \quad \text{per unit}$$

$$I_{H1} = I_A = \frac{1000/(345\sqrt{3})}{2.008}\underline{/-\cos^{-1}0.9} = 0.8333\underline{/-25.84°} \quad \text{per unit}$$

Figure 4.18 Per-unit positive-sequence network for Example 4.7

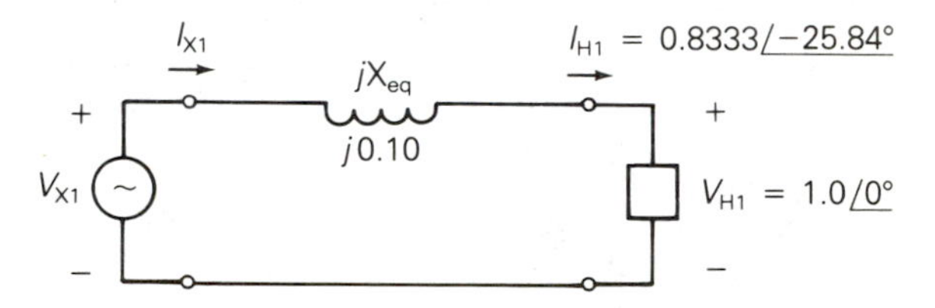

(a) Y-connected low-voltage windings

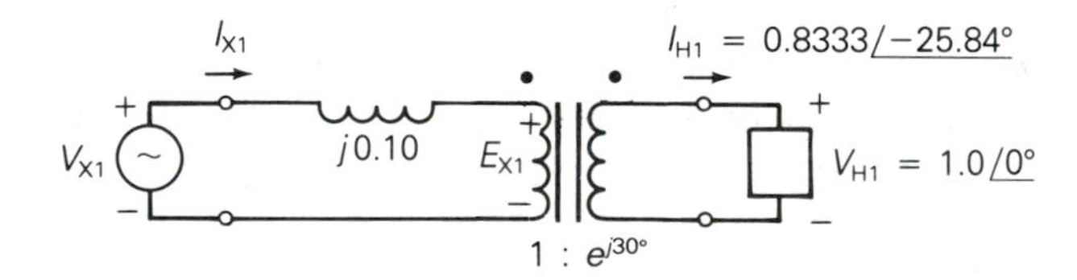

(b) Δ-connected low-voltage windings

a. For the Y–Y transformer, Figure 4.18(a),

$$I_{X1} = I_{H1} = 0.8333\underline{/-25.84°} \quad \text{per unit}$$

$$\begin{aligned} V_{X1} &= V_{H1} + (jX_{eq})I_{X1} \\ &= 1.0\underline{/0°} + (j0.10)(0.8333\underline{/-25.84°}) \\ &= 1.0 + 0.08333\underline{/64.16°} = 1.0363 + j0.0750 = 1.039\underline{/4.139°} \end{aligned}$$

Since $V_{X0} = V_{X2} = 0$,

$$V_{an} = V_{X1} = 1.039\underline{/4.139°} \quad \text{per unit}$$

Further, since $V_{baseXLN} = 13.8$ kV for the low-voltage Y windings, $V_{an} = 1.039(13.8) = 14.34$ kV, and

$$V_{an} = 14.34\underline{/4.139°} \quad \text{kV}$$

b. For the Δ–Y transformer, Figure 4.18(b),

$$E_{X1} = e^{-j30^\circ}V_{H1} = 1.0\underline{/-30^\circ} \quad \text{per unit}$$

$$I_{X1} = e^{-j30^\circ}I_{H1} = 0.8333\underline{/-25.84^\circ - 30^\circ} = 0.8333\underline{/-55.84^\circ} \text{ per unit}$$

$$V_{X1} = E_{X1} + (jX_{eq})I_{X1}$$

$$= 1.0\underline{/-30^\circ} + (j0.10)(0.8333\underline{/-55.84^\circ})$$

$$= 1.039\underline{/-25.861^\circ} \quad \text{per unit}$$

Since $V_{X0} = V_{X2} = 0$,

$$V_{an} = V_{X1} = 1.039\underline{/-25.861^\circ} \quad \text{per unit}$$

Further, since $V_{baseXLN} = 13.8/\sqrt{3} = 7.967$ kV for the low-voltage Δ windings, $V_{an} = (1.039)(7.967) = 8.278$ kV, and

$$V_{an} = 8.278\underline{/-25.861^\circ} \quad \text{kV}$$ ◁

Example 4.8 A 75-kVA, 480-volt Δ/208-volt Y transformer with a solidly grounded neutral is connected between the source and line of Example 3.6. The transformer leakage reactance is $X_{eq} = 0.10$ per unit; winding resistances and exciting current are neglected. Using the transformer ratings as base quantities, draw the per-unit sequence networks and calculate the phase a source current I_a.

Solution The base quantities are $S_{base1\phi} = 75/3 = 25$ kVA, $V_{baseHLN} = 480/\sqrt{3} = 277.1$ volts, $V_{baseXLN} = 208/\sqrt{3} = 120.1$ volts, and $Z_{baseX} = (120.1)^2/25{,}000 = 0.5770\ \Omega$. The sequence components of the actual source voltages are given in Figure 3.15. In per-unit, these voltages are

$$V_0 = \frac{15.91\underline{/62.11^\circ}}{277.1} = 0.05742\underline{/62.11^\circ} \quad \text{per unit}$$

$$V_1 = \frac{277.1\underline{/-1.772^\circ}}{277.1} = 1.0\underline{/-1.772^\circ} \quad \text{per unit}$$

$$V_2 = \frac{9.218\underline{/216.59^\circ}}{277.1} = 0.03327\underline{/216.59^\circ} \quad \text{per unit}$$

The per-unit line and load impedances, which are located on the low-voltage side of the transformer, are

$$Z_{L0} = Z_{L1} = Z_{L2} = \frac{1\underline{/85^\circ}}{0.577} = 1.733\underline{/85^\circ} \quad \text{per unit}$$

$$Z_{load1} = Z_{load2} = \frac{Z_\Delta}{3} = \frac{10\underline{/40^\circ}}{0.577} = 17.33\underline{/40^\circ} \quad \text{per unit}$$

The per-unit sequence networks are shown in Figure 4.19. Note that the per-unit line and load impedances, when referred to the high-voltage side of the phase-shifting transformer, do not change [see (4.1.26)]. Therefore, from

Figure 4.19

Per-unit sequence networks for Example 4.8

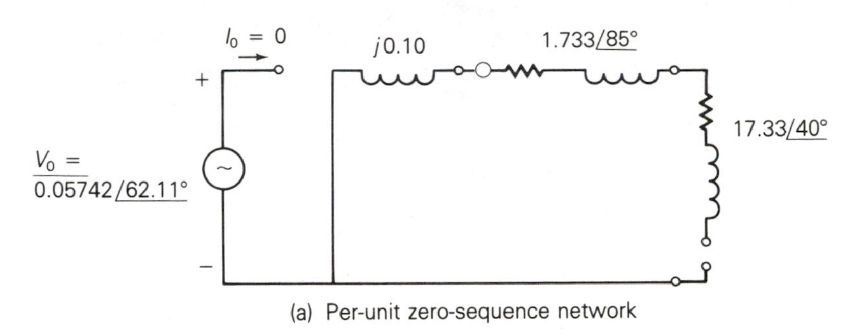

(a) Per-unit zero-sequence network

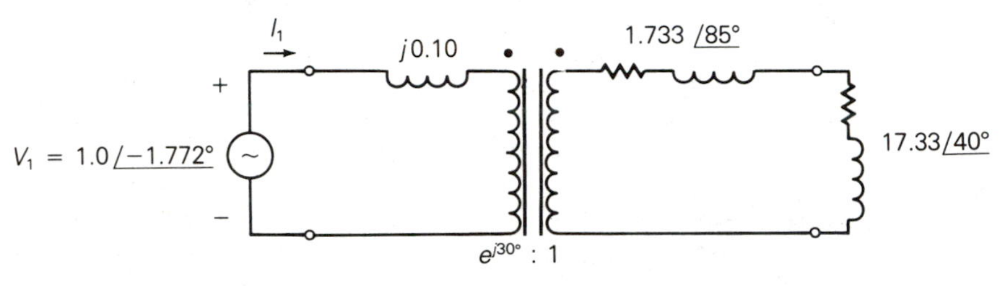

(b) Per-unit positive-sequence network

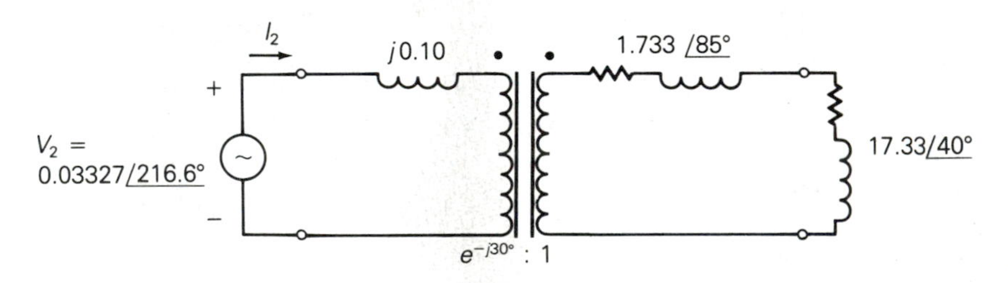

(c) Per-unit negative-sequence network

Figure 4.19, the sequence components of the source currents are

$$I_0 = 0$$

$$I_1 = \frac{V_1}{jX_{eq} + Z_{L1} + Z_{load1}} = \frac{1.0\underline{/-1.772°}}{j0.10 + 1.733\underline{/85°} + 17.33\underline{/40°}}$$

$$= \frac{1.0\underline{/-1.772°}}{13.43 + j12.97} = \frac{1.0\underline{/-1.772°}}{18.67\underline{/44.0°}} = 0.05356\underline{/-45.77°} \quad \text{per unit}$$

$$I_2 = \frac{V_2}{jX_{eq} + Z_{L2} + Z_{load2}} = \frac{0.03327\underline{/216.59°}}{18.67\underline{/44.0°}}$$

$$= 0.001782\underline{/172.59°} \quad \text{per unit}$$

The phase a source current is then, using (3.1.20),

$$\begin{aligned} I_a &= I_0 + I_1 + I_2 \\ &= 0 + 0.05356\underline{/-45.77^\circ} + 0.001782\underline{/172.59^\circ} \\ &= 0.03511 - j0.03764 = 0.05216\underline{/-46.19^\circ} \quad \text{per unit} \end{aligned}$$

$$\text{Using } \mathrm{I}_{\text{baseH}} = \frac{75{,}000}{480\sqrt{3}} = 90.21 \text{ A},$$

$$I_a = (0.05216)(90.21)\underline{/-46.19^\circ} = 4.705\underline{/-46.19^\circ} \quad \text{A} \qquad \triangleleft$$

Example 4.9 A 200-MVA, 345-kV Δ/34.5-kV Y substation transformer has an 8% leakage reactance. The transformer acts as a connecting link between 345-kV transmission and 34.5-kV distribution. Transformer winding resistances and exciting current are neglected. The high-voltage bus connected to the transformer is assumed to be an ideal 345-kV positive-sequence source with negligible source impedance. Using the transformer ratings as base values, determine:

a. The per-unit magnitudes of transformer voltage drop and voltage at the low-voltage terminals when rated transformer current at 0.8 p.f. lagging enters the high-voltage terminals

b. The per-unit magnitude of the fault current when a three-phase-to-ground bolted short circuit occurs at the low-voltage terminals

Solution In both parts (a) and (b), only positive-sequence current will flow, since there are no imbalances; only the positive-sequence network need be considered. Also, as we are interested only in voltage and current magnitudes, the Δ–Y transformer phase shift can be omitted.

a. As shown in Figure 4.20(a),

$$\mathrm{V}_{\text{drop}} = \mathrm{I}_{\text{rated}}\mathrm{X}_{\text{eq}} = (1.0)(0.08) = 0.08 \quad \text{per unit}$$

and

$$\begin{aligned} V_{\mathrm{X}} &= V_{\mathrm{H}} - (j\mathrm{X}_{\text{eq}})I_{\text{rated}} \\ &= 1.0\underline{/0^\circ} - (j0.08)(1.0\underline{/-36.87^\circ}) \\ &= 1.0 - (j0.08)(0.8 - j0.06) = 0.952 - j0.064 \\ &= 0.954\underline{/-3.85^\circ} \quad \text{per unit} \end{aligned}$$

b. As shown in Figure 4.20(b),

$$\mathrm{I}_{\text{SC}} = \frac{\mathrm{V}_{\mathrm{H}}}{\mathrm{X}_{\text{eq}}} = \frac{1.0}{0.08} = 12.5 \quad \text{per unit}$$

Under rated current conditions [part (a)], the 0.08 per-unit voltage drop across the transformer leakage reactance causes the voltage at the low-voltage terminals to be 0.954 per unit. Also, under three-phase short-circuit conditions [part (b)], the fault current is 12.5 times the rated transformer current. This

Figure 4.20

Circuits for Example 4.9

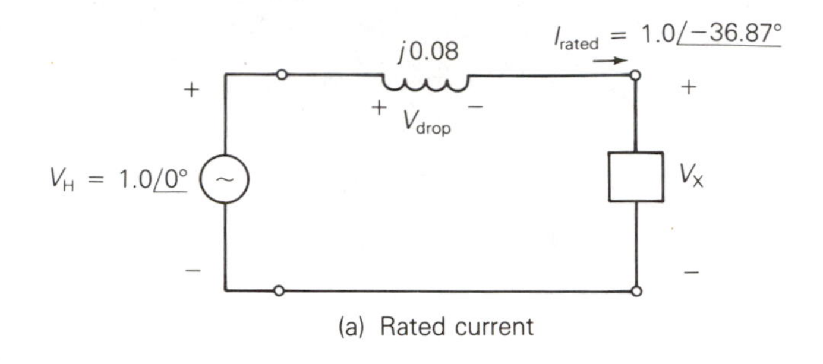

(a) Rated current

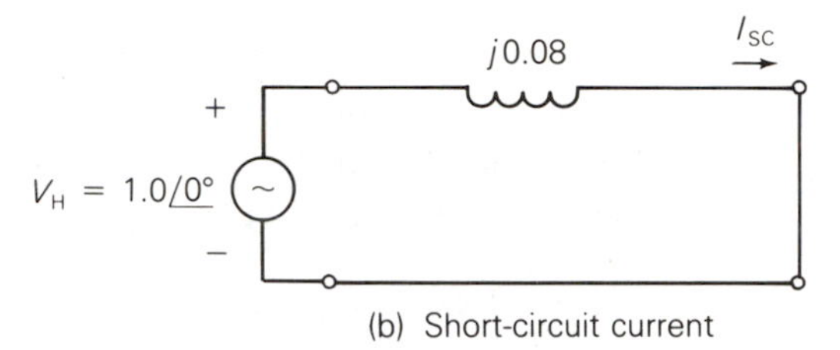

(b) Short-circuit current

example illustrates a compromise in the design or specification of transformer leakage reactance. A low value is desired to minimize voltage drops, but a high value is desired to limit fault currents. Typical transformer leakage reactances are given in Table A.2. ◁

Section 4.6

Three-winding Transformers

Figure 4.21(a) shows a basic single-phase three-winding transformer. The ideal transformer relations for a two-winding transformer, (4.1.8) and (4.1.14), can easily be extended to obtain corresponding relations for an ideal three-winding transformer. In actual units, these relations are:

$$N_1 I_1 = N_2 I_2 + N_3 I_3 \tag{4.6.1}$$

$$\frac{E_1}{N_1} = \frac{E_2}{N_2} = \frac{E_3}{N_3} \tag{4.6.2}$$

where I_1 enters the dotted terminal, I_2 and I_3 leave dotted terminals, and E_1, E_2, and E_3 have their + polarities at dotted terminals. In per-unit, (4.6.1) and (4.6.2) are:

$$I_{1\text{p.u.}} = I_{2\text{p.u.}} + I_{3\text{p.u.}} \tag{4.6.3}$$

$$E_{1\text{p.u.}} = E_{2\text{p.u.}} = E_{3\text{p.u.}} \tag{4.6.4}$$

where a common S_{base} is selected for all three windings, and voltage bases are

Figure 4.21

Single-phase three-winding transformer

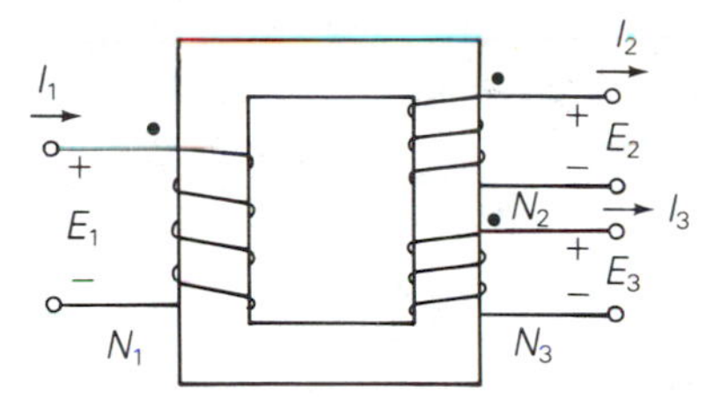

(a) Basic core and coil configuration

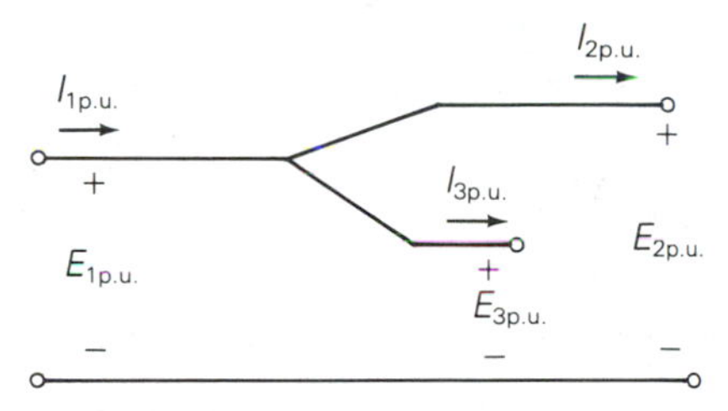

(b) Per-unit equivalent circuit—ideal transformer

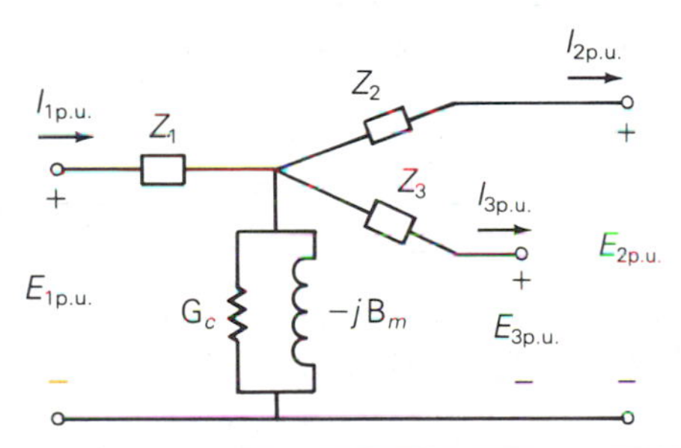

(c) Per-unit equivalent circuit—practical transformer

selected in proportion to the rated voltages of the windings. These two per-unit relations are satisfied by the per-unit equivalent circuit shown in Figure 4.21(b). Also, external series impedance and shunt admittance branches are included in the practical three-winding transformer circuit shown in Figure 4.21(c). The shunt admittance branch, a core loss resistor in parallel with a magnetizing inductor, can be evaluated from an open-circuit test. Also, when one winding is left open, the three-winding transformer behaves as a two-winding transformer, and standard short-circuit tests can be used to evaluate per-unit leakage impedances, which are defined as follows:

Z_{12} = per-unit leakage impedance measured from winding 1, with winding 2 shorted and winding 3 open

Z_{13} = per-unit leakage impedance measured from winding 1, with winding 3 shorted and winding 2 open

Z_{23} = per-unit leakage impedance measured from winding 2, with winding 3 shorted and winding 1 open

From Figure 4.21(c), with winding 2 shorted and winding 3 open, the leakage impedance measured from winding 1 is, neglecting the shunt admittance branch,

$$Z_{12} = Z_1 + Z_2 \tag{4.6.5}$$

Similarly,

$$Z_{13} = Z_1 + Z_3 \tag{4.6.6}$$

and

$$Z_{23} = Z_2 + Z_3 \tag{4.6.7}$$

Solving (4.6.5)–(4.6.7),

$$Z_1 = \tfrac{1}{2}(Z_{12} + Z_{13} - Z_{23}) \tag{4.6.8}$$

$$Z_2 = \tfrac{1}{2}(Z_{12} + Z_{23} - Z_{13}) \tag{4.6.9}$$

$$Z_3 = \tfrac{1}{2}(Z_{13} + Z_{23} - Z_{12}) \tag{4.6.10}$$

Equations (4.6.8)–(4.6.10) can be used to evaluate the per-unit series impedances Z_1, Z_2, and Z_3 of the three-winding transformer equivalent circuit from the per-unit leakage impedances Z_{12}, Z_{13}, and Z_{23}, which, in turn, are determined from short-circuit tests.

Note that each of the windings on a three-winding transformer may have a *different* kVA rating. If the leakage impedances from short-circuit tests are expressed in per-unit based on winding ratings, they must first be converted to per-unit on a common S_{base} *before* they are used in (4.6.8)–(4.6.10).

Example 4.10 The ratings of a single-phase three-winding transformer are:

winding 1: 300 MVA, 13.8 kV

winding 2: 300 MVA, 199.2 kV

winding 3: 50 MVA, 19.92 kV

The leakage reactances, from short-circuit tests, are:

$X_{12} = 0.10$ per unit on a 300-MVA, 13.8-kV base

$X_{13} = 0.16$ per unit on a 50-MVA, 13.8-kV base

$X_{23} = 0.14$ per unit on a 50-MVA, 199.2-kV base

Winding resistances and exciting current are neglected. Calculate the impedances of the per-unit equivalent circuit using a base of 300 MVA and 13.8 kV for terminal 1.

Solution $S_{base} = 300$ MVA is the same for all three terminals. Also, the specified voltage base for terminal 1 is $V_{base1} = 13.8$ kV. The base voltages for terminals 2 and 3 are then $V_{base2} = 199.2$ kV and $V_{base3} = 19.92$ kV, which are the rated voltages of these windings. From the data given, $X_{12} = 0.10$ per unit was measured from terminal 1 using the same base values as those specified for the circuit.

But $X_{13} = 0.16$ and $X_{23} = 0.14$ per unit on a 50-MVA base are first converted to the 300-MVA circuit base.

$$X_{13} = (0.16)\left(\frac{300}{50}\right) = 0.96 \quad \text{per unit}$$

$$X_{23} = (0.14)\left(\frac{300}{50}\right) = 0.84 \quad \text{per unit}$$

Then, from (4.6.8)–(4.6.10)

$$X_1 = \tfrac{1}{2}(0.10 + 0.96 - 0.84) = 0.11 \quad \text{per unit}$$

$$X_2 = \tfrac{1}{2}(0.10 + 0.84 - 0.96) = -0.01 \quad \text{per unit}$$

$$X_3 = \tfrac{1}{2}(0.84 + 0.96 - 0.10) = 0.85 \quad \text{per unit}$$

The per-unit equivalent circuit of this three-winding transformer is shown in Figure 4.22. Note that X_2 is negative. This illustrates the fact that X_1, X_2, and X_3 are *not* leakage reactances, but instead are equivalent reactances derived from the leakage reactances. Leakage reactances are always positive.

Figure 4.22

Circuit for Example 4.10

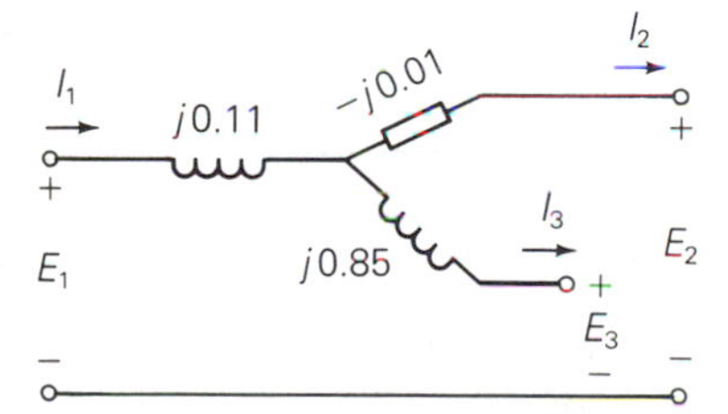

Note also that the node where the three equivalent circuit reactances are connected does not correspond to any physical location within the transformer. Rather, it is simply part of the equivalent circuit representation. ◁

Three identical single-phase three-winding transformers can be connected to form a three-phase bank. Figure 4.23 shows the general per-unit sequence networks of a three-phase three-winding transformer. Instead of labeling the windings 1, 2, and 3, as was done for the single-phase transformer, the letters H, M, and X are used to denote the high-, medium-, and low-voltage windings, respectively. By convention, a common S_{base} is selected for the H, M, and X terminals, and voltage bases V_{baseH}, V_{baseM}, and V_{baseX} are selected in proportion to the rated line-to-line voltages of the transformer.

For the general zero-sequence network, Figure 4.23(a), the connection between terminals H and H′ depends on how the high-voltage windings are connected, as follows:

1. Solidly grounded Y—Short H to H′.

Figure 4.23

Per-unit sequence networks of a three-phase three-winding transformer

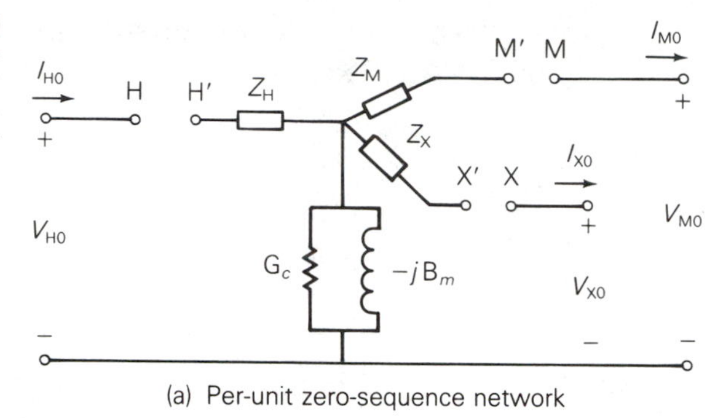

(a) Per-unit zero-sequence network

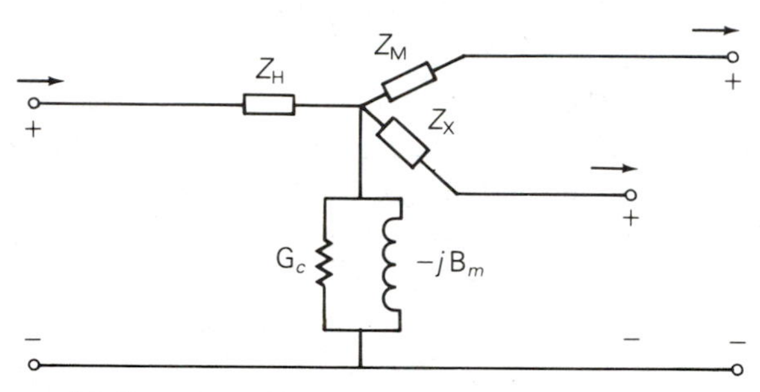

(b) Per-unit positive- or negative-sequence network (phase shift not shown)

2. Grounded Y through Z_N—Connect ($3Z_N$) from H to H′.
3. Ungrounded Y—Leave H–H′ open as shown.
4. Δ—Short H′ to the reference bus.

Terminals X–X′ and M–M′ are connected in a similar manner.

The impedances of the per-unit negative-sequence network are the same as those of the per-unit positive-sequence network, which is always true for nonrotating equipment. Phase-shifting transformers, not shown in Figure 4.23(b), can be included to model phase shift between Δ and Y windings.

Example 4.11 Three transformers, each identical to that described in Example 4.10, are connected as a three-phase bank in order to feed power from a 900-MVA, 13.8-kV generator to a 345-kV transmission line and to a 34.5-kV distribution line. The transformer windings are connected as follows:

13.8-kV windings (X): Δ, to generator

199.2-kV windings (H): solidly grounded Y, to 345-kV line

19.92-kV windings (M): grounded Y through $Z_n = j0.10\,\Omega$, to 34.5-kV line

The positive-sequence voltages and currents of the high- and medium-voltage Y windings lead the corresponding quantities of the low-voltage Δ winding by 30°. Draw the per-unit sequence networks, using a three-phase base of 900 MVA and 13.8 kV for terminal X.

Solution The per-unit sequence networks are shown in Figure 4.24. Since $V_{baseX} = 13.8$ kV is the rated line-to-line voltage of terminal X, $V_{baseM} = \sqrt{3}(19.92) = 34.5$ kV, which is the rated line-to-line voltage of terminal M. The base impedance of the medium-voltage terminal is then

$$Z_{baseM} = \frac{(34.5)^2}{900} = 1.3225 \quad \Omega$$

Figure 4.24

Per-unit sequence networks for Example 4.11

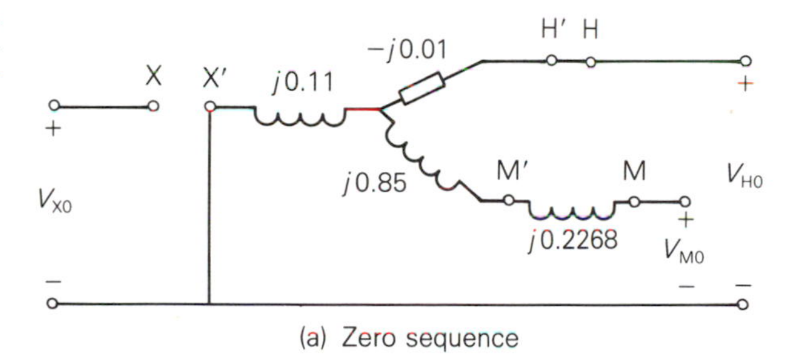

(a) Zero sequence

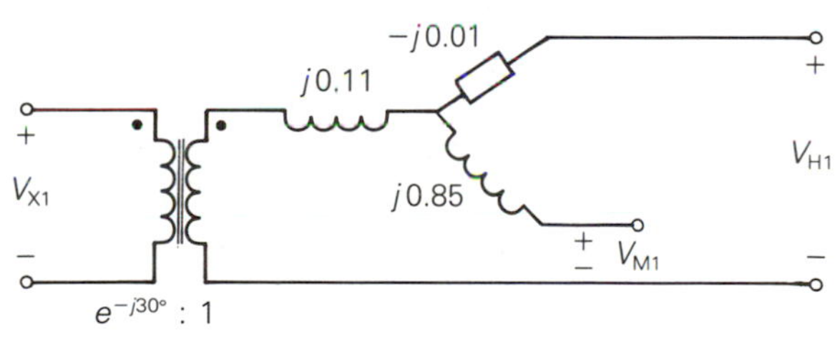

(b) Positive sequence

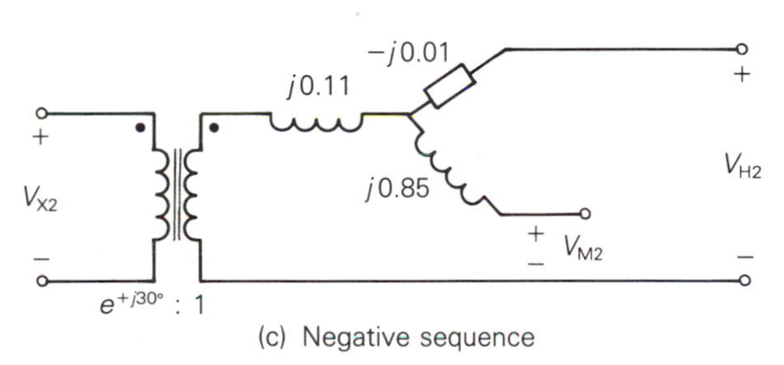

(c) Negative sequence

Therefore, the per-unit neutral impedance is

$$Z_n = \frac{j0.10}{1.3225} = j0.07561 \quad \text{per unit}$$

and $(3Z_n) = j0.2268$ is connected from terminal M to M′ in the per-unit zero-sequence network. Since the high-voltage windings have a solidly grounded neutral, H to H′ is shorted in the zero-sequence network. Also, phase-shifting transformers are included in the positive- and negative-sequence networks. ◁

Section 4.7

Autotransformers

A single-phase two-winding transformer is shown in Figure 4.25(a) with two separate windings, which is the usual two-winding transformer; the same transformer is shown in Figure 4.25(b) with the two windings connected in series, which is called an *autotransformer*. For the usual transformer [Figure 4.25(a)] the two windings are coupled magnetically via the mutual core flux. For the autotransformer [Figure 4.25(b)] the windings are both electrically and magnetically coupled. The autotransformer has smaller per-unit leakage impedances than the usual transformer; this results in both smaller series-voltage drops (an advantage) and higher short-circuit currents (a disadvantage). The autotransformer also has lower per-unit losses (higher efficiency), lower exciting current, and lower cost if the turns ratio is not too large. The electrical connection of the windings, however, allows transient overvoltages to pass through the autotransformer more easily.

Figure 4.25

Ideal single-phase transformers

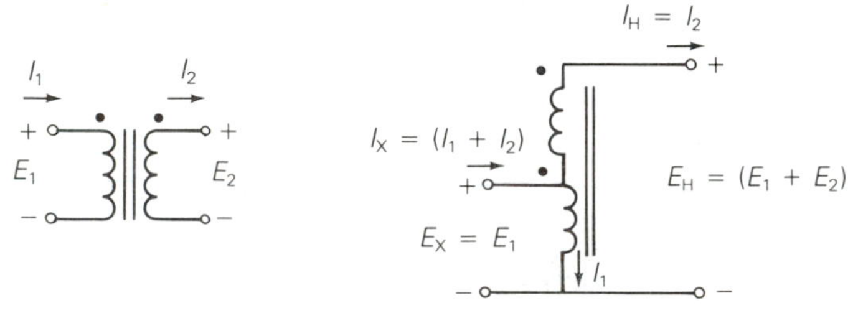

Example 4.12 The single-phase two-winding 20-kVA, 480/120-volt transformer of Example 4.3 is connected as an autotransformer, as in Figure 4.25(b), where winding 1 is the 120-volt winding. For this autotransformer, determine (a) the voltage

ratings E_X and E_H of the low- and high-voltage terminals, (b) the kVA rating, and (c) the per-unit leakage impedance.

Solution

a. Since the 120-volt winding is connected to the low-voltage terminal, $E_X = 120$ volts. When $E_X = E_1 = 120$ volts is applied to the low-voltage terminal, $E_2 = 480$ volts is induced across the 480-volt winding, neglecting the voltage drop across the leakage impedance. Therefore, $E_H = E_1 + E_2 = 120 + 480 = 600$ volts.

b. As a normal two-winding transformer rated 20 kVA, the rated current of the 480-volt winding is $I_2 = I_H = 20{,}000/480 = 41.667$ A. As an autotransformer, the 480-volt winding can carry the same current. Therefore, the kVA rating $S_H = E_H I_H = (600)(41.667) = 25$ kVA. Note also that when $I_H = I_2 = 41.667$ A, a current $I_1 = \frac{480}{120}(41.667) = 166.7$ A is induced in the 120-volt winding. Therefore, $I_X = I_1 + I_2 = 208.3$ A (neglecting exciting current) and $S_X = E_X I_X = (120)(208.3) = 25$ kVA, which is the same rating as calculated for the high-voltage terminal.

c. From Example 4.3, the leakage impedance is $0.0729\underline{/78.13^\circ}$ per unit as a normal, two-winding transformer. As an autotransformer, the leakage impedance *in ohms* is the same as for the normal transformer, since the core and windings are the same for both (only the external winding connections are different). But the base impedances are different. For the high-voltage terminal, using (4.3.4),

$$Z_{\text{baseHold}} = \frac{(480)^2}{20{,}000} = 11.52 \quad \Omega \qquad \text{as a normal transformer}$$

$$Z_{\text{baseHnew}} = \frac{(600)^2}{25{,}000} = 14.4 \quad \Omega \qquad \text{as an autotransformer}$$

Therefore, using (4.3.10),

$$Z_{\text{p.u.new}} = (0.0729\underline{/78.13^\circ})\left(\frac{11.52}{14.4}\right) = 0.05832\underline{/78.13^\circ} \quad \text{per unit}$$

For this example, the rating is 25 kVA, 120/600 volts as an autotransformer versus 20 kVA, 120/480 volts as a normal transformer. The autotransformer has both a larger kVA rating and a larger voltage ratio for the same cost. Also, the per-unit leakage impedance of the autotransformer is smaller. However, the increased high-voltage rating as well as the electrical connection of the windings may require more insulation for both windings. ◁

Section 4.8

Transformers with Off-nominal Turns Ratios

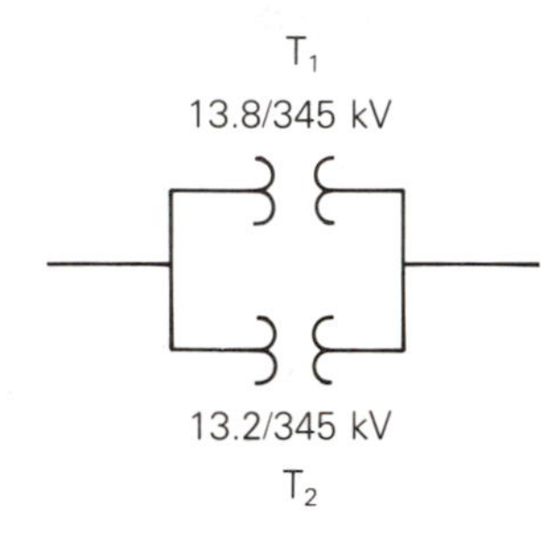

Figure 4.26

Two transformers connected in parallel

It has been shown that models of transformers that use per-unit quantities are simpler than those that use actual quantities. The ideal transformer winding is eliminated when the ratio of the selected voltage bases equals the ratio of the voltage ratings of the windings. In some cases, however, it is impossible to select voltage bases in this manner. For example, consider the two transformers connected in parallel in Figure 4.26. Transformer T_1 is rated 13.8/345 kV and T_2 is rated 13.2/345 kV. If we select $V_{baseH} = 345$ kV, then transformer T_1 requires $V_{baseX} = 13.8$ kV and T_2 requires $V_{baseX} = 13.2$ kV. It is clearly impossible to select the appropriate voltage bases for both transformers.

To accommodate this situation, we will develop a per-unit model of a transformer whose voltage ratings are not in proportion to the selected base voltages. Such a transformer is said to have an "off-nominal turns ratio." Figure 4.27(a) shows a transformer with rated voltages V_{1rated} and V_{2rated}, which satisfy

$$V_{1rated} = a_t V_{2rated} \tag{4.8.1}$$

Figure 4.27

Transformer with off-nominal turns ratio

1 2

$a_t : 1$

(a) Single-line diagram

1 2

$b : 1$ $c = \frac{a_t}{b} : 1$

(b) Represented as two transformers in series

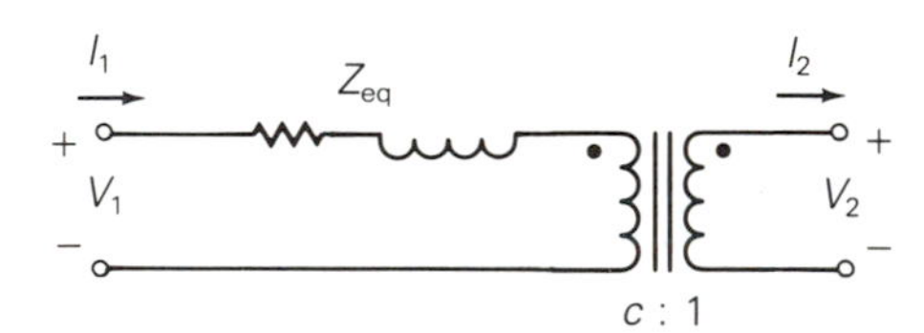

(c) Per-unit equivalent circuit
(Per-unit impedance is shown)

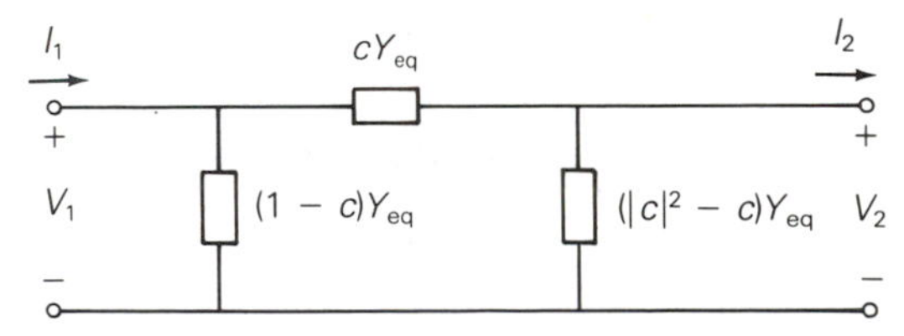

(d) π circuit representation for real c
(Per-unit admittances are shown; $Y_{eq} = \frac{1}{Z_{eq}}$)

where a_t is assumed, in general, to be either real or complex. Suppose the selected voltage bases satisfy

$$\mathrm{V}_{\mathrm{base}\,1} = b\,\mathrm{V}_{\mathrm{base}\,2} \tag{4.8.2}$$

Defining $c = a_t/b$, (4.8.1) can be rewritten as

$$\mathrm{V}_{1\mathrm{rated}} = b\left(\frac{a_t}{b}\right)\mathrm{V}_{2\mathrm{rated}} = bc\,\mathrm{V}_{2\mathrm{rated}} \tag{4.8.3}$$

Equation (4.8.3) can be represented by two transformers in series, as shown in Figure 4.27(b). The first transformer has the same ratio of rated winding voltages as the ratio of the selected base voltages, b. Therefore, this transformer has a standard per-unit model, as shown in Figure 4.9 or 4.17. We will assume that the second transformer is ideal, all real and reactive losses being associated with the first transformer. The resulting per-unit model is shown in Figure 4.27(c), where, for simplicity, the shunt-exciting branch is neglected. Note that if $a_t = b$, then the ideal transformer winding shown in this figure can be eliminated, since its turns ratio $c = (a_t/b) = 1$.

The per-unit model shown in Figure 4.27(c) is perfectly valid, but it is not suitable for some of the computer programs presented in later chapters because these programs do not accommodate ideal transformer windings. An alternative representation can be developed, however, by writing nodal equations for this figure as follows:

$$\begin{bmatrix} I_1 \\ -I_2 \end{bmatrix} = \begin{bmatrix} Y_{11} & Y_{12} \\ Y_{21} & Y_{22} \end{bmatrix} \begin{bmatrix} V_1 \\ V_2 \end{bmatrix} \tag{4.8.4}$$

where both I_1 and $-I_2$ are referenced *into* their nodes in accordance with the nodal equation method (Section 2.4). Recalling two-port network theory, the admittance parameters of (4.8.4) are, from Figure 4.27(c),

$$Y_{11} = \left.\frac{I_1}{V_1}\right|_{V_2 = 0} = \frac{1}{Z_{\mathrm{eq}}} = Y_{\mathrm{eq}} \tag{4.8.5}$$

$$Y_{22} = \left.\frac{-I_2}{V_2}\right|_{V_1 = 0} = \frac{1}{Z_{\mathrm{eq}}/|c|^2} = |c|^2 Y_{\mathrm{eq}} \tag{4.8.6}$$

$$Y_{12} = \left.\frac{I_1}{V_2}\right|_{V_1 = 0} = \frac{-cV_2/Z_{\mathrm{eq}}}{V_2} = -cY_{\mathrm{eq}} \tag{4.8.7}$$

$$Y_{21} = \left.\frac{-I_2}{V_1}\right|_{V_2 = 0} = \frac{-c^*I_1}{V_1} = -c^*Y_{\mathrm{eq}} \tag{4.8.8}$$

Equations (4.8.4)–(4.8.8) with real or complex c are convenient for representing transformers with off-nominal turns ratios in the computer programs presented later. Note that when c is complex, Y_{12} is not equal to Y_{21}, and the preceding admittance parameters cannot be synthesized with a passive RLC circuit. However, the π network shown in Figure 4.27(d), which has the same admittance parameters as (4.8.4)–(4.8.8), can be synthesized for real c. Note also that when $c = 1$, the shunt branches in this figure become open circuits (zero per unit mhos), and the series branch becomes Y_{eq} per unit mhos (or Z_{eq} per unit ohms).

Example 4.13 A three-phase generator step-up transformer is rated 1000 MVA, 13.8 kV Δ/345 kV Y with $Z_{eq} = j0.10$ per unit. The transformer high-voltage winding has $\pm 10\%$ taps. The system base quantities are

$$S_{\text{base}3\phi} = 500 \quad \text{MVA}$$

$$V_{\text{baseXLL}} = 13.8 \quad \text{kV}$$

$$V_{\text{baseHLL}} = 345 \quad \text{kV}$$

Determine the per-unit positive-sequence equivalent circuit for the following tap settings:

a. Rated tap

b. -10% tap (providing a 10% voltage decrease for the high-voltage winding)

Neglect transformer winding resistance, exciting current, and phase shift.

Solution **a.** Using (4.8.1) and (4.8.2) with the low-voltage winding denoted winding 1,

$$a_t = \frac{13.8}{345} = 0.04 \qquad b = \frac{V_{\text{baseXLL}}}{V_{\text{baseHLL}}} = \frac{13.8}{345} = a_t \qquad c = 1$$

From (4.3.11)

$$Z_{\text{p.u.new}} = (j0.10)\left(\frac{500}{1000}\right) = j0.05 \quad \text{per unit}$$

The positive-sequence equivalent circuit, not including winding resistance, exciting current, and phase shift is:

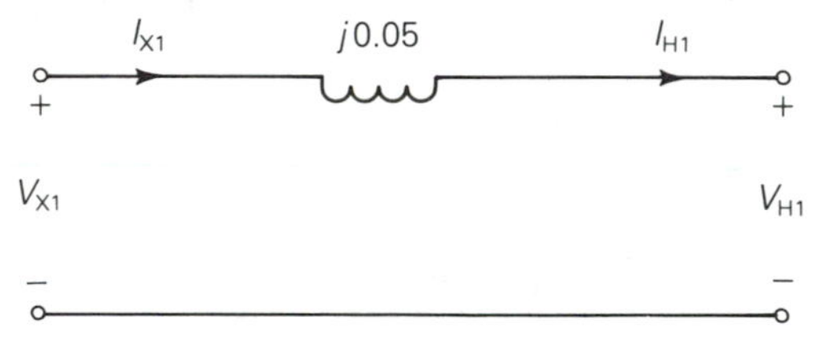

(Per-unit impedance is shown)

b. Using (4.8.1) and (4.8.2),

$$a_t = \frac{13.8}{345(0.9)} = 0.04444 \quad b = \frac{13.8}{345} = 0.04 \quad c = \frac{a_t}{b} = \frac{0.04444}{0.04} = 1.1111$$

From Figure 4.27(d),

$$cY_{\text{eq}} = 1.1111\left(\frac{1}{j0.05}\right) = -j22.22 \quad \text{per unit}$$

$$(1 - c)Y_{\text{eq}} = (-0.11111)(-j20) = +j2.222 \quad \text{per unit}$$

$$(|c|^2 - c)Y_{\text{eq}} = (1.2346 - 1.1)(-j20) = -j2.469 \quad \text{per unit}$$

The per-unit positive-sequence network is:

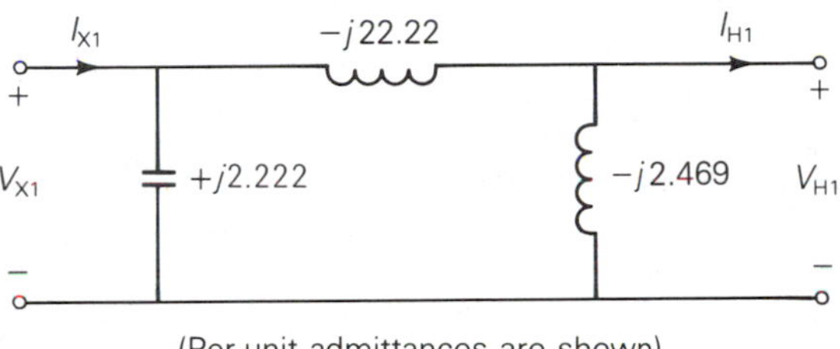

(Per-unit admittances are shown) ◁

The three-phase regulating transformers shown in Figures 4.28 and 4.29 can be modeled as transformers with off-nominal turns ratios. For the voltage-magnitude-regulating transformer shown in Figure 4.28, adjustable voltages ΔV_{an}, ΔV_{bn}, and ΔV_{cn}, which have equal magnitudes ΔV and which are in phase with the phase voltages V_{an}, V_{bn}, and V_{cn}, are placed in the series link between buses a–a', b–b', and c–c'. Modeled as a transformer with an off-nominal turns ratio (see Figure 4.27), $c = (1 + \Delta V)$ for a voltage-magnitude increase toward bus abc, or $c = (1 + \Delta V)^{-1}$ for an increase toward bus $a'b'c'$.

Figure 4.28

An example of a voltage-magnitude-regulating transformer

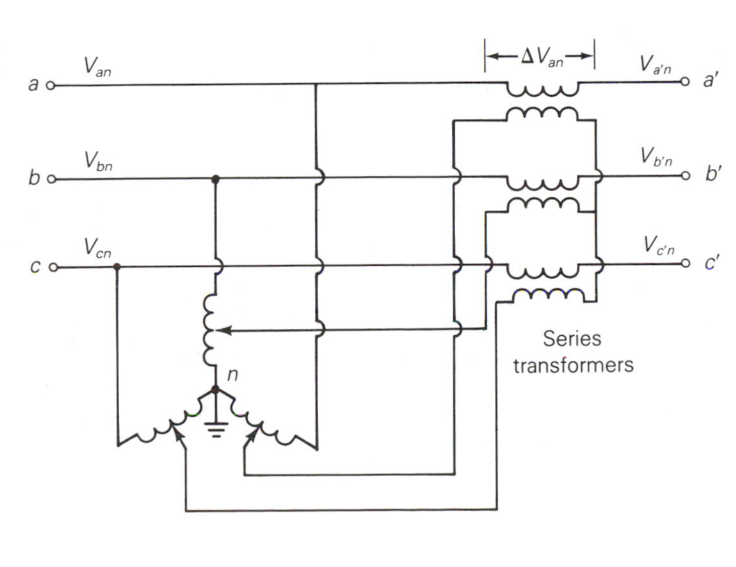

For the phase-angle-regulating transformer in Figure 4.29, the series voltages ΔV_{an}, ΔV_{bn}, and ΔV_{cn} are $\pm 90°$ out of phase with the phase voltages V_{an}, V_{bn}, and V_{cn}. The phasor diagram in Figure 4.29 indicates that each of the bus voltages $V_{a'n}$, $V_{b'n}$, and $V_{c'n}$ has a phase shift that is approximately proportional to the magnitude of the added series voltage. Modeled as a transformer with an off-nominal turns ratio (see Figure 4.27), $c \approx 1\underline{/\alpha}$ for a phase increase toward bus *abc* or $c \approx 1\underline{/-\alpha}$ for a phase increase toward bus *a'b'c'*.

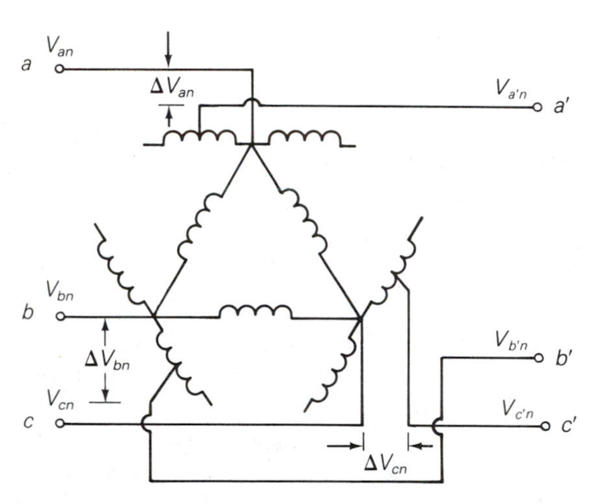

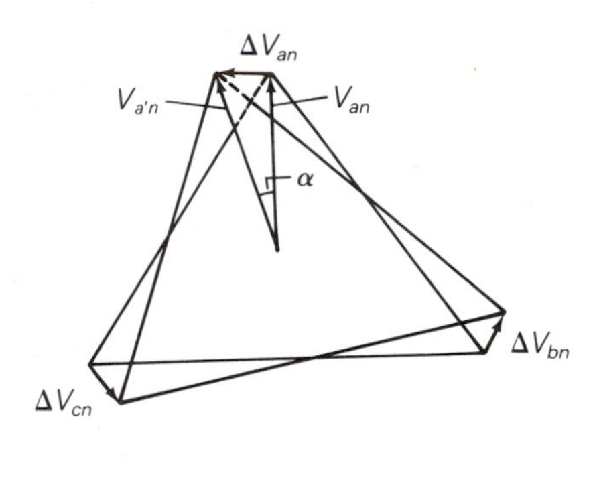

Figure 4.29 An example of a phase-angle-regulating transformer. Windings drawn in parallel are on the same core

Example 4.14 Two buses *abc* and *a'b'c'* are connected by two parallel lines L1 and L2 with positive-sequence series reactances $X_{L1} = 0.25$ and $X_{L2} = 0.20$ per unit. A regulating transformer is placed in series with line L1 at bus *a'b'c'*. Determine the 2×2 positive-sequence bus admittance matrix when the regulating transformer (a) provides a 0.05 per-unit increase in voltage magnitude toward bus *a'b'c'* and (b) advances the phase 3° toward bus *a'b'c'*. Assume that the regulating transformer is ideal. Also, the series resistance and shunt admittance of the lines are neglected.

Solution The circuit is shown in Figure 4.30.

a. For the voltage-magnitude-regulating transformer, $c = (1 + \Delta V)^{-1} = (1.05)^{-1} = 0.9524$ per unit. From (4.8.5)–(4.8.8), the admittance parameters of the regulating transformer in series with line L1 are

$$Y_{11L1} = \frac{1}{j0.25} = -j4.0$$

$$Y_{22L1} = (0.9524)^2(-j4.0) = -j3.628$$

$$Y_{12L1} = Y_{21L1} = (-0.9524)(-j4.0) = j3.810$$

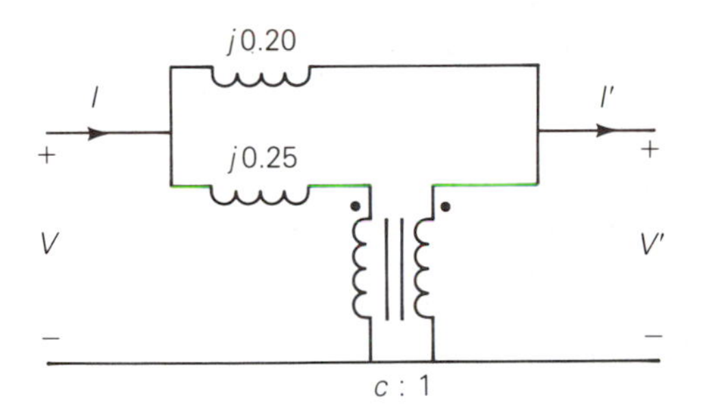

Figure 4.30

Positive-sequence circuit for Example 4.14

For line L2 alone,

$$Y_{11L2} = Y_{22L2} = \frac{1}{j0.20} = -j5.0$$

$$Y_{12L2} = Y_{21L2} = -(-j5.0) = j5.0$$

Combining the above admittances in parallel,

$$Y_{11} = Y_{11L1} + Y_{11L2} = -j4.0 - j5.0 = -j9.0$$

$$Y_{22} = Y_{22L1} + Y_{22L2} = -j3.628 - j5.0 = -j8.628$$

$$Y_{12} = Y_{21} = Y_{12L1} + Y_{12L2} = j3.810 + j5.0 = j8.810 \quad \text{per unit}$$

b. For the phase-angle-regulating transformer, $c = 1\underline{/-\alpha} = 1\underline{/-3^\circ}$. Then, for this regulating transformer in series with line L1,

$$Y_{11L1} = \frac{1}{j0.25} = -j4.0$$

$$Y_{22L1} = |1.0\underline{/-3^\circ}|^2(-j4.0) = -j4.0$$

$$Y_{12L1} = -(1.0\underline{/-3^\circ})(-j4.0) = 4.0\underline{/87^\circ} = 0.2093 + j3.9945$$

$$Y_{21L1} = -(1.0\underline{/-3^\circ})^*(-j4.0) = 4.0\underline{/93^\circ} = -0.2093 + j3.9945$$

The admittance parameters for line L2 alone are given in part (a) above. Combining the admittances in parallel,

$$Y_{11} = Y_{22} = -j4.0 - j5.0 = -j9.0$$

$$Y_{12} = 0.2093 + j3.9945 + j5.0 = 0.2093 + j8.9945$$

$$Y_{21} = -0.2093 + j3.9945 + j5.0 = -0.2093 + j8.9945 \quad \text{per unit} \quad \triangleleft$$

Note that a voltage-magnitude-regulating transformer controls the *reactive* power flow in the series link in which it is installed, whereas a phase-angle-regulating transformer controls the *real* power flow (see Problem 4.27).

Problems

Section 4.1

4.1 Consider an ideal transformer with $N_1 = 3000$ and $N_2 = 1000$ turns. Let winding 1 be connected to a source whose voltage is $e_1(t) = 100(1 - |t|)$ volts for $-1 \leqslant t \leqslant 1$ and $e_1(t) = 0$ for $|t| > 1$ second. A 2-farad capacitor is connected across winding 2. Sketch $e_1(t)$, $e_2(t)$, $i_1(t)$, and $i_2(t)$ versus time t.

4.2 A single-phase 50-kVA, 2400/240-volt, 60-Hz distribution transformer is used as a step-down transformer. The load, which is connected to the 240-volt secondary winding, absorbs 40 kVA at 0.8 power factor lagging and is at 230 volts. Assuming an ideal transformer, calculate the following: (a) primary voltage, (b) load impedance, (c) load impedance referred to the primary, and (d) the real and reactive power supplied to the primary winding.

4.3 For a conceptual single-phase, phase-shifting transformer, the primary voltage leads the secondary voltage by 30°. A load connected to the secondary winding absorbs 100 kVA at 0.9 power factor leading and at a voltage $E_2 = 277\underline{/0°}$ volts. Determine (a) the primary voltage, (b) primary and secondary currents, (c) load impedance referred to the primary winding, and (d) complex power supplied to the primary winding.

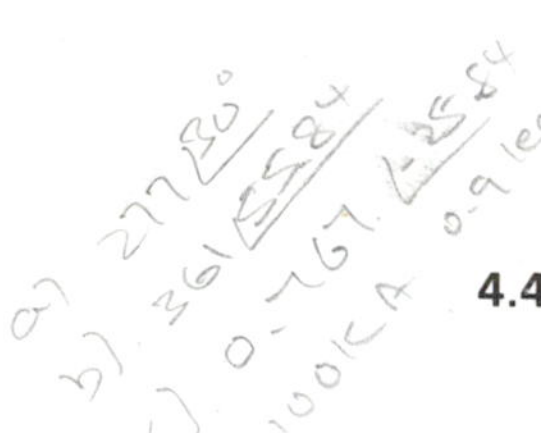

Section 4.2

4.4 The following data are obtained when open-circuit and short-circuit tests are performed on a single-phase, 50-kVA, 2400/240-volt, 60-Hz distribution transformer.

	VOLTAGE (volts)	CURRENT (amperes)	POWER (watts)
Measurements on low-voltage side with high-voltage winding open	240	4.85	173
Measurements on high-voltage side with low-voltage winding shorted	52.0	20.8	650

(a) Neglecting the series impedance, determine the exciting admittance referred to the high-voltage side. (b) Neglecting the exciting admittance, determine the equivalent series impedance referred to the high-voltage side. (c) Assuming equal series impedances for the primary and referred secondary, obtain an equivalent T-circuit referred to the high-voltage side.

4.5 A single-phase 50-kVA, 2400/240-volt, 60-Hz distribution transformer has a 1-ohm equivalent leakage reactance and a 5000-ohm magnetizing reactance referred to the high-voltage side. If rated voltage is applied to the high-voltage winding, calculate the open-circuit secondary voltage. Neglect I^2R and G_c^2V losses. Assume equal series leakage reactances for the primary and referred secondary.

4.6 A single-phase 50-kVA, 2400/240-volt, 60-Hz distribution transformer is used as a step-down transformer at the load end of a 2400-volt feeder whose series impedance is $(0.5 + j3.0)$ ohms. The equivalent series impedance of the transformer is $(1.5 + j2.0)$ ohms referred to the high-voltage (primary) side. The transformer is delivering rated load at 0.8 power factor lagging and at rated secondary voltage. Neglecting the transformer exciting current, determine (a) the voltage at the transformer primary terminals, (b) the voltage at the sending end of the feeder, and (c) the real and reactive power delivered to the sending end of the feeder.

Section 4.3

4.7 Using the transformer ratings as base quantities, work Problem 4.5 in per-unit.

4.8 Using the transformer ratings as base quantities, work Problem 4.6 in per-unit.

4.9 Using base values of 20 kVA and 115 volts in zone 3, rework Example 4.4.

4.10 A balanced-Y-connected voltage source with $E_{ag} = 277\underline{/0^\circ}$ volts is applied to a balanced-Y load in parallel with a balanced-Δ load, where $Z_Y = 20 + j10$ and $Z_\Delta = 30 - j15$ ohms. The Y load is solidly grounded. Using base values of $S_{\text{base }1\phi} = 10$ kVA and $V_{\text{baseLN}} = 277$ volts, calculate the source current I_a in per-unit and in amperes.

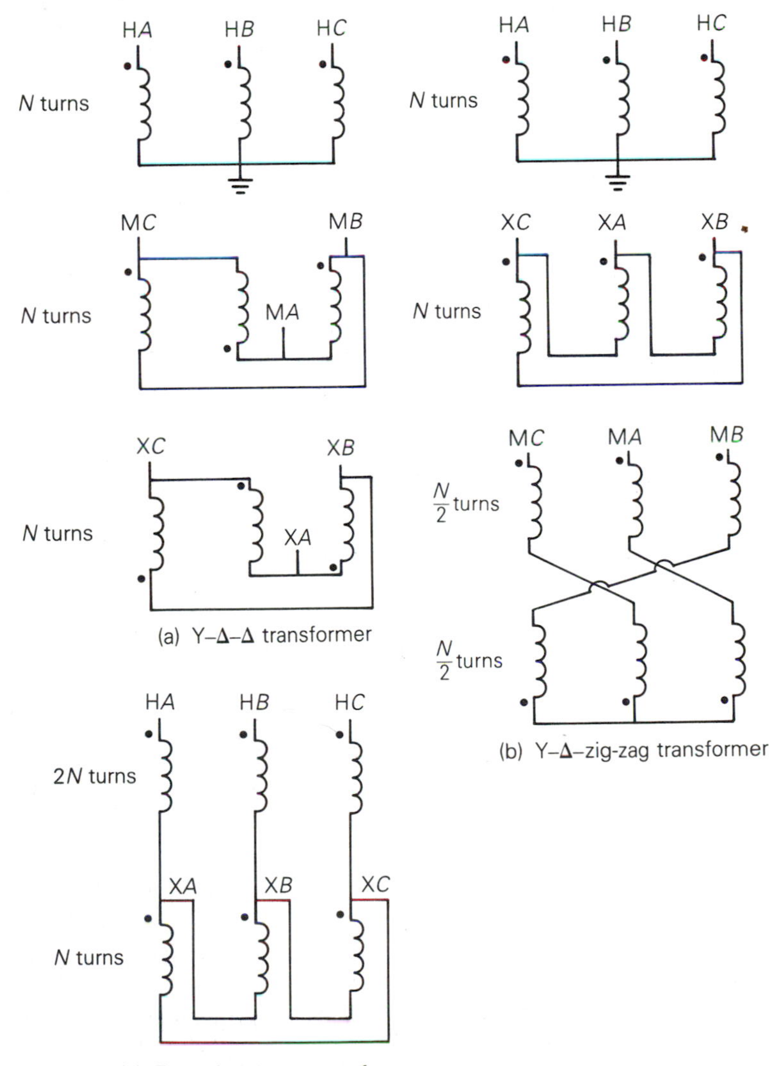

(a) Y–Δ–Δ transformer

(b) Y–Δ–zig-zag transformer

(c) Extended Δ autotransformer

Figure 4.31 Problems 4.11 and 4.24 (Coils drawn on the same vertical line are on the same core)

Section 4.4

4.11 Determine the positive- and negative-sequence phase shifts for the three-phase transformers shown in Figure 4.31.

4.12 Consider the three single-phase two-winding transformers shown in Figure 4.32. The high-voltage windings are connected in Y. (a) For the low-voltage side, connect the windings in Δ, place the polarity marks, and label the terminals a, b, and c in accordance with the American standard. (b) Relabel the terminals a', b', and c' such that V_{AN} is 90° out of phase with $V_{a'n}$ for positive sequence.

Figure 4.32

Problem 4.12

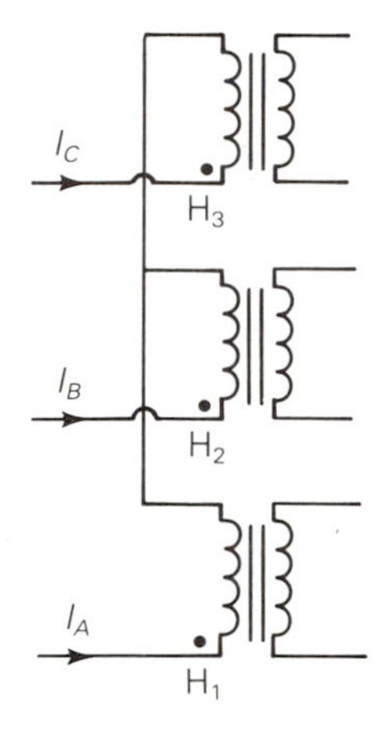

4.13 Consider a bank of three single-phase two-winding transformers whose high-voltage terminals are connected to a three-phase, 13.8-kV feeder. The low-voltage terminals are connected to a three-phase substation load rated 2.1 MVA and 2.3 kV. Determine the required voltage, current, and MVA ratings of both windings of each transformer, when the high-voltage/low-voltage windings are connected (a) Y–Δ, (b) Δ–Y, (c) Y–Y, and (d) Δ–Δ.

4.14 Three single-phase two-winding transformers, each rated 100 MVA, 34.5/13.8 kV, are connected to form a three-phase Δ–Δ bank. Balanced positive-sequence voltages are applied to the high-voltage terminals, and a balanced, resistive Y load connected to the low-voltage terminals absorbs 300 MW at 13.8 kV. If one of the single-phase transformers is removed (resulting in an open-Δ connection) and the balanced load is simultaneously reduced to 174 MW (58% of the original value), determine (a) the load voltages V_{an}, V_{bn}, and V_{cn}; (b) load currents I_a, I_b, and I_c; and (c) the MVA supplied by each of the remaining two transformers. Are balanced voltages still applied to the load? Is the open-Δ transformer overloaded?

Section 4.5

4.15 The leakage reactance of a three-phase, 300-MVA, 230 Y/23 Δ-kV transformer is 0.06 per unit based on its own ratings. The Y winding has a solidly grounded neutral. Draw the sequence networks. Neglect the exciting admittance and assume American standard phase shift.

4.16 Choosing system bases to be 240/24 kV and 100 MVA, redraw the sequence networks for Problem 4.15.

4.17 Consider the single-line diagram of the power system shown in Figure 4.33. Equipment ratings are:

generator 1:	1000 MVA, 18 kV, $X'' = 0.2$ per unit
generator 2:	1000 MVA, 18 kV, $X'' = 0.2$
synchronous motor 3:	1500 MVA, 20 kV, $X'' = 0.2$
3-phase Δ–Y transformers T_1, T_2, T_3, T_4:	1000 MVA, 500 kV Y/20 kV Δ, $X = 0.1$
3-phase Y–Y transformer T_5:	1500 MVA, 500 kV Y/20 kV Y, $X = 0.1$

Neglecting resistance, transformer phase shift, and magnetizing reactance, draw the positive-sequence reactance diagram. Use a base of 1000 MVA and 500 kV for the 50-ohm line. Determine the per-unit reactances.

Figure 4.33

Problems 4.17, 4.18, 4.19

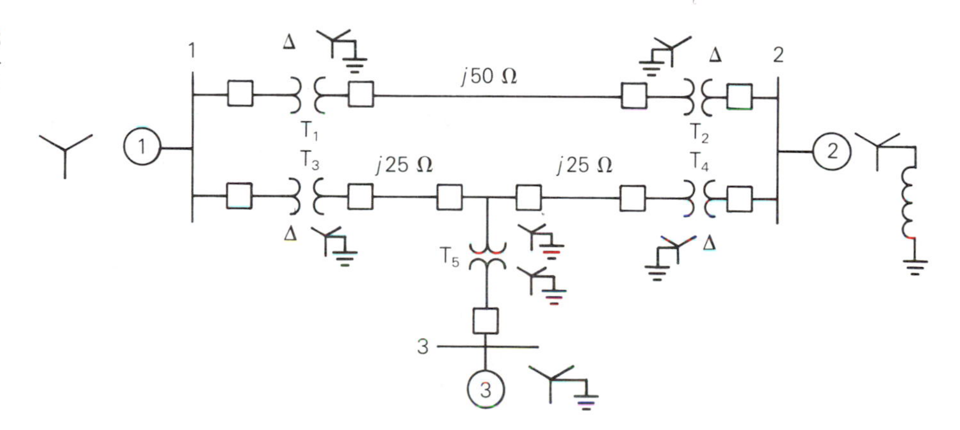

4.18 For the power system in Problem 4.17, the synchronous motor absorbs 1500 MW at 0.8 power factor leading with the bus 3 voltage at 18 kV. Determine the bus 1 and bus 2 voltages in kV. Assume that generators 1 and 2 deliver equal real powers and equal reactive powers. Also assume a balanced three-phase system with positive-sequence sources.

4.19 Draw the zero-sequence reactance diagram for the power system shown in Figure 4.33. The zero-sequence reactance of each generator and of the synchronous motor is 0.05 per unit based on equipment ratings. Generator 2 is grounded through a neutral reactor of 0.06 per unit on a 100-MVA, 18-kV base. The zero-sequence reactance of each transmission line is assumed to be three times its positive-sequence reactance. Use the same base as in Problem 4.17.

4.20 Three single-phase transformers, each rated 10 MVA, 66.4/12.5 kV, 60 Hz, with an equivalent series reactance of 0.1 per unit divided equally between primary and secondary, are connected in a three-phase bank. The high-voltage windings are Y connected and their terminals are directly connected to a 115-kV three-phase bus. The secondary terminals are all shorted together. Find the currents entering the high-voltage terminals and leaving the low-voltage terminals if the low-voltage windings are (a) Y connected, (b) Δ connected.

4.21 A 30-MVA, 13.2-kV three-phase generator, which has a positive-sequence reactance of 1.5 per unit on the generator base is connected to a 35-MVA, 13.2 Δ/115 Y-kV step-up transformer with a series impedance of $(0.005 + j0.1)$ per unit on its own base. (a) Calculate the per-unit generator reactance on the transformer base. (b) The load at the transformer terminals is 15 MW at unity power factor and at 115 kV. Choosing the transformer high-side voltage as the reference phasor, draw a phasor diagram for this condition. (c) For the condition of part (b), find the transformer low-side voltage and the generator internal voltage behind its reactance. Also compute the generator output power and power factor.

Section 4.6

4.22 A single-phase three-winding transformer has the following parameters: $Z_1 = Z_2 = Z_3 = 0 + j0.05$, $G_c = 0$, and $B_m = 0.2$ per unit. Three identical transformers, as described, are connected with their primaries in Y (solidly grounded neutral) and with their secondaries and tertiaries in Δ. Draw the per-unit sequence networks of this transformer bank.

4.23 The ratings of a three-phase three-winding transformer are:

Primary (1): Y connected, 66 kV, 15 MVA

Secondary (2): Y connected, 13.2 kV, 10 MVA

Tertiary (3): Δ connected, 2.3 kV, 5 MVA

Neglecting winding resistances and exciting current, the per-unit leakage reactances are:

$X_{12} = 0.08$ on a 15-MVA, 66-kV base

$X_{13} = 0.10$ on a 15-MVA, 66-kV base

$X_{23} = 0.09$ on a 10-MVA, 13.2-kV base

(a) Determine the per-unit reactances X_1, X_2, X_3 of the equivalent circuit on a 15-MVA, 66-kV base at the primary terminals. (b) Purely resistive loads of 7.5 MW at 13.2 kV and 5 MW at 2.3 kV are connected to the secondary and tertiary sides of the transformer, respectively. Draw the per-unit impedance diagram, showing the per-unit impedances on a 15-MVA, 66-kV base at the primary terminals.

4.24 Draw the positive-, negative-, and zero-sequence circuits for the transformers shown in Figure 4.31. Include ideal phase-shifting transformers showing phase shifts determined in Problem 4.11. Assume that all windings have the same kVA rating and that the equivalent leakage reactance of any two windings with the third winding open is 0.10 per unit. Neglect the exciting admittance.

Section 4.7

4.25 A single-phase 10-kVA, 2300/230-volt, 60-Hz two-winding distribution transformer is connected as an autotransformer to step up the voltage from 2300 to 2530 volts. (a) Draw a schematic diagram of this arrangement, showing all voltages and currents when delivering full load at rated voltage. (b) Find the permissible kVA rating of the autotransformer if the winding currents and voltages are not to exceed the rated values as a two-winding transformer. How much of this kVA rating is transformed by magnetic induction? (c) The following data are obtained from tests carried out on the transformer when it is connected as a two-winding transformer:

Open-circuit test with the low-voltage terminals excited:
Applied voltage = 230 V, Input current = 0.45 A, Input power = 70 W.

Short-circuit test with the high-voltage terminals excited:
Applied voltage = 120 V, Input current = 4.5 A, Input power = 240 W.

Based on this data, compute the efficiency of the autotransformer corresponding to full load, rated voltage, and 0.8 power factor lagging. Comment on why the efficiency is higher as an autotransformer than as a two-winding transformer.

4.26 Three single-phase two-winding transformers, each rated 3 kVA, 220/110 volts, 60 Hz, with a 0.10 per-unit leakage reactance are connected as a three-phase 110 Δ/571.6 Y-volt autotransformer bank. A three-phase load connected to the low-voltage terminals absorbs 6 kW at 110 volts and at 0.8 power factor lagging. Draw the per-unit impedance diagram and calculate the voltage and current at the high-voltage terminals. Assume positive-sequence operation.

Section 4.8

4.27 The two parallel lines in Example 4.14 supply a balanced load with a load current of $1.0\underline{/-30^\circ}$ per unit. Determine the real and reactive power supplied to the load bus from each parallel line with (a) no regulating transformer, (b) the voltage-magnitude-regulating transformer in Example 4.14(a), and (c) the phase-angle-regulating transformer in Example 4.14(b). Assume that the voltage at bus *abc* is adjusted so that the voltage at bus *a′b′c′* remains constant at $1.0\underline{/0^\circ}$ per unit. Also assume positive sequence. Comment on the effects of the regulating transformers.

References

1. R. Feinberg, *Modern Power Transformer Practice* (New York: John Wiley & Sons, 1979).
2. A. C. Franklin, and D. P. Franklin, *The J & P Transformer Book*, 11th ed. (London: Butterworths, 1983).
3. W. D. Stevenson Jr., *Elements of Power System Analysis*, 4th ed. (New York: McGraw-Hill, 1982).
4. J. R. Neuenswander, *Modern Power Systems* (Scranton, Pa.: International Textbook Company, 1971).
5. M. S. Sarma, *Electric Machines* (Dubuque, Ia.: William C. Brown, 1985).
6. A. E. Fitzgerald, C. Kingsley, and S. Umans, *Electric Machinery*, 4th ed. (New York: McGraw-Hill, 1983).
7. O. I. Elgerd, *Electric Energy Systems: An Introduction* (New York: McGraw-Hill, 1982).

CHAPTER 5

Transmission-Line Parameters

In this chapter we discuss the four basic transmission-line parameters: series resistance, series inductance, shunt capacitance, and shunt conductance. We also investigate transmission-line electric and magnetic fields.

Series resistance accounts for ohmic (I^2R) line losses. Series impedance, including resistance and inductive reactance, gives rise to series-voltage drops along the line. Shunt capacitance gives rise to line-charging currents. Shunt conductance accounts for V^2G line losses due to leakage currents between conductors or between conductors and ground. Shunt conductance of overhead lines is usually neglected.

Although the ideas developed in this chapter can be applied to underground transmission and distribution, the primary focus here is on overhead lines. Underground transmission in the United States presently accounts for less than 1% of total transmission, and is found mostly in large cities or under waterways. There is, however, a large application for underground cable in distribution systems.

Section 5.1

Transmission-Line Design Considerations

An overhead transmission line consists of conductors, insulators, support structures, and, in most cases, shield wires.

Conductors

Aluminum has replaced copper as the most common conductor metal for overhead transmission. Although a larger aluminum cross-sectional area is required to obtain the same loss as in a copper conductor, aluminum has a lower cost and lighter weight. Also, the supply of aluminum is abundant, while that of copper is limited.

One of the most common conductor types is aluminum conductor, steel-reinforced (ACSR), which consists of layers of aluminum strands surrounding a central core of steel strands (Figure 5.1). Stranded conductors are easier to manufacture, since larger conductor sizes can be obtained by simply adding successive layers of strands. Stranded conductors are also easier to handle and more flexible than solid conductors, especially in larger sizes. The use of steel strands gives ACSR conductors a high strength-to-weight ratio. For purposes of heat dissipation, overhead transmission-line conductors are bare (no insulating cover).

Figure 5.1

Typical ACSR conductor

54/7 Cardinal

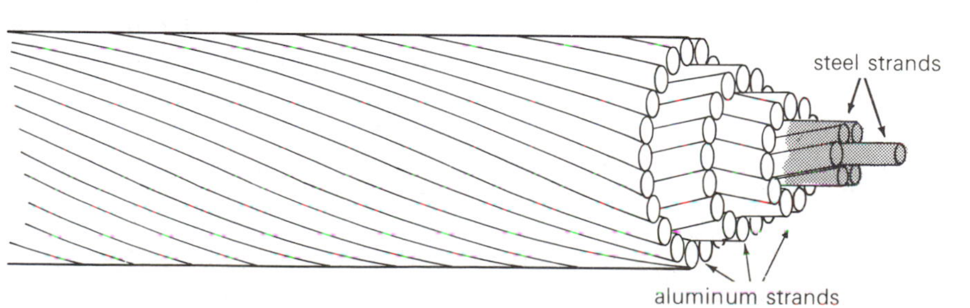

Figure 5.2

A 765 kV transmission line with self-supporting lattice steel towers

(Courtesy of the American Electric Power Company)

Other conductor types include the all-aluminum conductor (AAC), all-aluminum-alloy conductor (AAAC), aluminum conductor alloy-reinforced (ACAR), and aluminum-clad steel conductor (Alumoweld). There is also a conductor known as "expanded ACSR," which has a filler such as fiber or paper between the aluminum and steel strands. The filler increases the conductor diameter, which reduces the electric field at the conductor surface, to control corona.

EHV lines often have more than one conductor per phase; these conductors are called a *bundle*. The 765-kV line in Figure 5.2 has four conductors per phase, and the 345-kV double-circuit line in Figure 5.3 has two conductors per phase. Bundle conductors have a lower electric field strength at the conductor surfaces, thereby controlling corona. They also have a smaller series reactance.

Insulators

Insulators for transmission lines above 69 kV are suspension-type insulators, which consist of a string of discs, typically porcelain. The standard disc (Figure 5.5) has a 10-in. (0.254-m) diameter, $5\frac{3}{4}$-in. (0.146-m) spacing between centers of adjacent discs, and a mechanical strength of 7500 kg. The 765-kV line in Figure 5.2 has two strings per phase in a V-shaped arrangement, which helps to restrain conductor swings. The 345-kV line in Figure 5.4 has one vertical string per phase. The number of insulator discs in a string increases with line voltage (Table 5.1). Other types of discs include larger units with higher mechanical strength and fog insulators for use in contaminated areas.

Figure 5.3

A 345 kV double-circuit transmission line with self-supporting lattice steel towers

(Courtesy of Boston Edison Company)

Support Structures

Transmission lines employ a variety of support structures. Figure 5.2 shows a self-supporting, lattice steel tower typically used for 500- and 765-kV lines. Double-circuit 345-kV lines usually have self-supporting steel towers with the phases arranged either in a triangular configuration to reduce tower height or in a vertical configuration to reduce tower width (Figure 5.3). Wood frame configurations are commonly used for voltages of 345 kV and below (Figure 5.4).

Shield Wires

Shield wires located above the phase conductors protect the phase conductors against lightning. They are usually high- or extra-high-strength steel, Alumoweld, or ACSR with much smaller cross section than the phase conductors. The number and location of the shield wires are selected so that almost all lightning strokes terminate on the shield wires rather than on the phase conductors. Figures 5.2, 5.3, and 5.4 have two shield wires. Shield wires are grounded to the tower. As such, when lightning strikes a shield wire, it flows harmlessly to ground, provided the tower impedance and tower footing resistance are small.

The decision to build new transmission is based on power-system planning studies to meet future system requirements of load growth and new

Figure 5.4

Wood frame structure for a 345-kV line

(Courtesy of Boston Edison Company)

generation. The points of interconnection of each new line to the system, as well as the power and voltage ratings of each, are selected based on these studies. Thereafter, transmission-line design is based on optimization of electrical, mechanical, environmental, and economic factors.

Electrical Factors

Electrical design dictates the type, size, and number of bundle conductors per phase. Phase conductors are selected to have sufficient thermal capacity to meet continuous emergency overload and short-circuit current ratings. For EHV lines, the number of bundle conductors per phase is selected to control the voltage gradient at conductor surfaces, thereby reducing or eliminating corona.

Electrical design also dictates the number of insulator discs, vertical or V-shaped string arrangement, phase-to-phase clearance, and phase-to-tower clearance, all selected to provide adequate line insulation. Line insulation must withstand transient overvoltages due to lightning and switching surges, even when insulators are contaminated by fog, salt, or industrial pollution. Reduced clearances due to conductor swings during winds must also be accounted for.

Figure 5.5

Cut-away view of a standard insulator disc for suspension insulator strings

(Courtesy of Ohio Brass)

The number, type, and location of shield wires are selected to intercept lightning strokes that would otherwise hit the phase conductors. Also, tower footing resistance can be reduced by using driven ground rods or a buried conductor (called *counterpoise*) running parallel to the line. Line height is selected to satisfy prescribed conductor-to-ground clearances and to control ground-level electric field and its potential shock hazard.

Conductor spacings, types, and sizes also determine the series impedance and shunt admittance. Series impedance affects line-voltage drops, I^2R losses, and stability limits (Chapters 6, 12). Shunt admittance, primarily capacitive, affects line-charging currents, which inject reactive power into the power system. Shunt reactors (inductors) are often installed on lightly loaded EHV lines to absorb part of this reactive power, thereby reducing overvoltages.

NOMINAL VOLTAGE	PHASE CONDUCTORS				
	NUMBER OF CONDUCTORS PER BUNDLE	ALUMINUM CROSS-SECTION AREA PER CONDUCTOR (ACSR)	BUNDLE SPACING	MINIMUM CLEARANCES	
				PHASE-TO-PHASE	PHASE-TO-GROUND
kV		kcmil	cm	m	m
69	1	–	–	–	–
138	1	300–700	–	4 to 5	–
230	1	400–1000	–	6 to 9	–
345	1	2000–2500	–	6 to 9	7.6 to 11
345	2	800–2200	45.7	6 to 9	7.6 to 11
500	2	2000–2500	45.7	9 to 11	9 to 14
500	3	900–1500	45.7	9 to 11	9 to 14
765	4	900–1300	45.7	13.7	12.2

NOMINAL VOLTAGE	SUSPENSION INSULATOR STRING		SHIELD WIRES		
	NUMBER OF STRINGS PER PHASE	NUMBER OF STANDARD INSULATOR DISCS PER SUSPENSION STRING	TYPE	NUMBER	DIAMETER
kV					cm
69	1	4 to 6	Steel	0,1 or 2	–
138	1	8 to 11	Steel	0,1 or 2	–
230	1	12 to 21	Steel or ACSR	1 or 2	1.1 to 1.5
345	1	18 to 21	Alumoweld	2	0.87 to 1.5
345	1 and 2	18 to 21	Alumoweld	2	0.87 to 1.5
500	2 and 4	24 to 27	Alumoweld	2	0.98 to 1.5
500	2 and 4	24 to 27	Alumoweld	2	0.98 to 1.5
765	2 and 4	30 to 35	Alumoweld	2	0.98

Table 5.1 Typical transmission-line characteristics [1, 2]

Mechanical Factors

Mechanical design focuses on the strength of the conductors, insulator strings, and support structures. Conductors must be strong enough to support a specified thickness of ice and a specified wind in addition to their own weight. Suspension insulator strings must be strong enough to support the phase conductors with ice and wind loadings from tower to tower (span length).

Towers that satisfy minimum strength requirements, called suspension towers, are designed to support the phase conductors and shield wires with ice and wind loadings, and, in some cases, the unbalanced pull due to breakage of one or two conductors. Dead-end towers located every mile or so satisfy the maximum strength requirement of breakage of all conductors on one side of the tower. Angles in the line employ angle towers with intermediate strength. Conductor vibrations, which can cause conductor fatigue failure and damage to towers, are also of concern. Vibrations are controlled by adjustment of conductor tensions, use of vibration dampers, and—for bundle conductors—large bundle spacing and frequent use of bundle spacers.

Environmental Factors

Environmental factors include land usage and visual impact. When a line route is selected, the effect on local communities and population centers, land values, access to property, wildlife, and use of public parks and facilities must all be considered. Reduction in visual impact is obtained by aesthetic tower design and by blending the line with the countryside. Also, the biological effects of prolonged exposure to electric fields near transmission lines is of concern. Extensive research has been and continues to be done in this area.

Economic Factors

The optimum line design meets all the technical design criteria at lowest overall cost, which includes the total installed cost of the line as well as the cost of line losses over the operating life of the line. Many design factors affect cost. Utilities and consulting organizations use digital computer programs combined with specialized knowledge and physical experience to achieve optimum line design.

Section 5.2

Resistance

The dc resistance of a conductor at a specified temperature T is

$$R_{dc,T} = \frac{\rho_T l}{A} \quad \Omega \tag{5.2.1}$$

where ρ_T = conductor resistivity at temperature T

l = conductor length

A = conductor cross-sectional area

Two sets of units commonly used for calculating resistance, SI and English units, are summarized in Table 5.2. In English units conductor cross-sectional area is expressed in circular mils (cmil). One inch equals 1000 mils and

1 cmil equals $\pi/4$ sq mil. A circle with diameter D in., or (D in.) (1000 mil/in.) = 1000 D mil = d mil, has an area

$$A = \left(\frac{\pi}{4} D^2 \text{ in.}^2\right)\left(1000 \frac{\text{mil}}{\text{in.}}\right)^2 = \frac{\pi}{4}(1000\,D)^2 = \frac{\pi}{4} d^2 \quad \text{sq mil}$$

or

$$A = \left(\frac{\pi}{4} d^2 \text{ sq mil}\right)\left(\frac{1 \text{ cmil}}{\pi/4 \text{ sq mil}}\right) = d^2 \quad \text{cmil} \tag{5.2.2}$$

Table 5.2

Comparison of SI and English units for calculating conductor resistance

QUANTITY	SYMBOL	SI UNITS	ENGLISH UNITS
Resistivity	ρ	Ωm	Ω-cmil/ft
Length	ℓ	m	ft
Cross-sectional area	A	m^2	cmil
dc resistance	$R_{dc} = \frac{\rho \ell}{A}$	Ω	Ω

Resistivity depends on the conductor metal. Annealed copper is the international standard for measuring resistivity ρ (or conductivity σ, where $\sigma = 1/\rho$). Resistivity of conductor metals is listed in Table 5.3. As shown, hard-drawn aluminum, which has 61% of the conductivity of the international standard, has a resistivity at 20°C of 17.00 Ω-cmil/ft or 2.83×10^{-8} Ωm.

Table 5.3

% Conductivity, resistivity, and temperature constant of conductor metals

MATERIAL	% CONDUCTIVITY	$\rho_{20°C}$ RESISTIVITY AT 20°C		T TEMPERATURE CONSTANT
		$\Omega m \times 10^{-8}$	Ω-cmil/ft	°C
Copper:				
Annealed	100%	1.72	10.37	234.5
Hard-drawn	97.3%	1.77	10.66	241.5
Aluminum				
Hard-drawn	61%	2.83	17.00	228.1
Brass	20–27%	6.4–8.4	38–51	480
Bronze	9–13%	13–18	78–108	1980
Iron	17.2%	10	60	180
Silver	108%	1.59	9.6	243
Sodium	40%	4.3	26	207
Steel	2 to 14%	12 to 88	72–530	180–980

Conductor resistance depends on the following factors:

1. Spiraling
2. Temperature
3. Frequency ("skin effect")
4. Current magnitude—magnetic conductors

These are described in the following paragraphs.

For stranded conductors, alternate layers of strands are spiraled in opposite directions to hold the strands together. Spiraling makes the strands 1 or 2% longer than the actual conductor length. As a result, the dc resistance of a stranded conductor is 1 or 2% larger than that calculated from (5.2.1) for a specified conductor length.

Resistivity of conductor metals varies linearly over normal operating temperatures according to

$$\rho_{T2} = \rho_{T1}\left(\frac{T_2 + T}{T_1 + T}\right) \tag{5.2.3}$$

where ρ_{T2} and ρ_{T1} are resistivities at temperatures T_2 and T_1°C, respectively. T is a temperature constant that depends on the conductor material, and is listed in Table 5.3.

The ac resistance or *effective* resistance of a conductor is

$$R_{ac} = \frac{P_{loss}}{|I|^2} \quad \Omega \tag{5.2.4}$$

where P_{loss} is the conductor real power loss in watts and I is the rms conductor current. For dc, the current distribution is uniform throughout the conductor cross section, and (5.2.1) is valid. However, for ac, the current distribution is nonuniform. As frequency increases, the current in a solid cylindrical conductor tends to crowd toward the conductor surface, with smaller current density at the conductor center. This phenomenon is called *skin effect*. A conductor with a large radius can even have an oscillatory current density versus the radial distance from the conductor center.

With increasing frequency, conductor loss increases, which, from (5.2.4), causes the ac resistance to increase. At power frequencies (60 Hz) the ac resistance is at most a few percent higher than the dc resistance. Conductor manufacturers normally provide dc, 50-Hz, and 60-Hz conductor resistance based on test data (see Tables A.3 and A.4).

For magnetic conductors, such as steel conductors used for shield wires, resistance depends on current magnitude. The internal flux linkages, and therefore the iron or magnetic losses, depend on the current magnitude. For ACSR conductors, the steel core has a relatively high resistivity compared to the aluminum strands, and therefore the effect of current magnitude on ACSR conductor resistance is small. Tables on magnetic conductors list resistance at two current levels (see Table A.4).

Example 5.1 Table A.3 lists a 4/0 copper conductor with 12 strands. Strand diameter is 0.1328 in. For this conductor:

a. Verify the total copper cross-sectional area of 211,600 cmil.

b. Verify the dc resistance at 50°C of 0.302 Ω/mi. Assume a 2% increase in resistance due to spiraling.

c. From Table A.3, determine the percent increase in resistance at 60 Hz versus dc.

Solution **a.** The strand diameter is $d = (0.1328 \text{ in.})(1000 \text{ mil/in.}) = 132.8$ mil, and, from (5.2.2), the strand area is d^2 cmil. Using four significant figures, the cross-sectional area of the 12-strand conductor is

$$A = 12\,d^2 = 12(132.8)^2 = 211{,}600 \quad \text{cmil}$$

which agrees with the value given in Table A.3.

b. Using (5.2.3) and hard-drawn copper data from Table 5.3,

$$\rho_{50^\circ\text{C}} = 10.66\left(\frac{50 + 241.5}{20 + 241.5}\right) = 11.88 \quad \Omega\text{-cmil/ft}$$

From (5.2.1), the dc resistance at 50°C for a conductor length of 1 mile (5280 ft) is

$$R_{\text{dc},50^\circ\text{C}} = \frac{(11.88)(5280 \times 1.02)}{211{,}600} = 0.302 \quad \Omega/\text{mi}$$

which agrees with the value listed in Table A.3.

c. From Table A.3,

$$\frac{R_{60\text{Hz},50^\circ\text{C}}}{R_{\text{dc},50^\circ\text{C}}} = \frac{0.303}{0.302} = 1.003$$

$$\frac{R_{60\text{Hz},25^\circ\text{C}}}{R_{\text{dc},25^\circ\text{C}}} = \frac{0.278}{0.276} = 1.007$$

Thus the 60-Hz resistance of this conductor is about 0.3–0.7% higher than the dc resistance. The variation of these two ratios is due to the fact that resistance in Table A.3 is given to only three significant figures. ◁

Section 5.3

Conductance

Conductance accounts for real power loss between conductors or between conductors and ground. For overhead lines, this power loss is due to leakage currents at insulators and to corona. Insulator leakage current depends on the amount of dirt, salt, and other contaminants that have accumulated on insulators, as well as on meteorological factors, particularly the presence of moisture. Corona occurs when a high value of electric field strength at a conductor surface causes the air to become electrically ionized and to conduct. The real power loss due to corona, called *corona loss*, depends on meteorological conditions, particularly rain, and on conductor surface irregularities. Losses due to insulator leakage and corona are usually small compared to conductor I^2R loss. Conductance is usually neglected in power-system studies because it is a very small component of the shunt admittance.

Section 5.4

Inductance: Solid Cylindrical Conductor

The inductance of a magnetic circuit that has a constant permeability μ can be obtained by determining the following:

1. Magnetic field intensity H, from Ampere's law
2. Magnetic flux density B ($B = \mu H$)
3. Flux linkages λ
4. Inductance from flux linkages per ampere ($L = \lambda/I$)

As a step toward computing the inductances of more general conductors and conductor configurations, we first compute the internal, external, and total inductance of a solid cylindrical conductor. We also compute the flux linking one conductor in an array of current-carrying conductors.

Figure 5.6 shows a 1-meter section of a solid cylindrical conductor with radius r, carrying current I. For simplicity, assume that the conductor (1) is sufficiently long that end effects are neglected, (2) is nonmagnetic ($\mu = \mu_0 = 4\pi \times 10^{-7}$ H/m), and (3) has a uniform current density (skin effect is neglected). From (4.1.1), Ampere's law states that

$$\oint H_{\tan}\, dl = I_{\text{enclosed}} \tag{5.4.1}$$

Figure 5.6

Internal magnetic field of a solid cylindrical conductor

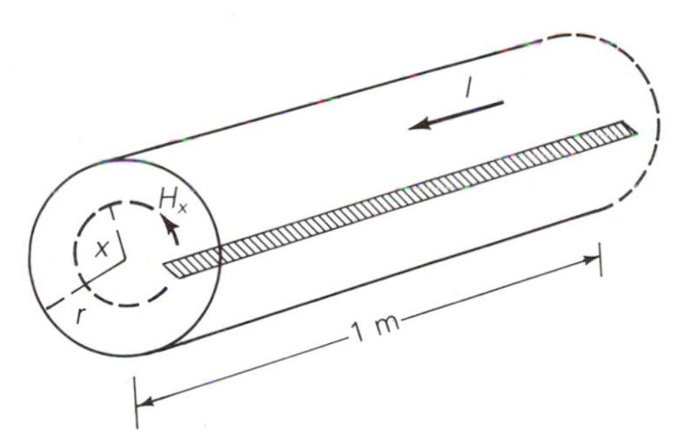

To determine the magnetic field inside the conductor, select the dashed circle of radius $x < r$ shown in Figure 5.6 as the closed contour for Ampere's law. Due to symmetry, H_x is constant along the contour. Also, there is no radial component of H_x, so H_x is tangent to the contour. That is, the conductor has a concentric magnetic field. From (5.4.1), the integral of H_x around the selected contour is

$$H_x(2\pi x) = I_x \qquad \text{for } x < r \tag{5.4.2}$$

where I_x is the portion of the total current enclosed by the contour. Solving (5.4.2)

$$H_x = \frac{I_x}{2\pi x} \quad \text{A/m} \tag{5.4.3}$$

Now assume a uniform current distribution within the conductor, that is

$$I_x = \left(\frac{x}{r}\right)^2 I \qquad \text{for } x < r \tag{5.4.4}$$

Using (5.4.4) in (5.4.3)

$$H_x = \frac{xI}{2\pi r^2} \quad \text{A/m} \tag{5.4.5}$$

For a nonmagnetic conductor, the magnetic flux density B_x is

$$B_x = \mu_0 H_x = \frac{\mu_0 x I}{2\pi r^2} \quad \text{Wb/m}^2 \tag{5.4.6}$$

The differential flux $d\phi$ per-unit length of conductor in the cross-hatched rectangle of width dx shown in Figure 5.6 is

$$d\phi = B_x\, dx \quad \text{Wb/m} \tag{5.4.7}$$

Computation of the differential flux linkage $d\lambda$ in the rectangle is tricky since only the fraction $(x/r)^2$ of the total current I is linked by the flux. That is,

$$d\lambda = \left(\frac{x}{r}\right)^2 d\phi = \frac{\mu_0 I}{2\pi r^4} x^3\, dx \quad \text{Wb-t/m} \tag{5.4.8}$$

Integrating (5.4.8) from $x = 0$ to $x = r$ determines the total flux linkages λ_{int} inside the conductor

$$\lambda_{\text{int}} = \int_0^r d\lambda = \frac{\mu_0 I}{2\pi r^4}\int_0^r x^3\, dx = \frac{\mu_0 I}{8\pi} = \frac{1}{2} \times 10^{-7} I \quad \text{Wb-t/m} \tag{5.4.9}$$

The internal inductance L_{int} per-unit length of conductor due to this flux linkage is then

$$L_{\text{int}} = \frac{\lambda_{\text{int}}}{I} = \frac{\mu_0}{8\pi} = \frac{1}{2} \times 10^{-7} \quad \text{H/m} \tag{5.4.10}$$

Next, in order to determine the magnetic field outside the conductor, select the dashed circle of radius $x > r$ shown in Figure 5.7 as the closed contour for Ampere's law. Noting that this contour encloses the entire current I, integration of (5.4.1) yields

$$H_x(2\pi x) = I \tag{5.4.11}$$

which gives

$$H_x = \frac{I}{2\pi x} \quad \text{A/m} \qquad x > r \tag{5.4.12}$$

Outside the conductor, $\mu = \mu_0$ and

$$B_x = \mu_0 H_x = (4\pi \times 10^{-7})\frac{I}{2\pi x} = 2 \times 10^{-7}\frac{I}{x} \quad \text{Wb/m}^2 \tag{5.4.13}$$

$$d\phi = B_x dx = 2 \times 10^{-7}\frac{I}{x}dx \quad \text{Wb/m} \tag{5.4.14}$$

Since the entire current I is linked by the flux outside the conductor

$$d\lambda = d\phi = 2 \times 10^{-7}\frac{I}{x}dx \quad \text{Wb-t/m} \tag{5.4.15}$$

Figure 5.7

External magnetic field of a solid cylindrical conductor

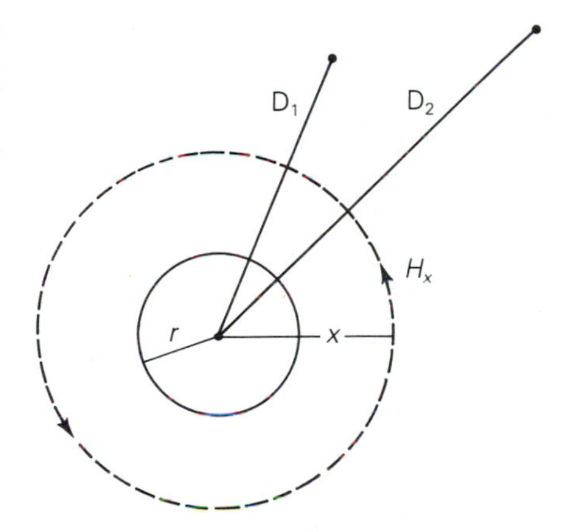

Integrating (5.4.15) between two external points at distances D_1 and D_2 from the conductor center gives the external flux linkage λ_{12} between D_1 and D_2:

$$\lambda_{12} = \int_{D_1}^{D_2} d\lambda = 2 \times 10^{-7} I \int_{D_1}^{D_2} \frac{dx}{x}$$

$$= 2 \times 10^{-7}\ I\ \ln\left(\frac{D_2}{D_1}\right) \quad \text{Wb-t/m} \tag{5.4.16}$$

The external inductance L_{12} per-unit length due to the flux linkages between D_1 and D_2 is then

$$L_{12} = \frac{\lambda_{12}}{I} = 2 \times 10^{-7} \ln\left(\frac{D_2}{D_1}\right) \quad \text{H/m} \tag{5.4.17}$$

The total flux λ_P linking the conductor out to external point P at distance D is the sum of the internal flux linkage, (5.4.9), and the external flux linkage, (5.4.16) from $D_1 = r$ to $D_2 = D$. That is

$$\lambda_P = \frac{1}{2} \times 10^{-7}\ I + 2 \times 10^{-7} I\ \ln\frac{D}{r} \tag{5.4.18}$$

Using the identity $\frac{1}{2} = 2 \ln e^{1/4}$ in (5.4.18), a more convenient expression for λ_P is obtained:

$$\lambda_P = 2 \times 10^{-7}\, I \left(\ln e^{1/4} + \ln \frac{D}{r} \right)$$

$$= 2 \times 10^{-7}\, I \ln \frac{D}{e^{-1/4} r}$$

$$= 2 \times 10^{-7}\, I \ln \frac{D}{r'} \quad \text{Wb-t/m} \tag{5.4.19}$$

where

$$r' = e^{-1/4} r = 0.7788\, r \tag{5.4.20}$$

Also, the total inductance L_P due to both internal and external flux linkages out to distance D is

$$L_P = \frac{\lambda_P}{I} = 2 \times 10^{-7} \ln\left(\frac{D}{r'}\right) \quad \text{H/m} \tag{5.4.21}$$

Finally, consider the array of M solid cylindrical conductors shown in Figure 5.8. Assume that each conductor m carries current I_m referenced out of the page. Also assume that the sum of the conductor currents is zero, that is,

$$I_1 + I_2 + \cdots + I_M = \sum_{m=1}^{M} I_m = 0 \tag{5.4.22}$$

The flux linkage λ_{kPk}, which links conductor k out to point P due to current I_k, is, from (5.4.19),

$$\lambda_{kPk} = 2 \times 10^{-7}\, I_k \ln \frac{D_{Pk}}{r'_k} \tag{5.4.23}$$

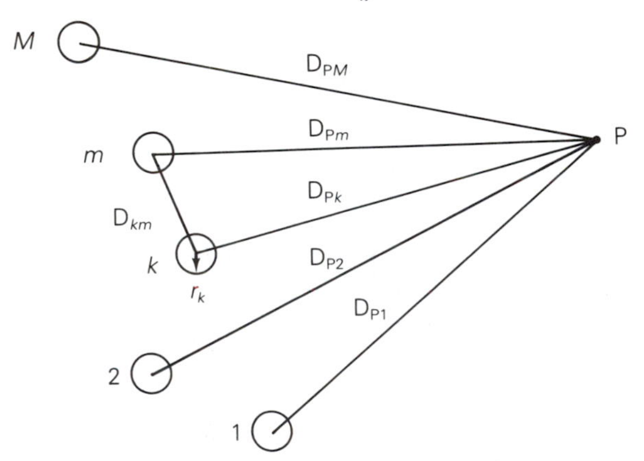

Figure 5.8 Array of M solid cylindrical conductors

Note that λ_{kPk} includes both internal and external flux linkages due to I_k. The flux linkage λ_{kPm}, which links conductor k out to P due to I_m, is, from (5.4.16),

$$\lambda_{kPm} = 2 \times 10^{-7}\, I_m \ln \frac{D_{Pm}}{D_{km}} \tag{5.4.24}$$

In (5.4.24) we use D_{km} instead of $(\mathrm{D}_{km} - r_k)$ or $(\mathrm{D}_{km} + r_k)$, which is a valid approximation when D_{km} is much greater than r_k. It can also be shown that this is a good approximation even when D_{km} is small. Using superposition, the total flux linkage $\lambda_{k\mathrm{P}}$, which links conductor k out to P due to all the currents, is

$$\lambda_{k\mathrm{P}} = \lambda_{k\mathrm{P}1} + \lambda_{k\mathrm{P}2} + \cdots + \lambda_{k\mathrm{P}M}$$

$$= 2 \times 10^{-7} \sum_{m=1}^{M} I_m \ln \frac{\mathrm{D}_{\mathrm{P}m}}{\mathrm{D}_{km}} \tag{5.4.25}$$

where we define $\mathrm{D}_{kk} = r'_k = e^{-1/4} r_k$ when $m = k$ in the above summation. Equation (5.4.25) is separated into two summations:

$$\lambda_{k\mathrm{P}} = 2 \times 10^{-7} \sum_{m=1}^{M} I_m \ln \frac{1}{\mathrm{D}_{km}} + 2 \times 10^{-7} \sum_{m=1}^{M} I_m \ln \mathrm{D}_{\mathrm{P}M} \tag{5.4.26}$$

Removing the last term from the second summation we get:

$$\lambda_{k\mathrm{P}} = 2 \times 10^{-7} \left[\sum_{m=1}^{M} I_m \ln \frac{1}{\mathrm{D}_{km}} + \sum_{m=1}^{M-1} I_m \ln \mathrm{D}_{\mathrm{P}m} + I_M \ln \mathrm{D}_{\mathrm{P}M} \right] \tag{5.4.27}$$

From (5.4.22),

$$I_M = -(I_1 + I_2 + \cdots + I_{M-1}) = -\sum_{m=1}^{M-1} I_m \tag{5.4.28}$$

Using (5.4.28) in (5.4.27)

$$\lambda_{k\mathrm{P}} = 2 \times 10^{-7} \left[\sum_{m=1}^{M} I_m \ln \frac{1}{\mathrm{D}_{km}} + \sum_{m=1}^{M-1} I_m \ln \mathrm{D}_{\mathrm{P}m} - \sum_{m=1}^{M-1} I_m \ln \mathrm{D}_{\mathrm{P}M} \right]$$

$$= 2 \times 10^{-7} \left[\sum_{m=1}^{M} I_m \ln \frac{1}{\mathrm{D}_{km}} + \sum_{m=1}^{M-1} I_m \ln \frac{\mathrm{D}_{\mathrm{P}m}}{\mathrm{D}_{\mathrm{P}M}} \right] \tag{5.4.29}$$

Now, let λ_k equal the total flux linking conductor k out to infinity. That is, $\lambda_k = \lim_{\mathrm{P} \to \infty} \lambda_{k\mathrm{P}}$. As $\mathrm{P} \to \infty$, all the distances $\mathrm{D}_{\mathrm{P}m}$ become equal, the ratios $\mathrm{D}_{\mathrm{P}m}/\mathrm{D}_{\mathrm{P}M}$ become unity, and $\ln(\mathrm{D}_{\mathrm{P}m}/\mathrm{D}_{\mathrm{P}M}) \to 0$. Therefore, the second summation in (5.4.29) becomes zero as $\mathrm{p} \to \infty$, and

$$\lambda_k = 2 \times 10^{-7} \sum_{m=1}^{M} I_m \ln \frac{1}{\mathrm{D}_{km}} \quad \text{Wb-t/m} \tag{5.4.30}$$

Equation (5.4.30) gives the total flux linking conductor k in an array of M conductors carrying currents $I_1, I_2, \ldots, I_M$, whose sum is zero. This equation is valid for either dc or ac currents. λ_k is a dc flux linkage when the currents are dc, and λ_k is a phasor flux linkage when the currents are phasor representations of sinusoids.

Section 5.5

Inductance: Single-phase Two-wire Line and Three-phase Three-wire Line with Equal Phase Spacing

The results of the previous section are used here to determine the inductances of two relatively simple transmission lines: a single-phase two-wire line and a three-phase three-wire line with equal phase spacing.

Figure 5.9(a) shows a single-phase two-wire line consisting of two solid cylindrical conductors x and y. Conductor x with radius r_x carries phasor current $I_x = I$ referenced out of the page. Conductor y with radius r_y carries return current $I_y = -I$. Since the sum of the two currents is zero, (5.4.30) is valid, from which the total flux linking conductor x is

$$\lambda_x = 2 \times 10^{-7}\left(I_x \ln\frac{1}{\mathrm{D}_{xx}} + I_y \ln\frac{1}{\mathrm{D}_{xy}}\right)$$

$$= 2 \times 10^{-7}\left(I \ln\frac{1}{r'_x} - I \ln\frac{1}{\mathrm{D}}\right)$$

$$= 2 \times 10^{-7} I \ln\frac{\mathrm{D}}{r'_x} \quad \text{Wb-t/m} \tag{5.5.1}$$

where $r'_x = e^{-1/4} r_x = 0.7788\, r_x$.

Figure 5.9

Single-phase two-wire line

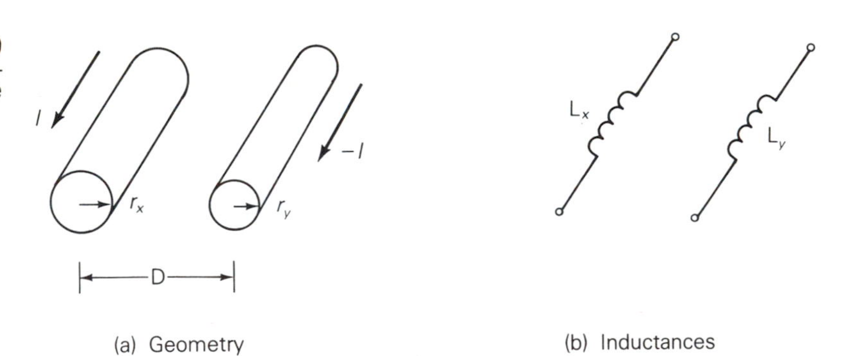

(a) Geometry (b) Inductances

The inductance of conductor x is then

$$\mathrm{L}_x = \frac{\lambda_x}{I_x} = \frac{\lambda_x}{I} = 2 \times 10^{-7} \ln\frac{\mathrm{D}}{r'_x} \quad \text{H/m per conductor} \tag{5.5.2}$$

Similarly, the total flux linking conductor y is

$$\lambda_y = 2 \times 10^{-7}\left(I_x \ln\frac{1}{D_{yx}} + I_y \ln\frac{1}{D_{yy}}\right)$$

$$= 2 \times 10^{-7}\left(I \ln\frac{1}{D} - I \ln\frac{1}{r'_y}\right)$$

$$= -2 \times 10^{-7} I \ln\frac{D}{r'_y} \tag{5.5.3}$$

and

$$L_y = \frac{\lambda_y}{I_y} = \frac{\lambda_y}{-I} = 2 \times 10^{-7} \ln\frac{D}{r'_y} \quad \text{H/m per conductor} \tag{5.5.4}$$

The total inductance of the single-phase circuit, also called *loop inductance*, is

$$L = L_x + L_y = 2 \times 10^{-7}\left(\ln\frac{D}{r'_x} + \ln\frac{D}{r'_y}\right)$$

$$= 2 \times 10^{-7} \ln\frac{D^2}{r'_x r'_y}$$

$$= 4 \times 10^{-7} \ln\frac{D}{\sqrt{r'_x r'_y}} \quad \text{H/m per circuit} \tag{5.5.5}$$

Also, if $r'_x = r'_y = r'$, the total circuit inductance is

$$L = 4 \times 10^{-7} \ln\frac{D}{r'} \quad \text{H/m per circuit} \tag{5.5.6}$$

Figure 5.10

Three-phase three-wire line with equal phase spacing

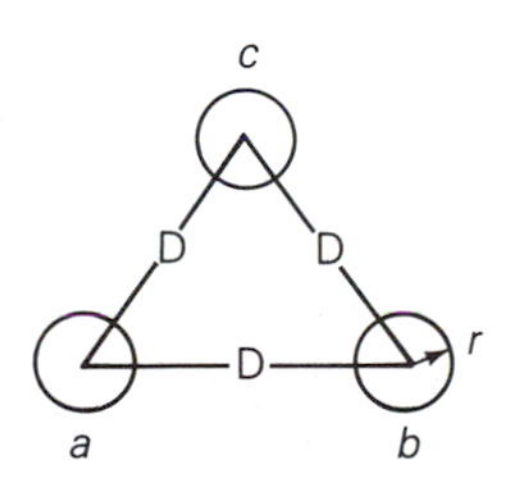

(a) Geometry

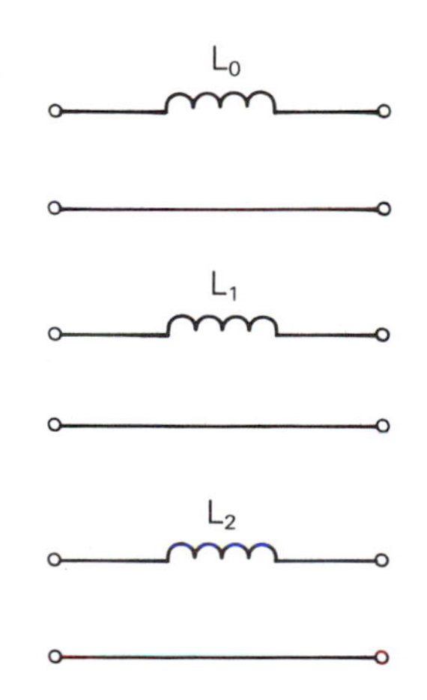

(b) Sequence inductances

The inductances of the single-phase two-wire line are shown in Figure 5.9(b).

Figure 5.10(a) shows a three-phase three-wire line consisting of three solid cylindrical conductors a, b, c, each with radius r, and with equal phase spacing D between any two conductors. To determine the positive-sequence inductance, assume positive-sequence currents I_a, I_b, I_c that satisfy $I_a + I_b + I_c = 0$. Then (5.4.30) is valid and the total flux linking the phase a conductor is

$$\lambda_a = 2 \times 10^{-7}\left(I_a \ln\frac{1}{r'} + I_b \ln\frac{1}{\mathrm{D}} + I_c \ln\frac{1}{\mathrm{D}}\right)$$

$$= 2 \times 10^{-7}\left[I_a \ln\frac{1}{r'} + (I_b + I_c)\ln\frac{1}{\mathrm{D}}\right] \tag{5.5.7}$$

Using $(I_b + I_c) = -I_a$,

$$\lambda_a = 2 \times 10^{-7}\left(I_a \ln\frac{1}{r'} - I_a \ln\frac{1}{\mathrm{D}}\right)$$

$$= 2 \times 10^{-7} I_a \ln\frac{\mathrm{D}}{r'} \quad \text{Wb-t/m} \tag{5.5.8}$$

The inductance of phase a is then

$$\mathrm{L}_a = \frac{\lambda_a}{I_a} = 2 \times 10^{-7} \ln\frac{\mathrm{D}}{r'} \quad \text{H/m per phase} \tag{5.5.9}$$

Due to symmetry, the same result is obtained for $\mathrm{L}_b = \lambda_b/I_b$ and for $\mathrm{L}_c = \lambda_c/I_c$. However, only one phase need be considered for positive-sequence operation of this line, since the flux linkages of each phase have equal magnitudes and 120° displacement. Thus the inductance of (5.5.9) is the positive-sequence inductance L_1. It is also the negative-sequence inductance L_2, since the positive- and negative-sequence inductances of nonrotating equipment are equal. That is

$$\mathrm{L}_1 = \mathrm{L}_2 = 2 \times 10^{-7} \ln\frac{\mathrm{D}}{r'} \quad \text{H/m} \tag{5.5.10}$$

Sequence inductances of the three-phase line with equal phase spacing are shown in Figure 5.10(b).

To determine the zero-sequence inductance L_0, the return paths including neutral conductors and earth return must be considered. In Section 5.7 we develop the equations for zero-sequence inductance as part of the series sequence impedances for more general three-phase lines.

Section 5.6

Inductance: Composite Conductors, Unequal Phase Spacing, Bundled Conductors

The results of Section 5.5 are extended here to include composite conductors, which consist of two or more solid cylindrical subconductors in parallel. A stranded conductor is one example of a composite conductor. For simplicity we assume that for each conductor, the subconductors are identical and share the conductor current equally.

Figure 5.11 shows a single-phase two-conductor line consisting of two composite conductors x and y. Conductor x has N identical subconductors, each with radius r_x and with current (I/N) referenced out of the page. Similarly, conductor y consists of M identical subconductors, each with radius r_y and with return current $(-I/M)$. Since the sum of all the currents is zero, (5.4.30) is valid and the total flux ϕ_k linking subconductor k of conductor x is

$$\phi_k = 2 \times 10^{-7}\left[\frac{I}{N}\sum_{m=1}^{N} \ln\frac{1}{D_{km}} - \frac{I}{M}\sum_{m=1'}^{M} \ln\frac{1}{D_{km}}\right] \tag{5.6.1}$$

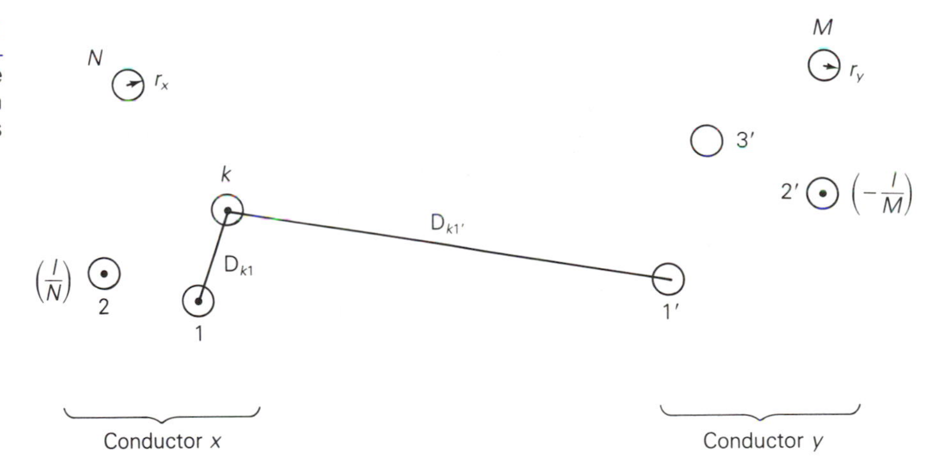

Figure 5.11 Single-phase two-conductor line with composite conductors

Since only the fraction $(1/N)$ of the total conductor current I is linked by this flux, the flux linkage λ_k of (the current in) subconductor k is

$$\lambda_k = \frac{\phi_k}{N} = 2 \times 10^{-7} I\left[\frac{1}{N^2}\sum_{m=1}^{N} \ln\frac{1}{D_{km}} - \frac{1}{NM}\sum_{m=1'}^{M} \ln\frac{1}{D_{km}}\right] \tag{5.6.2}$$

The total flux linkage of conductor x is

$$\lambda_x = \sum_{k=1}^{N} \lambda_k = 2 \times 10^{-7} I \sum_{k=1}^{N}\left[\frac{1}{N^2}\sum_{m=1}^{N} \ln\frac{1}{D_{km}} - \frac{1}{NM}\sum_{m=1'}^{M} \ln\frac{1}{D_{km}}\right] \tag{5.6.3}$$

Using $\ln A^{\alpha} = \alpha \ln A$ and $\sum \ln A_k = \ln \pi A_k$ (sum of lns = ln of products), (5.6.3) can be rewritten in the following form:

$$\lambda_x = 2 \times 10^{-7} I \ln \prod_{k=1}^{N} \frac{\left(\prod_{m=1'}^{M} D_{km}\right)^{1/NM}}{\left(\prod_{m=1}^{N} D_{km}\right)^{1/N^2}} \tag{5.6.4}$$

and the inductance of conductor x,

$L_x = \dfrac{\lambda_x}{I}$ can be written as

$$L_x = 2 \times 10^{-7} \ln \frac{D_{xy}}{D_{xx}} \quad \text{H/m per conductor} \tag{5.6.5}$$

where

$$D_{xy} = \sqrt[MN]{\prod_{k=1}^{N} \prod_{m=1'}^{M} D_{km}} \tag{5.6.6}$$

$$D_{xx} = \sqrt[N^2]{\prod_{k=1}^{N} \prod_{m=1}^{N} D_{km}} \tag{5.6.7}$$

D_{xy}, given by (5.6.6), is the MNth root of the product of the MN distances from the subconductors of conductor x to the subconductors of conductor y. Associated with each subconductor k of conductor x are the M distances $D_{k1'}$, $D_{k2'}, \ldots, D_{kM}$ to the subconductors of conductor y. For N subconductors in conductor x, there are therefore MN of these distances. D_{xy} is called the *geometric mean distance* or GMD between conductors x and y.

Also, D_{xx}, given by (5.6.7), is the N^2 root of the product of the N^2 distances between the subconductors of conductor x. Associated with each subconductor k are the N distances $D_{k1}, D_{k2}, \ldots, D_{kk} = r', \ldots, D_{kN}$. For N subconductors in conductor x, there are therefore N^2 of these distances. D_{xx} is called the *geometric mean radius* or GMR of conductor x.

Similarly, for conductor y,

$$L_y = 2 \times 10^{-7} \ln \frac{D_{xy}}{D_{yy}} \quad \text{H/m per conductor} \tag{5.6.8}$$

where

$$D_{yy} = \sqrt[M^2]{\prod_{k=1'}^{M} \prod_{m=1'}^{M} D_{km}} \tag{5.6.9}$$

D_{yy}, the GMR of conductor y, is the M^2 root of the product of the M^2

distances between the subconductors of conductor y. The total inductance L of the single-phase circuit is

$$L = L_x + L_y \quad \text{H/m per circuit} \tag{5.6.10}$$

Example 5.2 Expand (5.6.6), (5.6.7), and (5.6.9) for $N = 3$ and $M = 2'$. Then evaluate L_x, L_y, and L in H/m for the single-phase two-conductor line shown in Figure 5.12.

Solution For $N = 3$ and $M = 2'$, (5.6.6) becomes

$$D_{xy} = \sqrt[6]{\prod_{k=1}^{3} \prod_{m=1'}^{2'} D_{km}}$$

$$= \sqrt[6]{\prod_{k=1}^{3} D_{k1'} D_{k2'}}$$

$$= \sqrt[6]{(D_{11'}D_{12'})(D_{21'}D_{22'})(D_{31'}D_{32'})}$$

Similarly, (5.6.7) becomes

$$D_{xx} = \sqrt[9]{\prod_{k=1}^{3} \prod_{m=1}^{3} D_{km}}$$

$$= \sqrt[9]{\prod_{k=1}^{3} D_{k1} D_{k2} D_{k3}}$$

$$= \sqrt[9]{(D_{11}D_{12}D_{13})(D_{21}D_{22}D_{23})(D_{31}D_{32}D_{33})}$$

and (5.6.9) becomes

$$D_{yy} = \sqrt[4]{\prod_{k=1'}^{2'} \prod_{m=1'}^{2'} D_{km}}$$

$$= \sqrt[4]{\prod_{k=1'}^{2'} D_{k1'} D_{k2'}}$$

$$= \sqrt[4]{(D_{1'1'}D_{1'2'})(D_{2'1'}D_{2'2'})}$$

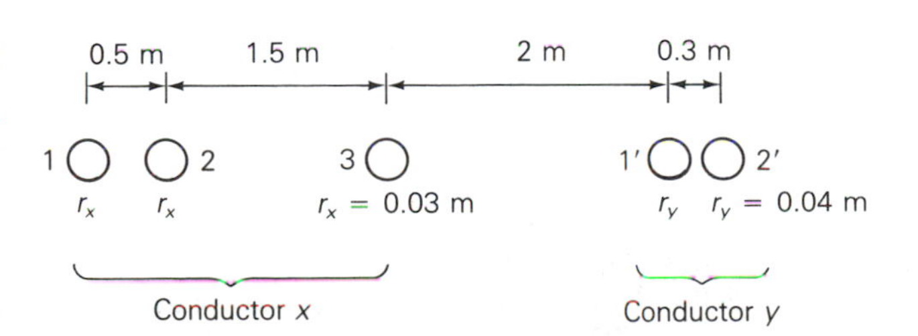

Figure 5.12 Single-phase two-conductor line for Example 5.2

Evaluating D_{xy}, D_{xx}, and D_{yy} for the single-phase two-conductor line shown in Figure 5.12,

$$D_{11'} = 4 \text{ m} \qquad D_{12'} = 4.3 \text{ m} \quad D_{21'} = 3.5 \text{ m}$$

$$D_{22'} = 3.8 \text{ m} \quad D_{31'} = 2 \text{ m} \qquad D_{32'} = 2.3 \text{ m}$$

$$D_{xy} = \sqrt[6]{(4)(4.3)(3.5)(3.8)(2)(2.3)} = 3.189 \text{ m}$$

$$D_{11} = D_{22} = D_{33} = r'_x = e^{-1/4} r_x = (0.7788)(0.03) = 0.02336 \text{ m}$$

$$D_{21} = D_{12} = 0.5 \text{ m}$$

$$D_{23} = D_{32} = 1.5 \text{ m}$$

$$D_{31} = D_{13} = 2.0 \text{ m}$$

$$D_{xx} = \sqrt[9]{(0.02336)^3(0.5)^2(1.5)^2(2.0)^2} = 0.3128 \text{ m}$$

$$D_{1'1'} = D_{2'2'} = r'_y = e^{-1/4} r_y = (0.7788)(0.04) = 0.03115 \text{ m}$$

$$D_{1'1'} = D_{2'1'} = 0.3 \text{ m}$$

$$D_{yy} = \sqrt[4]{(0.03115)^2(0.3)^2} = 0.09667 \text{ m}$$

Then, from (5.6.5), (5.6.8), and (5.6.10):

$$L_x = 2 \times 10^{-7} \ln\left(\frac{3.189}{0.3128}\right) = 4.644 \times 10^{-7} \quad \text{H/m per conductor}$$

$$L_y = 2 \times 10^{-7} \ln\left(\frac{3.189}{0.09667}\right) = 6.992 \times 10^{-7} \quad \text{H/m per conductor}$$

$$L = L_x + L_y = 1.164 \times 10^{-6} \quad \text{H/m per circuit}$$ ◁

It is seldom necessary to calculate GMR or GMD for standard lines. The GMR of standard conductors is provided by conductor manufacturers and can be found in various handbooks (see Tables A.3 and A.4). Also, if the distances between conductors are large compared to the distances between subconductors of each conductor, then the GMD between conductors is approximately equal to the distance between conductor centers.

Example 5.3 A single-phase line operating at 60 Hz consists of two 4/0 12-strand copper conductors with 5 ft spacing between conductor centers. The line length is 20 miles. Determine the total inductance in H and the total inductive reactance in Ω.

Solution The GMD between conductor centers is $D_{xy} = 5$ ft. Also, from Table A.3, the GMR of a 4/0 12-strand copper conductor is $D_{xx} = D_{yy} = 0.01750$ ft. From (5.6.5) and (5.6.8),

$$L_x = L_y = 2 \times 10^{-7} \ln\left(\frac{5}{0.01750}\right) \frac{\text{H}}{\text{m}} \times 1609 \frac{\text{m}}{\text{mi}} \times 20 \text{ mi}$$

$$= 0.03639 \quad \text{H per conductor}$$

The total inductance is

$$L = L_x + L_y = 2 \times 0.03639 = 0.07279 \quad \text{H per circuit}$$

and the total inductive reactance is

$$X_L = 2\pi f L = (2\pi)(60)(0.07279) = 27.44 \quad \Omega \text{ per circuit}$$ ◁

To calculate $L_1 = L_2$ for three-phase lines with stranded conductors and equal phase spacing, r' is replaced by the conductor GMR in (5.5.10). If the spacings between phases are unequal, then positive-sequence flux linkages are not obtained from positive-sequence currents. Instead, unbalanced flux linkages occur, and the phase inductances are unequal. However, balance can be restored by exchanging the conductor positions along the line, a technique called *transposition*.

Figure 5.13 shows a completely transposed three-phase line. The line is transposed at two locations such that each phase occupies each position for one-third of the line length. Conductor positions are denoted 1, 2, 3 with distances D_{12}, D_{23}, D_{31} between positions. The conductors are identical, each with GMR denoted DS. To calculate the positive-sequence inductance of this line, assume positive-sequence currents I_a, I_b, I_c, for which $I_a + I_b + I_c = 0$. Again, (5.4.30) is valid, and the total flux linking the phase a conductor while it is in position 1 is

$$\lambda_{a1} = 2 \times 10^{-7}\left[I_a \ln\frac{1}{D_S} + I_b \ln\frac{1}{D_{12}} + I_c \ln\frac{1}{D_{31}}\right] \quad \text{Wb-t/m} \tag{5.6.11}$$

Figure 5.13

Completely transposed three-phase line

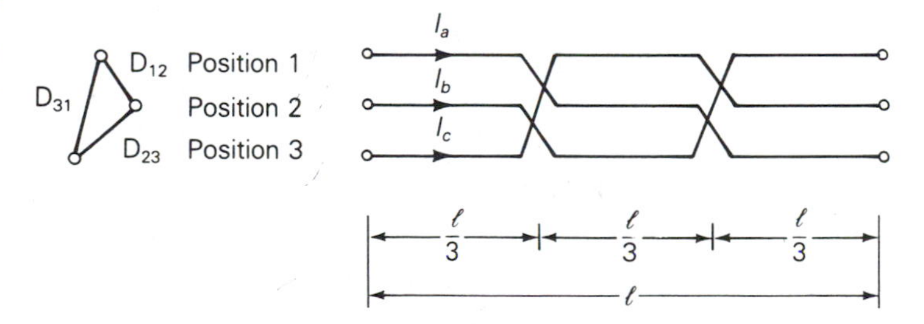

Similarly, the total flux linkage of this conductor while it is in positions 2 and 3 is

$$\lambda_{a2} = 2 \times 10^{-7}\left[I_a \ln\frac{1}{D_S} + I_b \ln\frac{1}{D_{23}} + I_c \ln\frac{1}{D_{12}}\right] \quad \text{Wb-t/m} \tag{5.6.12}$$

$$\lambda_{a3} = 2 \times 10^{-7}\left[I_a \ln\frac{1}{D_S} + I_b \ln\frac{1}{D_{31}} + I_c \ln\frac{1}{D_{23}}\right] \quad \text{Wb-t/m} \tag{5.6.13}$$

The average of the above flux linkages is

$$\lambda_a = \frac{\lambda_{a1}\left(\frac{l}{3}\right) + \lambda_{a2}\left(\frac{l}{3}\right) + \lambda_{a3}\left(\frac{l}{3}\right)}{l} = \frac{\lambda_{a1} + \lambda_{a2} + \lambda_{a3}}{3}$$

$$= \frac{2 \times 10^{-7}}{3}\left[3I_a \ln\frac{1}{D_S} + I_b \ln\frac{1}{D_{12}D_{23}D_{31}} + I_c \ln\frac{1}{D_{12}D_{23}D_{31}}\right] \tag{5.6.14}$$

Using $(I_b + I_c) = -I_a$ in (5.6.14),

$$\lambda_a = \frac{2 \times 10^{-7}}{3}\left[3I_a \ln\frac{1}{D_S} - I_a \ln\frac{1}{D_{12}D_{23}D_{31}}\right]$$

$$= 2 \times 10^{-7} I_a \ln\frac{\sqrt[3]{D_{12}D_{23}D_{31}}}{D_S} \quad \text{Wb-t/m} \tag{5.6.15}$$

and the average inductance of phase a is

$$L_a = \frac{\lambda_a}{I_a} = 2 \times 10^{-7} \ln\frac{\sqrt[3]{D_{12}D_{23}D_{31}}}{D_S} \quad \text{H/m per phase} \tag{5.6.16}$$

The same result is obtained for $L_b = \lambda_b/I_b$ and for $L_c = \lambda_c/I_c$. This inductance is the positive- (or negative-) sequence inductance of a completely transposed three-phase line. Defining

$$D_{eq} = \sqrt[3]{D_{12}D_{23}D_{31}} \tag{5.6.17}$$

we have

$$L_1 = L_2 = 2 \times 10^{-7} \ln\frac{D_{eq}}{D_S} \quad \text{H/m} \tag{5.6.18}$$

D_{eq}, the cube root of the product of the three-phase spacings, is the geometric mean distance between phases. Also, D_S is the conductor GMR for stranded conductors, or r' for solid cylindrical conductors.

Example 5.4 A completely transposed 60-Hz three-phase line has flat horizontal phase spacing with 10 m between adjacent conductors. The conductors are 1,590,000 cmil ACSR with 54/3 stranding. Line length is 200 km. Determine the positive-sequence inductance in H and the positive-sequence inductive reactance in Ω.

Solution From Table A.4 the GMR of a 1,590,000 cmil 54/3 ACSR conductor is

$$D_S = 0.0520 \text{ ft} \frac{1 \text{ m}}{3.28 \text{ ft}} = 0.0159 \text{ m}$$

Also, from (5.6.17) and (5.6.18),

$$D_{eq} = \sqrt[3]{(10)(10)(20)} = 12.6 \text{ m}$$

$$L_1 = 2 \times 10^{-7} \ln\left(\frac{12.6}{0.0159}\right) \frac{\text{H}}{\text{m}} \times \frac{1000 \text{ m}}{\text{km}} \times 200 \text{ km}$$

$$= 0.267 \quad \text{H}$$

The positive-sequence inductive reactance is

$$X_1 = 2\pi f L_1 = 2\pi(60)(0.267) = 101 \quad \Omega$$ ◁

It is common practice for EHV lines to use more than one conductor per phase, a practice called *bundling*. Bundling reduces the electric field strength at the conductor surfaces, which in turn reduces or eliminates corona and its results: undesirable power loss, communications interference and audible noise. Bundling also reduces the series reactance of the line by increasing the GMR of the bundle.

Figure 5.14 shows common EHV bundles consisting of two, three, or four conductors. The three-conductor bundle has its conductors on the vertices of an equilateral triangle, and the four-conductor bundle has its conductors on the corners of a square. To calculate $L_1 = L_2$, D_S in (5.6.18) is replaced by the GMR of the bundle. Since the bundle constitutes a composite conductor, calculation of bundle GMR is, in general, given by (5.6.7). If the conductors are stranded and the bundle spacing d is large compared to the conductor outside radius, each stranded conductor is first replaced by an equivalent solid cylindrical conductor with GMR $= D_S$. Then the bundle is replaced by one equivalent conductor with GMR $= D_{SL}$, given by (5.6.7) with $n = 2, 3,$ or 4 as follows:

Two-conductor bundle:

$$D_{SL} = \sqrt[4]{(D_S \times d)^2} = \sqrt{D_S d} \tag{5.6.19}$$

Three-conductor bundle:

$$D_{SL} = \sqrt[9]{(D_S \times d \times d)^3} = \sqrt[3]{D_S d^2} \tag{5.6.20}$$

Four-conductor bundle:

$$D_{SL} = \sqrt[16]{(D_S \times d \times d \times d\sqrt{2})^4} = 1.091\sqrt[4]{D_S d^3} \tag{5.6.21}$$

The positive- (or negative-) sequence inductance is then

$$L_1 = L_2 = 2 \times 10^{-7} \ln \frac{D_{eq}}{D_{SL}} \quad \text{H/m} \tag{5.6.22}$$

Figure 5.14 Bundle conductor configurations

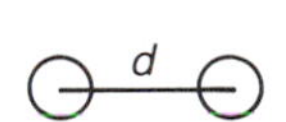

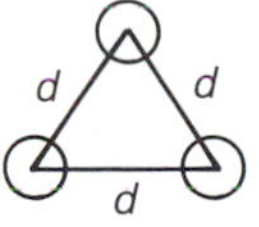

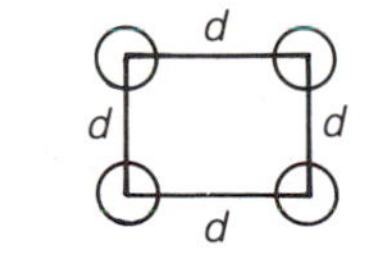

If the phase spacings are large compared to the bundle spacing, then sufficient accuracy for D_{eq} is obtained by using the distances between bundle centers.

Example 5.5 Each of the 1,590,000 cmil conductors in Example 5.4 is replaced by two 795,000 cmil ACSR 26/2 conductors, as shown in Figure 5.15. Bundle spacing is 0.40 m. Flat horizontal spacing is retained, with 10 m between adjacent bundle centers. Calculate the positive-sequence inductive reactance of the line and compare it with that of Example 5.4.

Solution From Table A.4, the GMR of a 795,000 cmil 26/2 ACSR conductor is

$$D_S = 0.0375 \text{ ft} \times \frac{1 \text{ m}}{3.28 \text{ ft}} = 0.0114 \text{ m}$$

From (5.6.19), the 2-conductor bundle GMR is

$$D_{SL} = \sqrt{(0.0114)(0.40)} = 0.0676 \quad \text{m}$$

Since $D_{eq} = 12.6$ m is the same as in Example 5.4,

$$L_1 = 2 \times 10^{-7} \ln\left(\frac{12.6}{0.0676}\right)(1000)(200) = 0.209 \quad \text{H}$$

$$X_1 = 2\pi f L_1 = (2\pi)(60)(0.209) = 78.8 \quad \Omega$$

The reactance of the bundled line, 78.8 Ω, is 22% less than that of Example 5.4, even though the 2-conductor bundle has the same amount of conductor material (that is, the same cmil per phase). One advantage of reduced series line reactance is smaller line-voltage drops. Also, the loadability of medium and long EHV lines is increased (see Chapter 6).

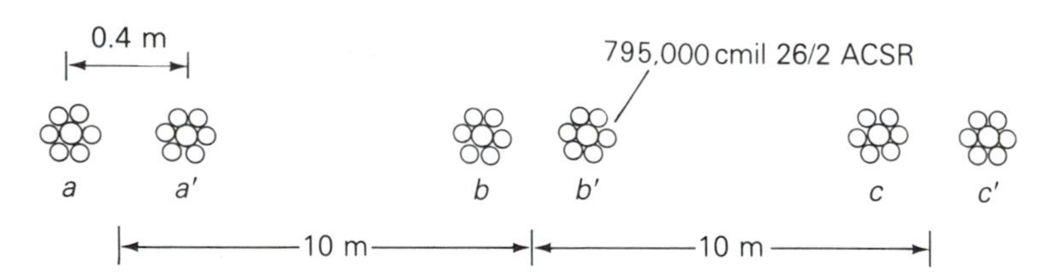

Figure 5.15 Three-phase bundled conductor line for Example 5.5

Section 5.7

Series Impedances: Three-phase Line with Neutral Conductors and Earth Return

In this section we develop equations suitable for computer calculation of the series impedances, including resistances and inductive reactances, for the three-phase line shown in Figure 5.16. This line has three phase conductors a, b, and c, where bundled conductors, if any, have already been replaced by equivalent conductors, as described in Section 5.6. The line also has N neutral conductors denoted $n1, n2, \ldots, nN$.* All the neutral conductors are connected in parallel and are grounded to the earth at regular intervals along the line. Any isolated neutral conductors that carry no current are omitted. The phase conductors are insulated from each other and from earth.

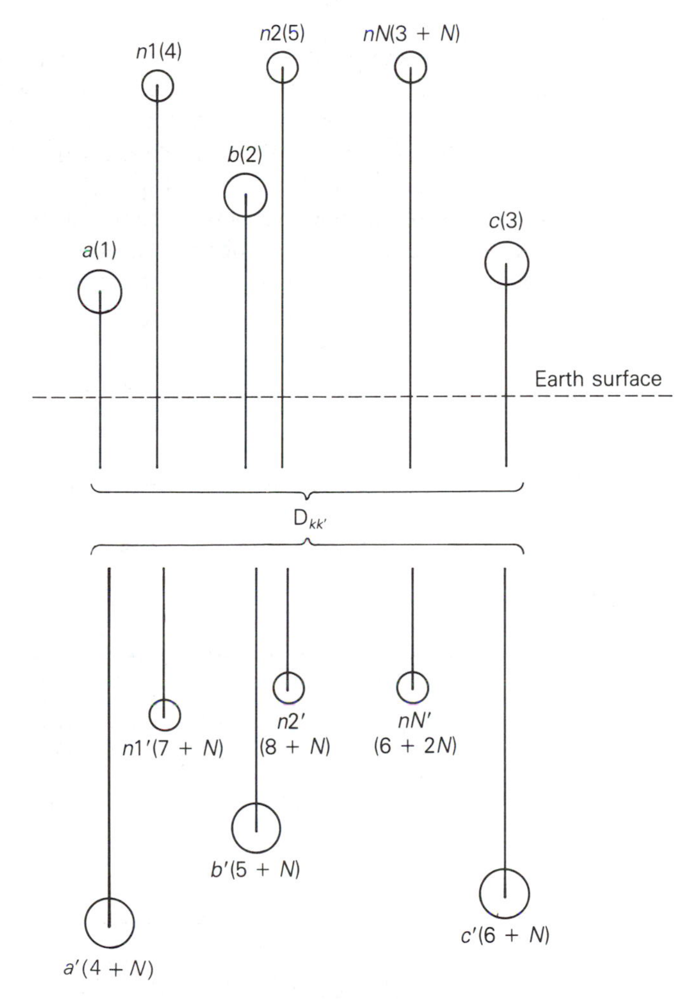

Figure 5.16

Three-phase transmission line with earth replaced by earth return conductors

If the phase currents are not balanced, there may be a return current in the grounded neutral wires and in the earth. The earth return current will spread out under the line, seeking the lowest impedance return path. A classic paper by Carson [4], later modified by others [5,6], shows that the earth can be replaced by a set of "earth return" conductors located directly under the overhead conductors, as shown in Figure 5.16. Each of these earth return conductors, denoted a', b', c', $n1'$, $n2'$, $\ldots$, nN', has a GMR denoted $D_{k'k'}$,

*Instead of *shield wire* we use the term *neutral conductor*, which applies to distribution as well as transmission lines.

distance $D_{kk'}$ from its overhead conductor, and resistance $R_{k'}$ given by:

$$D_{k'k'} = D_{kk} \text{ m} \tag{5.7.1}$$

$$D_{kk'} = [658.5\sqrt{\rho/f}]^{1/2} \quad \text{m} \tag{5.7.2}$$

$$R_{k'} = 9.869 \times 10^{-7} f \quad \Omega/\text{m} \tag{5.7.3}$$

where ρ is the earth resistivity in ohm-meters and f is frequency in hertz. Table 5.4 lists earth resistivities and 60-Hz equivalent conductor distances for various types of earth. It is common practice to select $\rho = 100\,\Omega\text{m}$ when actual data is unavailable. Note that all the earth return conductors have the same distance $D_{kk'}$ from their overhead conductors and the same resistance $R_{k'}$. For simplicity we renumber the conductors from 1 to $(6 + 2N)$, beginning with the overhead phase conductors, then overhead neutral conductors, then image phase conductors, then image neutral conductors, as shown in Figure 5.16.

Operating as a transmission line, the sum of the currents in all the conductors is zero. That is,

$$\sum_{k=1}^{(6+2N)} I_k = 0 \tag{5.7.4}$$

Equation (5.4.30) is therefore valid, and the flux linking conductor k is

$$\lambda_k = 2 \times 10^{-7} \sum_{m=1}^{(6+2N)} I_m \ln \frac{1}{D_{km}} \quad \text{Wb-t/m} \tag{5.7.5}$$

Table 5.4 Earth resistivities and 60-Hz equivalent conductor distances

TYPE OF EARTH	RESISTIVITY (Ωm)	$D_{kk'}$ (m)
Sea water	0.01–1.0	8.50–85.0
Swampy ground	10–100	269–850
Average damp earth	100	850
Dry earth	1000	2690
Pure slate	10^7	269,000
Sandstone	10^9	2,690,000

In matrix format, (5.7.5) becomes

$$\boldsymbol{\lambda} = \mathbf{L}\boldsymbol{I} \tag{5.7.6}$$

where

$\boldsymbol{\lambda}$ is a $(6 + 2N)$ vector

$\boldsymbol{I}$ is a $(6 + 2N)$ vector

$\mathbf{L}$ is a $(6 + 2N) \times (6 + 2N)$ matrix whose elements are:

$$L_{km} = 2 \times 10^{-7} \ln \frac{1}{D_{km}} \tag{5.7.7}$$

Figure 5.17

Circuit representation of series-phase impedances

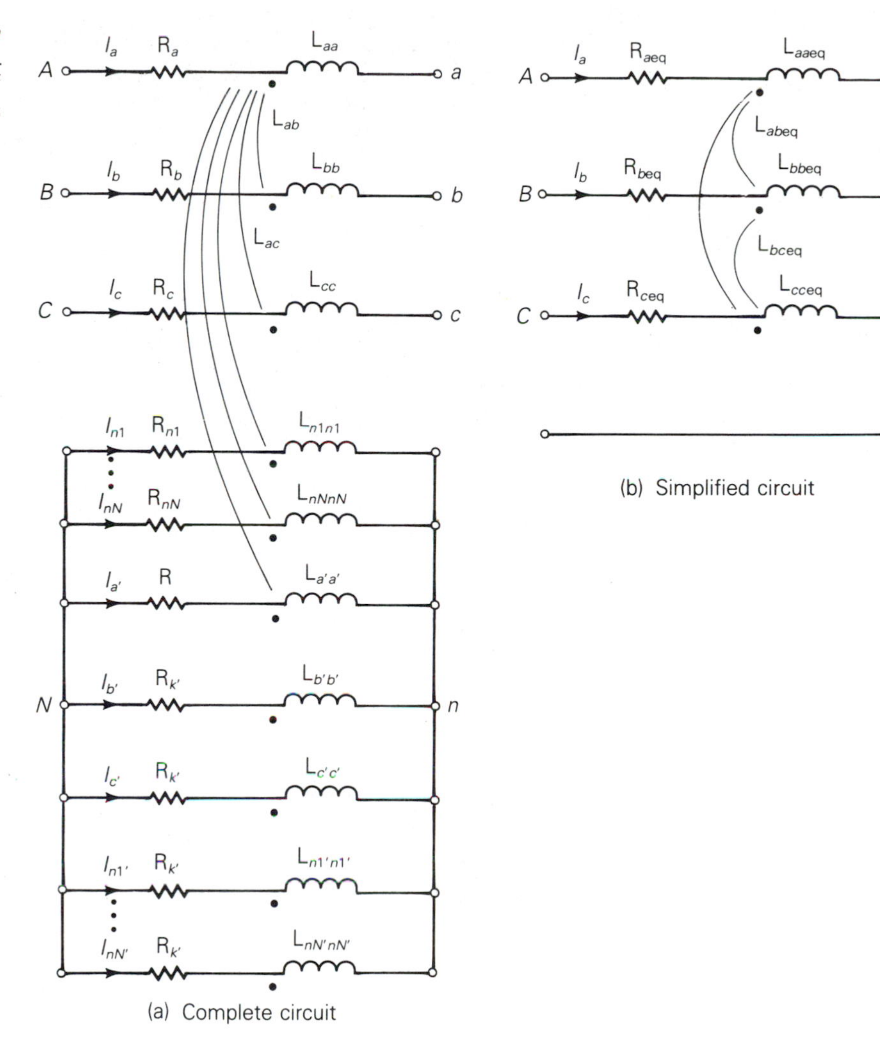

(a) Complete circuit

(b) Simplified circuit

When $k = m$, D_{kk} in (5.7.7) is the GMR of (bundled) conductor k. When $k \neq m$, D_{km} is the distance between conductors k and m.

A circuit representation of a 1-meter section of the line is shown in Figure 5.17(a). Using this circuit, the vector of voltage drops across the conductors is:

$$\begin{bmatrix} E_{Aa} \\ E_{Bb} \\ E_{Cc} \\ E_{Nn} \\ \vdots \\ E_{Nn} \\ \cdot \\ \cdot \\ \cdot \\ E_{Nn} \\ E_{Nn} \end{bmatrix} = (\mathbf{R} + j\omega\mathbf{L}) \begin{bmatrix} I_a \\ I_b \\ I_c \\ I_{n1} \\ \vdots \\ I_{Nn} \\ I_{a'} \\ I_{b'} \\ I_{c'} \\ I_{n1'} \\ I_{nN'} \end{bmatrix} \tag{5.7.8}$$

where $\mathbf{L}$ is given by (5.7.7) and $\mathbf{R}$ is a $(6 + 2N) \times (6 + 2N)$ diagonal matrix of conductor resistances.

$$\mathbf{R} = \begin{bmatrix} R_a & & & & & & & & & & & & \\ & R_b & & & & & & & & & \mathbf{0} & & \\ & & R_c & & & & & & & & & & \\ & & & R_{n1} & & & & & & & & & \\ & & & & \cdot & & & & & & & & \\ & & & & & \cdot & & & & & & & \\ & & & & & & \cdot & & & & & & \\ & & & & & & & R_{nN} & & & & & \\ & & & & & & & & R_{k'} & & & & \\ & & & & & & & & & R_{k'} & & & \\ & & & & & & & & & & \cdot & & \\ & \mathbf{0} & & & & & & & & & & \cdot & \\ & & & & & & & & & & & & R_{k'} \end{bmatrix} \Omega/\text{m} \tag{5.7.9}$$

The resistance R_k of each overhead conductor is obtained from conductor tables such as Table A.3 or A.4, for a specified frequency, temperature, and current. The resistance $R_{k'}$ of all the image conductors is the same, as given by (5.7.3).

Note that all the grounded neutral conductors as well as the equivalent earth conductors have the same voltage drop E_{Nn} in (5.7.8) since these conductors are connected in parallel. Our objective now is to reduce the $(6 + 2N)$ equations in (5.7.8) to three equations, thereby obtaining the simplified circuit representation shown in Figure 5.17(b). We do so by lumping the neutral voltage drop E_{Nn} into the voltage drops across the phase conductors and by eliminating the neutral and earth return currents from (5.7.8). First we subtract the last equation from all the other equations in (5.7.8), in order to lump E_{Nn} into E_{Aa}, E_{Bb}, and E_{Cc}. Also, we include (5.7.4) at the end of (5.7.8) to obtain:

$$\begin{bmatrix} E_{Aa} - E_{Nn} \\ E_{Bb} - E_{Nn} \\ E_{Cc} - E_{Nn} \\ \hline 0 \\ \cdots \\ \cdots \\ \cdots \\ \cdots \\ \cdots \\ \cdots \\ 0 \\ 0 \end{bmatrix} = \left[\begin{array}{ccc|ccc} \multicolumn{3}{c|}{\overbrace{\hphantom{Z_{11} \quad Z_{12} \quad Z_{13}}}^{Z_A}} & \multicolumn{3}{c}{\overbrace{\hphantom{Z_{14} \quad \cdots \quad Z_{1(6+2N)}}}^{Z_B}} \\ Z_{11} & Z_{12} & Z_{13} & Z_{14} & \cdots & Z_{1(6+2N)} \\ Z_{21} & Z_{22} & Z_{23} & Z_{24} & \cdots & Z_{2(6+2N)} \\ Z_{31} & Z_{32} & Z_{33} & Z_{34} & \cdots & Z_{3(6+2N)} \\ \hline Z_{41} & Z_{42} & Z_{43} & Z_{44} & \cdots & Z_{4(6+2N)} \\ & & & & & \\ & & & & & \\ & & & & & \\ & & & & & \\ & & & & & \\ Z_{(5+2N)1} & Z_{(5+2N)2} & Z_{(5+2N)3} & Z_{(5+2N)4} & \cdots & Z_{(5+2N)(6+2N)} \\ & & & & & \\ 1 & \cdots 1 \cdots & 1 & 1 & \cdots 1 \cdots & 1 \\ \multicolumn{3}{c|}{\underbrace{\hphantom{Z_{11} \quad Z_{12} \quad Z_{13}}}_{Z_C}} & \multicolumn{3}{c}{\underbrace{\hphantom{Z_{14} \quad \cdots \quad Z_{1(6+2N)}}}_{Z_D}} \end{array}\right] \begin{bmatrix} I_a \\ I_b \\ I_c \\ \hline I_{n1} \\ I_{nN} \\ I_{a'} \\ I_{b'} \\ I_{c'} \\ I_{n1'} \\ \vdots \\ I_{nN'} \end{bmatrix} \tag{5.7.10}$$

The diagonal elements of this matrix are

$$Z_{kk} = \begin{cases} R_k + j\omega L_{kk} - j\omega L_{(6+2N)k} = R_k + j\omega\, 2 \times 10^{-7} \ln \dfrac{D_{(6+2N)k}}{D_{kk}} \quad \Omega/\text{m} \quad k = 1,2,\ldots,(5+2N) \\ 1 \qquad k = (6+2N) \end{cases} \tag{5.7.11}$$

And the off-diagonal elements, for $k \neq m$, are

$$Z_{km} = \begin{cases} j\omega L_{km} - j\omega L_{(6+2N)m} = j\omega 2 \times 10^{-7} \ln \dfrac{D_{(6+2N)m}}{D_{km}} \quad \Omega/\text{m} \qquad \text{for } k \text{ and } m < (6+2N) \\[2ex] -R_{k'} + j\omega 2 \times 10^{-7} \ln \dfrac{D_{(6+2N)(6+2N)}}{D_{k(6+2N)}} \qquad \text{for } m = (6+2N) \\[2ex] 1 \qquad \text{for } k = (6+2N) \end{cases} \tag{5.7.12}$$

Next, (5.7.10) is partitioned as shown above to obtain

$$\begin{bmatrix} \boldsymbol{E}_{\mathrm{P}} \\ \hline \boldsymbol{0} \end{bmatrix} = \left[\begin{array}{c|c} \boldsymbol{Z}_A & \boldsymbol{Z}_B \\ \hline \boldsymbol{Z}_C & \boldsymbol{Z}_D \end{array}\right] \begin{bmatrix} \boldsymbol{I}_{\mathrm{P}} \\ \hline \boldsymbol{I}_n \end{bmatrix} \tag{5.7.13}$$

where

$$\boldsymbol{E}_{\mathrm{P}} = \begin{bmatrix} E_{Aa} - E_{Nn} \\ E_{Bb} - E_{Nn} \\ E_{Cc} - E_{Nn} \end{bmatrix}; \quad \boldsymbol{I}_{\mathrm{P}} = \begin{bmatrix} I_a \\ I_b \\ I_c \end{bmatrix}; \quad \boldsymbol{I}_n = \begin{bmatrix} I_{n1} \\ \vdots \\ I_{nN} \\ I_{a'} \\ I_{b'} \\ I_{c'} \\ I_{n1'} \\ \vdots \\ I_{nN'} \end{bmatrix}$$

$\boldsymbol{E}_{\mathrm{P}}$ is the 3 vector of voltage drops across the phase conductors (including the neutral voltage drop). $\boldsymbol{I}_{\mathrm{P}}$ is the 3 vector of phase currents and $\boldsymbol{I}_n$ is the $(3 + 2N)$ vector of neutral and earth return currents. Also, the $(6 + 2N) \times (6 + 2N)$ matrix in (5.7.10) is partitioned to obtain the following matrices:

$\boldsymbol{Z}_A$ with dimension 3×3

$\boldsymbol{Z}_B$ with dimension $3 \times (3 + 2N)$

$\boldsymbol{Z}_C$ with dimension $(3 + 2N) \times 3$

$\boldsymbol{Z}_D$ with dimension $(3 + 2N) \times (3 + 2N)$

Equation (5.7.13) is rewritten as two separate matrix equations:

$$\boldsymbol{E}_{\mathrm{P}} = \boldsymbol{Z}_A \boldsymbol{I}_{\mathrm{P}} + \boldsymbol{Z}_B \boldsymbol{I}_n \tag{5.7.14}$$

$$\boldsymbol{0} = \boldsymbol{Z}_C \boldsymbol{I}_{\mathrm{P}} + \boldsymbol{Z}_D \boldsymbol{I}_n \tag{5.7.15}$$

Solving (5.7.15) for $\boldsymbol{I}_n$,

$$\boldsymbol{I}_n = -\boldsymbol{Z}_D^{-1} \boldsymbol{Z}_C \boldsymbol{I}_{\mathrm{P}} \tag{5.7.16}$$

Using (5.7.16) in (5.7.14):

$$\boldsymbol{E}_{\mathrm{P}} = [\boldsymbol{Z}_A - \boldsymbol{Z}_B \boldsymbol{Z}_D^{-1} \boldsymbol{Z}_C]\, \boldsymbol{I}_{\mathrm{P}} \tag{5.7.17}$$

or

$$\boldsymbol{E}_{\mathrm{P}} = \boldsymbol{Z}_{\mathrm{P}} \boldsymbol{I}_{\mathrm{P}} \tag{5.7.18}$$

where

$$Z_P = Z_A - Z_B Z_D^{-1} Z_C \tag{5.7.19}$$

Equation (5.7.17), the desired result, relates the phase-conductor voltage drops (including neutral voltage drop) to the phase currents. Z_P given by (5.7.19) is the 3×3 series-phase impedance matrix, whose elements are denoted

$$Z_P = \begin{bmatrix} Z_{aaeq} & Z_{abeq} & Z_{aceq} \\ Z_{abeq} & Z_{bbeq} & Z_{bceq} \\ Z_{aceq} & Z_{bceq} & Z_{cceq} \end{bmatrix} \quad \Omega/\text{m} \tag{5.7.20}$$

In general, Z_P does not satisfy the conditions for a symmetrical impedance matrix, which require that the diagonal elements be equal and that the off-diagonal elements be equal. However, if the line is completely transposed, the diagonal and off-diagonal elements are averaged to obtain

$$\hat{Z}_P = \begin{bmatrix} \hat{Z}_{aaeq} & \hat{Z}_{abeq} & \hat{Z}_{abeq} \\ \hat{Z}_{abeq} & \hat{Z}_{aaeq} & \hat{Z}_{abeq} \\ \hat{Z}_{abeq} & \hat{Z}_{abeq} & \hat{Z}_{aaeq} \end{bmatrix} \quad \Omega/\text{m} \tag{5.7.21}$$

where

$$\hat{Z}_{aaeq} = \tfrac{1}{3}(Z_{aaeq} + Z_{bbeq} + Z_{cceq}) \tag{5.7.22}$$

$$\hat{Z}_{abeq} = \tfrac{1}{3}(Z_{abeq} + Z_{aceq} + Z_{bceq}) \tag{5.7.23}$$

$\hat{Z}_P$ is a symmetrical impedance matrix.

Equation (5.7.18) gives the series-voltage drops across the circuit in Figure 5.17(b), which is identical to the series impedance circuit shown in Figure 3.9. Therefore, (5.7.18) can be transformed to the sequence domain using the ideas of Section 3.3. From (3.3.3) and (3.3.4)

$$\begin{bmatrix} E_{00'} \\ E_{11'} \\ E_{22'} \end{bmatrix} = Z_S \begin{bmatrix} I_0 \\ I_1 \\ I_2 \end{bmatrix} \tag{5.7.24}$$

where

$$Z_S = A^{-1} Z_P A \tag{5.7.25}$$

and A is the 3 × 3 transformation matrix given by (3.1.8). $E_{00'}$, $E_{11'}$, and $E_{22'}$ are the zero-, positive-, and negative-sequence series-voltage drops, and I_0, I_1, I_2 are the sequence currents. $\boldsymbol{Z}_S$ is the 3 × 3 series sequence impedance matrix whose elements are

$$\boldsymbol{Z}_S = \begin{bmatrix} Z_0 & Z_{01} & Z_{02} \\ Z_{10} & Z_1 & Z_{12} \\ Z_{20} & Z_{21} & Z_2 \end{bmatrix} \quad \Omega/\text{m} \tag{5.7.26}$$

In general $\boldsymbol{Z}_S$ is not diagonal. However, if the line is completely transposed,

$$\hat{\boldsymbol{Z}}_S = A^{-1}\hat{\boldsymbol{Z}}_P A = \begin{bmatrix} \hat{Z}_0 & 0 & 0 \\ 0 & \hat{Z}_1 & 0 \\ 0 & 0 & \hat{Z}_2 \end{bmatrix} \tag{5.7.27}$$

where, from (3.3.7) and (3.3.8),

$$\hat{Z}_0 = \hat{Z}_{aaeq} + 2\hat{Z}_{abeq} \tag{5.7.28}$$

$$\hat{Z}_1 = \hat{Z}_2 = \hat{Z}_{aaeq} - \hat{Z}_{abeq} \tag{5.7.29}$$

A circuit representation of the series sequence impedances of a completely transposed three-phase line is shown in Figure 5.18.

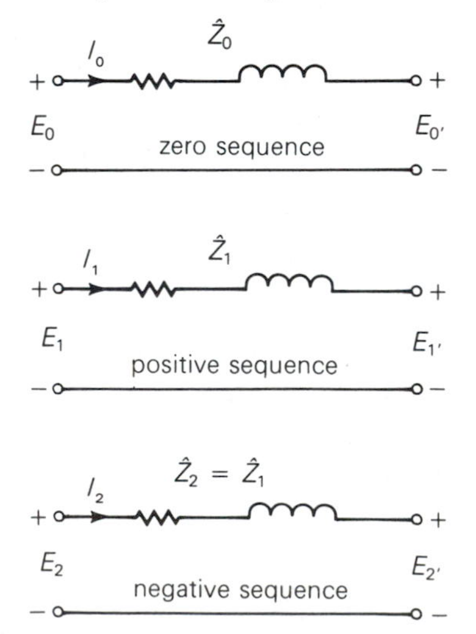

Figure 5.18 Circuit representation of the series sequence impedances of a completely transposed three-phase line

A personal computer program that computes the series phase impedance matrix and series sequence impedance matrix is described in Section 5.14.

Section 5.8

Electric Field and Voltage: Solid Cylindrical Conductor

The capacitance between conductors in a medium with constant permittivity ε can be obtained by determining the following:

1. Electric field strength E, from Gauss's law
2. Voltage between conductors
3. Capacitance from charge per unit volt ($C = q/V$)

As a step toward computing capacitances of general conductor configurations, we first compute the electric field of a uniformly charged, solid cylindrical conductor and the voltage between two points outside the conductor. We also compute the voltage between two conductors in an array of charged conductors.

Gauss's law states that the total electric flux leaving a closed surface equals the total charge within the volume enclosed by the surface. That is, the normal component of electric flux density integrated over a closed surface equals the charge enclosed:

$$\oiint D_\perp \, ds = \oiint \varepsilon E_\perp \, ds = Q_{\text{enclosed}} \tag{5.8.1}$$

where $D_\perp$ denotes the normal component of electric flux density, $E_\perp$ denotes the normal component of electric field strength, and ds denotes the differential surface area. From Gauss's law, electric charge is a source of electric fields. Electric field lines originate from positive charges and terminate at negative charges.

Figure 5.19 shows a solid cylindrical conductor with radius r and with charge q coulombs per meter (assumed positive in the figure), uniformly distributed on the conductor surface. For simplicity, assume that the conductor is (1) sufficiently long that end effects are negligible, and (2) a perfect conductor (that is, zero resistivity, $\rho = 0$).

Inside the perfect conductor, Ohm's law gives $E_{\text{int}} = \rho J = 0$. That is, the internal electric field E_{int} is zero. To determine the electric field outside the conductor, select the cylinder with radius $x > r$ and with 1-meter length, shown in Figure 5.19, as the closed surface for Gauss's law. Due to the uniform charge distribution, the electric field strength E_x is constant on the cylinder. Also, there is no tangential component of E_x, so the electric field is radial to the conductor. Then, integration of (5.8.1) yields

$$\varepsilon E_x(2\pi x)(1) = q(1)$$

$$E_x = \frac{q}{2\pi\varepsilon x} \quad \text{V/m} \tag{5.8.2}$$

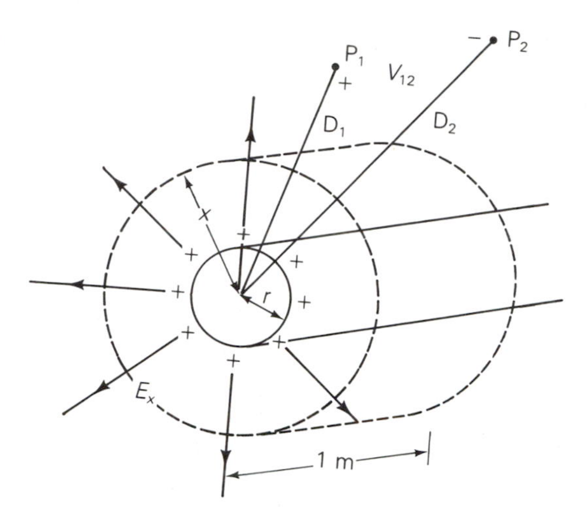

Figure 5.19

Perfectly conducting solid cylindrical conductor with uniform charge distribution

where, for a conductor in free space, $\varepsilon = \varepsilon_0 = 8.854 \times 10^{-12}$ F/m.

A plot of the electric field lines is also shown in Figure 5.19. The direction of the field lines, denoted by the arrows, is from the positive charges where the field originates, to the negative charges, which in this case are at infinity. If the charge on the conductor surface were negative, then the direction of the field lines would be reversed.

Concentric cylinders surrounding the conductor are constant potential surfaces. The potential difference between two concentric cylinders at distances D_1 and D_2 from the conductor center is

$$V_{12} = \int_{D_1}^{D_2} E_x \, dx \tag{5.8.3}$$

Using (5.8.2) in (5.8.1),

$$V_{12} = \int_{D_1}^{D_2} \frac{q}{2\pi\varepsilon x} dx = \frac{q}{2\pi\varepsilon} \ln \frac{D_2}{D_1} \quad \text{volts} \tag{5.8.4}$$

Equation (5.8.4) gives the voltage V_{12} between two points, P_1 and P_2, at distances D_1 and D_2 from the conductor center, as shown in Figure 5.19. Also, in accordance with our notation, V_{12} is the voltage at P_1 with respect to P_2. If q is positive and D_2 is greater than D_1, as shown in the figure, then V_{12} is positive; that is, P_1 is at a higher potential than P_2. Equation (5.8.4) is also valid for either dc or ac. For ac, V_{12} is a phasor voltage and q is a phasor representation of a sinusoidal charge.

Now apply (5.8.4) to the array of M solid cylindrical conductors shown in Figure 5.20. Assume that each conductor m has an ac charge q_m C/m uniformly distributed along the conductor. The voltage V_{kim} between conductors k and i due to the charge q_m acting alone is

$$V_{kim} = \frac{q_m}{2\pi\varepsilon} \ln \frac{D_{im}}{D_{km}} \quad \text{volts} \tag{5.8.5}$$

where $D_{mm} = r_m$ when $k = m$ or $i = m$. In (5.8.5) we have neglected the distortion of the electric field in the vicinity of the other conductors, caused by the fact that the other conductors themselves are constant potential surfaces. V_{kim} can be thought of as the voltage between cylinders with radii D_{km} and D_{im} concentric to conductor m at points on the cylinders remote from conductors, where there is no distortion.

Using superposition, the voltage V_{ki} between conductors k and i due to all the charges is

$$V_{ki} = \frac{1}{2\pi\varepsilon} \sum_{m=1}^{M} q_m \ln \frac{D_{im}}{D_{km}} \quad \text{volts} \tag{5.8.6}$$

Figure 5.20 Array of M solid cylindrical conductors

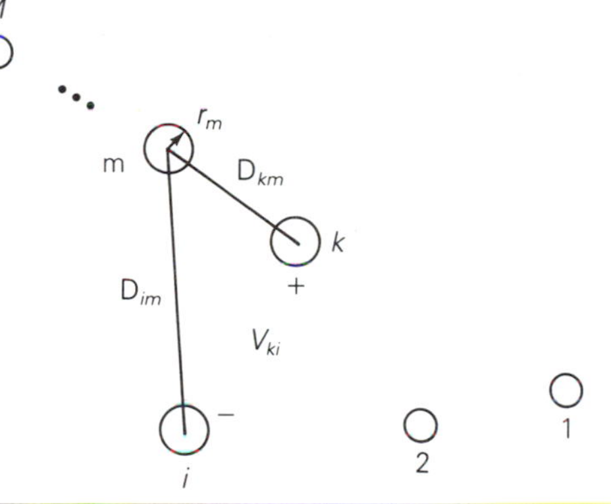

Section 5.9

Capacitance: Single-phase Two-wire Line and Three-phase Three-wire Line with Equal Phase Spacing

The results of the previous section are used here to determine the capacitances of the two relatively simple transmission lines considered in Section 5.5, a single-phase two-wire line and a three-phase three-wire line with equal phase spacing.

First we consider the single-phase two-wire line shown in Figure 5.9. Assume that the conductors are energized by a voltage source such that conductor x has a uniform charge q C/m and, assuming conservation of charge, conductor y has an equal quantity of negative charge $-q$. Using (5.8.6) with $k = x$, $i = y$, and $m = x, y$,

$$V_{xy} = \frac{1}{2\pi\varepsilon}\left[q \ln \frac{D_{yx}}{D_{xx}} - q \ln \frac{D_{yy}}{D_{xy}}\right]$$

$$= \frac{q}{2\pi\varepsilon} \ln \frac{D_{yx}D_{xy}}{D_{xx}D_{yy}} \tag{5.9.1}$$

Using $D_{xy} = D_{yx} = D$, $D_{xx} = r_x$, and $D_{yy} = r_y$, (5.9.1) becomes

$$V_{xy} = \frac{q}{\pi\varepsilon} \ln \frac{D}{\sqrt{r_x r_y}} \quad \text{volts} \tag{5.9.2}$$

For a 1-meter line length, the capacitance between conductors is

$$C_{xy} = \frac{q}{V_{xy}} = \frac{\pi\varepsilon}{\ln\left(\dfrac{D}{\sqrt{r_x r_y}}\right)} \quad \text{F/m line-to-line} \tag{5.9.3}$$

and if $r_x = r_y = r$,

$$C_{xy} = \frac{\pi\varepsilon}{\ln(D/r)} \quad \text{F/m line-to-line} \tag{5.9.4}$$

If the two-wire line is supplied by a transformer with a grounded center tap, then the voltage between each conductor and ground is one-half that given by (5.9.2). That is,

$$V_{xn} = V_{yn} = \frac{V_{xy}}{2} \tag{5.9.5}$$

and the capacitance from either line to the grounded neutral is

$$C_n = C_{xn} = C_{yn} = \frac{q}{V_{xn}} = 2C_{xy}$$

$$= \frac{2\pi\varepsilon}{\ln(D/r)} \quad \text{F/m line-to-neutral} \tag{5.9.6}$$

Circuit representations of the line-to-line and line-to-neutral capacitances are shown in Figure 5.21. Note that if the neutral is open in Figure 5.21(b), the two line-to-neutral capacitances combine in series to give the line-to-line capacitance.

Figure 5.21

Circuit representation of capacitances for a single-phase two-wire line

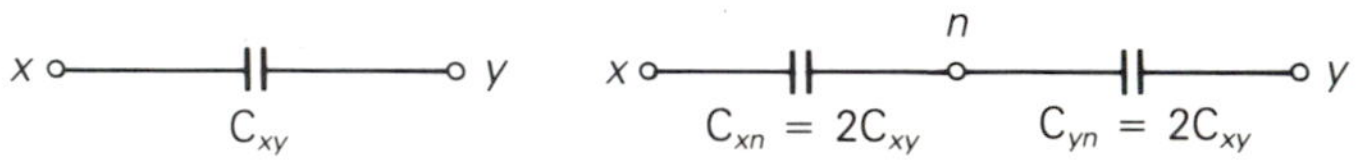

(a) Line-to-line capacitance (b) Line-to-neutral capacitances

Next consider the three-phase line with equal phase spacing shown in Figure 5.10. We shall neglect the effect of earth and neutral conductors here. To determine the positive-sequence capacitance, assume positive-sequence charges q_a, q_b, q_c such that $q_a + q_b + q_c = 0$. Using (5.8.6) with $k = a$, $i = b$, and $m = a, b, c$, the voltage V_{ab} between conductors a and b is

$$V_{ab} = \frac{1}{2\pi\varepsilon}\left[q_a \ln\frac{D_{ba}}{D_{aa}} + q_b \ln\frac{D_{bb}}{D_{ab}} + q_c \ln\frac{D_{bc}}{D_{ac}}\right] \tag{5.9.7}$$

Using $D_{aa} = D_{bb} = r$, and $D_{ab} = D_{ba} = D_{ca} = D_{cb} = D$, (5.9.7) becomes

$$V_{ab} = \frac{1}{2\pi\varepsilon}\left[q_a \ln\frac{D}{r} + q_b \ln\frac{r}{D} + q_c \ln\frac{D}{D}\right]$$

$$= \frac{1}{2\pi\varepsilon}\left[q_a \ln\frac{D}{r} + q_b \ln\frac{r}{D}\right] \quad \text{volts} \tag{5.9.8}$$

Note that the third term in (5.9.8) is zero because conductors a and b are equidistant from conductor c. Thus conductors a and b lie on a constant potential cylinder for the electric field due to q_c.

Similarly, using (5.8.6) with $k = a$, $i = c$, and $m = a, b, c$, the voltage V_{ac} is

$$V_{ac} = \frac{1}{2\pi\varepsilon}\left[q_a \ln\frac{D_{ca}}{D_{aa}} + q_b \ln\frac{D_{cb}}{D_{ab}} + q_c \ln\frac{D_{cc}}{D_{ac}}\right]$$

$$= \frac{1}{2\pi\varepsilon}\left[q_a \ln\frac{D}{r} + q_b \ln\frac{D}{D} + q_c \ln\frac{r}{D}\right]$$

$$= \frac{1}{2\pi\varepsilon}\left[q_a \ln\frac{D}{r} + q_c \ln\frac{r}{D}\right] \quad \text{volts} \tag{5.9.9}$$

Recall that for positive-sequence voltages,

$$V_{ab} = \sqrt{3}\, V_{an}\underline{/+30^\circ} = \sqrt{3}\, V_{an}\left[\frac{\sqrt{3}}{2} + j\frac{1}{2}\right] \tag{5.9.10}$$

$$V_{ac} = -V_{ca} = \sqrt{3}\, V_{an}\underline{/-30^\circ} = \sqrt{3}\, V_{an}\left[\frac{\sqrt{3}}{2} - j\frac{1}{2}\right] \tag{5.9.11}$$

Adding (5.9.10) and (5.9.11) yields

$$V_{ab} + V_{ac} = 3V_{an} \tag{5.9.12}$$

Using (5.9.8) and (5.9.9) in (5.9.12),

$$V_{an} = \frac{1}{3}\left(\frac{1}{2\pi\varepsilon}\right)\left[2q_a \ln\frac{D}{r} + (q_b + q_c)\ln\frac{r}{D}\right] \tag{5.9.13}$$

and with $q_b + q_c = -q_a$,

$$V_{an} = \frac{1}{2\pi\varepsilon} q_a \ln\frac{D}{r} \quad \text{volts} \tag{5.9.14}$$

The capacitance-to-neutral per line length is

$$C_{an} = \frac{q_a}{V_{an}} = \frac{2\pi\varepsilon}{\ln(D/r)} \quad \text{F/m line-to-neutral} \tag{5.9.15}$$

Due to symmetry the same result is obtained for $C_{bn} = q_b/V_{bn}$ and $C_{cn} = q_c/V_{cn}$. For positive sequence, however, only one phase need be considered. Thus

(5.9.15) gives the positive-sequence capacitance, which also equals the negative-sequence capacitance:

$$C_1 = C_2 = \frac{2\pi\varepsilon}{\ln(D/r)} \quad \text{F/m} \tag{5.9.16}$$

Circuit representations of the sequence capacitances are shown in Figure 5.22.

To determine the zero-sequence capacitance C_0, we must consider the neutral conductors and earth. The neutral conductors and earth also affect the positive- (and negative-) sequence capacitance, since they alter the electric field. We develop equations for the sequence capacitance matrix of more general three-phase lines in Section 5.11.

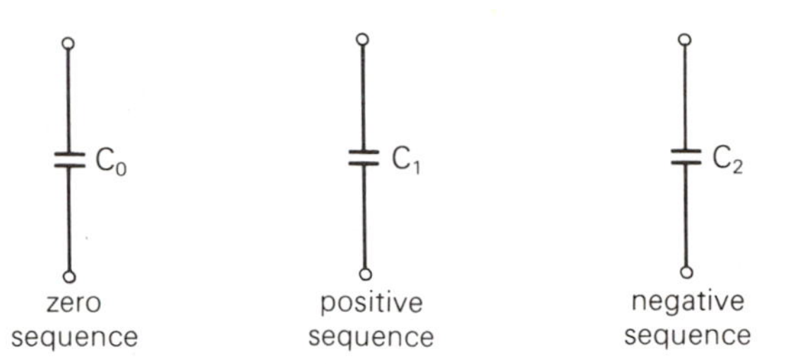

Figure 5.22 Circuit representation of the sequence capacitances of a three-phase line with equal phase spacing

Section 5.10

Capacitance: Stranded Conductors, Unequal Phase Spacing, Bundled Conductors

Equations (5.9.6) and (5.9.16) are based on the assumption that the conductors are solid cylindrical conductors with zero resistivity. The electric field inside these conductors is zero, and the external electric field is perpendicular to the conductor surfaces. Practical conductors with resistivities similar to those listed in Table 5.3 have a small internal electric field. As a result, the external electric field is slightly altered near the conductor surfaces. Also, the electric field near the surface of a stranded conductor is not the same as that of a solid cylindrical conductor. However, it is normal practice when calculating line capacitance to replace a stranded conductor by a perfectly conducting solid cylindrical conductor whose radius equals the outside radius of the stranded conductor. The resulting error in capacitance is small since only the electric field near the conductor surfaces is affected.

Also, (5.8.2) is based on the assumption that there is uniform charge distribution. But conductor charge distribution is nonuniform in the presence of other charged conductors. Therefore (5.9.6) and (5.9.16), which are derived from (5.8.2), are not exact. However, the nonuniformity of conductor charge distribution can be shown to have a negligible effect on line capacitance.

For three-phase lines with unequal phase spacing, positive-sequence voltages are not obtained with positive-sequence charges. Instead, unbalanced line-to-neutral voltages occur, and the phase-to-neutral capacitances are unequal. Balance can be restored by transposing the line such that each phase occupies each position for one-third of the line length. If equations similar to (5.9.7) for V_{ab} as well as for V_{ac} are written for each position in the transposition cycle, and are then averaged and used in (5.9.12)–(5.9.14), the resulting positive- and negative-sequence capacitance becomes

$$C_1 = C_2 = \frac{2\pi\varepsilon}{\ln(D_{eq}/r)} \quad \text{F/m} \tag{5.10.1}$$

where

$$D_{eq} = \sqrt[3]{D_{ab}D_{bc}D_{ac}} \tag{5.10.2}$$

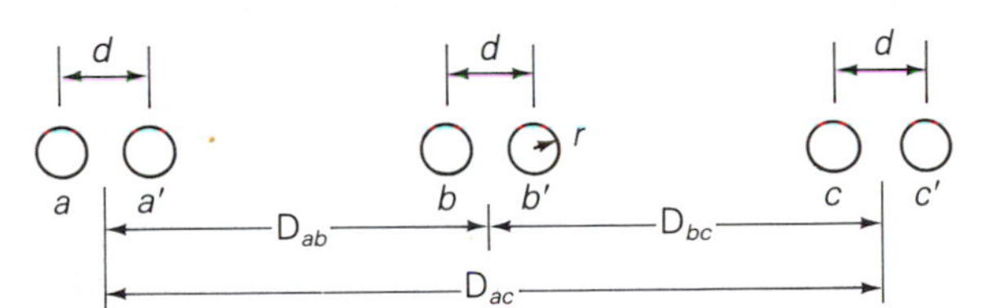

Figure 5.23

Three-phase line with two conductors per bundle

Figure 5.23 shows a bundled conductor line with two conductors per bundle. To determine the positive-sequence capacitance of this line, assume positive-sequence charges q_a, q_b, q_c for each phase such that $q_a + q_b + q_c = 0$. Assume that the conductors in each bundle, which are in parallel, share the charges equally. Thus conductors a and a' each have the charge $q_a/2$. Also assume that the phase spacings are much larger than the bundle spacings so that D_{ab} may be used instead of $(D_{ab} - d)$ or $(D_{ab} + d)$. Then, using (5.8.6) with $k = a$, $i = b$, $m = a, a', b, b', c, c'$,

$$\begin{aligned} V_{ab} &= \frac{1}{2\pi\varepsilon}\left[\frac{q_a}{2}\ln\frac{D_{ba}}{D_{aa}} + \frac{q_a}{2}\ln\frac{D_{ba'}}{D_{aa'}} + \frac{q_b}{2}\ln\frac{D_{bb}}{D_{ab}}\right. \\ &\qquad \left. + \frac{q_b}{2}\ln\frac{D_{bb'}}{D_{ab'}} + \frac{q_c}{2}\ln\frac{D_{bc}}{D_{ac}} + \frac{q_c}{2}\ln\frac{D_{bc'}}{D_{ac'}}\right] \\ &= \frac{1}{2\pi\varepsilon}\left[\frac{q_a}{2}\left(\ln\frac{D_{ab}}{r} + \ln\frac{D_{ab}}{d}\right) + \frac{q_b}{2}\left(\ln\frac{r}{D_{ab}} + \ln\frac{d}{D_{ab}}\right)\right. \\ &\qquad \left. + \frac{q_c}{2}\left(\ln\frac{D_{bc}}{D_{ac}} + \ln\frac{D_{bc}}{D_{ac}}\right)\right] \\ &= \frac{1}{2\pi\varepsilon}\left[q_a\ln\frac{D_{ab}}{\sqrt{rd}} + q_b\ln\frac{\sqrt{rd}}{D_{ab}} + q_c\ln\frac{D_{bc}}{D_{ac}}\right] \end{aligned} \tag{5.10.3}$$

Equation (5.10.3) is the same as (5.9.7), except that D_{aa} and D_{bb} in (5.9.7) are replaced by $\sqrt{rd}$ in this equation. Therefore, for a transposed line, derivation of the positive- and negative-sequence capacitance would yield

$$C_1 = C_2 = \frac{2\pi\varepsilon}{\ln(D_{eq}/D_{SC})} \quad \text{F/m} \tag{5.10.4}$$

where

$$D_{SC} = \sqrt{rd} \qquad \text{for a two-conductor bundle} \tag{5.10.5}$$

Similarly,

$$D_{SC} = \sqrt[3]{rd^2} \qquad \text{for a three-conductor bundle} \tag{5.10.6}$$

$$D_{SC} = 1.091\sqrt[4]{rd^3} \qquad \text{for a four-conductor bundle} \tag{5.10.7}$$

Equation (5.10.4) for capacitance is analogous to (5.6.22) for inductance. In both cases D_{eq}, given by (5.6.17) or (5.10.2), is the geometric mean of the distances between phases. Also, (5.10.5)–(5.10.7) for D_{SC} are analogous to (5.6.19)–(5.6.21) for D_{SL}, except that the conductor outside radius r replaces the conductor GMR D_S.

The current supplied to the transmission-line capacitance is called *charging current*. For a single-phase circuit operating at line-to-line voltage $V_{xy} = V_{xy}\underline{/0°}$, the charging current is

$$I_{chg} = Y_{xy}V_{xy} = j\omega C_{xy}V_{xy} \quad \text{A} \tag{5.10.8}$$

As shown in Chapter 2, a capacitor delivers reactive power. From (2.3.5), the reactive power delivered by the line-to-line capacitance is

$$Q_C = \frac{V_{xy}^2}{X_c} = Y_{xy}V_{xy}^2 = \omega C_{xy}V_{xy}^2 \quad \text{var} \tag{5.10.9}$$

For a completely transposed three-phase line that has positive-sequence voltages with $V_{an} = V_{LN}\underline{/0°}$, the phase a charging current is

$$I_{chg} = YV_{an} = j\omega C_1 V_{LN} \quad \text{A} \tag{5.10.10}$$

and the reactive power delivered by phase a is

$$Q_{C1\phi} = YV_{an}^2 = \omega C_1 V_{LN}^2 \quad \text{var} \tag{5.10.11}$$

The total reactive power supplied by the three-phase line is

$$Q_{C3\phi} = 3Q_{C1\phi} = 3\omega C_1 V_{LN}^2 = \omega C_1 V_{LL}^2 \quad \text{var} \tag{5.10.12}$$

Example 5.6 For the single-phase line in Example 5.3, determine the line-to-line capacitance in F and the line-to-line admittance in S. If the line voltage is 20 kV, determine the reactive power in kvar supplied by this capacitance.

Solution From Table A.3, the outside radius of a 4/0 12-strand copper conductor is

$$r = \frac{0.552}{2}\ \text{in.} \times \frac{1\ \text{ft}}{12\ \text{in.}} = 0.023\ \text{ft}$$

and from (5.9.4),

$$C_{xy} = \frac{\pi(8.854 \times 10^{-12})}{\ln\left(\frac{5}{0.023}\right)} = 5.169 \times 10^{-12} \quad \text{F/m}$$

or

$$C_{xy} = 5.169 \times 10^{-12} \frac{\text{F}}{\text{m}} \times 1609 \frac{\text{m}}{\text{mi}} \times 20 \text{ mi} = 1.66 \times 10^{-7} \quad \text{F}$$

and the shunt admittance is

$$\begin{aligned} Y_{xy} = j\omega C_{xy} &= j(2\pi 60)(1.66 \times 10^{-7}) \\ &= j6.27 \times 10^{-5} \quad \text{S line-to-line} \end{aligned}$$

From (5.10.9),

$$Q_C = (6.27 \times 10^{-5})(20 \times 10^3)^2 = 25.1 \quad \text{kvar}$$

◁

Example 5.7 For the three-phase line in Example 5.5, determine the positive-sequence capacitance in F and the shunt admittance in S. If the line voltage is 345 kV, determine the charging current in kA per phase and the total reactive power in Mvar supplied by the line capacitance. Assume positive-sequence voltages.

Solution From Table A.4, the outside radius of a 795,000 cmil 26/2 ACSR conductor is

$$r = \frac{1.108}{2} \text{ in.} \times 0.0254 \frac{\text{m}}{\text{in.}} = 0.0141 \quad \text{m}$$

From (5.10.5), the equivalent radius of the two-conductor bundle is

$$D_{SC} = \sqrt{(0.0141)(0.40)} = 0.0750 \quad \text{m}$$

$D_{eq} = 12.6$ m is the same as in Example 5.5. Therefore, from (5.10.4),

$$\begin{aligned} C_1 &= \frac{(2\pi)(8.854 \times 10^{-12})}{\ln\left(\dfrac{12.6}{0.0750}\right)} \frac{\text{F}}{\text{m}} \times 1000 \frac{\text{m}}{\text{km}} \times 200 \text{ km} \\ &= 2.17 \times 10^{-6} \quad \text{F} \end{aligned}$$

The positive-sequence shunt admittance is

$$\begin{aligned} Y_1 = j\omega C_1 &= j(2\pi 60)(2.17 \times 10^{-6}) \\ &= j8.19 \times 10^{-4} \quad \text{S} \end{aligned}$$

From (5.10.10),

$$I_{chg} = |I_{chg}| = (8.19 \times 10^{-4})\left(\frac{345}{\sqrt{3}}\right) = 0.163 \quad \text{kA/phase}$$

and from (5.10.12),

$$Q_{C3\phi} = (8.19 \times 10^{-4})(345)^2 = 97.5 \quad \text{Mvar}$$

◁

Section 5.11

Shunt Admittances: Lines with Neutral Conductors and Earth Return

In this section we develop equations suitable for computer calculation of the shunt admittances for the three-phase line shown in Figure 5.16. We approximate the earth surface as a perfectly conducting horizontal plane, even though the earth under the line may have irregular terrain and resistivities as shown in Table 5.4.

The effect of the earth plane is accounted for by the *method of images*, described as follows. Consider a single conductor with uniform charge distribution and with height H above a perfectly conducting earth plane, as shown in Figure 5.24(a). When the conductor has a positive charge, an equal quantity of negative charge is induced on the earth. The electric field lines will originate from the positive charges on the conductor and terminate at the negative charges on the earth. Also, the electric field lines are perpendicular to the surfaces of the conductor and earth.

Now replace the earth by the image conductor shown in Figure 5.24(b), which has the same radius as the original conductor, lies directly below the original conductor with conductor separation $H_{11} = 2H$, and has an equal quantity of negative charge. The electric field above the dotted line representing the location of the removed earth plane in Figure 5.24(b) is identical to the electric field above the earth plane in Figure 5.24(a). Therefore, the voltage between any two points above the earth is the same in both figures.

Figure 5.24

Method of images

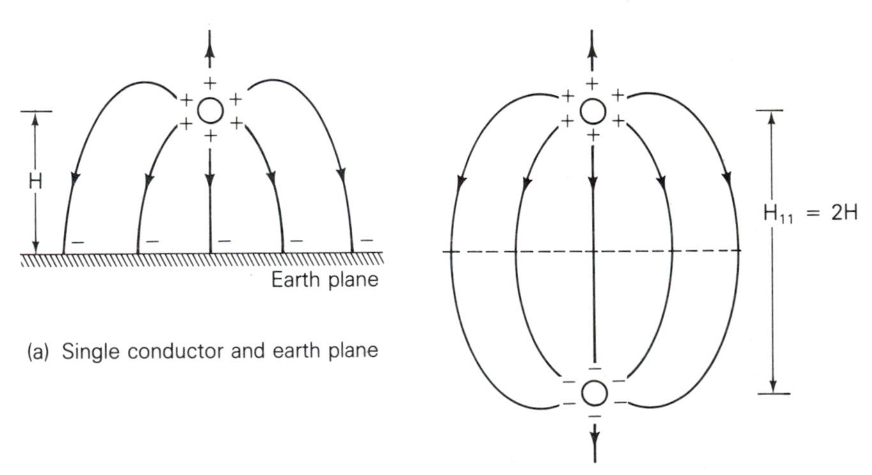

(a) Single conductor and earth plane

(b) Earth plane replaced by image conductor

Example 5.8

If the single-phase line in Example 5.6 has flat horizontal spacing with 18-ft average line height, determine the effect of the earth on capacitance. Assume a perfectly conducting earth plane.

Solution The earth plane is replaced by a separate image conductor for each overhead conductor, and the conductors are charged as shown in Figure 5.25. From (5.8.6), the voltage between conductors x and y is

$$V_{xy} = \frac{q}{2\pi\varepsilon}\left[\ln\frac{D_{yx}}{D_{xx}} - \ln\frac{D_{yy}}{D_{xy}} - \ln\frac{H_{yx}}{H_{xx}} + \ln\frac{H_{yy}}{H_{xy}}\right]$$

$$= \frac{q}{2\pi\varepsilon}\left[\ln\frac{D_{yx}D_{xy}}{D_{xx}D_{yy}} - \ln\frac{H_{yx}H_{xy}}{H_{xx}H_{yy}}\right]$$

$$= \frac{q}{\pi\varepsilon}\left[\ln\frac{D}{r} - \ln\frac{H_{xy}}{H_{xx}}\right]$$

The line-to-line capacitance is

$$C_{xy} = \frac{q}{V_{xy}} = \frac{\pi\varepsilon}{\ln\dfrac{D}{r} - \ln\dfrac{H_{xy}}{H_{xx}}} \quad \text{F/m}$$

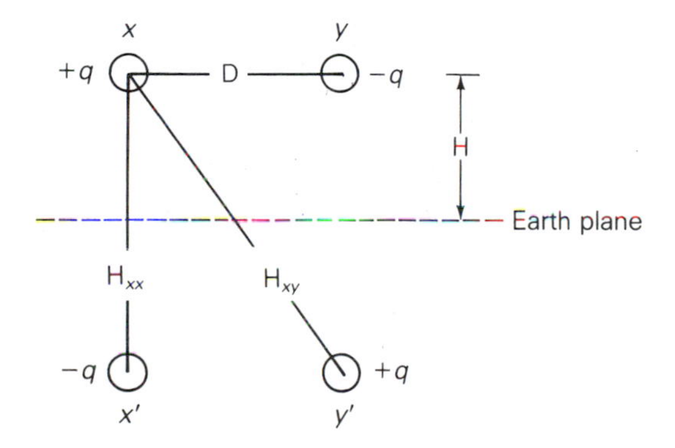

Figure 5.25 Single-phase line for Example 5.8

Using $D = 5\,\text{ft}$, $r = 0.023\,\text{ft}$, $H_{xx} = 2H = 36\,\text{ft}$, and $H_{xy} = \sqrt{(36)^2 + (5)^2} = 36.346\,\text{ft}$,

$$C_{xy} = \frac{\pi(8.854 \times 10^{-12})}{\ln\dfrac{5}{0.023} - \ln\dfrac{36.346}{36}} = 5.178 \times 10^{-12} \quad \text{F/m}$$

compared with 5.169×10^{-12} F/m in Example 5.6. The effect of the earth plane is to slightly increase the capacitance. Note that as the line height H increases, the ratio H_{xy}/H_{xx} approaches 1, $\ln(H_{xy}/H_{xx}) \to 0$, and the effect of the earth becomes negligible. ◁

For the three-phase line with N neutral conductors shown in Figure 5.26, the perfectly conducting earth plane is replaced by a separate image conductor for each overhead conductor. The overhead conductors a, b, c, $n1$, $n2, \ldots, nN$

carry charges $q_a, q_b, q_c, q_{n1}, \ldots, q_{nN}$, and the image conductors $a', b', c', n1', \ldots, nN'$ carry charges $-q_a, -q_b, -q_c, -q_{n1}, \ldots, -q_{nN}$. Applying (5.8.6) to determine the voltage $V_{kk'}$ between any conductor k and its image conductor k',

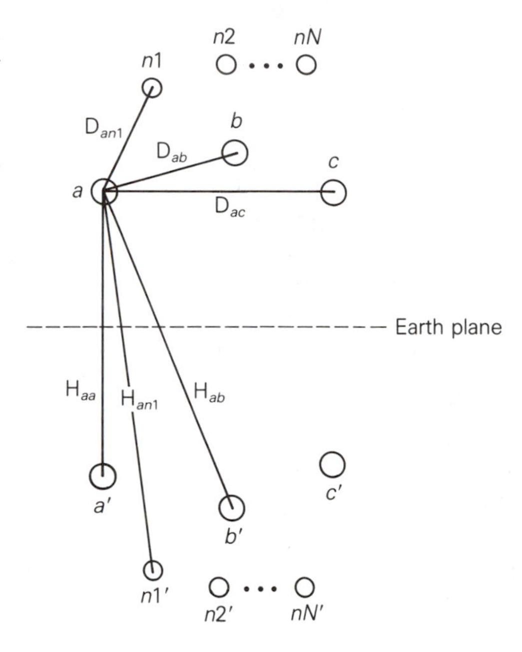

Figure 5.26 Three-phase line with neutral conductors and with earth plane replaced by image conductors

$$V_{kk'} = \frac{1}{2\pi\varepsilon}\left[\sum_{m=a}^{nN} q_m \ln\frac{\mathrm{H}_{km}}{\mathrm{D}_{km}} - \sum_{m=a}^{nN} q_m \ln\frac{\mathrm{D}_{km}}{\mathrm{H}_{km}}\right]$$

$$= \frac{2}{2\pi\varepsilon}\sum_{m=a}^{nN} q_m \ln\frac{\mathrm{H}_{km}}{\mathrm{D}_{km}} \tag{5.11.1}$$

where $\mathrm{D}_{kk} = r_k$ and D_{km} is the distance between overhead conductors k and m. H_{km} is the distance between overhead conductor k and image conductor m. By symmetry, the voltage V_{kn} between conductor k and the earth is one-half of $V_{kk'}$.

$$V_{kn} = \frac{1}{2}V_{kk'} = \frac{1}{2\pi\varepsilon}\sum_{m=a}^{nN} q_m \ln\frac{\mathrm{H}_{km}}{\mathrm{D}_{km}} \tag{5.11.2}$$

where

$$k = a, b, c, n1, n2, \ldots, nN$$

$$m = a, b, c, n1, n2, \ldots, nN$$

Since all the neutral conductors are grounded to the earth,

$$V_{kn} = 0 \qquad \text{for } k = n1, n2, \ldots, nN \tag{5.11.3}$$

In matrix format, (5.11.2) and (5.11.3) are

$$\begin{bmatrix} V_{an} \\ V_{bn} \\ V_{cn} \\ \hline 0 \\ \vdots \\ 0 \end{bmatrix} = \left[\begin{array}{ccc|ccc} \overbrace{P_{aa} \quad P_{ab} \quad P_{ac}}^{\mathbf{P}_A} & & & \overbrace{P_{an1} \quad \cdots \quad P_{anN}}^{\mathbf{P}_B} & & \\ P_{ba} & P_{bb} & P_{bc} & P_{bn1} & \cdots & P_{bnN} \\ P_{ca} & P_{cb} & P_{cc} & P_{cn1} & \cdots & P_{cnN} \\ \hline P_{n1a} & P_{n1b} & P_{n1c} & P_{n1n1} & \cdots & P_{n1nN} \\ \vdots & & & & & \\ \underbrace{P_{nNa} \quad P_{nNb} \quad P_{nNc}}_{\mathbf{P}_C} & & & \underbrace{P_{nNn1} \quad \cdots \quad P_{nNnN}}_{\mathbf{P}_D} & & \end{array}\right] \begin{bmatrix} q_a \\ q_b \\ q_c \\ \hline q_{n1} \\ \vdots \\ q_{nN} \end{bmatrix} \tag{5.11.4}$$

The elements of the $(3 + N) \times (3 + N)$ matrix $\mathbf{P}$ are

$$P_{km} = \frac{1}{2\pi\varepsilon} \ln \frac{H_{km}}{D_{km}} \quad \text{m/F} \tag{5.11.5}$$

where

$$k = a, b, c, n1, \ldots, nN$$

$$m = a, b, c, n1, \ldots, nN$$

Equation (5.11.4) is now partitioned as shown above to obtain

$$\begin{bmatrix} \boldsymbol{V}_P \\ \hline \mathbf{0} \end{bmatrix} = \left[\begin{array}{c|c} \mathbf{P}_A & \mathbf{P}_B \\ \hline \mathbf{P}_C & \mathbf{P}_D \end{array}\right] \begin{bmatrix} \boldsymbol{q}_P \\ \hline \boldsymbol{q}_n \end{bmatrix} \tag{5.11.6}$$

$\boldsymbol{V}_P$ is the 3 vector of phase-to-neutral voltages. $\boldsymbol{q}_P$ is the 3 vector of phase-conductor charges and $\boldsymbol{q}_n$ is the N vector of neutral conductor charges. The $(3 + N) \times (3 + N)$ $\mathbf{P}$ matrix is partitioned as shown in (5.11.4) to obtain:

$\mathbf{P}_A$ with dimension 3×3

$\mathbf{P}_B$ with dimension $3 \times N$

$\mathbf{P}_C$ with dimension $N \times 3$

$\mathbf{P}_D$ with dimension $N \times N$

Equation (5.11.6) is rewritten as two separate equations:

$$\boldsymbol{V}_P = \mathbf{P}_A \boldsymbol{q}_P + \mathbf{P}_B \boldsymbol{q}_n \tag{5.11.7}$$

$$\mathbf{0} = \mathbf{P}_C \boldsymbol{q}_P + \mathbf{P}_D \boldsymbol{q}_n \tag{5.11.8}$$

Then (5.11.8) is solved for q_n, which is used in (5.11.7) to obtain

$$V_P = (\mathbf{P}_A - \mathbf{P}_B\mathbf{P}_D^{-1}\mathbf{P}_C)\, q_P \tag{5.11.9}$$

or

$$q_P = \mathbf{C}_P V_P \tag{5.11.10}$$

where

$$\mathbf{C}_P = (\mathbf{P}_A - \mathbf{P}_B\mathbf{P}_D^{-1}\mathbf{P}_C)^{-1} \quad \text{F/m} \tag{5.11.11}$$

Equation (5.11.10), the desired result, relates the phase-conductor charges to the phase-to-neutral voltages. $\mathbf{C}_P$ is the 3×3 matrix of phase capacitances whose elements are denoted

$$\mathbf{C}_P = \begin{bmatrix} C_{aa} & C_{ab} & C_{ac} \\ C_{ab} & C_{bb} & C_{bc} \\ C_{ac} & C_{bc} & C_{cc} \end{bmatrix} \quad \text{F/m} \tag{5.11.12}$$

It can be shown that $\mathbf{C}_P$ is a symmetric matrix whose diagonal terms C_{aa}, C_{bb}, C_{cc} are positive, and whose off-diagonal terms C_{ab}, C_{bc}, C_{ac} are negative. This indicates that when a positive line-to-neutral voltage is applied to one phase, a positive charge is induced on that phase and negative charges are induced on the other phases, which is physically correct.

In general, $\mathbf{C}_P$ does not satisfy the conditions for a symmetrical capacitance matrix. However, if the line is completely transposed, the diagonal and off-diagonal elements of $\mathbf{C}_P$ are averaged to obtain

$$\hat{\mathbf{C}}_P = \begin{bmatrix} \hat{C}_{aa} & \hat{C}_{ab} & \hat{C}_{ab} \\ \hat{C}_{ab} & \hat{C}_{aa} & \hat{C}_{ab} \\ \hat{C}_{ab} & \hat{C}_{ab} & \hat{C}_{aa} \end{bmatrix} \quad \text{F/m} \tag{5.11.13}$$

where

$$\hat{C}_{aa} = \tfrac{1}{3}(C_{aa} + C_{bb} + C_{cc}) \quad \text{F/m} \tag{5.11.14}$$

$$\hat{C}_{ab} = \tfrac{1}{3}(C_{ab} + C_{bc} + C_{ac}) \quad \text{F/m} \tag{5.11.15}$$

$\hat{\mathbf{C}}_P$ is a symmetrical capacitance matrix.

The shunt phase admittance matrix is given by

$$Y_P = j\omega \mathbf{C}_P = j(2\pi f)\mathbf{C}_P \quad \text{S/m} \tag{5.11.16}$$

or, for a completely transposed line,

$$\hat{Y}_P = j\omega \hat{\mathbf{C}}_P = j(2\pi f)\hat{\mathbf{C}}_P \quad \text{S/m} \tag{5.11.17}$$

Equation (5.11.16) can also be transformed to the sequence domain to obtain

$$Y_S = A^{-1} Y_P A \tag{5.11.18}$$

where

$$Y_S = \mathbf{G}_S + j(2\pi f)\mathbf{C}_S \tag{5.11.19}$$

$$\mathbf{C}_S = \begin{bmatrix} C_0 & C_{01} & C_{02} \\ C_{10} & C_1 & C_{12} \\ C_{20} & C_{21} & C_2 \end{bmatrix} \quad \text{F/m} \tag{5.11.20}$$

In general, $\mathbf{C}_S$ is not diagonal. However, for the completely transposed line,

$$\hat{Y}_S = A^{-1}\hat{Y}_P A = \begin{bmatrix} \hat{y}_0 & 0 & 0 \\ 0 & \hat{y}_1 & 0 \\ 0 & 0 & \hat{y}_2 \end{bmatrix} = j(2\pi f)\begin{bmatrix} \hat{C}_0 & 0 & 0 \\ 0 & \hat{C}_1 & 0 \\ 0 & 0 & \hat{C}_2 \end{bmatrix} \tag{5.11.21}$$

where

$$\hat{C}_0 = \hat{C}_{aa} + 2\hat{C}_{ab} \quad \text{F/m} \tag{5.11.22}$$

$$\hat{C}_1 = \hat{C}_2 = \hat{C}_{aa} - \hat{C}_{ab} \quad \text{F/m} \tag{5.11.23}$$

Since $\hat{C}_{ab}$ is negative, the zero-sequence capacitance $\hat{C}_0$ is usually much less than the positive- or negative-sequence capacitance.

Circuit representations of the phase and sequence capacitances of a completely transposed three-phase line are shown in Figure 5.27. A personal computer program that computes the shunt admittance matrices is described in Section 5.14.

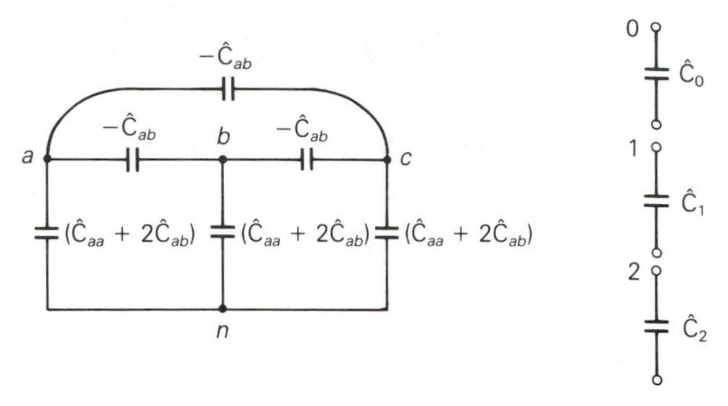

Figure 5.27

Circuit representations of the capacitances of a completely transposed three-phase line

Section 5.12

Electric Field Strength at Conductor Surfaces and at Ground Level

When the electric field strength at a conductor surface exceeds the breakdown strength of air, current discharges occur. This phenomenon, called corona, causes additional line losses (corona loss), communications interference, and audible noise. Although breakdown strength depends on many factors, a rough value is 30 kV/cm in a uniform electric field for dry air at atmospheric pressure. The presence of water droplets or rain can lower this value significantly. To control corona, transmission lines are usually designed to maintain calculated values of conductor surface electric field strength below 20 kV_{rms}/cm.

When line capacitances are determined and conductor voltages are known, the conductor charges can be calculated from (5.9.3) for a single-phase line or from (5.11.10) for a three-phase line. Then the electric field strength at the surface of one phase conductor, neglecting the electric fields due to charges on other phase conductors and neutral wires, is, from (5.8.2),

$$E_r = \frac{q}{2\pi\varepsilon r} \quad \text{V/m} \tag{5.12.1}$$

where r is the conductor outside radius.

For bundled conductors with N_b conductors per bundle and with charge q C/m per phase, the charge per conductor is q/N_b and

$$E_{r\text{ave}} = \frac{q/N_b}{2\pi\varepsilon r} \quad \text{V/m} \tag{5.12.2}$$

Equation (5.12.2) represents an average value for an individual conductor in a bundle. The maximum electric field strength at the surface of one conductor due to all charges in a bundle, obtained by the vector addition of electric fields (as shown in Figure 5.28), is as follows:

Two-conductor bundle ($N_b = 2$):

$$E_{r\max} = \frac{q/2}{2\pi\varepsilon r} + \frac{q/2}{2\pi\varepsilon d} = \frac{q/2}{2\pi\varepsilon r}\left(1 + \frac{r}{d}\right)$$

$$= E_{r\text{ave}}\left(1 + \frac{r}{d}\right) \tag{5.12.3}$$

Three-conductor bundle ($N_b = 3$):

$$E_{r\max} = \frac{q/3}{2\pi\varepsilon}\left(\frac{1}{r} + \frac{2\cos 30^\circ}{d}\right) = E_{r\text{ave}}\left(1 + \frac{r\sqrt{3}}{d}\right) \tag{5.12.4}$$

Four-conductor bundle ($N_b = 4$):

$$E_{r\max} = \frac{q/4}{2\pi\varepsilon}\left(\frac{1}{r} + \frac{1}{d\sqrt{2}} + \frac{2\cos 45^\circ}{d}\right) = E_{r\text{ave}}\left[1 + \frac{r}{d}(2.1213)\right] \tag{5.12.5}$$

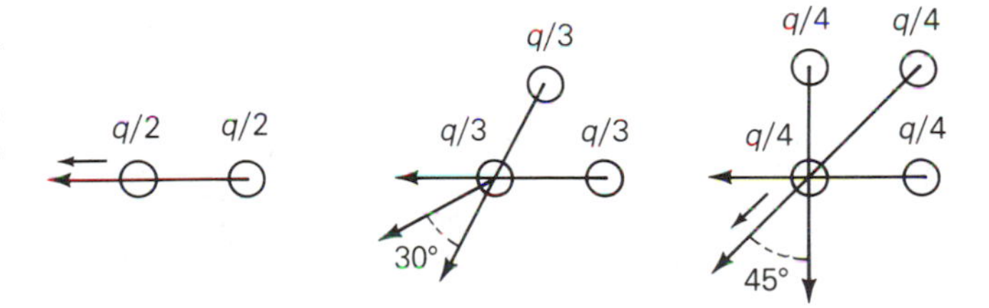

Figure 5.28 Vector addition of electric fields at the surface of one conductor in a bundle

Although the electric field strength at ground level is much less than at conductor surfaces where corona occurs, there are still capacitive coupling effects. Charges are induced on ungrounded equipment such as vehicles with rubber tires located near a line. If a person contacts the vehicle and ground, a discharge current will flow to ground. Transmission-line heights are designed to maintain discharge currents below prescribed levels for any equipment that may be on the right-of-way. Table 5.5 shows examples of maximum ground-level electric field strength.

As shown in Figure 5.29, the ground-level electric field strength due to charged conductor k and its image conductor is perpendicular to the earth plane, with value

$$E_k(w) = \left(\frac{q_k}{2\pi\varepsilon}\right)\frac{2\cos\theta}{\sqrt{y_k^2 + (w - x_k)^2}}$$

$$= \left(\frac{q_k}{2\pi\varepsilon}\right)\frac{2y_k}{y_k^2 + (w - x_k)^2} \quad \text{V/m} \tag{5.12.6}$$

where (x_k, y_k) are the horizontal and vertical coordinates of conductor k with respect to reference point R, w is the horizontal coordinate of the ground-level point where the electric field strength is to be determined, and q_k is the charge on conductor k. The total ground-level electric field is the phasor sum of terms $E_k(w)$ for all overhead conductors. A lateral profile of ground-level electric field strength is obtained by varying w from the center of the line to the edge of the right-of-way.

Figure 5.29

Ground-level electric field strength due to an overhead conductor and its image

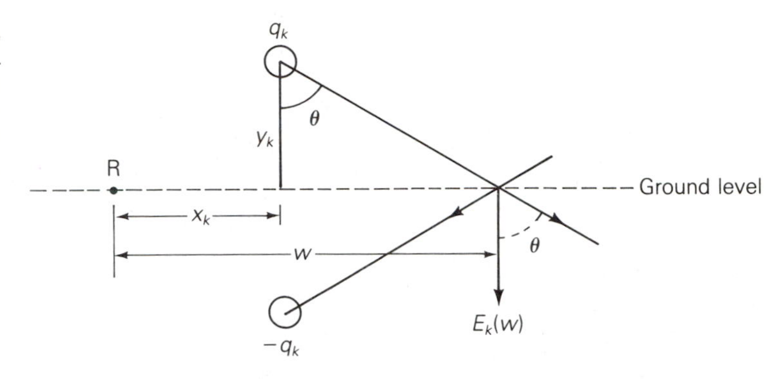

Table 5.5

Examples of maximum ground-level electric field strength versus transmission-line voltage [1]

LINE VOLTAGE kV_{rms}	MAXIMUM GROUND-LEVEL ELECTRIC FIELD STRENGTH kV_{rms}/m
23 (1ϕ)	0.01–0.025
23 (3ϕ)	0.01–0.05
115	0.1 –0.2
345	2.3 –5.0
345 (double circuit)	5.6
500	8.0
765	10.0

Example 5.9

For the single-phase line of Example 5.8, calculate the conductor surface electric field strength in kV_{rms}/cm. Also calculate the ground-level electric field in kV_{rms}/m directly under conductor x. The line voltage is 20 kV.

Solution

From Example 5.8, $C_{xy} = 5.178 \times 10^{-12}$ F/m. Using (5.9.3) with $V_{xy} = 20\underline{/0^\circ}$ kV,

$$q_x = -q_y = (5.178 \times 10^{-12})(20 \times 10^3\underline{/0^\circ}) = 1.036 \times 10^{-7}\underline{/0^\circ} \quad \text{C/m}$$

From (5.12.1), the conductor surface electric field strength is, with $r = 0.023\text{ ft} = 0.00701\text{ m}$,

$$E_r = \frac{1.036 \times 10^{-7}}{(2\pi)(8.854 \times 10^{-12})(0.00701)}\frac{\text{V}}{\text{m}} \times \frac{\text{kV}}{1000\text{ V}} \times \frac{\text{m}}{100\text{ cm}}$$

$$= 2.66 \quad kV_{rms}/cm$$

Selecting the center of the line as the reference point R, the coordinates (x_x, y_x) for conductor x are $(-2.5\,\text{ft}, 18\,\text{ft})$ or $(-0.762\,\text{m}, 5.49\,\text{m})$ and $(+0.762\,\text{m}, 5.49\,\text{m})$ for conductor y. The ground-level electric field directly under conductor x, where $w = -0.762\,\text{m}$, is, from (5.12.6),

$$E(-0.762) = E_x(-0.762) + E_y(-0.762)$$

$$= \frac{1.036 \times 10^{-7}}{(2\pi)(8.85 \times 10^{-12})}\left[\frac{(2)(5.49)}{(5.49)^2} - \frac{(2)(5.49)}{(5.49)^2 + (0.762 + 0.762)^2}\right]$$

$$= 1.862 \times 10^3(0.364 - 0.338) = 48.5\underline{/0^\circ} \quad \text{V/m} = 0.0485 \quad \text{kV/m}$$

For this 20-kV line, the electric field strengths at the conductor surface and at ground level are low enough to be of relatively small concern. For EHV lines, electric field strengths and the possibility of corona and shock hazard are of more concern. ◁

Section 5.13

Parallel Circuit Three-phase Lines

If two parallel three-phase circuits are close together, either on the same tower as in Figure 5.3, or on the same right-of-way, there are mutual inductive and capacitive couplings between the two circuits. When calculating the equivalent series impedance and shunt admittance matrices, these couplings should not be neglected unless the spacing between the circuits is large.

Consider the double-circuit line shown in Figure 5.30. For simplicity, assume that the lines are not transposed. Since both are connected in parallel, they have the same series-voltage drop for each phase. Following the same procedure as in Section 5.7, we can write $2(6 + N)$ equations similar to (5.7.6)–(5.7.9): six equations for the overhead phase conductors, N equations for the overhead neutral conductors, and $(6 + N)$ equations for the earth return conductors. After lumping the neutral voltage drop into the voltage drops across the phase conductors, and eliminating the neutral and earth return currents, we obtain

$$\begin{bmatrix} \boldsymbol{E}_P \\ \boldsymbol{E}_P \end{bmatrix} = \boldsymbol{Z}_P \begin{bmatrix} \boldsymbol{I}_{P1} \\ \boldsymbol{I}_{P2} \end{bmatrix} \tag{5.13.1}$$

where $\boldsymbol{E}_P$ is the vector of phase-conductor voltage drops (including the neutral

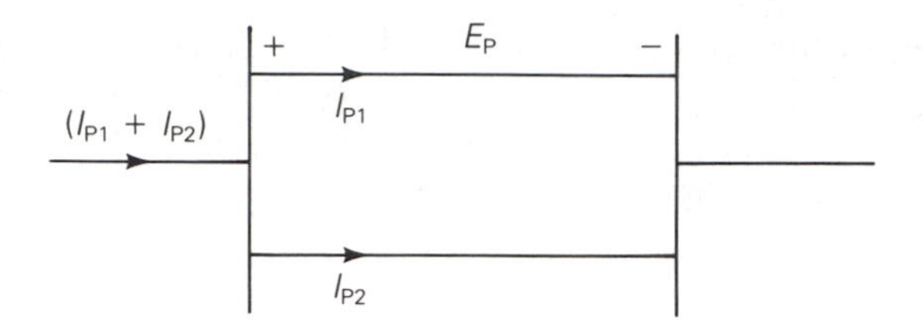

Figure 5.30

Single-line diagram of a double-circuit line

voltage drop), and $\boldsymbol{I}_{P1}$ and $\boldsymbol{I}_{P2}$ are the vectors of phase currents for lines 1 and 2. $\boldsymbol{Z}_P$ is a 6×6 impedance matrix. Solving (5.13.1)

$$\left[\begin{array}{c} \boldsymbol{I}_{P1} \\ -- \\ \boldsymbol{I}_{P2} \end{array}\right] = \boldsymbol{Z}_P^{-1} \left[\begin{array}{c} \boldsymbol{E}_P \\ -- \\ \boldsymbol{E}_P \end{array}\right] = \left[\begin{array}{c|c} \boldsymbol{Y}_A & \boldsymbol{Y}_B \\ \hline \boldsymbol{Y}_C & \boldsymbol{Y}_D \end{array}\right] \left[\begin{array}{c} \boldsymbol{E}_P \\ -- \\ \boldsymbol{E}_P \end{array}\right] = \left[\begin{array}{c} (\boldsymbol{Y}_A + \boldsymbol{Y}_B) \\ \\ (\boldsymbol{Y}_C + \boldsymbol{Y}_D) \end{array}\right] \boldsymbol{E}_P \tag{5.13.2}$$

where $\boldsymbol{Y}_A$, $\boldsymbol{Y}_B$, $\boldsymbol{Y}_C$, and $\boldsymbol{Y}_D$ are obtained by partitioning $\boldsymbol{Z}_P^{-1}$ into four 3×3 matrices. Adding $\boldsymbol{I}_{P1}$ and $\boldsymbol{I}_{P2}$,

$$(\boldsymbol{I}_{P1} + \boldsymbol{I}_{P2}) = (\boldsymbol{Y}_A + \boldsymbol{Y}_B + \boldsymbol{Y}_C + \boldsymbol{Y}_D)\boldsymbol{E}_P \tag{5.13.3}$$

and solving for $\boldsymbol{E}_P$,

$$\boldsymbol{E}_P = \boldsymbol{Z}_{Peq}(\boldsymbol{I}_{P1} + \boldsymbol{I}_{P2}) \tag{5.13.4}$$

where

$$\boldsymbol{Z}_{Peq} = (\boldsymbol{Y}_A + \boldsymbol{Y}_B + \boldsymbol{Y}_C + \boldsymbol{Y}_D)^{-1} \tag{5.13.5}$$

$\boldsymbol{Z}_{Peq}$ is the equivalent 3×3 series phase impedance matrix of the double-circuit line. Note that in (5.13.5) the matrices $\boldsymbol{Y}_B$ and $\boldsymbol{Y}_C$ account for the inductive coupling between the two circuits.

An analogous procedure can be used to obtain the shunt admittance matrix. Following the ideas of Section 5.11, we can write $(6 + N)$ equations similar to (5.11.4). After eliminating the neutral wire charges, we obtain

$$\left[\begin{array}{c} \boldsymbol{q}_{P1} \\ -- \\ \boldsymbol{q}_{P2} \end{array}\right] = \mathbf{C}_P \left[\begin{array}{c} \boldsymbol{V}_P \\ -- \\ \boldsymbol{V}_P \end{array}\right] = \left[\begin{array}{c|c} \mathbf{C}_A & \mathbf{C}_B \\ \hline \mathbf{C}_C & \mathbf{C}_D \end{array}\right] \left[\begin{array}{c} \boldsymbol{V}_P \\ -- \\ \boldsymbol{V}_P \end{array}\right] = \left[\begin{array}{c} (\mathbf{C}_A + \mathbf{C}_B) \\ \\ (\mathbf{C}_C + \mathbf{C}_D) \end{array}\right] \boldsymbol{V}_P \tag{5.13.6}$$

where $\boldsymbol{V}_P$ is the vector of phase-to-neutal voltages, and $\boldsymbol{q}_{P1}$ and $\boldsymbol{q}_{P2}$ are the vectors of phase-conductor charges for lines 1 and 2. $\mathbf{C}_P$ is a 6×6 capacitance matrix that is partitioned into four 3×3 matrices $\mathbf{C}_A$, $\mathbf{C}_B$, $\mathbf{C}_C$, and $\mathbf{C}_D$. Adding $\boldsymbol{q}_{P1}$ and $\boldsymbol{q}_{P2}$

$$(\boldsymbol{q}_{P1} + \boldsymbol{q}_{P2}) = \mathbf{C}_{Peq} \boldsymbol{V}_P \tag{5.13.7}$$

where

$$\mathbf{C}_{\text{Peq}} = (\mathbf{C}_A + \mathbf{C}_B + \mathbf{C}_C + \mathbf{C}_D) \tag{5.13.8}$$

Also,

$$\boldsymbol{Y}_{\text{Peq}} = j\omega\mathbf{C}_{\text{Peq}} \tag{5.13.9}$$

$\boldsymbol{Y}_{\text{Peq}}$ is the equivalent 3 × 3 shunt admittance matrix of the double-circuit line. The matrices $\mathbf{C}_B$ and $\mathbf{C}_C$ in (5.13.8) account for the capacitive coupling between the two circuits.

The corresponding sequence matrices of the double-circuit line are computed in the same manner as for the single-circuit line using (5.7.25) and (5.11.18). Also, these ideas can be extended in a straightforward fashion to more than two parallel circuits.

Section 5.14

Personal Computer Program: LINE CONSTANTS

The software package that accompanies this text includes a computer program entitled "LINE CONSTANTS" that calculates the series impedance and shunt admittance matrices of single- and double-circuit three-phase transmission lines. The program also calculates electric field strength at the surface of the phase conductors and a lateral profile of ground-level electric field strength.

Input data for the program consist of: (1) line voltage; (2) frequency; (3) earth resistivity; (4) right-of-way width; and, for each circuit, (5) outside radius, GMR, and resistance of each phase conductor and each neutral wire; (6) number of phase conductors in bundle and bundle spacing; and (7) horizontal and vertical position of each phase and of each neutral wire. Resistance and GMR data are given for a specified temperature, frequency, and current.

The output data consist of: the series phase impedance matrix $\boldsymbol{Z}_{\text{P}}$, (5.7.19); the series sequence impedance matrix $\boldsymbol{Z}_{\text{S}}$, (5.7.25); shunt phase admittance matrix $\boldsymbol{Y}_{\text{P}}$, (5.11.16); shunt sequence admittance matrix $\boldsymbol{Y}_{\text{S}}$, (5.11.18); conductor surface electric field strength (5.12.1)–(5.12.5); and the ground-level electric field strength from the center to the edge of the right-of-way, (5.12.6). The output also includes sequence impedances and admittances for completely transposed lines, (5.7.27) and (5.11.21). Output data for double-circuit lines are computed from the equations in Section 5.13.

Example 5.10 Run the LINE CONSTANTS program for the 765-kV single-circuit line shown in Figure 5.31.

Solution Program output data are given in Table 5.6.

Figure 5.31

Three-phase 765-kV line for Example 5.10

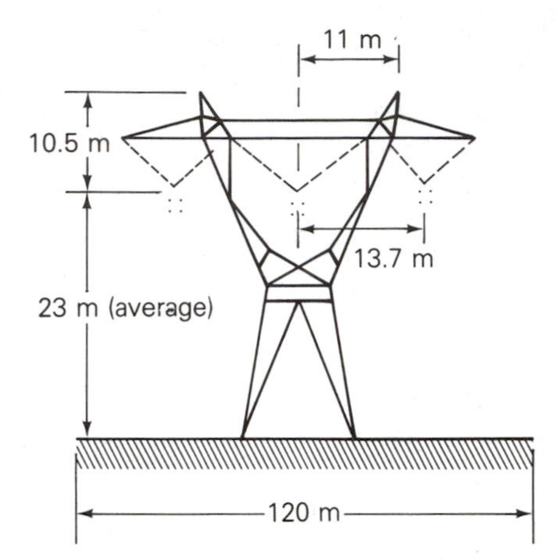

Neutrals:
2—Alumoweld 7 no. 8
Radius = 0.489 cm
GMR = 0.0636 cm
Resistance = 1.52 Ω/km

Phase conductors:
4—ACSR 954 kcmil, 54/7
Radius = 1.519 cm
GMR = 1.229 cm
Resistance = 0.0701 Ω/km
Bundle spacing = 45.7 cm

Earth resistivity = 100 Ωm
Frequency = 60 Hz
Voltage = 765 kV

Series phase impedance matrix $\boldsymbol{Z}_{\text{P}}$ Eq. 5.7.19 Units:Ohms/km

$$\begin{bmatrix} 2.283E-01+j7.189E-01 & 2.122E-01+j3.979E-01 & 2.084E-01+j3.499E-01 \\ 2.122E-01+j3.979E-01 & 2.323E-01+j7.121E-01 & 2.122E-01+j3.979E-01 \\ 2.084E-01+j3.499E-01 & 2.122E-01+j3.979E-01 & 2.283E-01+j7.189E-01 \end{bmatrix}$$

Series sequence impedance matrix $\boldsymbol{Z}_{\text{S}}$ Eq. 5.7.25 Units:Ohms/km

$$\begin{bmatrix} 6.514E-01+j1.480E+00 & 1.060E-02-j9.113E-03 & -1.319E-02-j4.620E-03 \\ -1.319E-02-j4.620E-03 & 1.875E-02+j3.348E-01 & -2.904E-02+j1.815E-02 \\ 1.060E-02-j9.113E-03 & 3.024E-02+j1.608E-02 & 1.875E-02+j3.348E-01 \end{bmatrix}$$

Shunt phase admittance matrix $\boldsymbol{Y}_{\text{P}}$ Eq. 5.11.16 Units:S/km

$$\begin{bmatrix} +j4.898E-06 & -j9.777E-07 & -j2.541E-07 \\ -j9.777E-07 & +j5.083E-06 & -j9.777E-07 \\ -j2.541E-07 & -j9.777E-07 & +j4.898E-06 \end{bmatrix}$$

Shunt sequence admittance matrix $\boldsymbol{Y}_{\text{S}}$ Eq. 5.11.20 Units:S/km

$$\begin{bmatrix} 0.000E+00 & +j3.487E-06 & -1.555E-07 & +j8.977E-08 & 1.555E-07 & +j8.977E-08 \\ 1.555E-07 & +j8.977E-08 & 0.000E+00 & +j5.696E-06 & 4.711E-07 & -j2.720E-07 \\ -1.555E-07 & +j8.977E-08 & -4.711E-07 & -j2.720E-07 & 0.000E+00 & +j5.696E-06 \end{bmatrix}$$

Conductor surface electric field strength Eqs. 5.12.1–5.12.5

$E_{\text{rmax}} = 22.5\,\text{kV}_{\text{rms}}/\text{cm}$

Lateral profile of ground level electric field strength Eq. 5.12.6

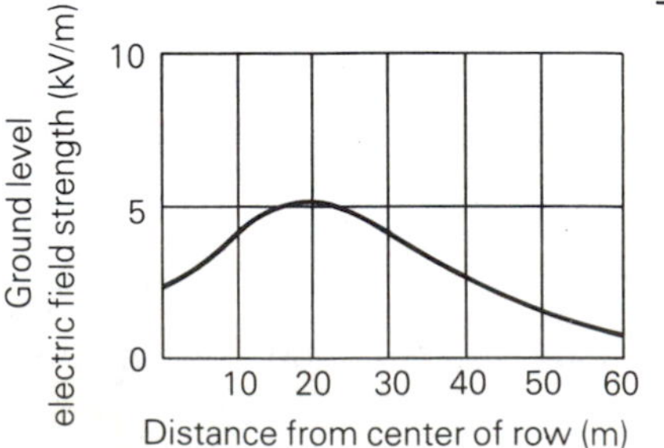

Table 5.6 Output data for Example 5.10

Problems

Section 5.2

5.1 The *Aluminum Electrical Conductor Handbook* lists a dc resistance of 0.01558 ohm per 1000 ft at 20°C and a 60-Hz resistance of 0.0956 ohm per mile at 50°C for the all-aluminum Marigold conductor, which has 61 strands and whose size is 1113 kcmil. Assuming an increase in resistance of 2% for spiraling, calculate and verify the dc resistance. Then calculate the dc resistance at 50°C, and determine the percentage increase due to skin effect.

5.2 One thousand circular mils or 1 kcmil is sometimes designated by the abbreviation MCM. Data for commercial bare aluminum electrical conductors lists a 60-Hz resistance of 0.0880 ohm per kilometer at 75°C for a 795-MCM AAC conductor.
(a) Determine the cross-sectional conducting area of this conductor in square meters.
(b) Find the 60-Hz resistance of this conductor in ohms per kilometer at 50°C.

5.3 A 60-Hz, 765-kV three-phase overhead transmission line has four ACSR 900 kcmil 54/3 conductors per phase. Determine the 60-Hz resistance of this line in ohms per kilometer per phase at 50°C.

Sections 5.4 and 5.5

5.4 A 60-Hz single-phase, two-wire overhead line has solid cylindrical copper conductors with 1.5 cm diameter. The conductors are arranged in a horizontal configuration with 1.2 m spacing. Calculate in mH/km (a) the inductance of each conductor due to internal flux linkages only, (b) the inductance of each conductor due to both internal and external flux linkages, and (c) the total inductance of the line.

5.5 A 60-Hz three-phase, three-wire overhead line has solid cylindrical conductors arranged in the form of an equilateral triangle with 4 ft conductor spacing. Conductor diameter is 0.5 in. Calculate the positive-sequence inductance in H/m and the positive-sequence inductive reactance in Ω/km.

Section 5.6

5.6 Find the GMR of a stranded conductor consisting of six outer strands surrounding and touching one central strand, all strands having the same radius r.

5.7 A bundle configuration for UHV lines (above 1000 kV) has identical conductors equally spaced around a circle, as shown in Figure 5.32. N_b is the number of conductors in the bundle, A is the circle radius, and D_S is the conductor GMR. Using the distance D_{1n}

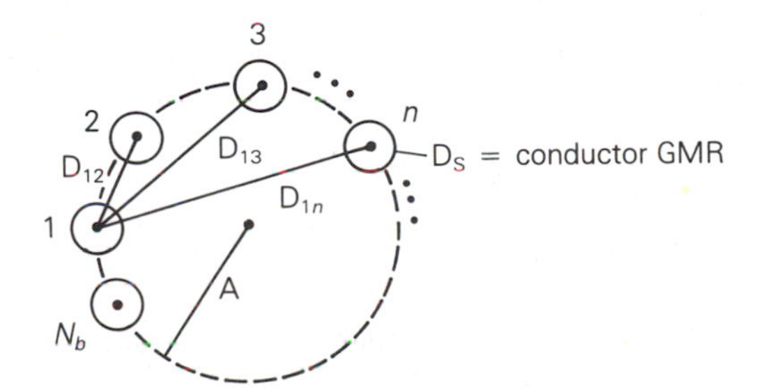

Figure 5.32 Bundle configuration for Problem 5.7

between conductors 1 and n given by $D_{1n} = 2A \sin[(n-1)\pi/N_b]$ for $n = 1, 2, \ldots, N_b$, and the following trigonometric identity:

$$[2\sin(\pi/N_b)][2\sin(2\pi/N_b)][2\sin(3\pi/N_b)]\ldots[2\sin\{(N_b-1)\pi/N_b\}] = N_b$$

show that the bundle GMR, denoted D_{SL}, is

$$D_{SL} = [N_b D_S A^{(N_b-1)}]^{(1/N_b)}$$

Also show that the above formula agrees with (5.6.19)–(5.6.21) for EHV lines with $N_b = 2, 3,$ and 4.

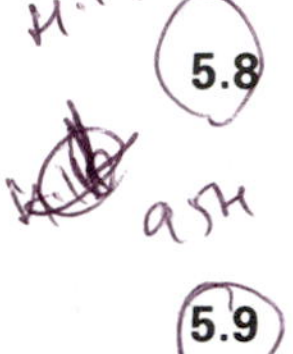

5.8 A 230-kV, 60-Hz, three-phase completely transposed overhead line has one ACSR 1113-kcmil conductor per phase and flat horizontal phase spacing, with 7 m between adjacent conductors. Determine the positive-sequence inductance in H/m and the positive-sequence inductive reactance in Ω/km.

5.9 Calculate the positive-sequence inductive reactance in Ω/km of a bundled 500-kV, 60-Hz, three-phase completely transposed overhead line having three ACSR 954-kcmil conductors per bundle, with 45.7 cm between conductors in the bundle. The horizontal phase spacings between bundle centers are 10, 10, and 20 m.

Section 5.9

5.10 Calculate the capacitance-to-neutral in F/m and the admittance-to-neutral in S/km for the single-phase line in Problem 5.4. Neglect the effect of the earth plane.

5.11 Calculate the positive-sequence shunt capacitance in F/m and the positive-sequence shunt admittance in S/km for the three-phase line in Problem 5.5. Neglect the effect of the earth plane.

Section 5.10

5.12 Calculate the positive-sequence shunt capacitance in F/m and the positive-sequence shunt admittance in S/km for the three-phase line in Problem 5.8. Also calculate the line-charging current in kA/phase if the line is 100 km in length and is operated at 230 kV. Neglect the effect of the earth plane.

5.13 Calculate the positive-sequence shunt capacitance in F/m and the positive-sequence shunt admittance in S/km for the line in Problem 5.9. Also calculate the total reactive power in Mvar/km supplied by the line capacitance when it is operated at 500 kV. Neglect the effect of the earth plane.

Section 5.11

5.14 For an average line height of 10 m, determine the effect of the earth on capacitance for the single-phase line in Problem 5.10. Assume a perfectly conducting earth plane.

Section 5.12

5.15 Calculate the conductor surface electric field strength in kV_{rms}/cm for the single-phase line in Problem 5.10 when the line is operating at 20 kV. Also calculate the ground-level electric field strength in kV_{rms}/m directly under one conductor. Assume a line height of 10 m.

Section 5.14

5.16 Run the LINE CONSTANTS program for Example 5.10 and verify the computed results.

5.17 The phase conductors in Example 5.10 are replaced by 1113 kcmil 54/19 ACSR conductors, four per bundle. Use the LINE CONSTANTS program to compute the positive-sequence series impedance, positive-sequence shunt capacitance, conductor surface electric field strength, and ground-level electric field strength profile. Compare the computed results with those of Problem 5.16.

5.18 Determine the effect of a 10% decrease as well as a 10% increase in bundle spacing on the positive-sequence series impedance and positive-sequence shunt admittance for the line in Example 5.10.

5.19 Determine the effect of a change in earth resistivity to 0.01, 1000, and 1.0×10^9 Ωm on the zero-sequence series impedance for the line in Example 5.10.

5.20 Using the LINE CONSTANTS program, compute the series sequence impedance matrix for the three-phase line in Problem 5.8. Assume an average line height of 12 m, an earth resistivity of 100 Ωm, 60-Hz conductor resistances at 50°C, and no neutral wires. Compare the computed positive-sequence inductive reactance with the result calculated in Problem 5.8.

5.21 Rework Problem 5.20 with one neutral wire located 6 m directly above the center phase conductor. Compare the series sequence impedance matrix with that of Problem 5.20.

5.22 Using the LINE CONSTANTS program, compute the shunt sequence admittance matrix for the line in Problem 5.9. Assume an average line height of 20 m and no neutral wires. Compare the computed positive-sequence shunt admittance with the result calculated in Problem 5.9.

5.23 Rework Problem 5.22 with two neutral wires located 7 m above and 8 m to the left and right of the center bundle.

5.24 Using the LINE CONSTANTS program, compute the conductor surface electric field strength and the ground-level electric field strength profile for the line in Problem 5.23. Assume a 100 m right-of-way width.

Figure 5.33 Double-circuit line for Problems 5.27 and 5.28

Phase conductors:
1—ACSR 2515 kcmil, 76/19
Radius = 2.388 cm
GMR = 1.893 cm
Resistance = 0.0280 Ω/km

Neutrals:
2—Alumoweld 7 no. 8
Radius = 0.489 cm
GMR = 0.0636 cm
Resistance = 1.52 Ω/km

Earth resistivity = 100 Ωm
Frequency = 60 Hz
Voltage = 345 kV

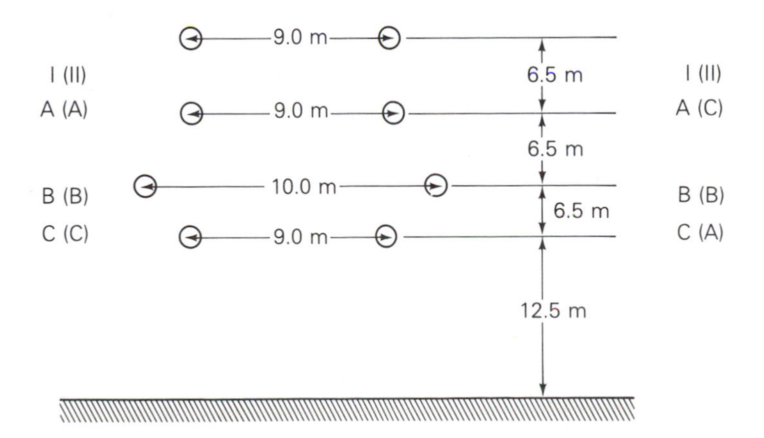

5.25 Determine the effect of a 10% decrease as well as a 10% increase in phase spacing on the conductor surface electric field strength and on the ground-level electric field strength profile for the line in Problem 5.24.

5.26 Determine the effect of a 10% decrease as well as a 10% increase in the average line height on the conductor surface electric field strength as well as the ground-level electric field strength profile for the line in Problem 5.24.

5.27 Using the LINE CONSTANTS program, calculate the equivalent series sequence impedance matrix and the equivalent shunt sequence admittance matrix for the double-circuit, three-phase line shown in Figure 5.33 with phase arrangement I.

5.28 Rework Problem 5.27 for phase arrangement II shown in parentheses in Figure 5.33. Compare the computed results of the two phase arrangements.

References

1. General Electric Company, *Transmission Line Reference Book–345 kV and Above*, 2d ed. (Palo Alto, Ca.: Electric Power Research Institute, 1982).
2. Westinghouse Electric Corporation, *Electrical Transmission and Distribution Reference Book*, 4th ed. (East Pittsburgh, Pa., 1964).
3. General Electric Company, *Electric Utility Systems and Practices*, 4th ed. (New York: John Wiley and Sons, 1983).
4. John R. Carson, "Wave Propagation in Overhead Wires with Ground Return," *Bell System Tech. J.* 5 (1926): 539–554.
5. C. F. Wagner, and R. D. Evans, *Symmetrical Components* (New York: McGraw-Hill, 1933).
6. Paul M. Anderson, *Analysis of Faulted Power Systems* (Ames, Iowa: Iowa State Press, 1973).
7. M. H. Hesse, "Electromagnetic and Electrostatic Transmission Line Parameters by Digital Computer," *Trans. IEEE* PAS-82 (1963): 282–291.
8. W. D. Stevenson, Jr., *Elements of Power System Analysis*, 4th ed. (New York: McGraw-Hill, 1982).
9. C. A. Gross, *Power System Analysis* (New York: John Wiley and Sons, 1979).

CHAPTER 6

Transmission Lines: Steady-State Operation

In this chapter we analyze the performance of single-phase and balanced three-phase transmission lines under normal steady-state operating conditions. Expressions for voltage and current at any point along a line are developed, where the distributed nature of the series impedance and shunt admittance is taken into account. A line is treated here as a two-port network for which the *ABCD* parameters and an equivalent π circuit are derived. Also, approximations are given for a medium-length line lumping the shunt admittance, for a short line neglecting the shunt admittance, and for a lossless line assuming zero series resistance and shunt conductance. The concepts of *surge impedance loading* and transmission-line *wavelength* are also presented.

An important issue discussed in this chapter is *voltage regulation*. Transmission-line voltages are generally high during light load periods and low during heavy load periods. Voltage regulation, defined in Section 6.1, refers to the change in line voltage as line loading varies from no-load to full load.

Another important issue discussed here is line loadability. Three major line-loading limits are: (1) the thermal limit; (2) the voltage-drop limit; and (3) the steady-state stability limit. Thermal and voltage-drop limits are discussed in Section 6.1. The theoretical steady-state stability limit, discussed in Section 6.4 for lossless lines and in Section 6.5 for lossy lines, refers to the ability of synchronous machines at the ends of a line to remain in synchronism. Practical line loadability is discussed in Section 6.6.

In Section 6.7 we discuss line compensation techniques for improving voltage regulation and for raising line loadings closer to the thermal limit. A personal computer program entitled "TRANSMISSION LINES—STEADY-STATE OPERATION," suitable for transmission-line analysis and design with respect to voltage regulation, line loadability, and transmission efficiency, is described in Section 6.8.

Section 6.1

Medium and Short Line Approximations

In this section we present short and medium-length transmission-line approximations as a means of introducing *ABCD* parameters. Some readers may prefer to start in Section 6.2, which presents the exact transmission-line equations.

It is convenient to represent a transmission line by the two-port network shown in Figure 6.1, where V_S and I_S are the sending-end voltage and current, and V_R and I_R are the receiving-end voltage and current.

Figure 6.1

Representation of two-port network

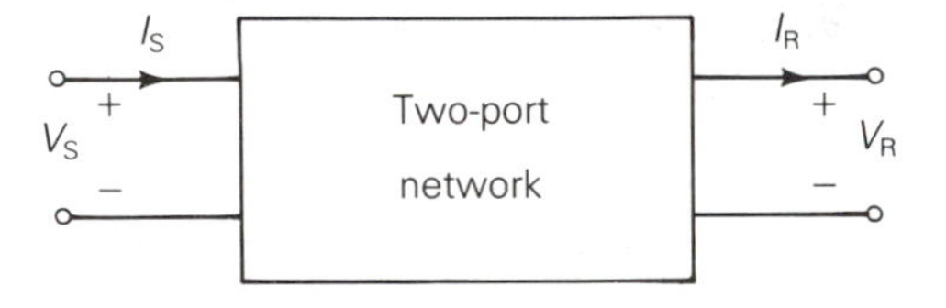

The relation between the sending-end and receiving-end quantities can be written as

$$V_S = AV_R + BI_R \quad \text{volts} \tag{6.1.1}$$

$$I_S = CV_R + DI_R \quad \text{A} \tag{6.1.2}$$

or, in matrix format,

$$\begin{bmatrix} V_S \\ I_S \end{bmatrix} = \begin{bmatrix} A & B \\ C & D \end{bmatrix} \begin{bmatrix} V_R \\ I_R \end{bmatrix} \tag{6.1.3}$$

where A, B, C, and D are parameters that depend on the transmission-line constants R, L, C, and G. The $ABCD$ parameters are, in general, complex numbers. A and D are dimensionless. B has units of ohms, and C has units of siemens. Network theory texts [5] show that $ABCD$ parameters apply to linear, passive, bilateral two-port networks, with the following general relation:

$$AD - BC = 1 \tag{6.1.4}$$

Figure 6.2

Short transmission line

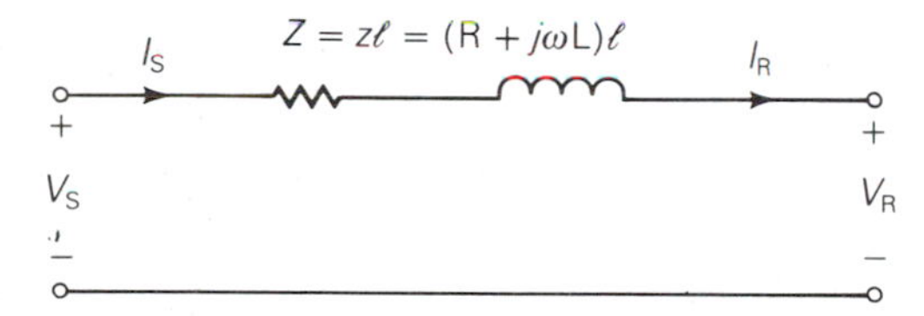

The circuit in Figure 6.2 represents a short transmission line, usually applied to overhead 60-Hz lines less than 80 km long. Only the series resistance and reactance are included. The shunt admittance is neglected. The circuit applies to either single-phase or completely transposed three-phase lines operating under balanced conditions. For a completely transposed three-phase line, Z is the positive-sequence series impedance, V_S and V_R are positive-sequence line-to-neutral voltages, and I_S and I_R are positive-sequence line currents. Although similar zero-sequence and negative-sequence circuits can be employed to represent unbalanced operating conditions, we consider only the positive-sequence network in this chapter.

To avoid confusion between total series impedance and series impedance per unit length, we use the following notation:

$z = R + j\omega L$ Ω/m, series impedance per unit length

$y = G + j\omega C$ S/m, shunt admittance per unit length

$Z = zl$ Ω, total series impedance

$Y = yl$ S, total shunt admittance

l = line length m

For three-phase lines, the subscript 1 indicating positive-sequence quantities is omitted in this chapter. Recall also that shunt conductance G is usually neglected for overhead transmission.

The $ABCD$ parameters for the short line in Figure 6.2 are easily obtained by writing a KVL and KCL equation as

$$V_S = V_R + ZI_R \tag{6.1.5}$$

$$I_S = I_R \tag{6.1.6}$$

or, in matrix format,

$$\begin{bmatrix} V_S \\ I_S \end{bmatrix} = \left[\begin{array}{c|c} 1 & Z \\ \hline 0 & 1 \end{array}\right] \begin{bmatrix} V_R \\ I_R \end{bmatrix} \tag{6.1.7}$$

Comparing (6.1.7) and (6.1.3), the *ABCD* parameters for a short line are

$$A = D = 1 \quad \text{per unit} \tag{6.1.8}$$

$$B = Z \quad \Omega \tag{6.1.9}$$

$$C = 0 \quad \text{S} \tag{6.1.10}$$

For medium-length lines, typically ranging from 80 to 250 km at 60 Hz, it is common to lump the total shunt capacitance and locate half at each end of the line. Such a circuit, called a *nominal π circuit*, is shown in Figure 6.3. To obtain the *ABCD* parameters of the nominal π circuit, note first that the current in the series branch in Figure 6.3 equals $I_R + \frac{V_R Y}{2}$. Then, writing a KVL equation,

$$\begin{aligned} V_S &= V_R + Z\left(I_R + \frac{V_R Y}{2}\right) \\ &= \left(1 + \frac{YZ}{2}\right) V_R + Z I_R \end{aligned} \tag{6.1.11}$$

Also, writing a KCL equation at the sending end,

$$I_S = I_R + \frac{V_R Y}{2} + \frac{V_S Y}{2} \tag{6.1.12}$$

Using (6.1.11) in (6.1.12),

$$\begin{aligned} I_S &= I_R + \frac{V_R Y}{2} + \left[\left(1 + \frac{YZ}{2}\right) V_R + Z I_R\right] \frac{Y}{2} \\ &= Y\left(1 + \frac{YZ}{4}\right) V_R + \left(1 + \frac{YZ}{2}\right) I_R \end{aligned} \tag{6.1.13}$$

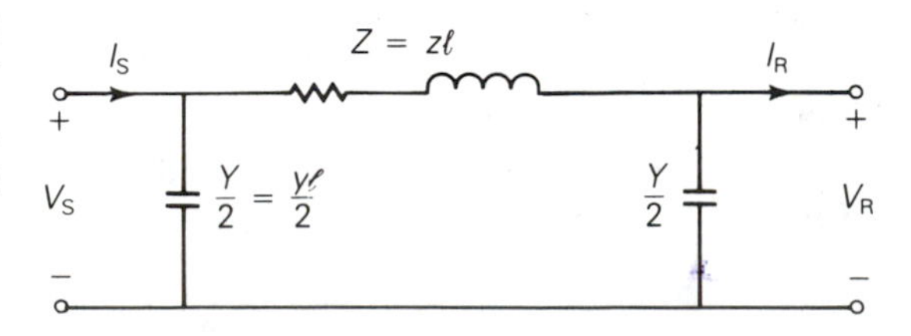

Figure 6.3 Medium-length transmission line—nominal π circuit

Writing (6.1.11) and (6.1.13) in matrix format,

$$\begin{bmatrix} V_S \\ \\ I_S \end{bmatrix} = \left[\begin{array}{c|c} \left(1 + \dfrac{YZ}{2}\right) & Z \\ \hline Y\left(1 + \dfrac{YZ}{4}\right) & \left(1 + \dfrac{YZ}{2}\right) \end{array}\right] \begin{bmatrix} V_R \\ \\ I_R \end{bmatrix} \tag{6.1.14}$$

Thus, comparing (6.1.14) and (6.1.3)

$$A = D = 1 + \frac{YZ}{2} \quad \text{per unit} \tag{6.1.15}$$

$$B = Z \quad \Omega \tag{6.1.16}$$

$$C = Y\left(1 + \frac{YZ}{4}\right) \quad \text{S} \tag{6.1.17}$$

Note that for both the short and medium-length lines, the relation $AD - BC = 1$ is verified. Note also that since the line is the same when viewed from either end, $A = D$.

Figure 6.4 gives the $ABCD$ parameters for some common networks, including a series impedance network that approximates a short line and a π circuit that approximates a medium-length line. A medium-length line could also be approximated by the T circuit shown in Figure 6.4, lumping half of the series impedance at each end of the line. Also given are the $ABCD$ parameters for networks in series, which are conveniently obtained by multiplying the $ABCD$ matrices of the individual networks.

$ABCD$ parameters can be used to describe the variation of line voltage with line loading. *Voltage regulation* is the change in voltage at the receiving end of the line when the load varies from no-load to a specified full load at a specified power factor, while the sending-end voltage is held constant. Expressed in percent of full-load voltage,

$$\text{percent VR} = \frac{|V_{RNL}| - |V_{RFL}|}{|V_{RFL}|} \times 100 \tag{6.1.18}$$

where percent VR is the percent voltage regulation, $|V_{RNL}|$ is the magnitude of the no-load receiving-end voltage, and $|V_{RFL}|$ is the magnitude of the full-load receiving-end voltage.

The effect of load power factor on voltage regulation is illustrated by the phasor diagrams in Figure 6.5 for short lines. The phasor diagrams are graphical representations of (6.1.5) for lagging and leading power factor loads. Note that, from (6.1.5) at no-load, $I_{RNL} = 0$ and $V_S = V_{RNL}$ for a short line. As

Figure 6.4

ABCD parameters of common networks

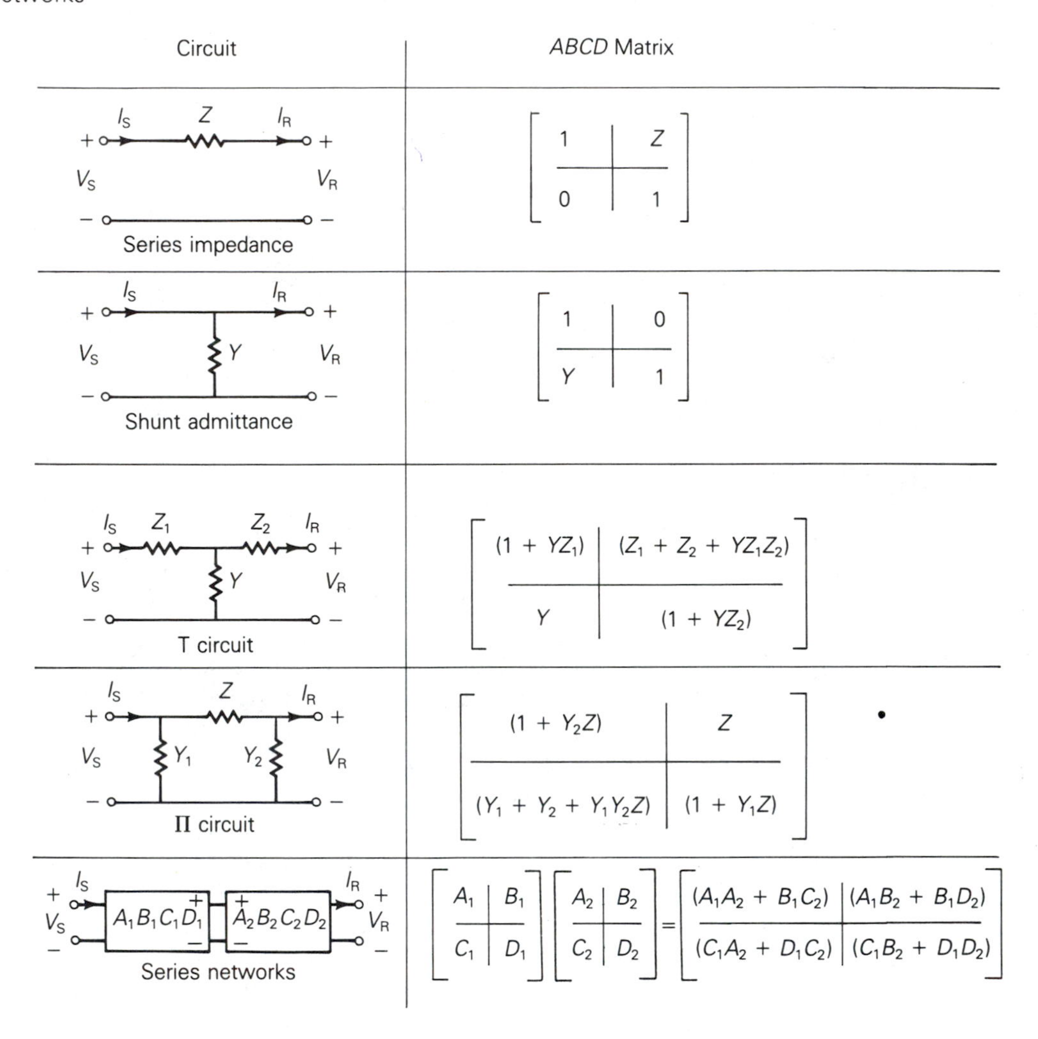

Figure 6.5

Phasor diagrams for a short transmission line

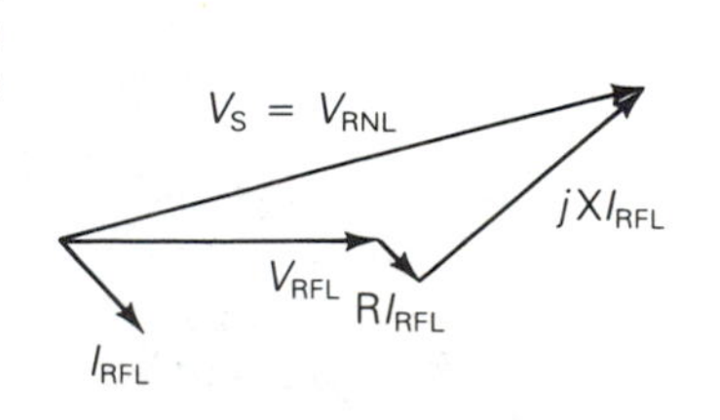

(a) Lagging p.f. load

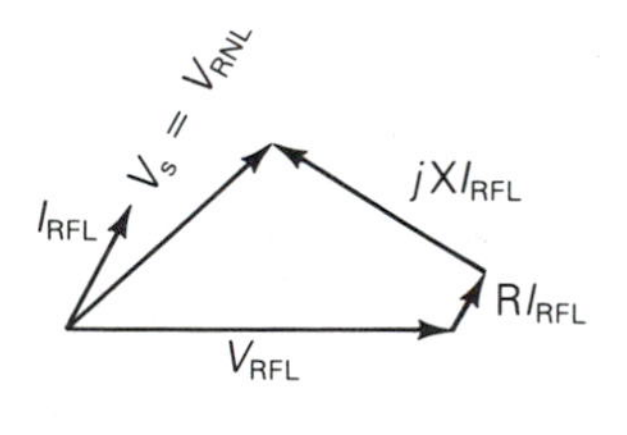

(b) Leading p.f. load

shown, the higher (worse) voltage regulation occurs for the lagging p.f. load, where V_{RNL} exceeds V_{RFL} by the larger amount. A smaller or even negative voltage regulation occurs for the leading p.f. load. In general, the no-load voltage is, from (6.1.1), with $I_{RNL} = 0$,

$$V_{RNL} = \frac{V_S}{A} \tag{6.1.19}$$

which can be used in (6.1.18) to determine voltage regulation.

In practice, transmission-line voltages decrease when heavily loaded and increase when lightly loaded. When voltages on EHV lines are maintained within $\pm 5\%$ of rated voltage, corresponding to about 10% voltage regulation, unusual operating problems are not encountered. Ten percent voltage regulation for lower voltage lines including transformer-voltage drops is also considered good operating practice.

In addition to voltage regulation, line loadability is an important issue. Three major line-loading limits are: (1) the thermal limit; (2) the voltage-drop limit; and (3) the steady-state stability limit.

The maximum temperature of a conductor determines its thermal limit. Conductor temperature affects the conductor sag between towers and the loss of conductor tensile strength due to annealing. If the temperature is too high, prescribed conductor-to-ground clearances may not be met, or the elastic limit of the conductor may be exceeded such that it cannot shrink to its original length when cooled. Conductor temperature depends on the current magnitude and its time duration, as well as on ambient temperature, wind velocity, and conductor surface conditions. Tables A.3 and A.4 give approximate current-carrying capacities of copper and ACSR conductors. The loadability of short transmission lines (less than 80 km in length for 60-Hz overhead lines) is usually determined by the conductor thermal limit or by ratings of line terminal equipment such as circuit breakers.

For longer line lengths (up to 300 km), line loadability is often determined by the voltage-drop limit. Although more severe voltage drops may be tolerated in some cases, a heavily loaded line with $V_R/V_S \geqslant 0.95$ is usually considered safe operating practice. For line lengths over 300 km, steady-state stability becomes a limiting factor. Stability, discussed in Section 6.4, refers to the ability of synchronous machines on either end of a line to remain in synchronism.

Example 6.1 A three-phase, 60-Hz, completely transposed 345-kV, 200-km line has two 795,000-cmil 26/2 ACSR conductors per bundle and the following positive-sequence line constants:

$$z = 0.032 + j0.35 \quad \Omega/\text{km}$$

$$y = j4.2 \times 10^{-6} \quad \text{S/km}$$

Full load at the receiving end of the line is 700 MW at 0.99 p.f. leading and at

95% of rated voltage. Assuming a medium-length line, determine the following:

a. *ABCD* parameters of the nominal π circuit

b. Sending-end voltage V_S, current I_S, and real power P_S

c. Percent voltage regulation

d. Thermal limit, based on the approximate current-carrying capacity listed in Table A.4

e. Transmission-line efficiency at full load

Solution **a.** The total series impedance and shunt admittance values are

$$Z = zl = (0.032 + j0.35)(200) = 6.4 + j70 = 70.29\underline{/84.78^\circ} \quad \Omega$$

$$Y = yl = (j4.2 \times 10^{-6})(200) = 8.4 \times 10^{-4}\underline{/90^\circ} \quad \text{S}$$

From (6.1.15)–(6.1.17),

$$A = D = 1 + (8.4 \times 10^{-4}\underline{/90^\circ})\,(70.29\underline{/84.78^\circ})(\tfrac{1}{2})$$

$$= 1 + 0.02952\underline{/174.48^\circ}$$

$$= 0.9706 + j0.00269 = 0.9706\underline{/0.159^\circ} \quad \text{per unit}$$

$$B = Z = 70.29\underline{/84.78^\circ} \quad \Omega$$

$$C = (8.4 \times 10^{-4}\underline{/90^\circ})(1 + 0.01476\underline{/174.78^\circ})$$

$$= (8.4 \times 10^{-4}\underline{/90^\circ})(0.9853 + j0.00134)$$

$$= 8.277 \times 10^{-4}\underline{/90.08^\circ} \quad \text{S}$$

b. The receiving-end voltage and current quantities are

$$\text{V}_R = (0.95)(345) = 327.8 \quad \text{kV}_{LL}$$

$$V_R = \frac{327.8}{\sqrt{3}}\underline{/0^\circ} = 189.2\underline{/0^\circ} \quad \text{kV}_{LN}$$

$$I_R = \frac{700\underline{/\cos^{-1} 0.99}}{(\sqrt{3})(0.95 \times 345)(0.99)} = 1.246\underline{/8.11^\circ} \quad \text{kA}$$

From (6.1.1) and (6.1.2), the sending-end quantities are

$$V_S = (0.9706\underline{/0.159^\circ})(189.2\underline{/0^\circ}) + (70.29\underline{/84.78^\circ})(1.246\underline{/8.11^\circ})$$

$$= 183.6\underline{/0.159^\circ} + 87.55\underline{/92.89^\circ}$$

$$= 179.2 + j87.95 = 199.6\underline{/26.14^\circ} \quad \text{kV}_{LN}$$

$$\text{V}_S = 199.6\sqrt{3} = 345.8 \text{ kV}_{LL} \approx 1.00 \quad \text{per unit}$$

$$I_S = (8.277 \times 10^{-4}\underline{/90.08^\circ})(189.2\underline{/0^\circ}) + (0.9706\underline{/0.159^\circ})(1.246\underline{/8.11^\circ})$$

$$= 0.1566\underline{/90.08^\circ} + 1.209\underline{/8.27^\circ}$$

$$= 1.196 + j0.331 = 1.241\underline{/15.5^\circ} \quad \text{kA}$$

and the real power delivered to the sending end is

$$P_S = (\sqrt{3})(345.8)(1.241)\cos(26.14° - 15.5°)$$
$$= 730.5 \quad \text{MW}$$

c. From (6.1.19), the no-load receiving-end voltage is

$$V_{RNL} = \frac{V_S}{A} = \frac{345.8}{0.9706} = 356.3 \quad kV_{LL}$$

and, from (6.1.18),

$$\text{percent VR} = \frac{356.3 - 327.8}{327.8} \times 100 = 8.7\%$$

d. From Table A.4, the approximate current-carrying capacity of two 795,000-cmil 26/2 ACSR conductors is $2 \times 0.9 = 1.8$ kA.

e. The full-load line losses are $P_S - P_R = 730.5 - 700 = 30.5$ MW and the full-load transmission efficiency is

$$\text{percent EFF} = \frac{P_R}{P_S} \times 100 = \frac{700}{730.5} \times 100 = 95.8\%$$

Since $V_S = 1.00$ per unit, the full-load receiving-end voltage of 0.95 per unit corresponds to $V_R/V_S = 0.95$, considered in practice to be the lowest operating voltage possible without encountering operating problems. Thus, for this 345-kV 200-km uncompensated line, voltage drop limits the full-load current to 1.246 kA at 0.99 p.f. leading, well below the thermal limit of 1.8 kA. ◁

Section 6.2

Transmission-Line Differential Equations

The line constants R, L, and C are derived in Chapter 5 as per-length values having units of Ω/m, H/m, and F/m. They are not lumped, but rather are uniformly distributed along the length of the line. In order to account for the distributed nature of transmission-line constants, consider the circuit shown in Figure 6.6, which represents a line section of length Δx. $V(x)$ and $I(x)$ denote the voltage and current at position x, which is measured in meters from the right, or receiving end of the line. Similarly, $V(x + \Delta x)$ and $I(x + \Delta x)$ denote the voltage and current at position $(x + \Delta x)$. The circuit constants are

$$z = R + j\omega L \quad \Omega/m \tag{6.2.1}$$

$$y = G + j\omega C \quad S/m \tag{6.2.2}$$

Figure 6.6

Transmission-line section of length Δx

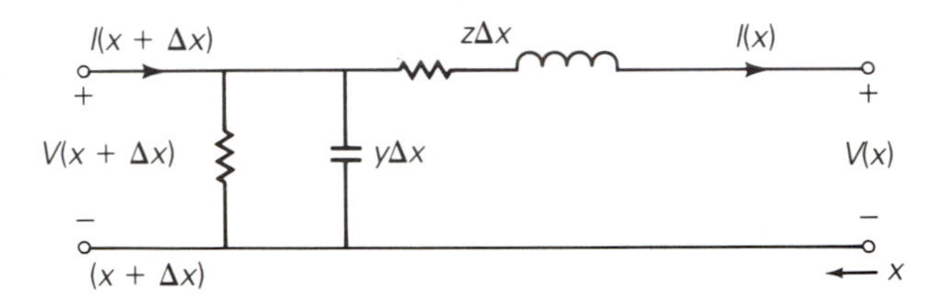

where G is usually neglected for overhead 60-Hz lines. Writing a KVL equation for the circuit

$$V(x + \Delta x) = V(x) + (z\Delta x)I(x) \quad \text{volts} \tag{6.2.3}$$

Rearranging (6.2.3),

$$\frac{V(x + \Delta x) - V(x)}{\Delta x} = z\,I(x) \tag{6.2.4}$$

and taking the limit as Δx approaches zero,

$$\frac{dV(x)}{dx} = z\,I(x) \tag{6.2.5}$$

Similarly, writing a KCL equation for the circuit,

$$I(x + \Delta x) = I(x) + (y\Delta x)V(x + \Delta x) \quad \text{A} \tag{6.2.6}$$

Rearranging,

$$\frac{I(x + \Delta x) - I(x)}{\Delta x} = y\,V(x) \tag{6.2.7}$$

and taking the limit as Δx approaches zero,

$$\frac{dI(x)}{dx} = y\,V(x) \tag{6.2.8}$$

Equations (6.2.5) and (6.2.8) are two linear, first-order, homogeneous differential equations with two unknowns, $V(x)$ and $I(x)$. We can eliminate $I(x)$ by differentiating (6.2.5) and using (6.2.8) as follows:

$$\frac{d^2V(x)}{dx^2} = z\frac{dI(x)}{dx} = zy\,V(x) \tag{6.2.9}$$

or

$$\frac{d^2V(x)}{dx^2} - zy\,V(x) = 0 \tag{6.2.10}$$

Equation (6.2.10) is a linear, second-order, homogeneous differential equation with one unknown, $V(x)$. By inspection, its solution is

$$V(x) = A_1 e^{\gamma x} + A_2 e^{-\gamma x} \quad \text{volts} \tag{6.2.11}$$

where A_1 and A_2 are integration constants and

$$\gamma = \sqrt{zy} \quad \text{m}^{-1} \tag{6.2.12}$$

γ, whose units are m^{-1}, is called the *propagation constant*. By inserting (6.2.11) and (6.2.12) into (6.2.10), the solution to the differential equation can be verified.

Next, using (6.2.11) in (6.2.5),

$$\frac{dV(x)}{dx} = \gamma A_1 e^{\gamma x} - \gamma A_2 e^{-\gamma x} = z\, I(x) \tag{6.2.13}$$

Solving for $I(x)$,

$$I(x) = \frac{A_1 e^{\gamma x} - A_2 e^{-\gamma x}}{z/\gamma} \tag{6.2.14}$$

Using (6.2.12), $z/\gamma = z/\sqrt{zy} = \sqrt{z/y}$, (6.2.14) becomes

$$I(x) = \frac{A_1 e^{\gamma x} - A_2 e^{-\gamma x}}{Z_c} \tag{6.2.15}$$

where

$$Z_c = \sqrt{\frac{z}{y}} \quad \Omega \tag{6.2.16}$$

Z_c, whose units are Ω, is called the *characteristic impedance*.

Next, the integration constants A_1 and A_2 are evaluated from the boundary conditions. At $x = 0$, the receiving end of the line, the receiving-end voltage and current are

$$V_R = V(0) \tag{6.2.17}$$

$$I_R = I(0) \tag{6.2.18}$$

Also, at $x = 0$, (6.2.11) and (6.2.15) become

$$V_R = A_1 + A_2 \tag{6.2.19}$$

$$I_R = \frac{A_1 - A_2}{Z_c} \tag{6.2.20}$$

Solving for A_1 and A_2,

$$A_1 = \frac{V_R + Z_c I_R}{2} \tag{6.2.21}$$

$$A_2 = \frac{V_R - Z_c I_R}{2} \tag{6.2.22}$$

Substituting A_1 and A_2 into (6.2.11) and (6.2.15),

$$V(x) = \left(\frac{V_R + Z_c I_R}{2}\right) e^{\gamma x} + \left(\frac{V_R - Z_c I_R}{2}\right) e^{-\gamma x} \tag{6.2.23}$$

$$I(x) = \left(\frac{V_R + Z_c I_R}{2Z_c}\right) e^{\gamma x} - \left(\frac{V_R - Z_c I_R}{2Z_c}\right) e^{-\gamma x} \tag{6.2.24}$$

Rearranging (6.2.23) and (6.2.24),

$$V(x) = \left(\frac{e^{\gamma x} + e^{-\gamma x}}{2}\right) V_R + Z_c \left(\frac{e^{\gamma x} - e^{-\gamma x}}{2}\right) I_R \tag{6.2.25}$$

$$I(x) = \frac{1}{Z_c}\left(\frac{e^{\gamma x} - e^{-\gamma x}}{2}\right) V_R + \left(\frac{e^{\gamma x} + e^{-\gamma x}}{2}\right) I_R \tag{6.2.26}$$

Recognizing the hyperbolic functions cosh and sinh,

$$V(x) = \cosh(\gamma x) V_R + Z_c \sinh(\gamma x) I_R \tag{6.2.27}$$

$$I(x) = \frac{1}{Z_c} \sinh(\gamma x) V_R + \cosh(\gamma x) I_R \tag{6.2.28}$$

Equations (6.2.27) and (6.2.28) give the *ABCD* parameters of the distributed line. In matrix format,

$$\begin{bmatrix} V(x) \\ I(x) \end{bmatrix} = \left[\begin{array}{c|c} A(x) & B(x) \\ \hline C(x) & D(x) \end{array}\right] \begin{bmatrix} V_R \\ I_R \end{bmatrix} \tag{6.2.29}$$

where

$$A(x) = D(x) = \cosh(\gamma x) \quad \text{per unit} \tag{6.2.30}$$

$$B(x) = Z_c \sinh(\gamma x) \quad \Omega \tag{6.2.31}$$

$$C(x) = \frac{1}{Z_c} \sinh(\gamma x) \quad \text{S} \tag{6.2.32}$$

Equation (6.2.29) gives the current and voltage at any point x along the line in terms of the receiving-end voltage and current. At the sending end, where $x = l$, $V(l) = V_S$ and $I(l) = I_S$. That is,

$$\begin{bmatrix} V_S \\ I_S \end{bmatrix} = \left[\begin{array}{c|c} A & B \\ \hline C & D \end{array}\right] \begin{bmatrix} V_R \\ I_R \end{bmatrix} \tag{6.2.33}$$

where

$$A = D = \cosh(\gamma l) \quad \text{per unit} \tag{6.2.34}$$

$$B = Z_c \sinh(\gamma l) \quad \Omega \tag{6.2.35}$$

$$C = \frac{1}{Z_c} \sinh(\gamma l) \quad \text{S} \tag{6.2.36}$$

Equations (6.2.34)–(6.2.36) give the *ABCD* parameters of the distributed line. In these equations, the propagation constant γ is a complex quantity with real and imaginary parts denoted α and β. That is,

$$\gamma = \alpha + j\beta \quad \text{m}^{-1} \tag{6.2.37}$$

The quantity γl is dimensionless. Also

$$e^{\gamma l} = e^{(\alpha l + j\beta l)} = e^{\alpha l}\, e^{j\beta l} = e^{\alpha l}\underline{/\beta l} \tag{6.2.38}$$

Using (6.2.38) the hyperbolic functions cosh and sinh can be evaluated as follows:

$$\cosh(\gamma l) = \frac{e^{\gamma l} + e^{-\gamma l}}{2} = \frac{1}{2}\left(e^{\alpha l}\underline{/\beta l} + e^{-\alpha l}\underline{/-\beta l}\right) \tag{6.2.39}$$

and

$$\sinh(\gamma l) = \frac{e^{\gamma l} - e^{-\gamma l}}{2} = \frac{1}{2}\left(e^{\alpha l}\underline{/\beta l} - e^{-\alpha l}\underline{/-\beta l}\right) \tag{6.2.40}$$

Alternatively, the following identities can be used:

$$\cosh(\alpha l + j\beta l) = \cosh(\alpha l)\cos(\beta l) + j\sinh(\alpha l)\sin(\beta l) \tag{6.2.41}$$

$$\sinh(\alpha l + j\beta l) = \sinh(\alpha l)\cos(\beta l) + j\cosh(\alpha l)\sin(\beta l) \tag{6.2.42}$$

Note that in (6.2.39)–(6.2.42), the dimensionless quantity βl is in radians, not degrees.

The *ABCD* parameters given by (6.2.34)–(6.2.36) are exact parameters valid for any line length. For accurate calculations, these equations must be used for overhead 60-Hz lines longer than 250 km. The *ABCD* parameters derived in Section 6.1 are approximate parameters that are more conveniently used for hand calculations involving short and medium-length lines. Table 6.1 summarizes the *ABCD* parameters for short, medium, long, and lossless (see Section 6.4) lines.

Table 6.1

Summary: Transmission-line *ABCD* parameters

PARAMETER	$A = D$	B	C
UNITS	Per Unit	Ω	S
Short line (less than 80 km	1	Z	0
Medium line—nominal π circuit (80 to 250 km)	$1 + \frac{YZ}{2}$	Z	$Y\left(1 + \frac{YZ}{4}\right)$
Long line—equivalent π circuit (more than 250 km)	$\cosh(\gamma\ell) = 1 + \frac{Y'Z'}{2}$	$Z_c \sinh(\gamma\ell) = Z'$	$(1/Z_c)\sinh(\gamma\ell) = Y'\left(1 + \frac{Y'Z'}{4}\right)$
Lossless line ($R = G = 0$)	$\cos(\beta\ell)$	$jZ_c \sin(\beta\ell)$	$\frac{j\sin(\beta\ell)}{Z_c}$

Example 6.2

A three-phase 765-kV, 60-Hz, 300-km, completely transposed line has the following positive-sequence impedance and admittance:

$$z = 0.0165 + j0.3306 = 0.3310\underline{/87.14^\circ} \quad \Omega/\text{km}$$

$$y = j4.674 \times 10^{-6} \quad \text{S/km}$$

Assuming positive-sequence operation, calculate the exact *ABCD* parameters of the line. Compare the exact *B* parameter with that of the nominal π circuit.

Solution From (6.2.12) and (6.2.16):

$$Z_c = \sqrt{\frac{0.3310\underline{/87.14^\circ}}{4.674 \times 10^{-6}\underline{/90^\circ}}} = \sqrt{7.082 \times 10^4\underline{/-2.86^\circ}}$$

$$= 266.1\underline{/-1.43^\circ} \quad \Omega$$

and

$$\gamma l = \sqrt{(0.3310\underline{/87.14^\circ})(4.674 \times 10^{-6}\underline{/90^\circ})} \times (300)$$

$$= \sqrt{1.547 \times 10^{-6}\underline{/177.14^\circ}} \times (300)$$

$$= 0.3731\underline{/88.57^\circ} = 0.00931 + j0.3730 \quad \text{per unit}$$

From (6.2.38),

$$e^{\gamma l} = e^{0.00931}e^{+j0.3730} = 1.0094\underline{/0.3730} \quad \text{radians}$$

$$= 0.9400 + j0.3678$$

and

$$e^{-\gamma l} = e^{-0.00931}e^{-j0.3730} = 0.9907\underline{/-0.3730} \quad \text{radians}$$

$$= 0.9226 - j0.3610$$

Then, from (6.2.39) and (6.2.40),

$$\cosh(\gamma l) = \frac{(0.9400 + j0.3678) + (0.9226 - j0.3610)}{2}$$

$$= 0.9313 + j0.0034 = 0.9313\underline{/0.209^\circ}$$

$$\sinh(\gamma l) = \frac{(0.9400 + j0.3678) - (0.9226 - j0.3610)}{2}$$

$$= 0.0087 + j0.3644 = 0.3645\underline{/88.63^\circ}$$

Finally, from (6.2.34)–(6.2.36),

$$A = D = \cosh(\gamma l) = 0.9313\underline{/0.209^\circ} \quad \text{per unit}$$

$$B = (266.1\underline{/-1.43^\circ})(0.3645\underline{/88.63^\circ}) = 97.0\underline{/87.2^\circ} \quad \Omega$$

$$C = \frac{0.3645\underline{/88.63^\circ}}{266.1\underline{/-1.43^\circ}} = 1.37 \times 10^{-3}\underline{/90.06^\circ} \quad \text{S}$$

Using (6.1.16), the B parameter for the nominal π circuit is

$$B_{\text{nominal}\,\pi} = Z = (0.3310\underline{/87.14^\circ})(300) = 99.3\underline{/87.14^\circ} \quad \Omega$$

which is 2% larger than the exact value. ◁

Section 6.3

Equivalent π Circuit

Many computer programs used in power-system analysis and design assume circuit representations of components such as transmission lines and transformers (see the power-flow program described in Chapter 7 as an example).

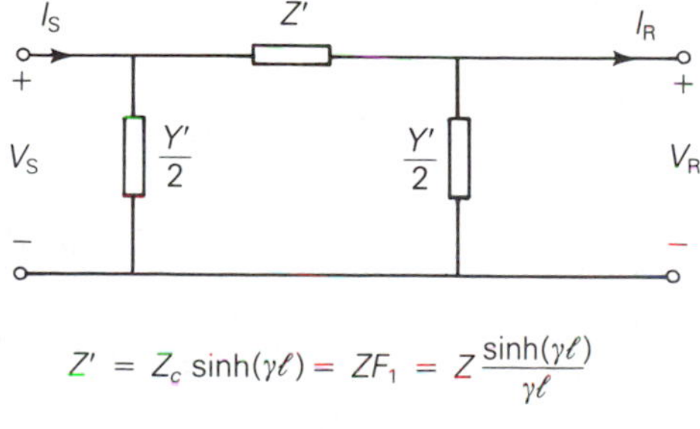

Figure 6.7

Transmission-line equivalent π circuit

$$Z' = Z_c \sinh(\gamma \ell) = ZF_1 = Z\frac{\sinh(\gamma \ell)}{\gamma \ell}$$

$$\frac{Y'}{2} = \frac{\tanh(\gamma \ell/2)}{Z_c} = \frac{Y}{2}F_2 = \frac{Y}{2}\frac{\tanh(\gamma \ell/2)}{(\gamma \ell/2)}$$

It is therefore convenient to represent the terminal characteristics of a transmission line by an equivalent circuit instead of its *ABCD* parameters.

The circuit shown in Figure 6.7. is called an *equivalent* π *circuit*. It is identical in structure to the nominal π circuit of Figure 6.3, except that Z' and Y' are used instead of Z and Y. Our objective is to determine Z' and Y' such that the equivalent π circuit has the same *ABCD* parameters as those of the distributed line, (6.2.34)–(6.2.36). The *ABCD* parameters of the equivalent π circuit, which has the same structure as the nominal π, are

$$A = D = 1 + \frac{Y'Z'}{2} \quad \text{per unit} \tag{6.3.1}$$

$$B = Z' \quad \Omega \tag{6.3.2}$$

$$C = Y'\left(1 + \frac{Y'Z'}{4}\right) \quad \text{S} \tag{6.3.3}$$

where we have replaced Z and Y in (6.1.15)–(6.1.17) with Z' and Y' in (6.3.1)–(6.3.3). Equating (6.3.2) to (6.2.35),

$$Z' = Z_c \sinh(\gamma l) = \sqrt{\frac{z}{y}}\sinh(\gamma l) \tag{6.3.4}$$

Rewriting (6.3.4) in terms of the nominal π circuit impedance $Z = zl$,

$$Z' = zl\left[\sqrt{\frac{z}{y}}\,\frac{\sinh(\gamma l)}{zl}\right] = zl\left[\frac{\sinh(\gamma l)}{\sqrt{zy}\,l}\right]$$

$$= ZF_1 \quad \Omega \tag{6.3.5}$$

where

$$F_1 = \frac{\sinh(\gamma l)}{\gamma l} \quad \text{per unit} \tag{6.3.6}$$

Similarly, equating (6.3.1) to (6.2.34),

$$1 + \frac{Y'Z'}{2} = \cosh(\gamma l)$$

$$\frac{Y'}{2} = \frac{\cosh(\gamma l) - 1}{Z'} \tag{6.3.7}$$

Using (6.3.4) and the identity $\tanh\left(\frac{\gamma l}{2}\right) = \frac{\cosh(\gamma l) - 1}{\sinh(\gamma l)}$, (6.3.7) becomes

$$\frac{Y'}{2} = \frac{\cosh(\gamma l) - 1}{Z_c \sinh(\gamma l)} = \frac{\tanh(\gamma l/2)}{Z_c} = \frac{\tanh(\gamma l/2)}{\sqrt{\frac{z}{y}}} \tag{6.3.8}$$

Rewriting (6.3.8) in terms of the nominal π circuit admittance $Y = yl$,

$$\frac{Y'}{2} = \frac{yl}{2}\left[\frac{\tanh(\gamma l/2)}{\sqrt{\frac{z}{y}}\frac{yl}{2}}\right] = \frac{yl}{2}\left[\frac{\tanh(\gamma l/2)}{\sqrt{zy}\ l/2}\right]$$

$$= \frac{Y}{2}F_2 \quad \text{S} \tag{6.3.9}$$

where

$$F_2 = \frac{\tanh(\gamma l/2)}{\gamma l/2} \quad \text{per unit} \tag{6.3.10}$$

Equations (6.3.6) and (6.3.10) give the correction factors F_1 and F_2 to convert Z and Y for the nominal π circuit to Z' and Y' for the equivalent π circuit.

Example 6.3 Compare the equivalent and nominal π circuits for the line in Example 6.2.

Solution For the nominal π circuit,

$$Z = zl = (0.3310\underline{/87.14^\circ})(300) = 99.3\underline{/87.14^\circ} \quad \Omega$$

$$\frac{Y}{2} = \frac{yl}{2} = \left(\frac{j4.674 \times 10^{-6}}{2}\right)(300) = 7.011 \times 10^{-4}\underline{/90^\circ} \quad \text{S}$$

From (6.3.6) and (6.3.10), the correction factors are

$$F_1 = \frac{0.3645\underline{/88.63^\circ}}{0.3731\underline{/88.57^\circ}} = 0.9769\underline{/0.06^\circ} \quad \text{per unit}$$

$$F_2 = \frac{\tanh(\gamma l/2)}{\gamma l/2} = \frac{\cosh(\gamma l) - 1}{(\gamma l/2)\sinh(\gamma l)}$$

$$= \frac{0.9313 + j0.0034 - 1}{\left(\frac{0.3731}{2}\underline{/88.57^\circ}\right)(0.3645\underline{/88.63^\circ})}$$

$$= \frac{-0.0687 + j0.0034}{0.06800\underline{/177.20^\circ}}$$

$$= \frac{0.06878\underline{/177.17^\circ}}{0.06800\underline{/177.20^\circ}} = 1.012\underline{/-0.03^\circ} \quad \text{per unit}$$

Then, from (6.3.5) and (6.3.9), for the equivalent π circuit,

$$Z' = (99.3\underline{/87.14°})(0.9769\underline{/0.06°}) = 97.0\underline{/87.2°} \quad \Omega$$

$$\frac{Y'}{2} = (7.011 \times 10^{-4}\underline{/90°})(1.012\underline{/-0.03°}) = 7.095 \times 10^{-4}\underline{/89.97°} \quad \text{S}$$

$$= 3.7 \times 10^{-7} + j7.095 \times 10^{-4} \quad \text{S}$$

Comparing these nominal and equivalent π circuit values, Z' is about 2% smaller than Z, and $Y'/2$ is about 1% larger than $Y/2$. Although the circuit values are approximately the same for this line, the equivalent π circuit should be used for accurate calculations involving long lines. Note the small shunt conductance, $G' = 3.7 \times 10^{-7}$ S introduced in the equivalent π circuit. G' is often neglected. ◁

Section 6.4

Lossless Lines

In this section we discuss the following concepts for lossless lines: surge impedance, *ABCD* parameters, equivalent π circuit, wavelength, surge impedance loading, voltage profiles and steady-state stability limit.

When line losses are neglected, simpler expressions for the line parameters are obtained and the above concepts are more easily understood. Since transmission and distribution lines for power transfer generally are designed to have low losses, the equations and concepts developed here can be used for quick and reasonably accurate hand calculations leading to seat-of-the-pants analyses and to initial designs. More accurate calculations can then be made with the personal computer program given in Section 6.8 for follow-up analysis and design.

Surge Impedance

For a lossless line, R = G = 0, and

$$z = j\omega \text{L} \quad \Omega/\text{m} \tag{6.4.1}$$

$$y = j\omega \text{C} \quad \text{S/m} \tag{6.4.2}$$

From (6.2.12) and (6.2.16),

$$Z_c = \sqrt{\frac{z}{y}} = \sqrt{\frac{j\omega L}{j\omega C}} = \sqrt{\frac{L}{C}} \quad \Omega \tag{6.4.3}$$

and

$$\gamma = \sqrt{zy} = \sqrt{(j\omega L)(j\omega C)} = j\omega\sqrt{LC} = j\beta \quad m^{-1} \tag{6.4.4}$$

where

$$\beta = \omega\sqrt{LC} \quad m^{-1} \tag{6.4.5}$$

The characteristic impedance $Z_c = \sqrt{L/C}$, commonly called *surge* impedance for a lossless line, is pure real, that is, resistive. The propagation constant $\gamma = j\beta$ is pure imaginary.

ABCD Parameters

The *ABCD* parameters are, from (6.2.30)–(6.2.32),

$$A(x) = D(x) = \cosh(\gamma x) = \cosh(j\beta x)$$
$$= \frac{e^{j\beta x} + e^{-j\beta x}}{2} = \cos(\beta x) \quad \text{per unit} \tag{6.4.6}$$

$$\sinh(\gamma x) = \sinh(j\beta x) = \frac{e^{j\beta x} - e^{-j\beta x}}{2} = j\sin(\beta x) \quad \text{per unit} \tag{6.4.7}$$

$$B(x) = Z_c \sinh(\gamma x) = jZ_c \sin(\beta x) = j\sqrt{\frac{L}{C}}\sin(\beta x) \quad \Omega \tag{6.4.8}$$

$$C(x) = \frac{\sinh(\gamma x)}{Z_c} = \frac{j\sin(\beta x)}{\sqrt{\frac{L}{C}}} \quad S \tag{6.4.9}$$

$A(x)$ and $D(x)$ are pure real; $B(x)$ and $C(x)$ are pure imaginary.

A comparison of lossless versus lossy *ABCD* parameters is shown in Table 6.1.

Equivalent π Circuit

For the equivalent π circuit, using (6.3.4),

$$Z' = jZ_c \sin(\beta l) = jX' \quad \Omega \tag{6.4.10}$$

or, from (6.3.5) and (6.3.6),

$$Z' = (j\omega L l)\left(\frac{\sin(\beta l)}{\beta l}\right) = jX' \quad \Omega \tag{6.4.11}$$

Also, from (6.3.9) and (6.3.10),

$$\frac{Y'}{2} = \frac{Y}{2}\frac{\tanh(j\beta l/2)}{j\beta l/2} = \frac{Y}{2}\frac{\sinh(j\beta l/2)}{(j\beta l/2)\cosh(j\beta l/2)}$$

$$= \left(\frac{j\omega C l}{2}\right)\frac{j\sin(\beta l/2)}{(j\beta l/2)\cos(\beta l/2)} = \left(\frac{j\omega C l}{2}\right)\frac{\tan(\beta l/2)}{\beta l/2}$$

$$= \left(\frac{j\omega C' l}{2}\right) \quad \text{S} \tag{6.4.12}$$

Z' and Y' are both pure imaginary. Also, for βl less than π radians, Z' is pure inductive and Y' is pure capacitive. Thus the equivalent π circuit for a lossless line, shown in Figure 6.8, is also lossless.

Figure 6.8

Equivalent π circuit for a lossless line ($\beta\ell$ less than π)

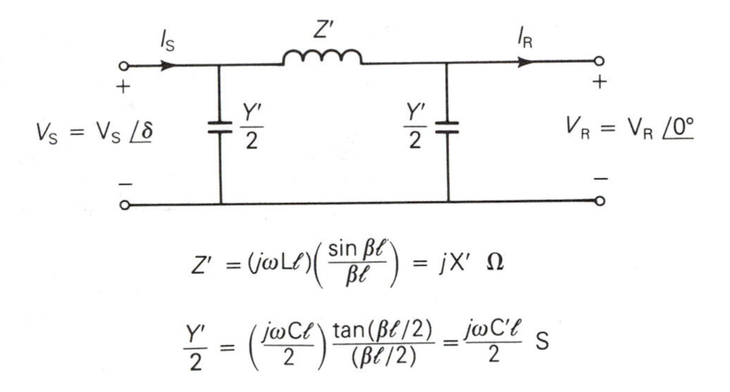

Wavelength

A *wavelength* is the distance required to change the phase of the voltage or current by 2π radians or 360°. For a lossless line, using (6.2.29),

$$V(x) = A(x)V_R + B(x)I_R$$

$$= \cos(\beta x)V_R + jZ_c\sin(\beta x)I_R \tag{6.4.13}$$

and

$$I(x) = C(x)V_R + D(x)I_R$$

$$= \frac{j\sin(\beta x)}{Z_c}V_R + \cos(\beta x)I_R \tag{6.4.14}$$

From (6.4.13) and (6.4.14), $V(x)$ and $I(x)$ change phase by 2π radians when $x = 2\pi/\beta$. Denoting wavelength by λ, and using (6.4.5),

$$\lambda = \frac{2\pi}{\beta} = \frac{2\pi}{\omega\sqrt{LC}} = \frac{1}{f\sqrt{LC}} \quad \text{m} \tag{6.4.15}$$

or

$$f\lambda = \frac{1}{\sqrt{\mathrm{LC}}} \tag{6.4.16}$$

It will be shown in Chapter 11 that the term $(1/\sqrt{\mathrm{LC}})$ in (6.4.16) is the velocity of propagation of voltage and current waves along a lossless line. For overhead lines, $(1/\sqrt{\mathrm{LC}}) \approx 3 \times 10^8$ m/s. And for $f = 60$ Hz, (6.4.14) gives

$$\lambda \approx \frac{3 \times 10^8}{60} = 5 \times 10^6 \text{ m} = 5000 \text{ km} = 3100 \text{ mi}$$

Typical power-line lengths are only a small fraction of the above 60-Hz wavelength.

Surge Impedance Loading

Surge impedance loading (SIL) is the power delivered by a lossless line to a load resistance equal to the surge impedance $Z_c = \sqrt{\mathrm{L/C}}$. Figure 6.9 shows a lossless line terminated by a resistance equal to its surge impedance. This line represents either a single-phase line or one phase-to-neutral of a balanced three-phase line. At SIL, from (6.4.13),

$$\begin{aligned} V(x) &= \cos(\beta x) V_{\mathrm{R}} + jZ_c \sin(\beta x) I_{\mathrm{R}} \\ &= \cos(\beta x) V_{\mathrm{R}} + jZ_c \sin(\beta x)\left(\frac{V_{\mathrm{R}}}{Z_c}\right) \\ &= (\cos\beta x + j\sin\beta x) V_{\mathrm{R}} \\ &= e^{j\beta x} V_{\mathrm{R}} \quad \text{volts} \end{aligned} \tag{6.4.17}$$

$$|V(x)| = |V_{\mathrm{R}}| \quad \text{volts} \tag{6.4.18}$$

Thus, at SIL, the voltage profile is flat. That is, the voltage magnitude at any point x along a lossless line at SIL is constant.

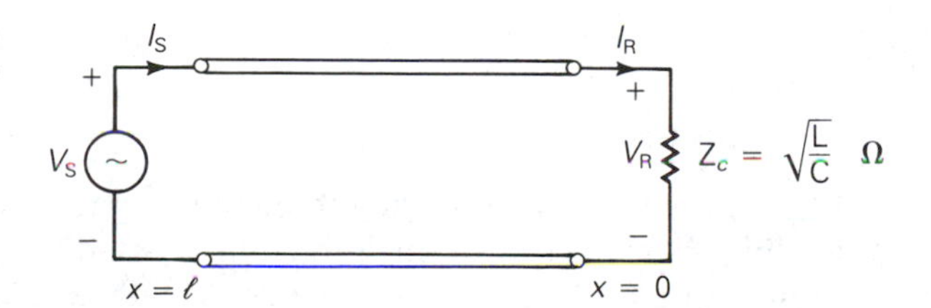

Figure 6.9 Lossless line terminated by its surge impedance

Also from (6.4.14) at SIL,

$$I(x) = \frac{j\sin(\beta x)}{Z_c} V_R + (\cos\beta x)\frac{V_R}{Z_c}$$

$$= (\cos\beta x + j\sin\beta x)\frac{V_R}{Z_c}$$

$$= (e^{j\beta x})\frac{V_R}{Z_c} \quad \text{A} \tag{6.4.19}$$

Using (6.4.17) and (6.4.19), the complex power flowing at any point x along the line is

$$S(x) = P(x) + jQ(x) = V(x)\,I^*(x)$$

$$= (e^{j\beta x}V_R)\left(\frac{e^{j\beta x}V_R}{Z_c}\right)^*$$

$$= \frac{|V_R|^2}{Z_c} \tag{6.4.20}$$

Thus the real power flow along a lossless line at SIL remains constant from the sending end to the receiving end. The reactive power flow is zero.

At rated line voltage, the real power delivered, or SIL, is, from (6.4.20),

$$\text{SIL} = \frac{V_{\text{rated}}^2}{Z_c} \quad \text{W} \tag{6.4.21}$$

where rated voltage is used for a single-phase line and rated line-to-line voltage is used for the total real power delivered by a three-phase line. Table 6.2 lists surge impedance and SIL values for typical overhead 60-Hz three-phase lines.

Table 6.2

Surge impedance and SIL values for typical 60-Hz overhead lines [1, 2]

V_{rated}	$Z_c = \sqrt{L/C}$	$\text{SIL} = V_{\text{rated}}^2/Z_c$
kV	Ω	MW
69	366–400	12–13
138	366–405	47–52
230	365–395	134–145
345	280–366	325–425
500	233–294	850–1075
765	254–266	2200–2300

Voltage Profiles

In practice, power lines are not terminated by their surge impedance. Instead, loadings can vary from a small fraction of SIL during light load conditions up to multiples of SIL, depending on line length and line compensation, during heavy load conditions. If a line is not terminated by its surge impedance, then the voltage profile is not flat. Figure 6.10 shows voltage profiles of lines with a fixed sending-end voltage magnitude V_S for line lengths l up to a quarter wavelength. This figure shows four loading conditions: (1) no-load, (2) SIL, (3) short circuit, and (4) full load, which are described as follows:

1. At no-load, $I_{RNL} = 0$ and (6.4.13) yields

$$V_{NL}(x) = (\cos \beta x) V_{RNL} \tag{6.4.22}$$

   The no-load voltage increases from $V_S = (\cos \beta l) V_{RNL}$ at the sending end to V_{RNL} at the receiving end (where $x = 0$).

2. From (6.4.18), the voltage profile at SIL is flat.

3. For a short circuit at the load, $V_{RSC} = 0$ and (6.4.13) yields

$$V_{SC}(x) = (Z_c \sin \beta x) I_{RSC} \tag{6.4.23}$$

   The voltage decreases from $V_S = (\sin \beta l)(Z_c I_{RSC})$ at the sending end to $V_{RSC} = 0$ at the receiving end.

4. The full-load voltage profile, which depends on the specification of full-load current, lies above the short-circuit voltage profile.

Figure 6.10

Voltage profiles of an uncompensated lossless line with fixed sending-end voltage for line lengths up to a quarter wavelength

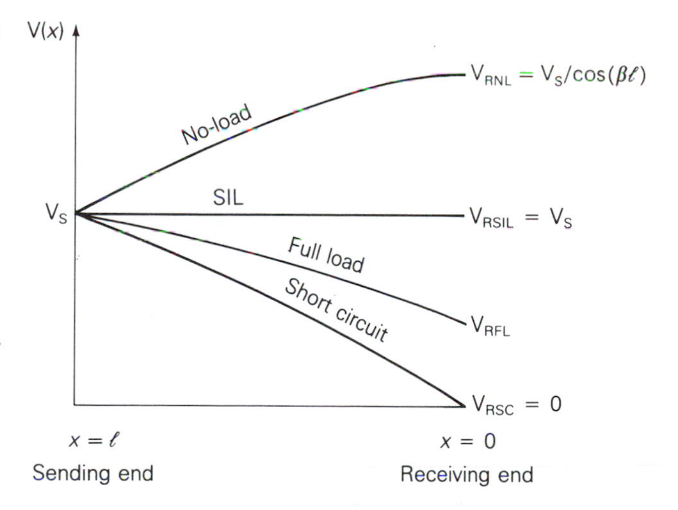

Figure 6.10 summarizes these results, showing a high receiving-end voltage at no-load and a low receiving-end voltage at full load. This voltage regulation problem becomes more severe as the line length increases. In Section 6.6, shunt compensation methods to reduce voltage fluctuations are discussed.

Steady-state Stability Limit

The equivalent π circuit of Figure 6.8 can be used to obtain an equation for the real power delivered by a lossless line. Assume that the voltage magnitudes

V_S and V_R at the ends of the line are held constant. Also, let δ denote the voltage-phase angle at the sending end with respect to the receiving end. From KVL, the receiving-end current I_R is

$$I_R = \frac{V_S - V_R}{Z'} - \frac{Y'}{2} V_R$$

$$= \frac{V_S e^{j\delta} - V_R}{jX'} - \frac{j\omega C' l}{2} V_R \tag{6.4.24}$$

and the complex power S_R delivered to the receiving end is

$$S_R = V_R I_R^* = V_R \left(\frac{V_S e^{j\delta} - V_R}{jX'} \right)^* + \frac{j\omega C' l}{2} V_R^2$$

$$= V_R \left(\frac{V_S e^{-j\delta} - V_R}{-jX'} \right) + \frac{j\omega C l}{2} V_R^2$$

$$= \frac{jV_R V_S \cos\delta + V_R V_S \sin\delta - jV_R^2}{X'} + \frac{j\omega C l}{2} V_R^2 \tag{6.4.25}$$

The real power delivered is

$$P = P_S = P_R = \text{Re}(S_R) = \frac{V_R V_S}{X'} \sin\delta \quad \text{W} \tag{6.4.26}$$

Note that since the line is lossless, $P_S = P_R$.

Equation (6.4.26) is plotted in Figure 6.11. For fixed voltage magnitudes V_S and V_R, the phase angle δ increases from 0 to 90° as the real power delivered increases. The maximum power that the line can deliver, which occurs when $\delta = 90°$, is given by

$$P_{max} = \frac{V_S V_R}{X'} \quad \text{W} \tag{6.4.27}$$

P_{max} represents the theoretical *steady-state stability* limit of a lossless line. If an attempt were made to exceed this steady-state stability limit, then synchronous machines at the sending end would lose synchronism with those at the receiving end. Stability is further discussed in Chapter 12.

It is convenient to express the steady-state stability limit in terms of SIL. Using (6.4.10) in (6.4.26),

$$P = \frac{V_S V_R \sin\delta}{Z_c \sin\beta l} = \left(\frac{V_S V_R}{Z_c} \right) \frac{\sin\delta}{\sin\left(\frac{2\pi l}{\lambda} \right)} \tag{6.4.28}$$

Figure 6.11

Real power delivered by a lossless line versus voltage angle across the line

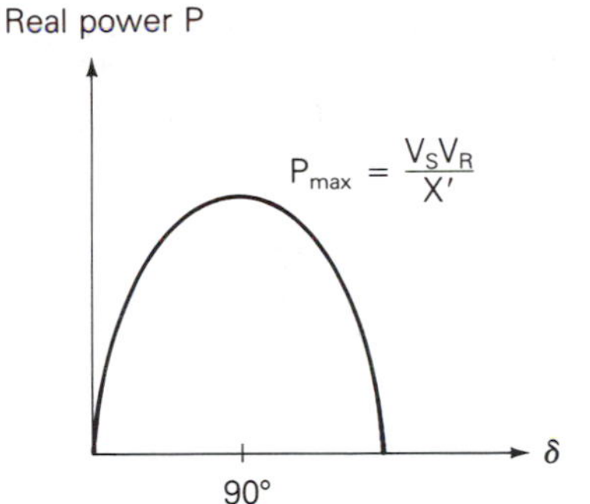

Expressing V_S and V_R in per-unit of rated line voltage,

$$P = \left(\frac{V_S}{V_{rated}}\right)\left(\frac{V_R}{V_{rated}}\right)\left(\frac{V_{rated}^2}{Z_c}\right)\frac{\sin\delta}{\sin\left(\frac{2\pi l}{\lambda}\right)}$$

$$= V_{s.p.u.}V_{R.p.u.}(\text{SIL})\frac{\sin\delta}{\sin\left(\frac{2\pi l}{\lambda}\right)} \quad \text{W} \tag{6.4.29}$$

And for $\delta = 90°$, the theoretical steady-state stability limit is

$$P_{max} = \frac{V_{s.p.u.}V_{R.p.u.}(\text{SIL})}{\sin\left(\frac{2\pi l}{\lambda}\right)} \quad \text{W} \tag{6.4.30}$$

Equations (6.4.27)–(6.4.30) reveal two important factors affecting the steady-state stability limit. First, from (6.4.27), it increases with the square of the line voltage. For example, a doubling of line voltage enables a fourfold increase in maximum power flow. Second, it decreases with line length. Equa-

Figure 6.12

Transmission-line loadability curve for 60-Hz overhead lines—no series or shunt compensation

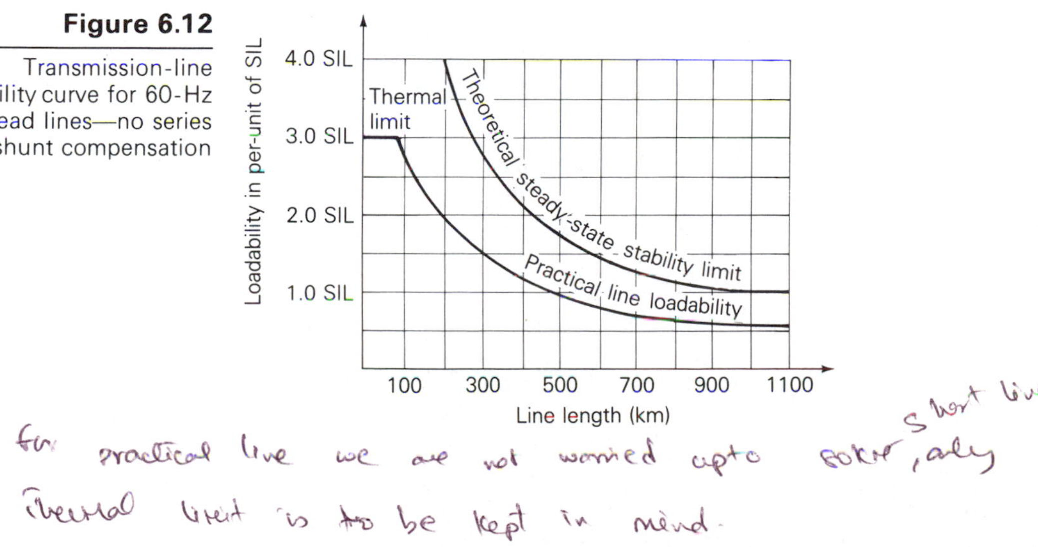

tion (6.4.30) is plotted in Figure 6.12 for $V_{s.p.u.} = V_{R.p.u.} = 1$, $\lambda = 5000$ km, and line lengths up to 1100 km. As shown, the theoretical steady-state stability limit decreases from 4(SIL) for a 200-km line to about 2(SIL) for a 400-km line.

Example 6.4 Neglecting line losses, find the theoretical steady-state stability limit for the 300-km line in Example 6.2. Assume a 266.1-Ω surge impedance, a 5000-km wavelength, and $V_S = V_R = 765$ kV.

Solution From (6.4.21),

$$\text{SIL} = \frac{(765)^2}{266.1} = 2199 \quad \text{MW}$$

From (6.4.30) with $l = 300$ km and $\lambda = 5000$ km,

$$P_{max} = \frac{(1)(1)(2199)}{\sin\left(\frac{2\pi \times 300}{5000}\right)} = (2.716)(2199) = 5974 \quad \text{MW}$$

Alternatively, from Figure 6.12, for a 300-km line, the theoretical steady-state stability limit is (2.72)SIL = (2.72)(2199) = 5980 MW, about the same as the above result. ◁

Section 6.5

Maximum Power Flow

Maximum power flow, discussed in Section 6.4 for lossless lines, is derived here in terms of the *ABCD* parameters for lossy lines. The following notation is used:

$$A = \cosh(\gamma l) = \text{A}\angle\theta_A$$

$$B = Z' = \text{Z}'\angle\theta_Z$$

$$V_S = \text{V}_S\angle\delta \qquad V_R = \text{V}_R\angle 0°$$

Solving (6.2.33) for the receiving-end current,

$$I_R = \frac{V_S - AV_R}{B} = \frac{\text{V}_S e^{j\delta} - \text{A}\text{V}_R e^{j\theta_A}}{\text{Z}' e^{j\theta_Z}} \tag{6.5.1}$$

The complex power delivered to the receiving end is

$$S_R = P_R + jQ_R = V_R I_R^* = V_R\left[\frac{V_S e^{j(\delta-\theta_Z)} - AV_R e^{j(\theta_A-\theta_Z)}}{Z'}\right]^*$$

$$= \frac{V_R V_S}{Z'} e^{j(\theta_Z-\delta)} - \frac{AV_R^2}{Z'} e^{j(\theta_Z-\theta_A)} \tag{6.5.2}$$

The real and reactive power delivered to the receiving end are thus

$$P_R = \text{Re}(S_R) = \frac{V_R V_S}{Z'}\cos(\theta_Z - \delta) - \frac{AV_R^2}{Z'}\cos(\theta_Z - \theta_A) \tag{6.5.3}$$

$$Q_R = \text{Im}(S_R) = \frac{V_R V_S}{Z'}\sin(\theta_Z - \delta) - \frac{AV_R^2}{Z'}\sin(\theta_Z - \theta_A) \tag{6.5.4}$$

Note that for a lossless line, $\theta_A = 0°$, $B = Z' = jX'$, $Z' = X'$, $\theta_Z = 90°$, and (6.5.3) reduces to

$$P_R = \frac{V_R V_S}{X'}\cos(90 - \delta) - \frac{AV_R^2}{X'}\cos(90°)$$

$$= \frac{V_R V_S}{X'}\sin\delta \tag{6.5.5}$$

which is the same as (6.4.26).

lossy line

The theoretical maximum real power delivered (or steady-state stability limit) occurs when $\delta = \theta_Z$ in (6.5.3):

$$P_{R\max} = \frac{V_R V_S}{Z'} - \frac{AV_R^2}{Z'}\cos(\theta_Z - \theta_A) \tag{6.5.6}$$

The second term in (6.5.6), and the fact that Z' is larger than X', reduce $P_{R\max}$ to a value somewhat less than that given by (6.4.27) for a lossless line.

Example 6.5 Determine the theoretical maximum power, in MW and in per-unit of SIL, that the line in Example 6.2 can deliver. Assume $V_S = V_R = 765$ kV.

Solution From Example 6.2,

$$A = 0.9313 \quad \text{per unit}; \quad \theta_A = 0.209°$$

$$B = Z' = 97.0 \quad \Omega; \quad \theta_Z = 87.2°$$

$$Z_c = 266.1 \quad \Omega$$

From (6.5.6) with $V_S = V_R = 765$ kV,

$$P_{R\max} = \frac{(765)^2}{97} - \frac{(0.9313)(765)^2}{97}\cos(87.2° - 0.209°)$$

$$= 6033 - 295 = 5738 \quad \text{MW}$$

From (6.4.20),

$$\text{SIL} = \frac{(765)^2}{266.1} = 2199 \quad \text{MW}$$

Thus

$$P_{R\max} = \frac{5738}{2199} = 2.61 \quad \text{per unit}$$

This value is about 4% less than that found in Example 6.4, where losses were neglected. ◁

Section 6.6

Line Loadability

In practice, power lines are not operated to deliver their theoretical maximum power, which is based on rated terminal voltages and an angular displacement $\delta = 90°$ across the line. Figure 6.12 shows a practical line loadability curve plotted below the theoretical steady-state stability limit. This curve is based on the voltage-drop limit $V_R/V_S \geqslant 0.95$ and on a maximum angular displacement of 30 to 35° across the line (or about 45° across the line and equivalent system reactances), in order to maintain stability during transient disturbances [1, 3]. The curve is valid for typical overhead 60-Hz lines with no compensation. Note that for short lines less than 80 km long, loadability is limited by the thermal rating of the conductors or by terminal equipment ratings, not by voltage drop or stability considerations. In Section 6.7 we investigate series and shunt compensation techniques to increase the loadability of longer lines toward their thermal limit.

Example 6.6 The 300-km uncompensated line in Example 6.2 has four 1,272,000-cmil 54/3 ACSR conductors per bundle. The sending-end voltage is held constant at 1.0 per-unit of rated line voltage. Determine the following:

a. The practical line loadability (Assume an approximate receiving-end voltage $V_R = 0.95$ per unit and $\delta = 35°$ maximum angle across the line.)

b. The full-load current at 0.986 p.f. leading based on the above practical line loadability

c. The exact receiving-end voltage for the full-load current found in part (b)

d. Percent voltage regulation for the above full-load current

e. Thermal limit of the line, based on the approximate current-carrying capacity given in Table A.4

Solution **a.** From (6.5.3), with $V_S = 765$, $V_R = 0.95 \times 765$ kV, and $\delta = 35°$, using the values of Z', θ_Z, A, and θ_A from Example 6.5,

$$P_R = \frac{(765)(0.95 \times 765)}{97.0}\cos(87.2° - 35°) - \frac{(0.9313)(0.95 \times 765)^2}{97.0}\cos(87.2° - 0.209°)$$

$$= 3513 - 266 = 3247 \quad \text{MW}$$

$P_R = 3247$ MW is the practical line loadability, provided the thermal and voltage-drop limits are not exceeded. Alternatively, from Figure 6.12 for a 300-km line, the practical line loadability is (1.49)SIL = (1.49)(2199) = 3277 MW, about the same as the above result.

b. For the above loading at 0.986 p.f. leading and at 0.95 × 765 kV, the full-load receiving-end current is

$$I_{RFL} = \frac{P}{\sqrt{3}V_R(\text{p.f.})} = \frac{3247}{(\sqrt{3})(0.95 \times 765)(0.986)} = 2.616 \quad \text{kA}$$

c. From (6.1.1) with $I_{RFL} = 2.616\underline{/\cos^{-1} 0.986} = 2.616\underline{/9.599°}$ kA, using the A and B parameters from Example 6.2,

$$V_S = AV_{RFL} + BI_{RFL}$$

$$\frac{765}{\sqrt{3}}\underline{/\delta} = (0.9313\underline{/0.209°})\,(V_{RFL}\underline{/0°}) + (97.0\underline{/87.2°})\,(2.616\underline{/9.599°})$$

$$441.7\underline{/\delta} = (0.9313V_{RFL} - 30.04) + j(0.0034V_{RFL} + 251.97)$$

Taking the squared magnitude of the above equation,

$$(441.7)^2 = 0.8673\,V_{RFL}^2 - 54.24V_{RFL} + 64{,}391$$

Solving,

$$V_{RFL} = 420.7 \quad \text{kV}_{LN}$$

$$V_{RFL} = 420.7\sqrt{3} = 728.7 \quad \text{kV}_{LL} = 0.953 \quad \text{per unit}$$

d. From (6.1.19), the receiving-end no-load voltage is

$$V_{RNL} = \frac{V_S}{A} = \frac{765}{0.9313} = 821.4 \quad \text{kV}_{LL}$$

And from (6.1.18),

$$\text{percent VR} = \frac{821.4 - 728.7}{728.7} \times 100 = 12.72\%$$

e. From Table A.4, the approximate current-carrying capacity of four 1,272,000-cmil 54/3 ACSR conductors is $4 \times 1.2 = 4.8$ kA.

Since the voltages $V_S = 1.0$ and $V_{RFL} = 0.953$ per unit satisfy the voltage-drop limit $V_R/V_S \geqslant 0.95$, the factor that limits line loadability is steady-state stability for this 300-km uncompensated line. The full-load current of 2.616 kA corresponding to loadability is also well below the thermal limit of 4.8 kA. The 12.7% voltage regulation is too high because the no-load voltage is too high. Compensation techniques to reduce no-load voltages are discussed in Section 6.7. ◁

Example 6.7 From a hydroelectric power plant 9000 MW are to be transmitted to a load center located 500 km from the plant. Based on practical line loadability criteria, determine the number of three-phase, 60-Hz lines required to transmit this power, with one line out of service, for the following cases: (a) 345-kV lines with $Z_c = 297\,\Omega$; (b) 500-kV lines with $Z_c = 277\,\Omega$; (c) 765-kV lines with $Z_c = 266\,\Omega$. Assume $V_S = 1.0$ per unit, $V_R = 0.95$ per unit, and $\delta = 35°$. Also assume that the lines are uncompensated and widely separated such that there is negligible mutual coupling between them.

Solution **a.** For 345-kV lines, (6.4.21) yields

$$\text{SIL} = \frac{(345)^2}{297} = 401 \quad \text{MW}$$

Neglecting losses, from (6.4.29), with $l = 500$ km and $\delta = 35°$,

$$P = \frac{(1.0)(0.95)(401)\sin(35°)}{\sin\left(\dfrac{2\pi \times 500}{5000}\right)} = (401)(0.927) = 372 \quad \text{MW/line}$$

Alternatively, the practical line loadability curve in Figure 6.12 can be used to obtain $P = (0.93)\text{SIL}$ for typical 500-km overhead 60-Hz uncompensated lines.

In order to transmit 9000 MW with one line out of service,

$$\#\,345\text{-kV lines} = \frac{9000\text{ MW}}{372\text{ MW/line}} + 1 = 24.2 + 1 \approx 26$$

b. For 500-kV lines,

$$\text{SIL} = \frac{(500)^2}{277} = 903 \quad \text{MW}$$

$$P = (903)(0.927) = 837 \quad \text{MW/line}$$

$$\#\,500\text{-kV lines} = \frac{9000}{837} + 1 = 10.8 + 1 \approx 12$$

c. For 765-kV lines,

$$\mathrm{SIL} = \frac{(765)^2}{266} = 2200 \quad \mathrm{MW}$$

$$\mathrm{P} = (2200)(0.927) = 2039 \quad \mathrm{MW/line}$$

$$\#765\text{-kV lines} = \frac{9000}{2039} + 1 = 4.4 + 1 \approx 6$$

Increasing the line voltage from 345 to 765 kV, a factor of 2.2, reduces the required number of lines from 26 to 6, a factor of 4.3. ◁

Example 6.8 Can five instead of six 765-kV lines transmit the required power in Example 6.7 if there are two intermediate substations that divide each line into three 167-km line sections, and if only one line section is out of service?

Solution The lines are shown in Figure 6.13. For simplicity, we neglect line losses. The equivalent π circuit of one 500-km, 765-kV line has a series reactance, from (6.4.10) and (6.4.15),

$$\mathrm{X}' = (266)\sin\left(\frac{2\pi \times 500}{5000}\right) = 156.35 \quad \Omega$$

Combining series/parallel reactances in Figure 6.13, the equivalent reactance of five lines with one line section out of service is

$$\mathrm{X_{eq}} = \frac{1}{5}\left(\frac{2}{3}\mathrm{X}'\right) + \frac{1}{4}\left(\frac{\mathrm{X}'}{3}\right) = 0.2167\ \mathrm{X}' = 33.88 \quad \Omega$$

Then, from (6.4.26) with $\delta = 35°$,

$$\mathrm{P} = \frac{(765)(765 \times 0.95)\sin(35°)}{33.88} = 9412 \quad \mathrm{MW}$$

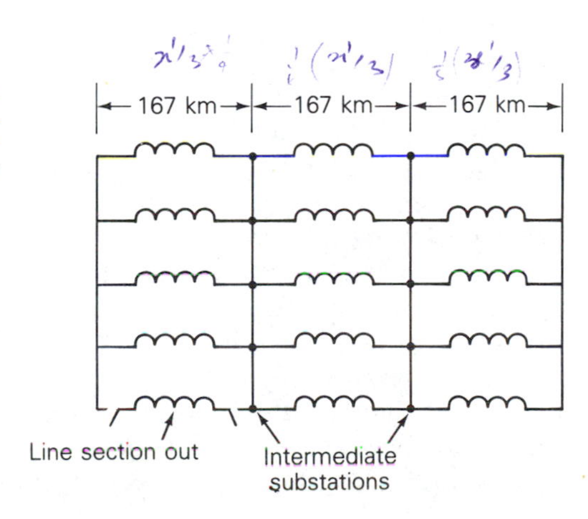

Figure 6.13 Transmission-line configuration for Example 6.8

Inclusion of line losses would reduce the above value by 3 or 4% to about 9100 MW. Therefore, the answer is yes. Five 765-kV, 500-km uncompensated lines with two intermediate substations and with one line section out of service will transmit 9000 MW. Intermediate substations are often economical if their costs do not outweigh the reduction in line costs. ◁

Section 6.7

Reactive Compensation Techniques

Inductors and capacitors are used on medium-length and long transmission lines to increase line loadability and to maintain voltages near rated values. Shunt reactors (inductors) are commonly installed at selected points along EHV lines from each phase to neutral. The inductors absorb reactive power and reduce overvoltages during light load conditions. They also reduce transient overvoltages due to switching and lightning surges. However, shunt reactors can reduce line loadability if they are not removed under full-load conditions.

In addition to shunt reactors, shunt capacitors are sometimes used to deliver reactive power and increase transmission voltages during heavy load conditions. Another type of shunt compensation includes thyristor-switched reactors in parallel with capacitors. These devices, called *static var systems*, can absorb reactive power during light loads and deliver reactive power during heavy loads. Through automatic control of the thyristor switches, voltage fluctuations are minimized and line loadability is increased. Synchronous condensors (synchronous motors with no mechanical load) can also control their reactive power output, although more slowly than static var systems.

Series capacitors are sometimes used on long lines to increase line loadability. Capacitor banks are installed in series with each phase conductor at selected points along a line. Their effect is to reduce the net series impedance of the line in series with the capacitor banks, thereby reducing line-voltage drops and increasing the steady-state stability limit. A disadvantage of series capacitor banks is that automatic protection devices must be installed to bypass high currents during faults and to reinsert the capacitor banks after fault clearing. Also, the addition of series capacitors can excite low-frequency oscillations, a phenomenon called *subsynchronous resonance*, which may damage turbine-generator shafts. Studies have shown, however, that series capacitive compensation can increase the loadability of long lines at only a fraction of the cost of new transmission [1].

Figure 6.14 shows a schematic and an equivalent circuit for a compensated line section, where N_C is the amount of series capacitive compensation expressed in percent of the positive-sequence line impedance and N_L is the

Figure 6.14

Compensated transmission-line section

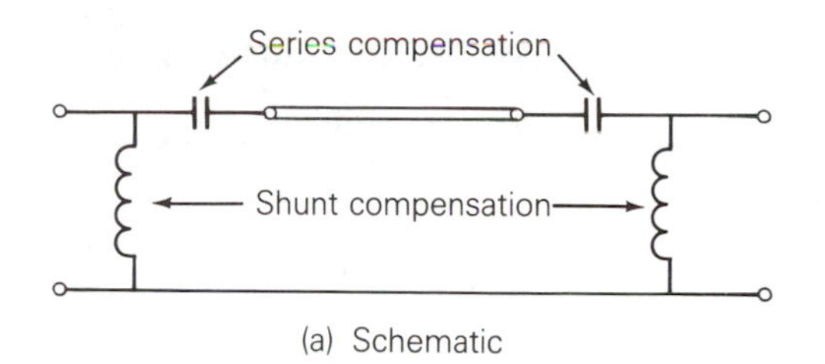

(a) Schematic

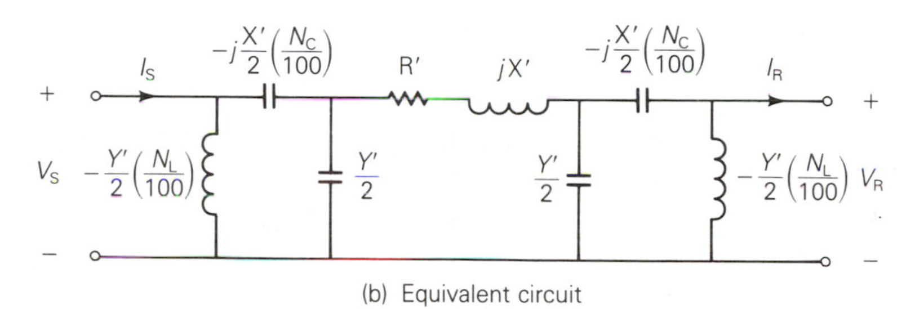

(b) Equivalent circuit

amount of shunt reactive compensation in percent of the positive-sequence line admittance. It is assumed in Figure 6.14 that half of the compensation is installed at each end of the line section. The following two examples illustrate the effect of compensation.

Example 6.9

Identical shunt reactors (inductors) are connected from each phase conductor to neutral at both ends of the 300-km line in Example 6.2 during light load conditions, providing 75% compensation. The reactors are removed during heavy load conditions. Full load is 1.90 kA at unity p.f. and at 730 kV. Assuming that the sending-end voltage is constant, determine the following:

a. Percent voltage regulation of the uncompensated line

b. The equivalent shunt admittance and series impedance of the compensated line

c. Percent voltage regulation of the compensated line

Solution

a. From (6.1.1) with $I_{RFL} = 1.9\underline{/0^\circ}$ kA, using the A and B parameters from Example 6.2,

$$V_S = AV_{RFL} + BI_{RFL}$$

$$= (0.9313\underline{/0.209^\circ})\left(\frac{730}{\sqrt{3}}\underline{/0^\circ}\right) + (97.0\underline{/87.2^\circ})(1.9\underline{/0^\circ})$$

$$= 392.5\underline{/0.209^\circ} + 184.3\underline{/87.2^\circ}$$

$$= 401.5 + j185.5$$

$$= 442.3\underline{/24.8^\circ} \quad \text{kV}_{LN}$$

$$V_S = 442.3\sqrt{3} = 766.0 \quad \text{kV}_{LL}$$

The no-load receiving-end voltage is, from (6.1.19),

$$V_{RNL} = \frac{766.0}{0.9313} = 822.6 \quad kV_{LL}$$

and the percent voltage regulation for the uncompensated line is, from (6.1.18),

$$\text{percent VR} = \frac{822.6 - 730}{730} \times 100 = 12.68\%$$

b. From Example 6.3, the shunt admittance of the equivalent π circuit without compensation is

$$Y' = 2(3.7 \times 10^{-7} + j7.094 \times 10^{-4})$$

$$= 7.4 \times 10^{-7} + j14.188 \times 10^{-4} \quad S$$

With 75% shunt compensation, the equivalent shunt admittance is

$$Y_{eq} = 7.4 \times 10^{-7} + j14.188 \times 10^{-4}(1 - \tfrac{75}{100})$$

$$= 3.547 \times 10^{-4}\underline{/89.88^\circ} \quad S$$

Since there is no series compensation, the equivalent series impedance is the same as without compensation:

$$Z_{eq} = Z' = 97.0\underline{/87.2^\circ} \quad \Omega$$

c. The equivalent A parameter for the compensated line is

$$A_{eq} = 1 + \frac{Y_{eq}Z_{eq}}{2}$$

$$= 1 + \frac{(3.547 \times 10^{-4}\underline{/89.88^\circ})(97.0\underline{/87.2^\circ})}{2}$$

$$= 1 + 0.0172\underline{/177.1^\circ}$$

$$= 0.9828\underline{/0.05^\circ} \quad \text{per unit}$$

Then, from (6.1.19),

$$V_{RNL} = \frac{766}{0.9828} = 779.4 \quad kV_{LL}$$

Since the shunt reactors are removed during heavy load conditions, $V_{RFL} = 730$ kV is the same as without compensation. Therefore

$$\text{percent VR} = \frac{779.4 - 730}{730} \times 100 = 6.77\%$$

The use of shunt reactors at light loads improves the voltage regulation from 12.68% to 6.77% for this line. ◁

Example 6.10 Identical series capacitors are installed in each phase at both ends of the line in Example 6.2, providing 30% compensation. Determine the theoretical maximum power that this compensated line can deliver and compare with that of the uncompensated line. Assume $V_S = V_R = 765$ kV.

Solution From Example 6.3, the equivalent series reactance without compensation is

$$X' = 97.0 \sin 87.2° = 96.88 \quad \Omega$$

Based on 30% series compensation, half at each end of the line, the impedance of each series capacitor is

$$Z_{cap} = -jX_{cap} = -j(\tfrac{1}{2})(0.30)(96.88) = -j14.53 \quad \Omega$$

From Figure 6.4, the *ABCD* matrix of this series impedance is

$$\left[\begin{array}{c|c} 1 & -j14.53 \\ \hline 0 & 1 \end{array}\right]$$

As also shown in Figure 6.4, the equivalent *ABCD* matrix of networks in series is obtained by multiplying the *ABCD* matrices of the individual networks. For this example there are three networks: the series capacitors at the sending end, the line, and the series capacitors at the receiving end. Therefore the equivalent *ABCD* matrix of the compensated line is, using the *ABCD* parameters, from Example 6.2,

$$\left[\begin{array}{c|c} 1 & -j14.53 \\ \hline 0 & 1 \end{array}\right] \left[\begin{array}{c|c} 0.9313\underline{/0.209°} & 97.0\underline{/87.2°} \\ \hline 1.37 \times 10^{-3}\underline{/90.06°} & 0.9313\underline{/0.209°} \end{array}\right] \left[\begin{array}{c|c} 1 & -j14.53 \\ \hline 0 & 1 \end{array}\right]$$

After performing these matrix multiplications, we obtain

$$\left[\begin{array}{c|c} A_{eq} & B_{eq} \\ \hline C_{eq} & D_{eq} \end{array}\right] = \left[\begin{array}{c|c} 0.9512\underline{/0.205°} & 69.70\underline{/86.02°} \\ \hline 1.37 \times 10^{-3}\underline{/90.06°} & 0.9512\underline{/0.205°} \end{array}\right]$$

Therefore

$$A_{eq} = 0.9512 \quad \text{per unit} \qquad \theta_{Aeq} = 0.205°$$

$$B_{eq} = Z'_{eq} = 69.70 \quad \Omega \qquad \theta_{Zeq} = 86.02°$$

From (6.5.6) with $V_S = V_R = 765$ kV,

$$P_{R\max} = \frac{(765)^2}{69.70} - \frac{(0.9512)(765)^2}{69.70}\cos(86.02° - 0.205°)$$

$$= 8396 - 583 = 7813 \quad \text{MW}$$

which is 36.2% larger than the value of 5738 MW found in Example 6.5 without compensation. We note that the practical line loadability of this series compensated line is also about 35% larger than the value of 3247 MW found in Example 6.6 without compensation. ◁

Section 6.8

Personal Computer Program: TRANSMISSION LINES—STEADY-STATE OPERATION

The software package that accompanies this text includes the program entitled "TRANSMISSION LINES—STEADY-STATE OPERATION," which calculates: the *ABCD* parameters, equivalent π circuit values, sending-end quantities for a given receiving-end load, and voltage regulation and maximum power flow for single-phase or balanced three-phase lines, with or without compensation, operating under steady-state conditions.

The input data to the program consist of: (1) line constants R, ωL, G, and ωC; (2) rated line voltage, and line length; (3) receiving-end voltage, full-load MVA and power factor; (4) the number and location of intermediate substations; and (5) the percent shunt reactive (inductive) and percent series capacitive compensation installed at each line terminal and at each intermediate substation. The shunt reactors are removed at heavy loads.

The output data consist of (a) without compensation: (1) the characteristic impedance, (6.2.16); (2) propagation constant, (6.2.12); (3) wavelength, (6.4.15); (4) surge impedance loading, (6.4.21); (5) the equivalent π circuit impedance, (6.3.5) and (6.3.6), and admittance, (6.3.9) and (6.3.10); and (6) the *ABCD* parameters, (6.3.1)–(6.3.3); and (b) with compensation: (1) the equivalent *ABCD* parameters; (2) sending-end voltage, current, and power, (6.1.3); (3) percent voltage regulation, (6.1.18); and (4) the theoretical maximum real power delivered to the receiving end with rated terminal voltages, (6.5.6).

The following example compares the equivalent π circuit quantities R$'$, X$'$, and Y$'$ with the nominal π circuit quantities R, X, and Y for some representative EHV lines.

Example 6.11 The following constants represent typical 345-, 500-, and 765-kV, three-phase, 60-Hz overhead transmission lines:

LINE VOLTAGE	SERIES IMPEDANCE*	SHUNT ADMITTANCE*
kV	z Ω/km	y S/km
345	$0.047 + j0.37$	$j4.1 \times 10^{-6}$
500	$0.020 + j0.34$	$j4.5 \times 10^{-6}$
765	$0.012 + j0.33$	$j4.7 \times 10^{-6}$

*Positive sequence quantities are shown.
Resistances are given at 25°C.

Figure 6.15

Equivalent π to nominal π circuit ratios (R′/R), (X′/X), and (Y′/Y) for typical 345-, 500-, and 765-kV overhead 60-Hz lines

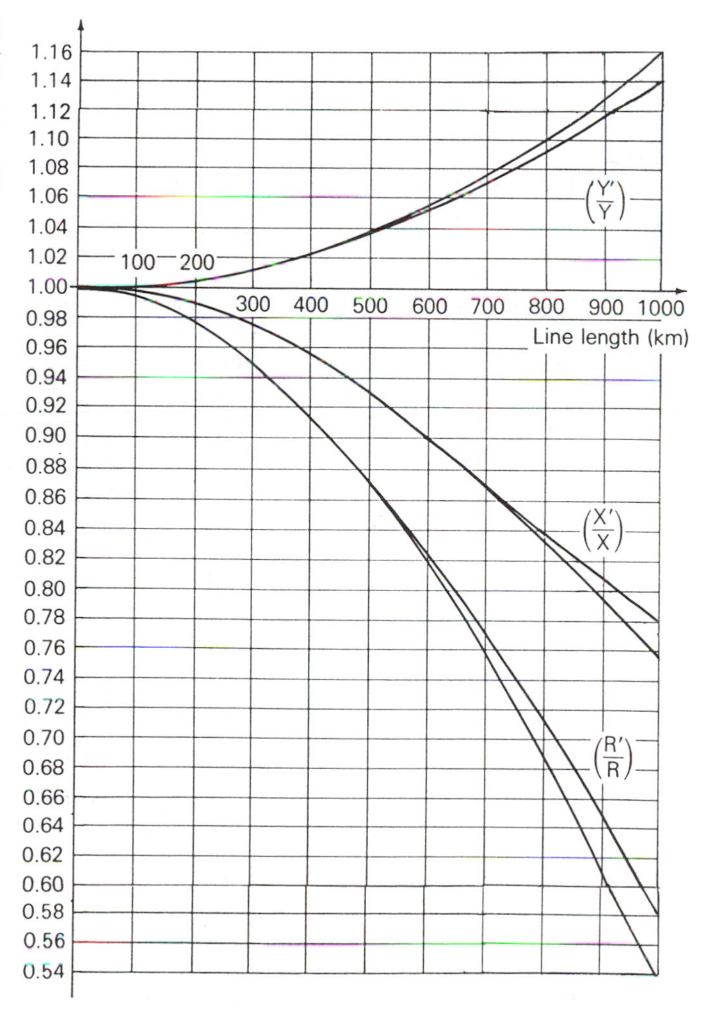

Compute and plot the ratios (R′/R), (X′/X), and (Y′/Y) for the above three lines with line lengths varying from 0 to 1000 km. Neglect the conductances G′ and G.

Solution The TRANSMISSION LINES—STEADY-STATE OPERATION program was run as a subprogram in a main program for the given three lines, with line lengths varying from 25 to 1000 km, in increments of 25 km. The outputs (R′/R), (X′/X), and (Y′/Y) are plotted in Figure 6.15. As shown in the figure, each curve falls within a reasonably narrow band for the three lines. These curves can be used for approximate hand calculations employing typical 60-Hz EHV lines (see Problem 6.25).

The curves also explain why the nominal π circuit gives reasonably accurate results for medum-length lines less than 250 km long. As the line length decreases, all the curves in Figure 6.15 approach unity; that is, the equivalent π circuit approaches the nominal π circuit. ◁

Problems

Section 6.1

6.1 A 25-km 34.5-kV, 60-Hz three-phase line has a positive-sequence series impedance $z = 0.19 + j0.34\ \Omega$/km. The load at the receiving end absorbs 10 MVA at 33 kV. Assuming a short line, calculate: (a) the $ABCD$ parameters, (b) the sending-end voltage for a load power factor of 0.9 lagging, (c) the sending-end voltage for a load power factor of 0.9 leading.

6.2 A 200-km 230-kV, 60-Hz three-phase line has a positive-sequence series impedance $z = 0.08 + j0.48\ \Omega$/km and a positive-sequence shunt admittance $y = j3.33 \times 10^{-6}$ S/km. At full load the line delivers 250 MW at 0.99 p.f. lagging and at 220 kV. Using the nominal π circuit, calculate: (a) the $ABCD$ parameters, (b) the sending-end voltage and current, and (c) the percent voltage regulation.

6.3 Rework Problem 6.2 in per-unit using 100-MVA (three-phase) and 230-kV (line-to-line) base values. Calculate: (a) the per-unit $ABCD$ parameters, (b) the per-unit sending-end voltage and current, and (c) the percent voltage regulation.

6.4 Derive the $ABCD$ parameters for the two networks in series, as shown in Figure 6.4.

6.5 Derive the $ABCD$ parameters for the T circuit shown in Figure 6.4.

Section 6.2

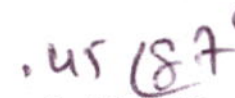

6.6 Evaluate $\cosh(\gamma l)$ and $\tanh(\gamma l/2)$ for $\gamma l = 0.40\underline{/85^\circ}$ per unit.

6.7 A 400-km 500-kV, 60-Hz three-phase uncompensated three-phase line has a positive-sequence series impedance $z = 0.03 + j0.35\ \Omega$/km and a positive-sequence shunt admittance $y = j4.4 \times 10^{-6}$ S/km. Calculate: (a) Z_c, (b) (γl), and (c) the exact $ABCD$ parameters for this line.

6.8 At full load the line in Problem 6.7 delivers 1000 MW at unity power factor and at

475 kV. Calculate: (a) the sending-end voltage, (b) the sending-end current, (c) the sending-end power factor, (d) the full-load line losses, and (e) the percent voltage regulation.

Section 6.3

6.9 Determine the equivalent π circuit for the line in Problem 6.7 and compare it with the nominal π circuit.

Section 6.4

6.10 A 300-km 500-kV, 60-Hz three-phase uncompensated line has a positive-sequence series reactance $x = 0.34\,\Omega/\text{km}$ and a positive-sequence shunt admittance $y = j4.5 \times 10^{-6}$ S/km. Neglecting losses, calculate: (a) Z_c, (b) (γl), (c) the *ABCD* parameters, (d) the wavelength λ of the line, in kilometers, and (e) the surge impedance loading in MW.

6.11 Determine the equivalent π circuit for the line in Problem 6.10.

6.12 Rated line voltage is applied to the sending end of the line in Problem 6.10. Calculate the receiving-end voltage when the receiving end is terminated by (a) an open circuit, (b) the surge impedance of the line, and (c) one-half of the surge impedance. (d) Also calculate the theoretical maximum real power that the line can deliver when rated voltage is applied to both ends of the line.

6.13 From (5.6.22) and (5.10.4), the positive-sequence series inductance and shunt capacitance of a three-phase overhead line are

$$L_1 = 2 \times 10^{-7} \ln (D_{eq}/D_{SL}) = \frac{\mu_0}{2\pi} \ln (D_{eq}/D_{SL}) \quad \text{H/m}$$

$$C_1 = \frac{2\pi\varepsilon_0}{\ln(D_{eq}/D_{SC})} \quad \text{F/m}$$

$$\text{where } \mu_0 = 4\pi \times 10^{-7} \ \text{H/m} \quad \text{and} \quad \varepsilon_0 = \left(\frac{1}{36\pi}\right) \times 10^{-9} \ \text{F/m}$$

Using these equations, determine formulas for surge impedance and velocity of propagation of an overhead lossless line. Then determine the surge impedance and velocity of propagation for the three-phase 765-kV line given in Example 5.10. Assume positive-sequence operation. Neglect line losses as well as the effects of the overhead neutral wires and the earth plane.

Section 6.5

6.14 The line in Problem 6.7 has three ACSR 1113-kcmil conductors per phase. Calculate the theoretical maximum real power that this line can deliver and compare with the thermal limit of the line. Assume $V_S = V_R = 1.0$ per unit and unity power factor at the receiving end.

6.15 Repeat Problems 6.7 and 6.14 if the line length is (a) 200 km, (b) 600 km.

Section 6.6

6.16 For the line in Problems 6.7 and 6.14, determine: (a) the practical line loadability in MW, assuming $V_S = 1.0$ per unit, $V_R \approx 0.95$ per unit, and $\delta_{max} = 35°$; (b) the full-load current at 0.99 p.f. leading, based on the above practical line loadability; (c) the exact receiving-end voltage for the full-load current in (b) above; and (d) the percent voltage regulation. For this line, is loadability determined by the thermal limit, the voltage-drop limit, or steady-state stability?

6.17 Determine the practical line loadability in MW and in per-unit of SIL for the line in

Problem 6.7 if the line length is (a) 200 km, (b) 600 km. Assume $V_S = 1.0$ per unit, $V_R = 0.95$ per unit, and $\delta_{max} = 35°$.

6.18 It is desired to transmit 2000 MW from a power plant to a load center located 300 km from the plant. Determine the number of 60-Hz three-phase, uncompensated transmission lines required to transmit this power with one line out of service for the following cases: (a) 345-kV lines, $Z_c = 300\,\Omega$, (b) 500-kV lines, $Z_c = 275\,\Omega$, (c) 765-kV lines, $Z_c = 260\,\Omega$. Assume that $V_S = 1.0$ per unit, $V_R = 0.95$ per unit, and $\delta_{max} = 35°$.

Section 6.7

6.19 Recalculate the percent voltage regulation in Problem 6.8 when identical shunt reactors are installed at both ends of the line during light loads, providing 65% total shunt compensation. The reactors are removed at full load. Also calculate the impedance of each shunt reactor.

6.20 Identical series capacitors are installed at both ends of the line in Problem 6.7, providing 40% total series compensation. Determine the equivalent *ABCD* parameters of this compensated line. Also calculate the impedance of each series capacitor.

6.21 Determine the theoretical maximum real power that the series compensated line in Problem 6.20 can deliver when $V_S = V_R = 1.0$ per unit. Compare your result with that of Problem 6.14.

6.22 What is the minimum amount of series capacitive compensation N_C in percent of the positive-sequence line reactance needed to reduce the number of 765-kV lines in Example 6.8 from five to four. Assume two intermediate substations with one line section out of service. Also, neglect line losses and assume that the series compensation is sufficiently distributed along the line so as to effectively reduce the series reactance of the equivalent π circuit to $X'(1 - N_C/100)$.

6.23 Determine the equivalent *ABCD* parameters for the line in Problem 6.7 if it has 65% shunt reactive (inductors) compensation and 40% series capacitive compensation. Half of this compensation is installed at each end of the line, as in Figure 6.14.

Section 6.8

6.24 Using the TRANSMISSION LINES—STEADY-STATE OPERATION program, verify the curves in Figure 6.15 for the 500-kV line in Example 6.11 with 100-, 300-, 500-, and 1000-km line lengths.

6.25 Using the curves in Figure 6.15, determine an approximate equivalent π circuit for the line in Problem 6.7 and compare it with the exact equivalent π circuit determined in Problem 6.9. Also determine approximate equivalent π circuits for 200- and 600-km line lengths.

6.26 Using the TRANSMISSION LINES—STEADY-STATE OPERATION program, rework Problems 6.7, 6.8, 6.9, and 6.14.

6.27 Using the TRANSMISSION LINES—STEADY-STATE OPERATION program, increase the receiving-end load at unity power factor and at $V_R = 0.95$ per unit for the line in Problem 6.7 until $V_R/V_S = 0.95$ or $\delta = 35°$, whichever occurs first. Determine whether voltage drop or steady-state stability limits the loadability of this line.

6.28 Repeat Problem 6.27 if the line has 35% total series capacitive compensation, half installed at each end of the line.

6.29 A 650-km 765-kV, 60-Hz three-phase transmission line has a positive-sequence series impedance $z = 0.0185 + j0.3170\,\Omega/\text{km}$ and a positive-sequence shunt admittance

$y = j4.875 \times 10^{-6}$ S/km. Full load at the receiving end is 2500 MVA at 0.99 p.f. leading and at 95% of rated voltage. Two intermediate substations divide the line into three equal line sections. Twenty-two percent total series capacitive compensation and 65% total shunt reactive (inductive) compensation is installed with equal compensation at the line ends and at the intermediate substations. The shunt reactors are removed during heavy loads. Run the TRANSMISSION LINES—STEADY-STATE OPERATION program to determine (a) without compensation: Z_c, γ, λ, the *ABCD* parameters, and the equivalent π circuit impedance and admittance; (b) with compensation: the equivalent *ABCD* parameters, sending-end voltage, sending-end current, sending-end power factor, percent voltage regulation, and the theoretical maximum real power delivered to the receiving end with rated terminal voltages.

References

1. General Electric Company, *Transmission Line Reference Book—345 kV and Above*, 2d ed. (Palo Alto, Ca.: Electric Power Research Institute, 1982).
2. Westinghouse Electric Corporation, *Electrical Transmission and Distribution Reference Book*, 4th ed. (East Pittsburgh, Pa., 1964).
3. R. D. Dunlop, R. Gutman, and P. P. Marchenko, "Analytical Development of Loadability Characteristics for EHV and UHV Lines," *IEEE TRANS. PAS*, Vol. PAS-98, No. 2 (March/April 1979): pp. 606–607.
4. W. D. Stevenson, Jr., *Elements of Power System Analysis*, 4th ed. (New York: McGraw-Hill, 1982).
5. W. H. Hayt, Jr., and J. E. Kemmerly, *Engineering Circuit Analysis*, 3d ed. (New York: McGraw-Hill, 1978).

CHAPTER 7

Power Flows

Successful power-system operation under normal balanced three-phase steady-state conditions requires the following:

1. Generation supplies the demand (load) plus losses.
2. Bus voltage magnitudes remain close to rated values.
3. Generators operate within specified real and reactive power limits.
4. Transmission lines and transformers are not overloaded.

The power-flow computer program (commonly called "load flow") is the basic tool for investigating these requirements. This program computes the voltage magnitude and angle at each bus in a power system under balanced three-phase steady-state conditions. Real and reactive power flows for all equipment interconnecting the buses, as well as equipment losses, are also computed. Both existing power systems and proposed changes including new generation and transmission to meet projected load growth are of interest.

Conventional nodal or loop analysis is not suitable for power-flow studies because the input data for loads is normally given in terms of power, not impedance. Also, generators are considered as power sources, not voltage or current sources. The power-flow problem is therefore formulated as a set of nonlinear algebraic equations suitable for computer solution.

In Sections 7.1–7.3 we review some basic methods, including direct and iterative techniques for solving algebraic equations. Then in Sections 7.4–7.6 the power-flow problem is formulated, computer input data are specified, and two solution methods, Gauss–Seidel and Newton–Raphson, are presented. Sparsity techniques for reducing computer storage and time requirements are introduced in Section 7.7, and the POWER FLOW personal computer program, which accompanies this text, is described in Section 7.8. A fast decoupled power-flow method is described in Section 7.9, and means for controlling power flows are discussed in Section 7.10.

Since balanced three-phase steady-state conditions are assumed, only positive-sequence networks are used in this chapter. Also, all power-flow equations and input/output data are given in per-unit.

Section 7.1

Direct Solutions to Linear Algebraic Equations: Gauss Elimination

Consider the following set of linear algebraic equations in matrix format:

$$\begin{bmatrix} A_{11} & A_{12} & \cdots & A_{1N} \\ A_{21} & A_{22} & \cdots & A_{2N} \\ \vdots & \vdots & & \\ A_{N1} & A_{N2} & \cdots & A_{NN} \end{bmatrix} \begin{bmatrix} x_1 \\ x_2 \\ \vdots \\ x_N \end{bmatrix} = \begin{bmatrix} y_1 \\ y_2 \\ \vdots \\ y_N \end{bmatrix} \tag{7.1.1}$$

or

$$\mathbf{Ax} = \mathbf{y} \tag{7.1.2}$$

where **x** and **y** are N vectors and **A** is an $N \times N$ square matrix. The components of **x**, **y**, and **A** may be real or complex. Given **A** and **y**, it is desired to solve for **x**. We assume the det (**A**) is nonzero, so a unique solution to (7.1.1) exists.

The solution **x** can easily be obtained when **A** is an upper triangular matrix with nonzero diagonal elements. Then (7.1.1) has the form

$$\begin{bmatrix} A_{11} & A_{12} & \cdots & & A_{1N} \\ 0 & A_{22} & \cdots & & A_{2N} \\ \vdots & & & & \\ 0 & 0 & \cdots & A_{N-1,N-1} & A_{N-1,N} \\ 0 & 0 & \cdots & 0 & A_{NN} \end{bmatrix} \begin{bmatrix} x_1 \\ x_2 \\ \vdots \\ x_{N-1} \\ x_N \end{bmatrix} = \begin{bmatrix} y_1 \\ y_2 \\ \vdots \\ y_{N-1} \\ y_N \end{bmatrix} \tag{7.1.3}$$

Since the last equation in (7.1.3) involves only x_N,

$$x_N = \frac{y_N}{A_{NN}} \tag{7.1.4}$$

After x_N is computed, the next-to-last equation can be solved:

$$x_{N-1} = \frac{y_{N-1} - A_{N-1,N}x_N}{A_{N-1,N-1}} \tag{7.1.5}$$

In general, with $x_N, x_{N-1}, \ldots, x_{k+1}$ already computed, the kth equation can be solved

$$x_k = \frac{y_k - \sum_{n=k+1}^{N} A_{kn}x_n}{A_{kk}} \qquad k = N, N-1, \ldots, 1 \tag{7.1.6}$$

This procedure for solving (7.1.3) is called *back substitution*.

If **A** is not upper triangular, (7.1.1) can be transformed to an equivalent equation with an upper triangular matrix. The transformation, called *Gauss elimination*, is described by the following $(N - 1)$ steps. During step 1, we use the first equation in (7.1.1) to eliminate x_1 from the remaining equations. That is, equation 1 is multiplied by A_{n1}/A_{11} and then subtracted from equation n, for $n = 2, 3, \ldots, N$. After completing step 1, we have

$$
\begin{bmatrix}
A_{11} & A_{12} & \cdots & A_{1N} \\
0 & \left(A_{22} - \dfrac{A_{21}}{A_{11}} A_{12}\right) & \cdots & \left(A_{2N} - \dfrac{A_{21}}{A_{11}} A_{1N}\right) \\
0 & \left(A_{32} - \dfrac{A_{31}}{A_{11}} A_{12}\right) & \cdots & \left(A_{3N} - \dfrac{A_{31}}{A_{11}} A_{1N}\right) \\
\vdots & \vdots & & \vdots \\
0 & \left(A_{N2} - \dfrac{A_{N1}}{A_{11}} A_{12}\right) & \cdots & \left(A_{NN} - \dfrac{A_{N1}}{A_{11}} A_{1N}\right)
\end{bmatrix}
\begin{bmatrix} x_1 \\ x_2 \\ x_3 \\ \vdots \\ x_N \end{bmatrix}
\begin{bmatrix} y_1 \\ y_2 - \dfrac{A_{21}}{A_{11}} y_1 \\ y_3 - \dfrac{A_{31}}{A_{11}} y_1 \\ \vdots \\ y_N - \dfrac{A_{N1}}{A_{11}} y_1 \end{bmatrix}
\tag{7.1.7}
$$

Equation (7.1.7) has the following form:

$$
\begin{bmatrix}
A_{11}^{(1)} & A_{12}^{(1)} & \cdots & A_{1N}^{(1)} \\
0 & A_{22}^{(1)} & \cdots & A_{2N}^{(1)} \\
0 & A_{32}^{(1)} & \cdots & A_{3N}^{(1)} \\
\vdots & \vdots & & \vdots \\
0 & A_{N2}^{(1)} & \cdots & A_{NN}^{(1)}
\end{bmatrix}
\begin{bmatrix} x_1 \\ x_2 \\ x_3 \\ \vdots \\ x_N \end{bmatrix}
=
\begin{bmatrix} y_1^{(1)} \\ y_2^{(1)} \\ y_3^{(1)} \\ \vdots \\ y_N^{(1)} \end{bmatrix}
\tag{7.1.8}
$$

where the superscript (1) denotes step 1 of Gauss elimination.

During step 2 we use the second equation in (7.1.8) to eliminate x_2 from the remaining (third, fourth, fifth, and so on) equations. That is, equation 2 is multiplied by $A_{n2}^{(1)}/A_{22}^{(1)}$ and subtracted from equation n, for $n = 3, 4, \ldots, N$. After step 2, we have

$$\begin{bmatrix} A_{11}^{(2)} & A_{12}^{(2)} & A_{13}^{(2)} & \cdots & A_{1N}^{(2)} \\ 0 & A_{22}^{(2)} & A_{23}^{(2)} & \cdots & A_{2N}^{(2)} \\ 0 & 0 & A_{33}^{(2)} & \cdots & A_{3N}^{(2)} \\ 0 & 0 & A_{43}^{(2)} & \cdots & A_{4N}^{(2)} \\ \vdots & \vdots & \vdots & & \vdots \\ 0 & 0 & A_{N3}^{(2)} & \cdots & A_{NN}^{(2)} \end{bmatrix} \begin{bmatrix} x_1 \\ x_2 \\ x_3 \\ x_4 \\ \vdots \\ x_N \end{bmatrix} = \begin{bmatrix} y_1^{(2)} \\ y_2^{(2)} \\ y_3^{(2)} \\ y_4^{(2)} \\ \vdots \\ y_N^{(2)} \end{bmatrix} \tag{7.1.9}$$

During step k, we start with $\mathbf{A}^{(k-1)}\mathbf{x} = \mathbf{y}^{(k-1)}$. The first k of these equations, already triangularized, are left unchanged. Also, equation k is multiplied by $A_{nk}^{(k-1)}/A_{kk}^{(k-1)}$ and then subtracted from equation n, for $n = k+1, k+2, \ldots, N$.

After $(N-1)$ steps, we arrive at the equivalent equation $\mathbf{A}^{(N-1)}\mathbf{x} = \mathbf{y}^{(N-1)}$, where $\mathbf{A}^{(N-1)}$ is upper triangular.

Example 7.1 Solve

$$\left[\begin{array}{c|c} 10 & 5 \\ \hline 2 & 9 \end{array}\right] \begin{bmatrix} x_1 \\ x_2 \end{bmatrix} = \begin{bmatrix} 6 \\ 3 \end{bmatrix}$$

using Gauss elimination and back substitution.

Solution Since $N = 2$ for this example, there is $(N-1) = 1$ Gauss elimination step. Multiplying the first equation by $A_{21}/A_{11} = 2/10$ and then subtracting from the second,

$$\left[\begin{array}{c|c} 10 & 5 \\ \hline 0 & 9 - \frac{2}{10}(5) \end{array}\right] \begin{bmatrix} x_1 \\ x_2 \end{bmatrix} = \begin{bmatrix} 6 \\ 3 - \frac{2}{10}(6) \end{bmatrix}$$

or

$$\left[\begin{array}{c|c} 10 & 5 \\ \hline 0 & 8 \end{array}\right] \begin{bmatrix} x_1 \\ x_2 \end{bmatrix} = \begin{bmatrix} 6 \\ 1.8 \end{bmatrix}$$

which has the form $\mathbf{A}^{(1)}\mathbf{x} = \mathbf{y}^{(1)}$, where $\mathbf{A}^{(1)}$ is upper triangular. Now, using back substitution, (7.1.6) gives, for $k = 2$:

$$x_2 = \frac{y_2^{(1)}}{A_{22}^{(1)}} = \frac{1.8}{8} = 0.225$$

and, for $k = 1$,

$$x_1 = \frac{y_1^{(1)} - A_{12}^{(1)} x_2}{A_{11}^{(1)}} = \frac{6 - (5)(0.225)}{10} = 0.4875 \qquad \triangleleft$$

Example 7.2 Use Gauss elimination to triangularize

$$\begin{bmatrix} 2 & 3 & -1 \\ -4 & 6 & 8 \\ 10 & 12 & 14 \end{bmatrix} \begin{bmatrix} x_1 \\ x_2 \\ x_3 \end{bmatrix} = \begin{bmatrix} 5 \\ 7 \\ 9 \end{bmatrix}$$

Solution There are $(N - 1) = 2$ Gauss elimination steps. During step 1, we subtract $A_{21}/A_{11} = -4/2 = -2$ times equation 1 from equation 2, and we subtract $A_{31}/A_{11} = 10/2 = 5$ times equation 1 from equation 3, to give

$$\left[\begin{array}{c|c|c} 2 & 3 & -1 \\ \hline 0 & 6 - (-2)(3) & 8 - (-2)(-1) \\ \hline 0 & 12 - (5)(3) & 14 - (5)(-1) \end{array}\right] \begin{bmatrix} x_1 \\ x_2 \\ x_3 \end{bmatrix} = \begin{bmatrix} 5 \\ 7 - (-2)(5) \\ 9 - (5)(5) \end{bmatrix}$$

or

$$\begin{bmatrix} 2 & 3 & -1 \\ 0 & 12 & 6 \\ 0 & -3 & 19 \end{bmatrix} \begin{bmatrix} x_1 \\ x_2 \\ x_3 \end{bmatrix} = \begin{bmatrix} 5 \\ 17 \\ -16 \end{bmatrix}$$

which is $\mathbf{A}^{(1)}\mathbf{x} = \mathbf{y}^{(1)}$. During step 2, we subtract $A_{32}^{(1)}/A_{22}^{(1)} = -3/12 = -0.25$ times equation 2 from equation 3, to give

$$\left[\begin{array}{c|c|c} 2 & 3 & -1 \\ \hline 0 & 12 & 6 \\ \hline 0 & 0 & 19-(-.25)(6) \end{array}\right]\begin{bmatrix} x_1 \\ x_2 \\ x_3 \end{bmatrix} = \begin{bmatrix} 5 \\ 17 \\ -16-(-.25)(17) \end{bmatrix}$$

or

$$\begin{bmatrix} 2 & 3 & -1 \\ 0 & 12 & 6 \\ 0 & 0 & 20.5 \end{bmatrix}\begin{bmatrix} x_1 \\ x_2 \\ x_3 \end{bmatrix} = \begin{bmatrix} 5 \\ 17 \\ -11.75 \end{bmatrix}$$

which is triangularized. The solution **x** can now be easily obtained via back substitution. ◁

Computer storage requirements for Gauss elimination and back substitution include N^2 memory locations for **A** and N locations for **y**. If there is no further need to retain **A** and **y**, then $\mathbf{A}^{(k)}$ can be stored in the location of **A**, and $\mathbf{y}^{(k)}$, as well as the solution **x**, can be stored in the location of **y**. Additional memory is also required for DO loops, arithmetic statements, and working space.

Computer time requirements can be evaluated by determining the number of arithmetic operations required for Gauss elimination and back substitution. One can show (see Problem 7.2) that Gauss elimination requires $(N^3 - N)/3$ multiplications, $(N)(N-1)/2$ divisions, and $(N^3 - N)/3$ subtractions. Also, back substitution (see Problem 7.3) requires $(N)(N-1)/2$ multiplications, N divisions, and $(N)(N-1)/2$ subtractions. Therefore, for very large N, the approximate computer time for solving (7.1.1) by Gauss elimination and back substitution is the time required to perform $N^3/3$ multiplications and $N^3/3$ subtractions.

For example, consider a digital computer with a 10^{-6} s multiplication or division time and 10^{-7} s addition or subtraction time. Solving $N = 1000$ equations would require approximately

$$\tfrac{1}{3}N^3(10^{-6}) + \tfrac{1}{3}N^3(10^{-7}) = \tfrac{1}{3}(1000)^3(10^{-6} + 10^{-7}) = 366.7 \quad \text{s}$$

plus some additional bookkeeping time for indexing and managing DO loops.

For matrices that have only a few nonzero elements, special techniques can be employed to significantly reduce computer storage and time requirements. These techniques are discussed in Section 7.7.

Section 7.2

Iterative Solutions to Linear Algebraic Equations: Jacobi and Gauss–Seidel

A general iterative solution to (7.1.1) proceeds as follows. First select an initial guess $\mathbf{x}(0)$. Then use

$$\mathbf{x}(i+1) = \mathbf{g}[\mathbf{x}(i)] \qquad i = 0, 1, 2, \ldots \tag{7.2.1}$$

where $\mathbf{x}(i)$ is the ith guess and $\mathbf{g}$ is an N vector of functions that specify the iteration method. The procedure continues until the following stopping condition is satisfied:

$$\left| \frac{x_k(i+1) - x_k(i)}{x_k(i)} \right| < \varepsilon \qquad \text{for all } k = 1, 2, \ldots, N \tag{7.2.2}$$

where $x_k(i)$ is the kth component of $\mathbf{x}(i)$ and ε is a specified *tolerance level.*

The following questions are pertinent:

1. Will the iteration procedure converge to the unique solution?
2. What is convergence rate (how many iterations are required)?
3. When using a digital computer, what are the computer storage and time requirements?

These questions are addressed for two specific iteration methods: *Jacobi* and *Gauss–Seidel.** The Jacobi method is obtained by considering the kth equation of (7.1.1), as follows:

$$y_k = A_{k1}x_1 + A_{k2}x_2 + \cdots + A_{kk}x_k + \cdots + A_{kN}x_N \tag{7.2.3}$$

Solving for x_k,

$$x_k = \frac{1}{A_{kk}}[y_k - (A_{k1}x_1 + \cdots + A_{k,k-1}x_{k-1} + A_{k,k+1}x_{k+1} + \cdots + A_{kN}x_N)]$$

$$= \frac{1}{A_{kk}}\left[y_k - \sum_{n=1}^{k-1} A_{kn}x_n - \sum_{n=k+1}^{N} A_{kn}x_n\right] \tag{7.2.4}$$

The Jacobi method uses the "old" values of $\mathbf{x}(i)$ at iteration i on the right side of (7.2.4) to generate the "new" value $x_k(i+1)$ on the left side of (7.2.4). That is,

$$x_k(i+1) = \frac{1}{A_{kk}}\left[y_k - \sum_{n=1}^{k-1} A_{kn}x_n(i) - \sum_{n=k+1}^{N} A_{kn}x_n(i)\right] \qquad k = 1, 2, \ldots, N \tag{7.2.5}$$

*The Jacobi method is also called the Gauss method.

The Jacobi method given by (7.2.5) can also be written in the following matrix format:

$$\mathbf{x}(i+1) = \mathbf{M}\,\mathbf{x}(i) + \mathbf{D}^{-1}\mathbf{y} \tag{7.2.6}$$

where

$$\mathbf{M} = \mathbf{D}^{-1}(\mathbf{D} - \mathbf{A}) \tag{7.2.7}$$

and

$$\mathbf{D} = \begin{bmatrix} A_{11} & 0 & 0 & \cdots & 0 \\ 0 & A_{22} & 0 & \cdots & 0 \\ 0 & \vdots & \vdots & & \vdots \\ \vdots & & & & 0 \\ 0 & 0 & 0 & \cdots & A_{NN} \end{bmatrix} \tag{7.2.8}$$

For Jacobi, **D** consists of the diagonal elements of the **A** matrix.

Example 7.3 Solve Example 7.1 using the Jacobi method. Start with $x_1(0) = x_2(0) = 0$ and continue until (7.2.2) is satisfied for $\varepsilon = 10^{-4}$.

Solution From (7.2.5) with $N = 2$,

$$k = 1 \qquad x_1(i+1) = \frac{1}{A_{11}}[y_1 - A_{12}x_2(i)] = \tfrac{1}{10}[6 - 5x_2(i)]$$

$$k = 2 \qquad x_2(i+1) = \frac{1}{A_{22}}[y_2 - A_{21}x_1(i)] = \tfrac{1}{9}[3 - 2x_1(i)]$$

Alternatively, in matrix format, using (7.2.6)–(7.2.8)

$$\mathbf{D}^{-1} = \left[\begin{array}{c|c} 10 & 0 \\ \hline 0 & 9 \end{array}\right]^{-1} = \left[\begin{array}{c|c} \frac{1}{10} & 0 \\ \hline 0 & \frac{1}{9} \end{array}\right]$$

$$\mathbf{M} = \left[\begin{array}{c|c} \frac{1}{10} & 0 \\ \hline 0 & \frac{1}{9} \end{array}\right] \left[\begin{array}{c|c} 0 & -5 \\ \hline -2 & 0 \end{array}\right] = \left[\begin{array}{c|c} 0 & -\frac{5}{10} \\ \hline -\frac{2}{9} & 0 \end{array}\right]$$

$$\begin{bmatrix} x_1(i+1) \\ x_2(i+1) \end{bmatrix} = \left[\begin{array}{c|c} 0 & -\frac{5}{10} \\ \hline -\frac{2}{9} & 0 \end{array}\right] \begin{bmatrix} x_1(i) \\ x_2(i) \end{bmatrix} + \left[\begin{array}{c|c} \frac{1}{10} & 0 \\ \hline 0 & \frac{1}{9} \end{array}\right] \begin{bmatrix} 6 \\ 3 \end{bmatrix}$$

The above two formulations are identical. Starting with $x_1(0) = x_2(0) = 0$, the iterative solution is given in the following table:

Jacobi

i	0	1	2	3	4	5	6	7	8	9	10
$x_1(i)$	0	0.60000	0.43334	0.50000	0.48148	0.48889	0.48683	0.48766	0.48743	0.48752	0.48749
$x_2(i)$	0	0.33333	0.20000	0.23704	0.22222	0.22634	0.22469	0.22515	0.22496	0.22502	0.22500

As shown, the Jacobi method converges to the unique solution obtained in Example 7.1. The convergence criterion is satisfied at the 10th iteration, since

$$\left| \frac{x_1(10) - x_1(9)}{x_1(9)} \right| = \left| \frac{0.48749 - 0.48752}{0.48749} \right| = 6.2 \times 10^{-5} < \varepsilon$$

and

$$\left| \frac{x_2(10) - x_2(9)}{x_2(9)} \right| = \left| \frac{0.22500 - 0.22502}{0.22502} \right| = 8.9 \times 10^{-5} < \varepsilon \qquad \triangleleft$$

The Gauss–Seidel method is given by

$$x_k(i+1) = \frac{1}{A_{kk}} \left[y_k - \sum_{n=1}^{k-1} A_{kn} x_n(i+1) - \sum_{n=k+1}^{N} A_{kn} x_n(i) \right] \tag{7.2.9}$$

Comparing (7.2.9) with (7.2.5), note that Gauss–Seidel is similar to Jacobi except that during each iteration, the "new" values, $x_n(i+1)$, for $n < k$ are used on the right side of (7.2.9) to generate the "new" value $x_k(i + 1)$ on the left side.

The Gauss–Seidel method of (7.2.9) can also be written in the matrix format of (7.2.6) and (7.2.7), where

$$\mathbf{D} = \begin{bmatrix} A_{11} & 0 & 0 \cdots 0 \\ A_{21} & A_{22} & 0 \cdots 0 \\ \vdots & \vdots & \vdots \\ A_{N1} & A_{N2} & \cdots A_{NN} \end{bmatrix} \tag{7.2.10}$$

For Gauss–Seidel, **D** in (7.2.10) is the lower triangular portion of **A**, whereas for Jacobi, **D** in (7.2.8) is the diagonal portion of **A**.

Example 7.4 Rework Example 7.3 using the Gauss–Seidel method.

Solution From (7.2.9),

$$k = 1 \qquad x_1(i+1) = \frac{1}{A_{11}}[y_1 - A_{12}x_2(i)] = \tfrac{1}{10}[6 - 5x_2(i)]$$

$$k = 2 \qquad x_2(i+1) = \frac{1}{A_{22}}[y_2 - A_{21}x_1(i+1)] = \tfrac{1}{9}[3 - 2x_1(i+1)]$$

Using this equation for $x_1(i + 1)$, $x_2(i + 1)$ can also be written as

$$x_2(i+1) = \tfrac{1}{9}\{3 - \tfrac{2}{10}[6 - 5x_2(i)]\}$$

Alternatively, in matrix format, using (7.2.10), (7.2.6), and (7.2.7):

$$\mathbf{D}^{-1} = \left[\begin{array}{c|c} 10 & 0 \\ \hline 2 & 9 \end{array}\right]^{-1} = \left[\begin{array}{c|c} \frac{1}{10} & 0 \\ \hline -\frac{2}{90} & \frac{1}{9} \end{array}\right]$$

$$\mathbf{M} = \begin{bmatrix} \frac{1}{10} & 0 \\ -\frac{2}{90} & \frac{1}{9} \end{bmatrix} \begin{bmatrix} 0 & -5 \\ 0 & 0 \end{bmatrix} = \begin{bmatrix} 0 & -\frac{1}{2} \\ 0 & \frac{1}{9} \end{bmatrix}$$

$$\begin{bmatrix} x_1(i+1) \\ x_2(i+1) \end{bmatrix} = \begin{bmatrix} 0 & -\frac{1}{2} \\ 0 & \frac{1}{9} \end{bmatrix} \begin{bmatrix} x_1(i) \\ x_2(i) \end{bmatrix} + \begin{bmatrix} \frac{1}{10} & 0 \\ -\frac{2}{90} & \frac{1}{9} \end{bmatrix} \begin{bmatrix} 6 \\ 3 \end{bmatrix}$$

These two formulations are identical. Starting with $x_1(0) = x_2(0) = 0$, the solution is given in the following table:

Gauss–Seidel

i	0	1	2	3	4	5	6
$x_1(i)$	0	0.60000	0.50000	0.48889	0.48765	0.48752	0.48750
$x_2(i)$	0	0.20000	0.22222	0.22469	0.22497	0.22500	0.22500

For this example, Gauss–Seidel converges in 6 iterations, compared to 10 iterations with Jacobi. ◁

The convergence rate is faster with Gauss–Seidel for some **A** matrices, but faster with Jacobi for other **A** matrices. In some cases, one method diverges while the other converges. In other cases both methods diverge, as illustrated by the next example.

Example 7.5 Using the Gauss–Seidel method with $x_1(0) = x_2(0) = 0$, solve

$$\begin{bmatrix} 5 & 10 \\ 9 & 2 \end{bmatrix} \begin{bmatrix} x_1 \\ x_2 \end{bmatrix} = \begin{bmatrix} 6 \\ 3 \end{bmatrix}$$

Solution Note that these equations are the same as those in Example 7.1, except that x_1 and x_2 are interchanged. Using (7.2.9),

$$k = 1 \qquad x_1(i+1) = \frac{1}{A_{11}}[y_1 - A_{12}x_2(i)] = \tfrac{1}{5}[6 - 10x_2(i)]$$

$$k = 2 \qquad x_2(i+1) = \frac{1}{A_{22}}[y_2 - A_{21}x_2(i+1)] = \tfrac{1}{2}[3 - 9x_1(i+1)]$$

Successive calculations of x_1 and x_2 are shown in the following table:

Gauss–Seidel

i	0	1	2	3	4	5
$x_1(i)$	0	1.2	9	79.2	711	6397
$x_2(i)$	0	−3.9	−39	−354.9	−3198	−28786

The unique solution by matrix inversion is

$$\begin{bmatrix} x_1 \\ x_2 \end{bmatrix} = \begin{bmatrix} 5 & 10 \\ 9 & 2 \end{bmatrix}^{-1} \begin{bmatrix} 6 \\ 3 \end{bmatrix} = \frac{-1}{80} \begin{bmatrix} 2 & -10 \\ -9 & 5 \end{bmatrix} \begin{bmatrix} 6 \\ 3 \end{bmatrix} = \begin{bmatrix} 0.225 \\ 0.4875 \end{bmatrix}$$

As shown, Gauss–Seidel does not converge to the unique solution; instead it diverges. We could show that Jacobi also diverges for this example. ◁

If any diagonal element A_{kk} equals zero, then Jacobi and Gauss–Seidel are undefined, because the right-hand sides of (7.2.5) and (7.2.9) are divided by A_{kk}. Also, if any one diagonal element has too small a magnitude, these methods will diverge. In Examples 7.3 and 7.4 Jacobi and Gauss–Seidel converge, since the diagonals (10 and 9) are both large; in Example 7.5, however, the diagonals (5 and 2) are small compared to the off-diagonals, and the methods diverge.

In general, convergence of Jacobi or Gauss–Seidel can be evaluated by recognizing that (7.2.6) represents a digital filter with input **y** and output $\mathbf{x}(i)$. The z-transform of (7.2.6) may be employed to determine the filter transfer function and its poles. The output $\mathbf{x}(i)$ converges if and only if all the filter poles have magnitudes less than 1 (see Problems 7.8, 7.9, and 7.10).

Rate of convergence is also established by the filter poles. Fast convergence is obtained when the magnitudes of all the poles are small. In addition, experience with specific **A** matrices has shown that more iterations are required for Jacobi and Gauss–Seidel as the dimension N increases.

Computer storage requirements for Jacobi include N^2 memory locations for the **A** matrix and $3N$ locations for the vectors **y**, $\mathbf{x}(i)$ and $\mathbf{x}(i+1)$. Storage space is also required for DO loops, arithmetic statements, and working space to compute (7.2.5). Gauss–Seidel requires N fewer memory locations, since for (7.2.9) the new value $x_k(i+1)$ can be stored in the location of the old value $x_k(i)$.

Computer time per iteration is relatively small for Jacobi and Gauss–Seidel. Inspection of (7.2.5) or (7.2.9) shows that N^2 multiplications/divisions and $N(N-1)$ subtractions per iteration are required [one division, $(N-1)$ multiplications, and $(N-1)$ subtractions for each $k = 1, 2, \ldots, N$]. For example, consider a digital computer with a 10^{-6} s multiplication or division time and a 10^{-7} s addition or subtraction time. Solving $N = 1000$ equations by Jacobi or Gauss–Seidel would require $N^2 \times 10^{-6} + N(N-1) \times 10^{-7} = (1000)^2 \times 10^{-6} + 1000(999) \times 10^{-7} = 1.1$ s per iteration plus some additional bookkeeping time for indexing and managing DO loops.

Section 7.3

Iterative Solutions to Nonlinear Algebraic Equations: Newton–Raphson

A set of nonlinear algebraic equations in matrix format is given by

$$\mathbf{f}(\mathbf{x}) = \begin{bmatrix} f_1(\mathbf{x}) \\ f_2(\mathbf{x}) \\ \vdots \\ f_N(\mathbf{x}) \end{bmatrix} = \mathbf{y} \tag{7.3.1}$$

where $\mathbf{y}$ and $\mathbf{x}$ are N vectors and $\mathbf{f}(\mathbf{x})$ is an N vector of functions. Given $\mathbf{y}$ and $\mathbf{f}(\mathbf{x})$, we want to solve for $\mathbf{x}$. The iterative methods described in Section 7.2 can be extended to nonlinear equations as follows. Rewriting (7.3.1),

$$\mathbf{0} = \mathbf{y} - \mathbf{f}(\mathbf{x}) \tag{7.3.2}$$

Adding $\mathbf{Dx}$ to both sides of (7.3.2), where $\mathbf{D}$ is a square $N \times N$ invertible matrix,

$$\mathbf{Dx} = \mathbf{Dx} + \mathbf{y} - \mathbf{f}(\mathbf{x}) \tag{7.3.3}$$

Premultiplying by $\mathbf{D}^{-1}$,

$$\mathbf{x} = \mathbf{x} + \mathbf{D}^{-1}[\mathbf{y} - \mathbf{f}(\mathbf{x})] \tag{7.3.4}$$

The old values $\mathbf{x}(i)$ are used on the right side of (7.3.4) to generate the new values $\mathbf{x}(i+1)$ on the left side. That is,

$$\mathbf{x}(i+1) = \mathbf{x}(i) + \mathbf{D}^{-1}\{\mathbf{y} - \mathbf{f}[\mathbf{x}(i)]\} \tag{7.3.5}$$

For linear equations, $\mathbf{f}(\mathbf{x}) = \mathbf{Ax}$ and (7.3.5) reduces to

$$\mathbf{x}(i+1) = \mathbf{x}(i) + \mathbf{D}^{-1}[\mathbf{y} - \mathbf{Ax}(i)] = \mathbf{D}^{-1}(\mathbf{D} - \mathbf{A})\mathbf{x}(i) + \mathbf{D}^{-1}\mathbf{y} \tag{7.3.6}$$

which is identical to the Jacobi and Gauss–Seidel methods of (7.2.6). For nonlinear equations, the matrix $\mathbf{D}$ in (7.3.5) must be specified.

One method for specifying $\mathbf{D}$, called *Newton–Raphson*, is based on the following Taylor series expansion of $\mathbf{f}(\mathbf{x})$ about an operating point $\mathbf{x}_0$.

$$\mathbf{y} = \mathbf{f}(\mathbf{x}_0) + \left.\frac{d\mathbf{f}}{d\mathbf{x}}\right|_{\mathbf{x}=\mathbf{x}_0} (\mathbf{x} - \mathbf{x}_0) \quad \ldots \tag{7.3.7}$$

Neglecting the higher order terms in (7.3.7) and solving for **x**,

$$\mathbf{x} = \mathbf{x}_0 + \left[\frac{d\mathbf{f}}{d\mathbf{x}}\bigg|_{\mathbf{x}=\mathbf{x}_0}\right]^{-1} [\mathbf{y} - \mathbf{f}(\mathbf{x}_0)] \tag{7.3.8}$$

The Newton–Raphson method replaces $\mathbf{x}_0$ by the old value $\mathbf{x}(i)$ and $\mathbf{x}$ by the new value $\mathbf{x}(i+1)$ in (7.3.8), Thus,

$$\mathbf{x}(i+1) = \mathbf{x}(i) + \mathbf{J}^{-1}(i)\{\mathbf{y} - \mathbf{f}[\mathbf{x}(i)]\} \tag{7.3.9}$$

where

$$\mathbf{J}(i) = \frac{d\mathbf{f}}{d\mathbf{x}}\bigg|_{\mathbf{x}=\mathbf{x}(i)} = \begin{bmatrix} \frac{\partial f_1}{\partial x_1} & \frac{\partial f_1}{\partial x_2} & \cdots & \frac{\partial f_1}{\partial x_N} \\ \frac{\partial f_2}{\partial x_1} & \frac{\partial f_2}{\partial x_2} & \cdots & \frac{\partial f_2}{\partial x_N} \\ \vdots & \vdots & & \vdots \\ \frac{\partial f_N}{\partial x_1} & \frac{\partial f_N}{\partial x_2} & \cdots & \frac{\partial f_N}{\partial x_N} \end{bmatrix}_{\mathbf{x}=\mathbf{x}(i)} \tag{7.3.10}$$

The $N \times N$ matrix $\mathbf{J}(i)$, whose elements are the partial derivatives shown in (7.3.10), is called the Jacobian matrix. The Newton–Raphson method is similar to extended Gauss–Seidel, except that $\mathbf{D}$ in (7.3.5) is replaced by $\mathbf{J}(i)$ in (7.3.9).

Example 7.6 Solve the scalar equation $f(x) = y$, where $y = 9$ and $f(x) = x^2$. Starting with $x(0) = 1$, use (a) Newton–Raphson and (b) extended Gauss–Seidel with $D = 3$ until (7.2.2) is satisfied for $\varepsilon = 10^{-4}$. Compare the two methods.

Solution **a.** Using (7.3.10) with $f(x) = x^2$.

$$J(i) = \frac{d}{dx}(x^2)\bigg|_{x=x(i)} = 2x\bigg|_{x=x(i)} = 2x(i)$$

Using $J(i)$ in (7.3.9),

$$x(i+1) = x(i) + \frac{1}{2x(i)}[9 - x^2(i)]$$

Starting with $x(0) = 1$, successive calculations of the Newton–Raphson equation are shown in the following table:

Newton–Raphson

i	0	1	2	3	4	5
$x(i)$	1	5.00000	3.40000	3.02353	3.00009	3.00000

b. Using (7.3.5) with D = 3, the Gauss–Seidel method is

$$x(i+1) = x(i) + \tfrac{1}{3}[9 - x^2(i)]$$

The corresponding Gauss–Seidel calculations are as follows:

Gauss–Seidel (D = 3)

i	0	1	2	3	4	5	6
$x(i)$	1	3.66667	2.18519	3.59351	2.28908	3.54245	2.35945

As shown, Gauss–Seidel oscillates about the solution, slowly converging, whereas Newton–Raphson converges in five iterations to the solution $x = 3$. Note that if $x(0)$ is negative, Newton–Raphson converges to the negative solution $x = -3$. Also, it is assumed that the matrix inverse $\mathbf{J}^{-1}$ exists. Thus the initial value $x(0) = 0$ should be avoided for this example. ◁

Example 7.7 Solve

$$\begin{bmatrix} x_1 + x_2 \\ x_1 x_2 \end{bmatrix} = \begin{bmatrix} 15 \\ 50 \end{bmatrix} \qquad \mathbf{x}(0) = \begin{bmatrix} 4 \\ 9 \end{bmatrix}$$

Use the Newton–Raphson method starting with the above $\mathbf{x}(0)$ and continue until (7.2.2) is satisfied with $\varepsilon = 10^{-4}$.

Solution Using (7.3.10) with $f_1 = (x_1 + x_2)$ and $f_2 = x_1 x_2$,

$$\mathbf{J}(i)^{-1} = \begin{bmatrix} \dfrac{\partial f_1}{\partial x_1} & \dfrac{\partial f_1}{\partial x_2} \\ \dfrac{\partial f_2}{\partial x_1} & \dfrac{\partial f_2}{\partial x_2} \end{bmatrix}^{-1}_{\mathbf{x}=\mathbf{x}(i)} = \begin{bmatrix} 1 & 1 \\ x_2(i) & x_1(i) \end{bmatrix}^{-1} = \frac{\begin{bmatrix} x_1(i) & -1 \\ -x_2(i) & 1 \end{bmatrix}}{x_1(i) - x_2(i)}$$

Using $\mathbf{J}(i)^{-1}$ in (7.3.9),

$$\begin{bmatrix} x_1(i+1) \\ x_2(i+1) \end{bmatrix} = \begin{bmatrix} x_1(i) \\ x_2(i) \end{bmatrix} + \frac{\begin{bmatrix} x_1(i) & -1 \\ -x_2(i) & 1 \end{bmatrix}}{x_1(i) - x_2(i)} \begin{bmatrix} 15 - x_1(i) - x_2(i) \\ 50 - x_1(i)x_2(i) \end{bmatrix}$$

Writing the preceding as two separate equations,

$$x_1(i+1) = x_1(i) + \frac{x_1(i)[15 - x_1(i) - x_2(i)] - [50 - x_1(i)x_2(i)]}{x_1(i) - x_2(i)}$$

$$x_2(i+1) = x_2(i) + \frac{-x_2(i)[15 - x_1(i) - x_2(i)] + [50 - x_1(i)x_2(i)]}{x_1(i) - x_2(i)}$$

Successive calculations of these equations are shown in the following table:

Newton–Raphson

i	0	1	2	3	4
$x_1(i)$	4	5.20000	4.99130	4.99998	5.00000
$x_2(i)$	9	9.80000	10.00870	10.00002	10.00000

Newton–Raphson converges in four iterations for this example. ◁

Equation (7.3.9) contains the matrix inverse $\mathbf{J}^{-1}$. Instead of computing $\mathbf{J}^{-1}$, (7.3.9) can be rewritten as follows:

$$\mathbf{J}(i)\,\Delta\mathbf{x}(i) = \Delta\mathbf{y}(i) \tag{7.3.11}$$

where

$$\Delta\mathbf{x}(i) = \mathbf{x}(i+1) - \mathbf{x}(i) \tag{7.3.12}$$

and

$$\Delta\mathbf{y}(i) = \mathbf{y} - \mathbf{f}[\mathbf{x}(i)] \tag{7.3.13}$$

Then, during each iteration, the following four steps are completed:

Step 1 Compute $\Delta\mathbf{y}(i)$ from (7.3.13).

Step 2 Compute $\mathbf{J}(i)$ from (7.3.10).

Step 3 Using Gauss elimination and back substitution, solve (7.3.11) for $\Delta\mathbf{x}(i)$.

Step 4 Compute $\mathbf{x}(i+1)$ from (7.3.12).

Example 7.8 Complete the above four steps for the first iteration of Example 7.7.

Solution

Step 1 $\Delta\mathbf{y}(0) = \mathbf{y} - \mathbf{f}[\mathbf{x}(0)] = \begin{bmatrix} 15 \\ 50 \end{bmatrix} - \begin{bmatrix} 4+9 \\ (4)(9) \end{bmatrix} = \begin{bmatrix} 2 \\ 14 \end{bmatrix}$

Step 2 $\mathbf{J}(0) = \left[\begin{array}{c|c} 1 & 1 \\ \hline x_2(0) & x_1(0) \end{array}\right] = \left[\begin{array}{c|c} 1 & 1 \\ \hline 9 & 4 \end{array}\right]$

Step 3 Using $\Delta\mathbf{y}(0)$ and $\mathbf{J}(0)$, (7.3.11) becomes

$$\left[\begin{array}{c|c} 1 & 1 \\ \hline 9 & 4 \end{array}\right]\begin{bmatrix} \Delta x_1(0) \\ \Delta x_2(0) \end{bmatrix} = \begin{bmatrix} 2 \\ 14 \end{bmatrix}$$

Using Gauss elimination, subtract $J_{21}/J_{11} = 9/1 = 9$ times the first equation from the second equation, giving

$$\left[\begin{array}{c|c} 1 & 1 \\ \hline 0 & -5 \end{array}\right]\begin{bmatrix} \Delta x_1(0) \\ \Delta x_2(0) \end{bmatrix} = \begin{bmatrix} 2 \\ -4 \end{bmatrix}$$

Solving by back substitution,

$$\Delta x_2(0) = \frac{-4}{-5} = 0.8$$

$$\Delta x_1(0) = 2 - 0.8 = 1.2$$

Step 4 $\mathbf{x}(1) = \mathbf{x}(0) + \Delta\mathbf{x}(0) = \begin{bmatrix} 4 \\ 9 \end{bmatrix} + \begin{bmatrix} 1.2 \\ 0.8 \end{bmatrix} = \begin{bmatrix} 5.2 \\ 9.8 \end{bmatrix}$

This is the same as computed in Example 7.7. ◁

Experience from power-flow studies has shown that Newton–Raphson converges in many cases where Jacobi and Gauss–Seidel diverge. Furthermore, the number of iterations required for convergence is independent of the dimension N for Newton–Raphson, but increases with N for Jacobi and Gauss–Seidel. Most Newton–Raphson power-flow problems converge in less than ten iterations [1].

The disadvantage of Newton–Raphson compared to Jacobi and Gauss–Seidel is that more computer storage and computer time per iteration are required. Newton–Raphson requires the storage space of Jacobi plus N^2 memory locations for $\mathbf{J}(i)$ and working space to solve (7.3.11). Also, as shown in Section 7.1, Gauss elimination and back substitution require approximately $\frac{1}{3}N^3$ multiplications and $\frac{1}{3}N^3$ subtractions to solve (7.3.11) for large N. For example, solving 1000 equations on a digital computer with a 10^{-6}-s multiplication or division time and a 10^{-7}-s addition or subtraction time would require 366.7 s per iteration plus some additional bookkeeping time for

Newton–Raphson. In Section 7.7, we discuss sparsity techniques that significantly reduce the storage and time requirements of Newton–Raphson.

Section 7.4

The Power-Flow Problem

The power-flow problem is the computation of voltage magnitude and phase angle at each bus in a power system under balanced three-phase steady-state conditions. As a by-product of this calculation, real and reactive power flows in equipment such as transmission lines and transformers, as well as equipment losses, can be computed.

The starting point for a power-flow problem is a single-line diagram of the power system, from which the input data for computer solutions can be obtained. Input data consist of bus data, transmission line data, and transformer data.

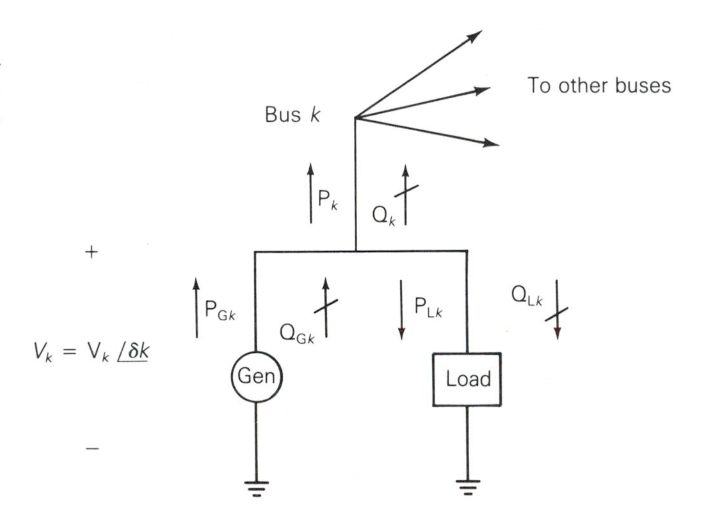

Figure 7.1 Bus variables V_k, δ_k, P_k, and Q_k

As shown in Figure 7.1, the following four variables are associated with each bus k: voltage magnitude V_k, phase angle δ_k, net real power P_k, and reactive power Q_k supplied to the bus. At each bus, two of these variables are specified as input data, and the other two are unknowns to be computed by the power-flow program. For convenience, the power delivered to bus k in Figure

7.1 is separated into generator and load terms. That is,

$$P_k = P_{Gk} - P_{Lk}$$
$$Q_k = Q_{Gk} - Q_{Lk} \qquad (7.4.1)$$

Each bus k is categorized into one of the following three bus types:

1. Swing bus—There is only one swing bus, which for convenience is numbered bus 1 in this text. The swing bus is a reference bus for which $V_1\angle\delta_1$, typically $1.0\angle 0°$ per unit, is input data. The power-flow program computes P_1 and Q_1.
2. Load bus—P_k and Q_k are input data. The power-flow program computes V_k and δ_k. Most buses in a typical power-flow program are load buses.
3. Voltage controlled bus—P_k and V_k are input data. The power-flow program computes Q_k and δ_k. Examples are buses to which generators, switched shunt capacitors, or static var systems are connected. Maximum and minimum var limits Q_{Gkmax} and Q_{Gkmin} that this equipment can supply are also input data. Another example is a bus to which a tap-changing transformer is connected; the power-flow program then computes the tap setting.

Note that when bus k is a load bus with no generation, $P_k = -P_{Lk}$ is negative; that is, the real power supplied to bus k in Figure 7.1 is negative. If the load is inductive, $Q_k = -Q_{Lk}$ is negative.

Transmission lines are represented by the equivalent π circuit, shown in Figure 6.7. Transformers are also represented by equivalent circuits, as shown in Figure 4.9 for a two-winding transformer, Figure 4.21 for a three-winding transformer, or Figure 4.27 for a tap-changing transformer.

Input data for each transmission line include the per-unit equivalent π circuit series impedance Z' and shunt admittance Y', the two buses to which the line is connected, and maximum MVA rating. Similarly, input data for each transformer include per-unit winding impedances Z, the per-unit exciting branch admittance Y, the buses to which the windings are connected, and maximum MVA ratings. Input data for tap-changing transformers also include maximum tap settings.

The bus admittance matrix $\boldsymbol{Y}_{bus}$ can be constructed from the line and transformer input data. From (2.4.3) and (2.4.4), the elements of $\boldsymbol{Y}_{bus}$ are:

Diagonal elements: Y_{kk} = sum of admittances connected to bus k

Off-diagonal elements: Y_{kn} = −(sum of admittances connected between buses k and n) $\quad k \neq n \qquad (7.4.2)$

Example 7.9 Figure 7.2 shows a single-line diagram of a five-bus power system. Input data are given in Tables 7.1, 7.2, and 7.3. As shown in Table 7.1, bus 1, to which a generator is connected, is the swing bus. Bus 3, to which a generator and a

load are connected, is a voltage-controlled bus. Buses 2, 4, and 5 are load buses. Note that the loads at buses 2 and 3 are inductive since $Q_2 = -Q_{L2} = -0.7$ and $-Q_{L3} = -0.1$ are negative.

For each bus k, determine which of the variables V_k, δ_k, P_k, and Q_k are input data and which are unknowns. Also, compute the elements of the second row of $\boldsymbol{Y}_{\text{bus}}$.

Solution The input data and unknowns are listed in Table 7.4. For bus 1, the swing bus, P_1 and Q_1 are unknowns. For bus 3, a voltage-controlled bus, Q_3 and δ_3 are unknowns. For buses 2, 4, and 5, load buses, V_2, V_4, V_5 and δ_2, δ_4, δ_5 are unknowns.

The elements of $\boldsymbol{Y}_{\text{bus}}$ are computed from (7.4.2). Since buses 1 and 3 are not directly connected to bus 2,

$$Y_{21} = Y_{23} = 0$$

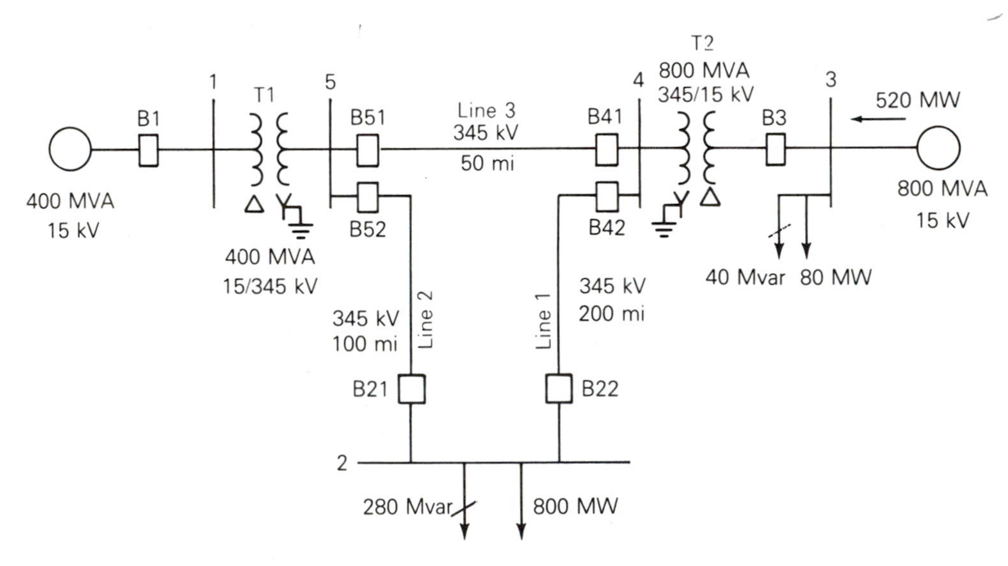

Figure 7.2 Single-line diagram for Example 7.9

Table 7.1 Bus input data for Example 7.9*

BUS	TYPE	V	δ	P_G	Q_G	P_L	Q_L	Q_{Gmax}	Q_{Gmin}
		per unit	degrees	per unit	per unit	per unit	per unit	per unit	per unit
1	Swing	1.0	0	—	—	0	0	—	—
2	Load	—	—	0	0	2.0	0.7	—	—
3	Constant voltage	1.05	—	1.3	—	0.2	0.1	1.0	−0.7
4	Load	—	—	0	0	0	0	—	—
5	Load	—	—	0	0	0	0	—	—

*$S_{\text{base}} = 400$ MVA, $V_{\text{base}} = 15$ kV at buses 1, 3, and 345 kV at buses 2, 4, 5

Table 7.2

Line input data for Example 7.9

BUS-TO-BUS	R′	X′	G′	B′	MAXIMUM MVA
	per unit	per unit	per unit	per unit	per unit
2–4	0.036	0.4	0	0.43	3.0
2–5	0.018	0.2	0	0.22	3.0
4–5	0.009	0.1	0	0.11	3.0

Table 7.3

Transformer input data for Example 7.9

BUS-TO-BUS	R	X	G_c	B_m	MAXIMUM MVA	MAXIMUM TAP SETTING
	per unit	per unit	per unit	per unit	per unit	per unit
1–5	0.006	0.08	0	0	1.5	—
3–4	0.003	0.04	0	0	2.5	—

Table 7.4

Input data and unknowns —Example 7.9

BUS	INPUT DATA	UNKNOWNS
1	$V_1 = 1.0,\ \delta_1 = 0$	P_1, Q_1
2	$P_2 = P_{G2} - P_{L2} = -2$ $Q_2 = Q_{G2} - Q_{L2} = -0.7$	V_2, δ_2
3	$V_3 = 1.05$ $P_3 = P_{G3} - P_{L3} = 1.1$	Q_3, δ_3
4	$P_4 = 0,\ Q_4 = 0$	V_4, δ_4
5	$P_5 = 0,\ Q_5 = 0$	V_5, δ_5

Using (7.4.2),

$$Y_{24} = \frac{-1}{R'_{24} + jX'_{24}} = \frac{-1}{0.036 + j0.4} = -0.22319 + j2.47991 \text{ per unit}$$

$$= 2.48993\underline{/95.143^\circ} \text{ per unit}$$

$$Y_{25} = \frac{-1}{R'_{25} + jX'_{25}} = \frac{-1}{0.018 + j0.2} = -0.44638 + j4.95983 \text{ per unit}$$

$$= 4.97988\underline{/95.143^\circ} \text{ per unit}$$

$$Y_{22} = \frac{1}{R'_{24} + jX'_{24}} + \frac{1}{R'_{25} + jX'_{25}} + j\frac{B'_{24}}{2} + j\frac{B'_{25}}{2}$$

$$= (0.22319 - j2.47991) + (0.44638 - j4.95983) + j\frac{0.43}{2} + j\frac{0.22}{2}$$

$$= 0.66957 - j7.11474 = 7.14618\underline{/-84.624^\circ} \text{ per unit}$$

where half of the shunt admittance of each line connected to bus 2 is included in Y_{22} (the other half is located at the other ends of these lines). ◁

Using $\boldsymbol{Y}_{\text{bus}}$, we can write nodal equations for a power-system network, as follows:

$$\boldsymbol{I} = \boldsymbol{Y}_{\text{bus}} \boldsymbol{V} \tag{7.4.3}$$

where $\boldsymbol{I}$ is the N vector of source currents injected into each bus and $\boldsymbol{V}$ is the N vector of bus voltages. For bus k, the kth equation in (7.4.3) is

$$I_k = \sum_{n=1}^{N} Y_{kn} V_n \tag{7.4.4}$$

The complex power delivered to bus k is

$$S_k = \mathrm{P}_k + j\mathrm{Q}_k = V_k I_k^* \tag{7.4.5}$$

Power-flow solutions by Gauss–Seidel are based on nodal equations, (7.4.4), where each current source I_k is calculated from (7.4.5). Using (7.4.4) in (7.4.5),

$$\mathrm{P}_k + j\mathrm{Q}_k = V_k \left[\sum_{n=1}^{N} Y_{kn} V_n \right]^* \qquad k = 1, 2, \ldots, N \tag{7.4.6}$$

With the following notation,

$$V_n = \mathrm{V}_n e^{j\delta_n} \tag{7.4.7}$$

$$Y_{kn} = \mathrm{Y}_{kn} e^{j\theta_{kn}} \qquad k, n = 1, 2, \ldots, N \tag{7.4.8}$$

(7.4.6) becomes

$$\mathrm{P}_k + j\mathrm{Q}_k = \mathrm{V}_k \sum_{n=1}^{N} \mathrm{Y}_{kn} \mathrm{V}_n e^{j(\delta_k - \delta_n - \theta_{kn})} \tag{7.4.9}$$

Taking the real and imaginary parts of (7.4.9)

$$\mathrm{P}_k = \mathrm{V}_k \sum_{n=1}^{N} \mathrm{Y}_{kn} \mathrm{V}_n \cos(\delta_k - \delta_n - \theta_{kn}) \tag{7.4.10}$$

$$\mathrm{Q}_k = \mathrm{V}_k \sum_{n=1}^{N} \mathrm{Y}_{kn} \mathrm{V}_n \sin(\delta_k - \delta_n - \theta_{kn}) \qquad k = 1, 2, \ldots, N \tag{7.4.11}$$

Power-flow solutions by Newton–Raphson are based on the nonlinear power-flow equations given by (7.4.10) and (7.4.11).

Section 7.5

Power-Flow Solution by Gauss–Seidel

Nodal equations $\boldsymbol{I} = \boldsymbol{Y}_{\text{bus}} \boldsymbol{V}$ are a set of linear equations analogous to $\boldsymbol{y} = \boldsymbol{A}\boldsymbol{x}$, solved in Section 7.2 using Gauss–Seidel. Since power-flow bus data consist of P_k and Q_k for load buses or P_k and V_k for voltage-controlled buses, nodal equations do not directly fit the linear equation format; the current source vector $\boldsymbol{I}$ is unknown and the equations are actually nonlinear. For each load bus, I_k can be calculated from (7.4.5), giving

$$I_k = \frac{\mathrm{P}_k - j\mathrm{Q}_k}{V_k^*} \tag{7.5.1}$$

Applying the Gauss–Seidel method, (7.2.9), to nodal equations, with I_k given above, we obtain

$$V_k(i+1) = \frac{1}{Y_{kk}}\left[\frac{\mathrm{P}_k - j\mathrm{Q}_k}{V_k^*(i)} - \sum_{n=1}^{k-1} Y_{kn}V_n(i+1) - \sum_{n=k+1}^{N} Y_{kn}V_n(i)\right] \tag{7.5.2}$$

Equation (7.5.2) can be applied twice during each iteration for load buses, first using $V_k^*(i)$, then replacing $V_k^*(i)$ by $V_k^*(i+1)$ on the right side of (7.5.2).

For a voltage-controlled bus, Q_k is unknown, but can be calculated from (7.4.11), giving

$$\mathrm{Q}_k = \mathrm{V}_k(i) \sum_{n=1}^{N} \mathrm{Y}_{kn}\mathrm{V}_n(i) \sin[\delta_k(i) - \delta_n(i) - \theta_{kn}] \tag{7.5.3}$$

Also,

$$\mathrm{Q}_{Gk} = \mathrm{Q}_k + \mathrm{Q}_{Lk}$$

If the calculated value of Q_{Gk} does not exceed its limits, then Q_k is used in (7.5.2) to calculate $V_k(i+1) = \mathrm{V}_k(i+1)\underline{/\delta_k(i+1)}$. Then the magnitude $\mathrm{V}_k(i+1)$ is changed to V_k, which is input data for the voltage-controlled bus. Thus we use (7.5.2) to compute only the angle $\delta_k(i+1)$ for voltage-controlled buses.

If the calculated value exceeds its limit $\mathrm{Q}_{Gk\max}$ or $\mathrm{Q}_{Gk\min}$ during any iteration, then the bus type is changed from a voltage-controlled bus to a load bus, with Q_{Gk} set to its limit value. Under this condition, the voltage-controlling device (capacitor bank, static var system, and so on) is not capable of maintaining V_k as specified by the input data. The power-flow program then calculates a new value of V_k.

For the swing bus, denoted bus 1, V_1 and δ_1 are input data. As such, no iterations are required for bus 1. After the iteration process has converged, one pass through (7.4.10) and (7.4.11) can be made to compute P_1 and Q_1.

Example 7.10 For the power system of Example 7.9, use Gauss–Seidel to calculate $V_2(1)$, the phasor voltage at bus 2 after the first iteration. Use zero initial phase angles and 1.0 per-unit initial voltage magnitudes (except at bus 3, where $\mathrm{V}_3 = 1.05$) to start the iteration procedure.

Solution Bus 2 is a load bus. Using the input data and bus admittance values from Example 7.9 in (7.5.2),

$$V_2(1) = \frac{1}{Y_{22}}\left\{\frac{P_2 - jQ_2}{V_2^*(0)} - [Y_{21}V_1(1) + Y_{23}V_3(0) + Y_{24}V_4(0) + Y_{25}V_5(0)]\right\}$$

$$= \frac{1}{7.14618\underline{/-84.624^\circ}}\left\{\frac{-2 - j(-0.7)}{1.0\underline{/0^\circ}}\right.$$

$$\left. - [(-0.44638 + j4.95983)(1.0) + (-0.22319 + j2.47991)(1.0)]\right\}$$

$$= \frac{(-2 + j0.7) - (-0.66957 + j7.43974)}{7.14618\underline{/-84.624^\circ}}$$

$$= 0.96132\underline{/-16.543^\circ} \quad \text{per unit}$$

Next, the above value is used in (7.5.2) to recalculate $V_2(1)$:

$$V_2(1) = \frac{1}{7.14618\underline{/-84.624^\circ}}$$

$$\left\{\frac{-2 + j0.7}{0.96132\underline{/16.543^\circ}} - [-0.66957 + j7.43974]\right\}$$

$$= \frac{-1.11745 - j6.14933}{7.14618\underline{/-84.624^\circ}} = 0.87460\underline{/-15.675^\circ} \quad \text{per unit}$$

Computations are next performed at buses 3, 4, and 5 to complete the first Gauss–Seidel iteration. ◁

Section 7.6

Power-Flow Solution by Newton–Raphson

Equations (7.4.10) and (7.4.11) are analogous to the nonlinear equation $\mathbf{y} = \mathbf{f}(\mathbf{x})$, solved in Section 7.3 by Newton–Raphson. We define the $\mathbf{x}$, $\mathbf{y}$, and $\mathbf{f}$ vectors for the power-flow problem as

$$\mathbf{x} = \begin{bmatrix} \boldsymbol{\delta} \\ \\ \mathbf{V} \end{bmatrix} = \begin{bmatrix} \delta_2 \\ \vdots \\ \delta_N \\ \\ V_2 \\ \vdots \\ V_N \end{bmatrix}; \quad \mathbf{y} = \begin{bmatrix} \mathbf{P} \\ \\ \mathbf{Q} \end{bmatrix} = \begin{bmatrix} P_2 \\ \vdots \\ P_N \\ \\ Q_2 \\ \vdots \\ Q_N \end{bmatrix}; \quad \mathbf{f(x)} = \begin{bmatrix} \mathbf{P(x)} \\ \\ \mathbf{Q(x)} \end{bmatrix} = \begin{bmatrix} P_2(\mathbf{x}) \\ \vdots \\ P_N(\mathbf{x}) \\ \\ Q_2(\mathbf{x}) \\ \vdots \\ Q_N(\mathbf{x}) \end{bmatrix} \tag{7.6.1}$$

where all V, P, and Q terms are in per-unit and δ terms are in radians. The swing bus variables δ_1 and V_1 are omitted from (7.6.1), since they are already known. Equations (7.4.10) and (7.4.11) then have the following form:

$$y_k = P_k = P_k(\mathbf{x}) = V_k \sum_{n=1}^{N} Y_{kn} V_n \cos(\delta_k - \delta_n - \theta_{kn}) \tag{7.6.2}$$

$$y_{k+N} = Q_k = Q_k(\mathbf{x}) = V_k \sum_{n=1}^{N} Y_{kn} V_n \sin(\delta_k - \delta_n - \theta_{kn}) \tag{7.6.3}$$

$$k = 2, 3, \ldots, N$$

The Jacobian matrix of (7.3.10) has the form

$$\mathbf{J} = \left[\begin{array}{ccc|ccc} \multicolumn{3}{c|}{\mathbf{J1}} & \multicolumn{3}{c}{\mathbf{J2}} \\ \frac{\partial P_2}{\partial \delta_2} & \cdots & \frac{\partial P_2}{\partial \delta_N} & \frac{\partial P_2}{\partial V_2} & \cdots & \frac{\partial P_2}{\partial V_N} \\ \vdots & & & \vdots & & \\ \frac{\partial P_N}{\partial \delta_2} & \cdots & \frac{\partial P_N}{\partial \delta_N} & \frac{\partial P_N}{\partial V_2} & \cdots & \frac{\partial P_N}{\partial V_N} \\ \hline \frac{\partial Q_2}{\partial \delta_2} & \cdots & \frac{\partial Q_2}{\partial \delta_N} & \frac{\partial Q_2}{\partial V_2} & \cdots & \frac{\partial Q_2}{\partial V_N} \\ \vdots & & & \vdots & & \\ \frac{\partial Q_N}{\partial \delta_2} & \cdots & \frac{\partial Q_N}{\partial \delta_N} & \frac{\partial Q_N}{\partial V_2} & \cdots & \frac{\partial Q_N}{\partial V_N} \\ \multicolumn{3}{c|}{\mathbf{J3}} & \multicolumn{3}{c}{\mathbf{J4}} \end{array}\right] \tag{7.6.4}$$

Equation (7.6.4) is partitioned into four blocks. The partial derivatives in each block, derived from (7.6.2) and (7.6.3), are given in Table 7.5.

Table 7.5

Elements of the Jacobian matrix

$n \neq k$

$$J1_{kn} = \frac{\partial P_k}{\partial \delta_n} = V_k Y_{kn} V_n \sin(\delta_k - \delta_n - \theta_{kn})$$

$$J2_{kn} = \frac{\partial P_k}{\partial V_n} = V_k Y_{kn} \cos(\delta_k - \delta_n - \theta_{kn})$$

$$J3_{kn} = \frac{\partial Q_k}{\partial \delta_n} = -V_k Y_{kn} V_n \cos(\delta_k - \delta_n - \theta_{kn})$$

$$J4_{kn} = \frac{\partial Q_k}{\partial V_n} = V_k Y_{kn} \sin(\delta_k - \delta_n - \theta_{kn})$$

$n = k$

$$J1_{kk} = \frac{\partial P_k}{\partial \delta_k} = -V_k \sum_{\substack{n=1\\ n\neq k}}^{N} Y_{kn} V_n \sin(\delta_k - \delta_n - \theta_{kn})$$

$$J2_{kk} = \frac{\partial P_k}{\partial V_k} = V_k Y_{kk} \cos\theta_{kk} + \sum_{n=1}^{N} Y_{kn} V_n \cos(\delta_k - \delta_n - \theta_{kn})$$

$$J3_{kk} = \frac{\partial Q_k}{\partial \delta_k} = V_k \sum_{\substack{n=1\\ n\neq k}}^{N} Y_{kn} V_n \cos(\delta_k - \delta_n - \theta_{kn})$$

$$J4_{kk} = \frac{\partial Q_k}{\partial V_k} = -V_k Y_{kk} \sin\theta_{kk} + \sum_{n=1}^{N} Y_{kn} V_n \sin(\delta_k - \delta_n - \theta_{kn})$$

$k, n = 2, 3, \ldots, N$

We now apply to the power-flow problem the four Newton–Raphson steps outlined in Section 7.3, starting with $\mathbf{x}(i) = \begin{bmatrix} \boldsymbol{\delta}(i) \\ \mathbf{V}(i) \end{bmatrix}$ at the ith iteration.

Step 1 Use (7.6.2) and (7.6.3) to compute

$$\Delta\mathbf{y}(i) = \begin{bmatrix} \Delta\mathbf{P}(i) \\ \Delta\mathbf{Q}(i) \end{bmatrix} = \begin{bmatrix} \mathbf{P} - \mathbf{P}[\mathbf{x}(i)] \\ \mathbf{Q} - \mathbf{Q}[\mathbf{x}(i)] \end{bmatrix} \tag{7.6.5}$$

Step 2 Use the equations in Table 7.5 to calculate the Jacobian matrix.

Step 3 Use Gauss elimination and back substitution to solve

$$\left[\begin{array}{c|c} \mathbf{J1}(i) & \mathbf{J2}(i) \\ \hline \mathbf{J3}(i) & \mathbf{J4}(i) \end{array}\right] \begin{bmatrix} \Delta\boldsymbol{\delta}(i) \\ \Delta\mathbf{V}(i) \end{bmatrix} = \begin{bmatrix} \Delta\mathbf{P}(i) \\ \Delta\mathbf{Q}(i) \end{bmatrix} \tag{7.6.6}$$

Step 4 Compute

$$\mathbf{x}(i+1) = \begin{bmatrix} \boldsymbol{\delta}(i+1) \\ \\ \mathbf{V}(i+1) \end{bmatrix} = \begin{bmatrix} \boldsymbol{\delta}(i) \\ \\ \mathbf{V}(i) \end{bmatrix} + \begin{bmatrix} \Delta\boldsymbol{\delta}(i) \\ \\ \Delta\mathbf{V}(i) \end{bmatrix} \tag{7.6.7}$$

Starting with initial value $\mathbf{x}(0)$, the procedure continues until convergence is obtained or until the number of iterations exceeds a specified maximum. Convergence criteria are often based on $\Delta\mathbf{y}(i)$ (called *power mismatches*) rather than on $\Delta\mathbf{x}(i)$ (phase angle and voltage magnitude mismatches).

For each voltage-controlled bus, the magnitude V_k is already known, and the function $Q_k(\mathbf{x})$ is not needed. Therefore we omit V_k from the $\mathbf{x}$ vector and Q_k from the $\mathbf{y}$ vector. We also omit from the Jacobian matrix the column corresponding to partial derivatives with respect to V_k and the row corresponding to partial derivatives of $Q_k(\mathbf{x})$.

At the end of each iteration, we compute $Q_k(\mathbf{x})$ from (7.6.3) and $Q_{Gk} = Q_k(\mathbf{x}) + Q_{Lk}$ for each voltage-controlled bus. If the computed value of Q_{Gk} exceeds its limits, then the bus type is changed to a load bus with Q_{Gk} set to its limit value. Also, the corresponding row and column are reinserted into the Jacobian matrix, and the power-flow program computes a new value for V_k.

Example 7.11 Determine the dimension of the Jacobian matrix for the power system in Example 7.9. Also calculate $\Delta P_2(0)$ in Step 1 and $J1_{24}(0)$ in Step 2 of the first Newton–Raphson iteration. Assume zero initial phase angles and 1.0 per-unit initial voltage magnitudes (except $V_3 = 1.05$).

Solution Since there are $N = 5$ buses for Example 7.9, (7.6.2) and (7.6.3) constitute $2(N-1) = 8$ equations, for which $\mathbf{J}(i)$ would have dimension 8×8. However, there is one voltage-controlled bus, bus 3. Therefore V_3 and the equation for $Q_3(\mathbf{x})$ are eliminated, and $\mathbf{J}(i)$ is reduced to a 7×7 matrix.

From Step 1 and (7.6.2),

$$\begin{aligned}\Delta P_2(0) = P_2 - P_2(\mathbf{x}) = P_2 - V_2(0)\{&Y_{21}V_1 \cos[\delta_2(0) - \delta_1(0) - \theta_{21}] \\ &+ Y_{22}V_2\cos[-\theta_{22}] + Y_{23}V_3\cos[\delta_2(0) - \delta_3(0) - \theta_{23}] \\ &+ Y_{24}V_4\cos[\delta_2(0) - \delta_4(0) - \theta_{24}] \\ &+ Y_{25}V_5\cos[\delta_2(0) - \delta_5(0) - \theta_{25}]\}\end{aligned}$$

$$\begin{aligned}\Delta P_2(0) = -2.0 - 1.0\{&7.14618(1.0)\cos(84.624°) \\ &+ 2.48993(1.0)\cos(-95.143°) \\ &+ 4.97988(1.0)\cos(-95.143°)\} \\ = -2.0 - (&-7.24 \times 10^{-5}) = -1.99993 \quad \text{per unit}\end{aligned}$$

From Step 2 and J1 given in Table 7.5

$$\begin{aligned} \mathrm{J1}_{24}(0) &= \mathrm{V}_2(0)\mathrm{Y}_{24}\mathrm{V}_4(0)\sin[\delta_2(0) - \delta_4(0) - \theta_{24}] \\ &= (1.0)(2.48993)(1.0)\sin[-95.143°] \\ &= -2.47991 \quad \text{per unit} \end{aligned}$$

◁

Voltage-controlled buses to which tap-changing or voltage-regulating transformers are connected can be handled by various methods. One method is to treat each of these buses as a load bus. The equivalent π circuit parameters (Figure 4.27) are first calculated with tap setting $c = 1.0$ for starting. During each iteration the computed bus voltage magnitude is compared with the desired value specified by the input data. If the computed voltage is low (or high), c is increased (or decreased) to its next setting, and the parameters of the equivalent π circuit as well as $\boldsymbol{Y}_{\text{bus}}$ are recalculated. The procedure continues until the computed bus voltage magnitude equals the desired value within a specified tolerance, or until the high or low tap-setting limit is reached. Phase-shifting transformers can be handled in a similar way by using a complex turns ratio $c = 1.0\underline{/\alpha}$, and by varying the phase-shift angle α.

A method with faster convergence makes c a variable and includes it in the $\mathbf{x}$ vector of (7.6.1). An equation is then derived to enter into the Jacobian matrix [4].

Section 7.7

Sparsity Techniques

A typical power system has an average of fewer than three lines connected to each bus. As such, each row of $\boldsymbol{Y}_{\text{bus}}$ has an average of fewer than four nonzero elements, one off-diagonal for each line and the diagonal. Such a matrix, which has only a few nonzero elements, is said to be *sparse*.

Newton–Raphson power-flow programs employ sparse matrix techniques to reduce computer storage and time requirements. These techniques include compact storage of $\boldsymbol{Y}_{\text{bus}}$ and $\mathbf{J}(i)$, and reordering of buses to avoid fill-in of $\mathbf{J}(i)$ during Gauss elimination steps. Consider the following matrix:

$$\mathbf{S} = \begin{bmatrix} 1.0 & -1.1 & -2.1 & -3.1 \\ -4.1 & 2.0 & 0 & -5.1 \\ -6.1 & 0 & 3.0 & 0 \\ -7.1 & 0 & 0 & 4.0 \end{bmatrix} \tag{7.7.1}$$

One method for compact storage of **S** consists of the following four vectors:

$$\mathbf{DIAG} = [1.0 \quad 2.0 \quad 3.0 \quad 4.0] \tag{7.7.2}$$

$$\mathbf{OFFDIAG} = [-1.1 \quad -2.1 \quad -3.1 \quad -4.1 \quad -5.1 \quad -6.1 \quad -7.1] \tag{7.7.3}$$

$$\mathbf{COL} = [2 \quad 3 \quad 4 \quad 1 \quad 4 \quad 1 \quad 1] \tag{7.7.4}$$

$$\mathbf{ROW} = [3 \quad 2 \quad 1 \quad 1] \tag{7.7.5}$$

DIAG contains the ordered diagonal elements and **OFFDIAG** contains the non-zero off-diagonal elements of **S. COL** contains the column number of each off-diagonal element. For example, the *fourth* element in **COL** is 1, indicating that the *fourth* element of **OFFDIAG**, -4.1, is located in column 1. **ROW** indicates the number of off-diagonal elements in each row of **S**. For example, the *first* element of **ROW** is 3, indicating the *first* three elements of **OFFDIAG**, -1.1, -2.1, and -3.1, are located in the *first* row. The *second* element of **ROW** is 2, indicating the next two elements of **OFFDIAG**, -4.1 and -5.1, are located in the *second* row. The **S** matrix can be completely reconstructed from these four vectors. Note that the dimension of **DIAG** and **ROW** equals the number of diagonal elements of **S**, whereas the dimension of **OFFDIAG** and **COL** equals the number of non-zero off-diagonals.

Now assume that computer storage requirements are 4 bytes to store each magnitude and 4 bytes to store each phase of Y_{bus} in an N-bus power system. Also assume Y_{bus} has an average of $3N$ nonzero off-diagonals (three lines per bus) along with its N diagonals. Using the preceding compact storage technique, we need $(4 + 4)3N = 24N$ bytes for **OFFDIAG** and $(4 + 4)N = 8N$ bytes for **DIAG.** Also, assuming 2 bytes to store each integer, we need $6N$ bytes for **COL** and $2N$ bytes for **ROW**. Total computer memory required is then $(24 + 8 + 6 + 2)N = 40N$ bytes with compact storage of Y_{bus}, compared to $8N^2$ bytes without compact storage. For a 1000-bus power system, this means 40 instead of 8000 kilobytes to store Y_{bus}. Further storage reduction could be obtained by storing only the upper triangular portion of the symmetric Y_{bus} matrix.

The Jacobian matrix is also sparse. From Table 7.5, whenever $Y_{kn} = 0$, $J1_{kn} = J2_{kn} = J3_{kn} = J4_{kn} = 0$. Compact storage of **J** for a 1000-bus power system requires less than 100 kilobytes with the above assumptions.

The other sparsity technique is to reorder buses. Suppose Gauss elimination is used to triangularize **S** in (7.7.1). After one Gauss elimination step, as described in Section 7.1, we have

$$\mathbf{S}^{(1)} = \begin{bmatrix} 1.0 & -1.1 & -2.1 & -3.1 \\ 0 & -2.51 & -8.61 & -7.61 \\ 0 & -6.71 & -9.81 & -18.91 \\ 0 & -7.81 & -14.91 & -18.01 \end{bmatrix} \tag{7.7.6}$$

It can be seen that the zeros in columns 2, 3, and 4 of $\mathbf{S}$ are filled in with nonzero elements in $\mathbf{S}^{(1)}$. The original degree of sparsity is lost.

One simple reordering method is to start with those buses having the fewest connected branches and to end with those having the most connected branches. For example, $\mathbf{S}$ in (7.7.1) has three branches connected to bus 1 (three off-diagonals in row 1), two branches connected to bus 2, and one branch connected to buses 3 and 4. Reordering the buses 4, 3, 2, 1 instead of 1, 2, 3, 4 we have

$$\mathbf{S}_{\text{reordered}} = \begin{bmatrix} 4.0 & 0 & 0 & -7.1 \\ 0 & 3.0 & 0 & -6.1 \\ -5.1 & 0 & 2.0 & -4.1 \\ -3.1 & -2.1 & -1.1 & 1.0 \end{bmatrix} \tag{7.7.7}$$

Now, after one Gauss elimination step,

$$\mathbf{S}^{(1)}_{\text{reordered}} = \begin{bmatrix} 4.0 & 0 & 0 & -7.1 \\ 0 & 3.0 & 0 & -6.1 \\ 0 & 0 & 2.0 & -13.15 \\ 0 & -2.1 & -1.1 & -4.5025 \end{bmatrix} \tag{7.7.8}$$

Note that the original degree of sparsity is not lost in (7.7.8).

Reordering buses according to the fewest connected branches can be performed once, before the Gauss elimination process begins. Alternatively, buses can be renumbered during each Gauss elimination step in order to account for changes during the elimination process.

Sparsity techniques similar to those described in this section are a standard feature of today's Newton–Raphson power-flow programs. As a result of these techniques, typical 1000-bus power-flow solutions require less than 200 kilobytes of storage, less than 25 seconds per iteration of computer time, and less than 10 iterations to converge.

Section 7.8

Personal Computer Program: POWER FLOW

The software package that accompanies this text includes the program "POWER FLOW," which computes voltage magnitude and angle at each bus in a power system under balanced three-phase steady-state operation. Bus voltages are then used to compute generator, line, and transformer loadings.

Input data for the program include bus, line, and transformer data, as shown in Tables 7.1, 7.2, and 7.3. All input data are given in per-unit on a common MVA base.

The bus admittance matrix is calculated from the input data using (7.4.2). The program then uses the Newton–Raphson method described in Section 7.6.

For single runs, starting values of bus voltage magnitudes and angles are set equal to those of the swing bus, except for voltage-controlled buses, whose voltage magnitudes are known. For a series of runs with input data changes, the final values of each run are used as starting values for the next run.

The Newton–Raphson procedure is terminated when the magnitudes of all power mismatches $\Delta P(i)$ and $\Delta Q(i)$ are less than a tolerance level (typically 0.001 per unit) or when the number of iterations exceeds a maximum (typically 10). Both the tolerance level and maximum number of iterations are selected by the program user.

Output data consist of bus, line, and transformer data. Bus output data include voltage magnitude and angle, real and reactive power of each generator and load, and identification of buses with voltage magnitudes more than 5% above or below that of the swing bus. Line (and transformer) output data include real and reactive power flows entering the line terminals and identification of lines with MVA flows above their maximum ratings. Other useful output include generation, load, and line loss totals, number of iterations to converge, and total mismatches ΔP and ΔQ after convergence. If desired, reactive power delivered by line capacitances or other shunt reactive elements can also be obtained.

Example 7.12 Run the POWER FLOW program for the power system given in Example 7.9.

Solution Bus, line, and transformer output data are shown in Tables 7.6, 7.7, and 7.8. Note that the bus 2 voltage magnitude is low (0.834 per unit).

			GENERATION		LOAD		
	VOLTAGE MAGNITUDE	PHASE ANGLE	PG	QG	PL	QL	0.95 > V > 1.05
bus #	per unit	degrees	per unit	per unit	per unit	per unit	
1	1.000	0.000	0.987	0.286	0.000	0.000	
2	0.834	−22.407	0.000	0.000	2.000	0.700	****
3	1.050	−0.597	1.300	0.844	0.200	0.100	
4	1.019	−2.834	0.000	0.000	0.000	0.000	
5	0.974	−4.548	0.000	0.000	0.000	0.000	
		TOTAL	2.287	1.129	2.200	0.800	

Table 7.6 Example 7.12 bus output data

LINE #	BUS TO BUS		P	Q	S	RATING EXCEEDED
1	2	4	−0.730	−0.348	0.808	
	4	2	0.759	0.304	0.818	
2	2	5	−1.270	−0.352	1.318	
	5	2	1.314	0.658	1.469	
3	4	5	0.336	0.376	0.504	
	5	4	−0.333	−0.456	0.565	

Table 7.7 Example 7.12 line output data

TRAN. #	BUS TO BUS		P	Q	S	RATING EXCEEDED
1	1	5	0.987	0.286	1.028	
	5	1	−0.981	−0.201	1.001	
2	3	4	1.100	0.744	1.328	
	4	3	−1.095	−0.680	1.289	

Table 7.8 Example 7.12 transformer output data

◁

Section 7.9

Fast Decoupled Power Flow

Contingencies are a major concern in power-system operations. For example, operating personnel need to know what power-flow changes will occur due to a particular generator outage or transmission-line outage. Contingency information, when obtained in real time, can be used to anticipate problems caused by such outages, and can be used to develop operating strategies to overcome the problems.

Fast power-flow algorithms have been developed to give power-flow solutions in seconds [8]. These algorithms are based on the following simplification of the Jacobian matrix. Neglecting $\mathbf{J}_2(i)$ and $\mathbf{J}_3(i)$, (7.6.6) reduces to two sets of decoupled equations:

$$\mathbf{J}_1(i)\,\Delta\boldsymbol{\delta}(i) = \Delta\mathbf{P}(i) \tag{7.9.1}$$

$$\mathbf{J}_4(i)\,\Delta\mathbf{V}(i) = \Delta\mathbf{Q}(i) \tag{7.9.2}$$

The computer time required to solve (7.9.1) and (7.9.2) is significantly less than that required to solve (7.6.6). Further reduction in computer time can be obtained from additional simplification of the Jacobian matrix. For example, assume $V_k \approx V_n \approx 1.0$ per unit and $\delta_k \approx \delta_n$. Then $\mathbf{J}_1$ and $\mathbf{J}_4$ are constant matrices whose elements in Table 7.5 are the imaginary components of Y_{bus}. As such, $\mathbf{J}_1$ and $\mathbf{J}_4$ do not have to be recalculated during successive iterations.

Simplifications similar to these enable rapid power-flow solutions. For a fixed number of iterations, the fast decoupled algorithm given by (7.9.1) and (7.9.2) is not as accurate as the exact Newton–Raphson algorithm. But the savings in computer time is considered more important.

Section 7.10

Control of Power Flow

The following means are used to control system power flows:

1. Prime mover and excitation control of generators
2. Switching of shunt capacitor banks, shunt reactors, and static var systems
3. Control of tap-changing and regulating transformers

A simple model of a generator operating under balanced steady-state conditions is the Thévenin equivalent shown in Figure 7.3. V_t is the generator terminal voltage, E_g is the excitation voltage, δ is the power angle, and X_g is the positive-sequence synchronous reactance. From the figure, the generator current is

$$I = \frac{E_g e^{j\delta} - V_t}{jX_g} \tag{7.10.1}$$

and the complex power delivered by the generator is

$$S = P + jQ = V_t I^* = V_t\left(\frac{E_g e^{-j\delta} - V_t}{-jX_g}\right)$$

$$= \frac{V_t E_g (j\cos\delta + \sin\delta) - jV_t^2}{X_g} \tag{7.10.2}$$

Figure 7.3

Generator Thévenin equivalent

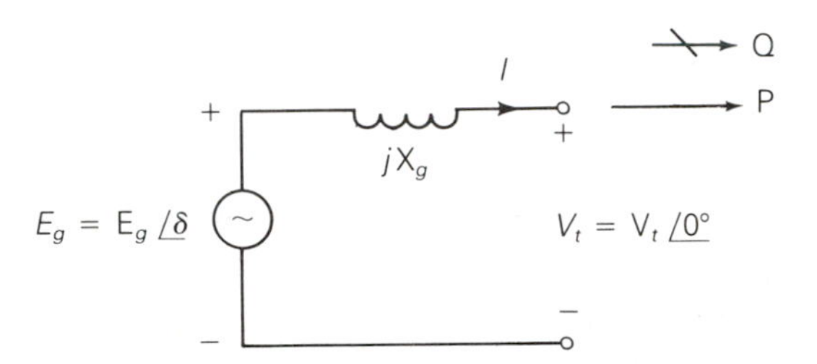

The real and reactive powers delivered are then

$$P = \operatorname{Re} S = \frac{V_t E_g}{X_g}\sin\delta \tag{7.10.3}$$

$$Q = \operatorname{Im} S = \frac{V_t}{X_g}(E_g\cos\delta - V_t) \tag{7.10.4}$$

Equation (7.10.3) shows that the real power P increases when the power angle δ increases. From an operational standpoint, when the prime mover increases the power input to the generator while the excitation voltage is held constant, the rotor speed increases. As the rotor speed increases, the power angle δ also increases, causing an increase in generator real power output P. There is also a decrease in reactive power output Q, given by (7.10.4). However, when δ is less than 15°, the increase in P is much larger than the decrease in Q. From the power-flow standpoint, an increase in prime-mover power corresponds to an increase in P at the constant-voltage bus to which the generator is connected. The power-flow program computes the increase in δ along with the small change in Q.

Equation (7.10.4) shows that reactive power output Q increases when the excitation voltage E_g increases. From the operational standpoint, when the generator exciter output increases while holding the prime-mover power constant, the rotor current increases. As the rotor current increases, the excitation voltage E_g also increases, causing an increase in generator reactive power output Q. There is also a small decrease in δ required to hold P constant in (7.10.3). From the power-flow standpoint, an increase in generator excitation corresponds to an increase in voltage magnitude at the constant-voltage bus to which the generator is connected. The power-flow program computes the increase in reactive power Q supplied by the generator along with the small change in δ.

Figure 7.4 shows the effect of adding a shunt capacitor bank to a power-system bus. The system is modeled by its Thévenin equivalent. Before the capacitor bank is connected, the switch SW is open and the bus voltage equals E_{Th}. After the bank is connected, SW is closed, and the capacitor current I_C leads the bus voltage V_t by 90°. The phasor diagram shows that V_t is larger than E_{Th} when SW is closed. From the power-flow standpoint, the addition of a shunt capacitor bank to a load bus corresponds to the addition of a negative reactive load, since a capacitor absorbs negative reactive power. The power-flow program computes the increase in bus voltage magnitude along with the small change in δ. Similarly, the addition of a shunt reactor corresponds to the addition of a positive reactive load, wherein the power-flow program computes the decrease in voltage magnitude.

Figure 7.4

Effect of adding a shunt capacitor bank to a power-system bus

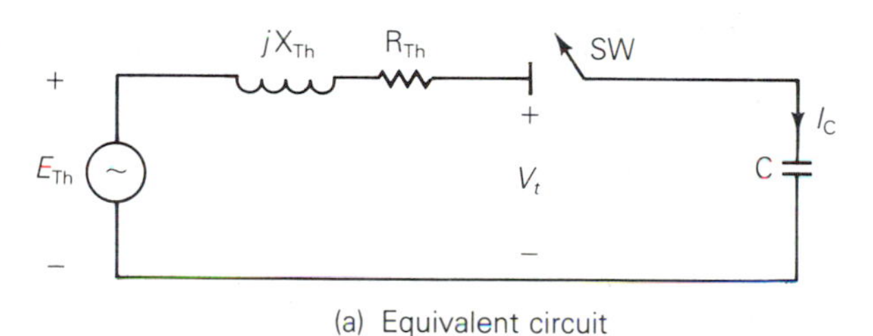

(a) Equivalent circuit

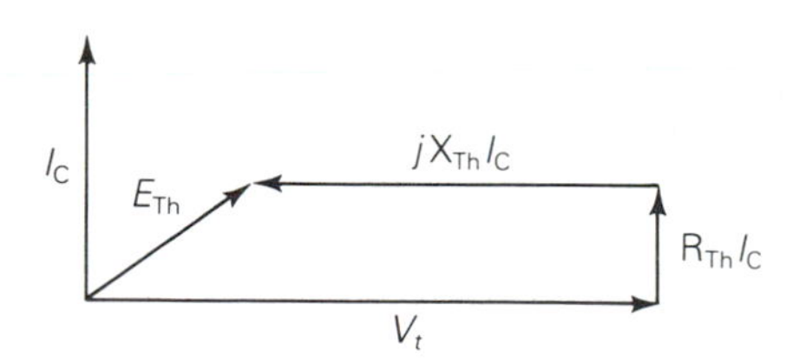

(b) Phasor diagram with switch SW closed

Tap-changing and voltage magnitude regulating transformers are used to control bus voltages as well as reactive power flows on lines to which they are connected. Similarly, phase-angle regulating transformers are used to control bus angles as well as real power flows on lines to which they are connected. Both tap-changing and regulating transformers are modeled by a transformer with an off-nominal turns ratio c (Figure 4.27). From the power-flow standpoint, a change in tap setting or voltage regulation corresponds to a change in c. The power-flow program computes the changes in $\boldsymbol{Y}_{\text{bus}}$, bus voltage magnitudes and angles, and branch flows.

Besides the above controls, the power-flow program can be used to investigate the effect of switching in or out lines, transformers, loads, and generators. Proposed system changes to meet future load growth, including new transmission, new transformers, and new generation can also be investigated. Power-flow design studies are normally conducted by trial and error. Using engineering judgment, adjustments in generation levels and controls are made until the desired equipment loadings and voltage profile are obtained.

Example 7.13 Determine the effect of adding a 200-Mvar shunt capacitor bank at bus 2 on the power system in Example 7.9.

Solution The addition of a 200-Mvar (0.5 per-unit) shunt capacitor bank corresponds to adding -0.5 per unit of reactive load. Therefore, Q_L at bus 2 is changed from 0.7 to $(0.7 - 0.5) = 0.2$ per unit, and the POWER FLOW program is rerun. Bus, line, and transformer output data are shown in Tables 7.9, 7.10, and 7.11. As shown, the capacitor bank increases the voltage at bus 2 to an acceptable value of 0.968 per unit.

			GENERATION		LOAD		
	VOLTAGE MAGNITUDE	PHASE ANGLE	PG	QG	PL	QL	0.95 > V > 1.05
bus #	per unit	degrees	per unit	per unit	per unit	per unit	
1	1.000	0.000	0.963	−0.118	0.000	0.000	
2	0.968	−19.629	0.000	0.000	2.000	0.200	
3	1.050	−0.408	1.300	0.370	0.200	0.100	
4	1.037	−2.680	0.000	0.000	0.000	0.000	
5	1.007	−4.427	0.000	0.000	0.000	0.000	
		TOTAL	2.263	0.252	2.200	0.300	

Table 7.9 Example 7.13 bus output data

LINE #	BUS TO BUS		P	Q	S	RATING EXCEEDED
1	2	4	−0.731	−0.195	0.757	
	4	2	0.752	−0.010	0.752	
2	2	5	−1.269	−0.005	1.269	
	5	2	1.300	0.137	1.307	
3	4	5	0.345	0.234	0.416	
	5	4	−0.343	−0.329	0.476	

Table 7.10 Example 7.13 line output data

TRAN. #	BUS TO BUS		P	Q	S	RATING EXCEEDED
1	1	5	0.963	−0.118	0.970	
	5	1	−0.957	0.193	0.976	
2	3	4	1.100	0.270	1.133	
	4	3	−1.097	−0.224	1.119	

Table 7.11 Example 7.13 transformer output data ◁

Problems

Section 7.1

7.1 Using Gauss elimination and back substitution, solve

$$\begin{bmatrix} 10 & 3 & 0 \\ 4 & 12 & 2 \\ 5 & 2 & 14 \end{bmatrix} \begin{bmatrix} x_1 \\ x_2 \\ x_3 \end{bmatrix} = \begin{bmatrix} 1 \\ 2 \\ 3 \end{bmatrix}$$

7.2 Show that the Gauss elimination method, which transforms the set of N linear equations $\mathbf{Ax} = \mathbf{y}$ to $\mathbf{A}^{(N-1)}\mathbf{x} = \mathbf{y}^{(N-1)}$, where $\mathbf{A}^{(N-1)}$ is triangular, requires $(N^3 - N)/3$ multiplications, $N(N-1)/2$ divisions, and $(N^3 - N)/3$ subtractions. Assume that all the elements of $\mathbf{A}$ and $\mathbf{y}$ are nonzero and real. (*Hint:* Investigate (7.1.7). Note that during the first Gauss elimination step, *each* of the $(N-1)$ rows that are changed requires *one* division, N multiplications, and N subtractions.)

7.3 Show that, after triangularizing $\mathbf{Ax} = \mathbf{y}$, the back substitution method of solving $\mathbf{A}^{(N-1)}\mathbf{x} = \mathbf{y}^{(N-1)}$ requires N divisions, $N(N-1)/2$ multiplications, and $N(N-1)/2$ subtractions. Assume that all the elements of $\mathbf{A}^{(N-1)}$ and $\mathbf{y}^{(N-1)}$ are nonzero and real.

7.4 For a digital computer with a 10^{-6}-s multiplication or division time and a 10^{-7}-s addition or subtraction time, determine how much computer time would be required to solve (7.1.1) for $N = 100$, using Gauss elimination and back substitution. Assume that all the elements of $\mathbf{A}$ and $\mathbf{y}$ are nonzero and real.

7.5 If 4 bytes are used to store each floating-point number in computer memory, how many kilobytes are required to store an N vector $\mathbf{y}$ and an $N \times N$ matrix $\mathbf{A}$, where $N = 1000$. Assume that all the elements of $\mathbf{y}$ and $\mathbf{A}$ are nonzero and real.

Section 7.2

7.6 Solve Problem 7.1 using the Jacobi iterative method. Start with $x_1(0) = x_2(0) = x_3(0) = 0$, and continue until (7.2.2) is satisfied with $\varepsilon = 0.01$.

7.7 Repeat Problem 7.6 using the Gauss–Seidel iterative method. Which method converges more rapidly?

7.8 Take the z-transform of (7.2.6) and show that $\mathbf{X}(z) = \mathbf{G}(z)\mathbf{Y}(z)$, where $\mathbf{G}(z) = (z\mathbf{U} - \mathbf{M})^{-1}\mathbf{D}^{-1}$ and $\mathbf{U}$ is the unit matrix.

$\mathbf{G}(z)$ is the matrix transfer function of a digital filter that represents the Jacobi or Gauss–Seidel methods. The filter poles are obtained by solving $\det(z\mathbf{U} - \mathbf{M}) = 0$. The filter is stable if and only if all the poles have magnitudes less than 1.

7.9 Using the results of Problem 7.8, determine the filter poles for Examples 7.3 and 7.5. Note that in Example 7.3 both poles have magnitudes less than 1, which means the filter is stable and Jacobi converges for this example. However, in Example 7.5, one pole has a magnitude greater than 1, which means the filter is unstable and Gauss–Seidel diverges for this example.

7.10 Determine the poles of the Jacobi and Gauss–Seidel digital filters for the general two-dimensional problem ($N = 2$):

$$\left[\begin{array}{c|c} A_{11} & A_{12} \\ \hline A_{21} & A_{22} \end{array}\right]\begin{bmatrix} x_1 \\ x_2 \end{bmatrix} = \begin{bmatrix} y_1 \\ y_2 \end{bmatrix}$$

Then determine a necessary and sufficient condition for convergence of these filters when $N = 2$.

7.11 For a digital computer with a 10^{-6}-s multiplication or division time and a 10^{-7}-s addition or subtraction time, determine how much computer time per iteration would be required to solve (7.1.1) with $N = 100$, using Gauss–Seidel. Assume that all the elements of $\mathbf{A}$ and $\mathbf{y}$ are nonzero and real.

Section 7.3

7.12 Use Newton–Raphson to find one solution to the polynomial equation $f(x) = y$, where $y = 0$ and $f(x) = 3x^3 + 4x^2 + 5x + 6$. Start with $x(0) = 1.0$ and continue until (7.2.2) is satisfied with $\varepsilon = 0.001$.

7.13 Use Newton–Raphson to find a solution to

$$\begin{bmatrix} e^{x_1 x_2} \\ \cos(x_1 + x_2) \end{bmatrix} = \begin{bmatrix} 1.2 \\ 0.5 \end{bmatrix}$$

where x_1 and x_2 are in radians. (a) Start with $x_1(0) = 1.0$ and $x_2(0) = 0.5$ and continue until (7.2.2) is satisfied with $\varepsilon = 0.005$. (b) Show that Newton–Raphson diverges for this example if $x_1(0) = 1.0$ and $x_2(0) = 2.0$.

Section 7.4

7.14 Compute the elements of the third row of Y_{bus} for the power system in Example 7.9.

Section 7.5

7.15 Using Gauss–Seidel, calculate $V_3(1)$, the phasor voltage at bus 3 after the first iteration. Note that bus 3 is a voltage-controlled bus.

Section 7.6

7.16 For the power system in Example 7.9, calculate $\Delta P_3(0)$ in Step 1 and $J1_{33}(0)$ in Step 2 of the first Newton–Raphson iteration. Assume zero initial phase angles and 1.0 per-unit initial voltage magnitudes (except $V_3 = 1.05$).

Section 7.7

7.17 Using the compact storage technique described in Section 7.7, determine the vectors **DIAG, OFFDIAG, COL,** and **ROW** for the following matrix:

$$\mathbf{S} = \begin{bmatrix} 17 & -9.1 & 0 & 0 & -2.1 & -7.1 \\ -9.1 & 25 & -8.1 & -1.1 & -6.1 & 0 \\ 0 & -8.1 & 9 & 0 & 0 & 0 \\ 0 & -1.1 & 0 & 2 & 0 & 0 \\ -2.1 & -6.1 & 0 & 0 & 14 & -5.1 \\ -7.1 & 0 & 0 & 0 & -5.1 & 15 \end{bmatrix}$$

7.18 If 4 bytes of computer storage are used for each floating-point number and 2 bytes for each integer, determine the total number of bytes required to store the **S** matrix in Problem 7.17 (a) with compact storage and (b) without compact storage.

7.19 Reorder the rows of the matrix given in Problem 7.17 such that after one Gauss elimination step, $\mathbf{S}^{(1)}_{\text{reordered}}$ has the same number of zeros in columns 2–6 as the original matrix.

Sections 7.8–7.10

7.20 For the power system given in Example 7.9, use the POWER FLOW program to find the Mvar rating of a shunt capacitor bank at bus 2 that increases V_2 to 1.00 per unit. Also determine the effect of this capacitor bank on line loadings and on total I^2R line losses.

7.21 For the power system given in Example 7.9, run the POWER FLOW program with an additional line installed from bus 2 to bus 4. The line parameters of the added line equal those of existing line 2–4. Determine the effect on V_2, on line loadings, and on total I^2R line losses.

7.22 For the power system given in Example 7.9, run the POWER FLOW program with V_3 decreased to 1.03 per unit. Determine the effect on V_2 and on the reactive power supplied by the generator at bus 3.

7.23 Assume that the transformer between buses 1 and 5 in Example 7.9 is a tap-changing transformer whose taps can be varied from 0.85 to 1.15 in increments of 0.05 per unit. Use the POWER FLOW program to determine the tap setting required to increase V_5 to 1.05 per unit.

7.24 A new transformer is installed between buses 1 and 5, in parallel with the existing transformer in Example 7.9. The new transformer is identical to the existing transformer. The taps on the new transformer are set at 1.10 to give a 10% increase in voltage at bus 5, while the taps on the existing transformer are set at the nominal value 1.0 per unit. Use the POWER FLOW program to find the real power, reactive power, and MVA supplied by each of these transformers to bus 5.

References

1. W. F. Tinney, and C. E. Hart, "Power Flow Solutions by Newton's Method." *Trans. IEEE PAS.* Vol. 86 (November 1967): p. 1449.

2. W. F. Tinney, and J. W. Walker, "Direct Solution of Sparse Network Equations by Optimally Ordered Triangular Factorization," *Proc. IEEE.* Vol. 55 (November 1967): pp. 1801–1809.

3. Glenn W. Stagg, and Ahmed H. El-Abiad, *Computer Methods in Power System Analysis* (New York: McGraw-Hill, 1968).

4. N. M. Peterson, and W. S. Meyer, "Automatic Adjustment of Transformer and Phase Shifter Taps in Newton Power Flow," *IEEE Trans. PAS.* Vol. 90 (January–February 1971): pp. 103–108.

5. W. D. Stevenson, Jr., *Elements of Power Systems Analysis*, 4th ed. (New York: McGraw-Hill, 1982).

6. A. Bramellar, and R. N. Allan, *Sparsity* (London: Pitman, 1976).

7. C. A. Gross, *Power Systems Analysis* (New York: John Wiley and Sons, 1979).

8. B. Stott, "Fast Decoupled Load Flow," *Trans. IEEE PAS*, Vol. PAS 91 (September–October 1972): pp. 1955–1959.

CHAPTER 8

Symmetrical Faults

Short circuits occur in power systems when equipment insulation fails, due to system overvoltages caused by lightning or switching surges, to insulation contamination (salt spray or pollution), or to other mechanical causes. The resulting short circuit or "fault" current is determined by the internal voltages of the synchronous machines and by the system impedances between the machine voltages and the fault. Short-circuit currents may be several orders of magnitude larger than normal operating currents and, if allowed to persist, may cause equipment thermal damage. Windings and busbars may also suffer mechanical damage due to high magnetic forces during faults. It is therefore necessary to remove from service as soon as possible faulted sections of a power system. Standard EHV protective equipment is designed to clear faults within 3 cycles (50 ms at 60 Hz). Lower voltage protective equipment operates more slowly (for example, 5 to 20 cycles).

We begin this chapter by reviewing series R–L circuit transients in Section 8.1, followed in Section 8.2 by a description of three-phase short-circuit currents at unloaded synchronous machines. Both the ac component, including subtransient, transient, and steady-state currents, and the dc component of fault current are analyzed. We then extend these results in Sections 8.3 and 8.4 to power-system three-phase short circuits by means of the superposition principle. We observe that the bus impedance matrix is the key to calculating fault currents. In Sections 8.5 and 8.6 we present a systematic procedure for computing the bus impedance matrix and a personal computer program based on this procedure to compute symmetrical short-circuit currents in power systems. This program may be utilized in power-system design to select, set, and coordinate protective equipment such as circuit breakers, fuses, relays, and instrument transformers. Circuit breaker and fuse selection is discussed in Section 8.7.

Balanced three-phase power systems are assumed throughout this chapter. Since three-phase fault currents in balanced systems are themselves balanced, only positive-sequence networks are used. We also work in per-unit.

Section 8.1

Series R–L Circuit Transients

Consider the series R–L circuit shown in Figure 8.1. The closing of switch SW at $t = 0$ represents to a first approximation a three-phase short circuit at the terminals of an unloaded synchronous machine. For simplicity, assume zero fault impedance; that is, the short circuit is a solid or "bolted" fault. The current is assumed to be zero before SW closes, and the source angle α determines the source voltage at $t = 0$. Writing a KVL equation for the circuit,

$$\frac{\mathrm{L}di(t)}{dt} + \mathrm{R}i(t) = \sqrt{2}\mathrm{V}\sin(\omega t + \alpha) \qquad t \geq 0 \tag{8.1.1}$$

The solution to (8.1.1) is

$$i(t) = i_{\mathrm{ac}}(t) + i_{\mathrm{dc}}(t) = \frac{\sqrt{2}\mathrm{V}}{\mathrm{Z}}\left[\sin(\omega t + \alpha - \theta) - \sin(\alpha - \theta)e^{-t/\mathrm{T}}\right] \quad \mathrm{A} \tag{8.1.2}$$

where

$$i_{ac}(t) = \frac{\sqrt{2}V}{Z} \sin(\omega t + \alpha - \theta) \quad A \tag{8.1.3}$$

$$i_{dc}(t) = -\frac{\sqrt{2}V}{Z} \sin(\alpha - \theta) e^{-t/T} \quad A \tag{8.1.4}$$

$$Z = \sqrt{R^2 + (\omega L)^2} = \sqrt{R^2 + X^2} \quad \Omega \tag{8.1.5}$$

$$\theta = \tan^{-1}\frac{\omega L}{R} = \tan^{-1}\frac{X}{R} \tag{8.1.6}$$

$$T = \frac{L}{R} = \frac{X}{\omega R} = \frac{X}{2\pi f R} \quad s \tag{8.1.7}$$

Figure 8.1

Current in a series R–L circuit with ac voltage source

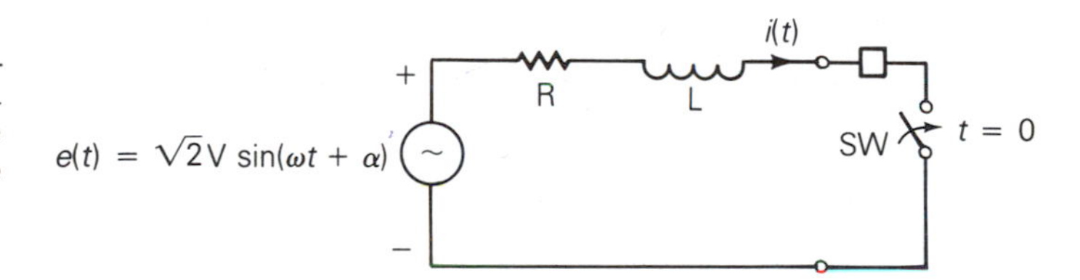

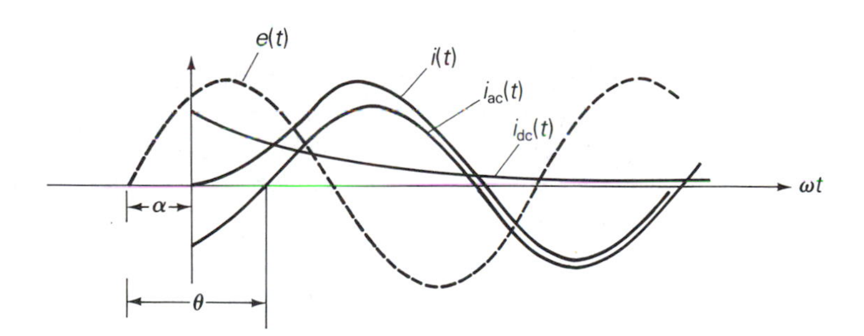

The total fault current in (8.1.2), called the *asymmetrical fault current*, is plotted in Figure 8.1 along with its two components. The ac fault current (also called *symmetrical* or *steady-state fault current*), given by (8.1.3), is a sinusoid. The *dc offset current*, given by (8.1.4), decays exponentially with time constant $T = L/R$.

The rms ac fault current is $I_{ac} = V/Z$. The magnitude of the dc offset, which depends on α, varies from 0 when $\alpha = \theta$ to $\sqrt{2}I_{ac}$ when $\alpha = (\theta \pm \pi/2)$. Note that a short circuit may occur at any instant during a cycle of the ac source; that is, α can have any value. Since we are primarily interested in the largest fault current, we choose $\alpha = (\theta - \pi/2)$. Then (8.1.2) becomes

$$i(t) = \sqrt{2}I_{ac}\,[\sin(\omega t - \pi/2) + e^{-t/T}] \quad A \tag{8.1.8}$$

where

$$I_{ac} = \frac{V}{Z} \quad A \tag{8.1.9}$$

The rms value of $i(t)$ is of interest. Since $i(t)$ in (8.1.8) is not strictly periodic, its rms value is not strictly defined. However, treating the exponential term as a constant, we stretch the rms concept to calculate the rms asymmetrical fault current with maximum dc offset, as follows:

$$\begin{aligned} I_{rms}(t) &= \sqrt{[I_{ac}]^2 + [I_{dc}(t)]^2} \\ &= \sqrt{[I_{ac}]^2 + [\sqrt{2}I_{ac}e^{-t/T}]^2} \\ &= I_{ac}\sqrt{1 + 2e^{-2t/T}} \quad A \end{aligned} \tag{8.1.10}$$

It is convenient to use $T = X/(2\pi f R)$ and $t = \tau/f$, where τ is time in cycles, and write (8.1.10) as

$$I_{rms}(\tau) = K(\tau)I_{ac} \quad A \tag{8.1.11}$$

where

$$K(\tau) = \sqrt{1 + 2e^{-4\pi\tau/(X/R)}} \quad \text{per unit} \tag{8.1.12}$$

From (8.1.11) and (8.1.12), the rms asymmetrical fault current equals the rms ac fault current times an "asymmetry factor," $K(\tau)$. $I_{rms}(\tau)$ decreases from $\sqrt{3}I_{ac}$ when $\tau = 0$ to I_{ac} when τ is large. Also, higher X to R ratios (X/R) give higher values of $I_{rms}(\tau)$. The above series R–L short-circuit currents are summarized in Table 8.1.

Table 8.1

Short-circuit current—series R–L circuit*

	INSTANTANEOUS CURRENT	rms CURRENT
COMPONENT \ UNITS	A	A
Symmetrical (ac)	$i_{ac}(t) = \frac{\sqrt{2}V}{Z}\sin(\omega t + \alpha - \theta)$	$I_{ac} = \frac{V}{Z}$
dc offset	$i_{dc}(t) = \frac{-\sqrt{2}V}{Z}\sin(\alpha - \theta)e^{-t/T}$	
Asymmetrical (total)	$i(t) = i_{ac}(t) + i_{dc}(t)$	$I_{rms}(t) = \sqrt{I_{ac}^2 + i_{dc}(t)^2}$ with maximum dc offset: $I_{rms}(\tau) = K(\tau)I_{ac}$

*See Figure 8.1 and (8.1.1)–(8.1.12).

Example 8.1 A bolted short circuit occurs in the series R–L circuit of Figure 8.1 with $V = 20\,kV$, $X = 8\,\Omega$, $R = 0.8\,\Omega$, and with maximum dc offset. The circuit breaker opens 3 cycles after fault inception. Determine (a) the rms ac fault current, (b) the rms "momentary" current at $\tau = 0.5$ cycle, which passes through the breaker before it opens, and (c) the rms asymmetrical fault current which the breaker interrupts.

Solution **a.** From (8.1.9),

$$I_{ac} = \frac{20 \times 10^3}{\sqrt{(8)^2 + (0.8)^2}} = \frac{20 \times 10^3}{8.040} = 2.488 \quad kA$$

b. From (8.1.11) and (8.1.12) with $(X/R) = 8/(0.8) = 10$ and $\tau = 0.5$ cycle,

$$K(0.5 \text{ cycle}) = \sqrt{1 + 2e^{-4\pi(0.5)/10}} = 1.438$$

$$I_{momentary} = K(0.5 \text{ cycle})I_{ac} = (1.438)(2.488) = 3.576 \quad kA$$

c. From (8.1.11) and (8.1.12) with $(X/R) = 10$ and $\tau = 3$ cycles,

$$K(3 \text{ cycles}) = \sqrt{1 + 2e^{-4\pi(3)/10}} = 1.023$$

$$I_{rms}(3 \text{ cycles}) = (1.023)(2.488) = 2.544 \quad kA$$

◁

Section 8.2

Three-phase Short Circuit—Unloaded Synchronous Machine

One way to investigate a three-phase short circuit at the terminals of a synchronous machine is to perform a test on an actual machine. Figure 8.2 shows an oscillogram of the ac fault current in one phase of an unloaded synchronous machine during such a test. The dc offset has been removed from the oscillogram. As shown, the amplitude of the sinusoidal waveform decreases from a high initial value to a lower steady-state value.

A physical explanation for this phenomenon is that the magnetic flux caused by the short-circuit armature currents (or by the resultant armature MMF) is initially forced to flow through high reluctance paths that do not link the field winding or damper circuits of the machine. This is a result of the theorem of constant flux linkages, which states that the flux linking a closed winding cannot change instantaneously. The armature inductance, which is inversely proportional to reluctance, is therefore initially low. As the flux then moves towards the lower reluctance paths, the armature inductance increases.

The ac fault current in a synchronous machine can be modeled by the series R–L circuit of Figure 8.1 if a time-varying inductance $L(t)$ or reactance $X(t) = \omega L(t)$ is employed. In standard machine theory texts [3,4], the following reactances are defined:

Figure 8.2

ac fault current in one phase of an unloaded synchronous machine during a three-phase short circuit (the dc offset current is removed)

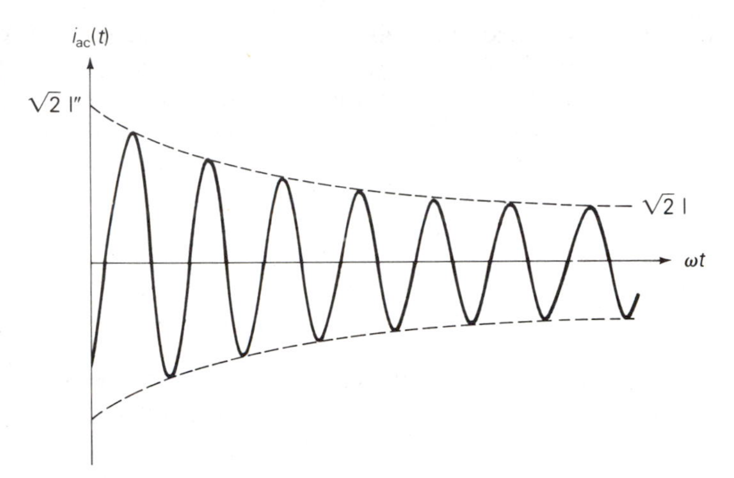

X''_d = direct axis subtransient reactance

X'_d = direct axis transient reactance

X_d = direct axis synchronous reactance

where $X''_d < X'_d < X_d$. The subscript d refers to the direct axis. There are similar quadrature axis reactances X''_q, X'_q, and X_q [3,4]. However, if the armature resistance is small, the quadrature axis reactances do not significantly affect the short-circuit current. Using the above direct axis reactances, the instantaneous ac fault current can be written as

$$i_{ac}(t) = \sqrt{2}E_g\left[\left(\frac{1}{X''_d} - \frac{1}{X'_d}\right)e^{-t/T''_d} + \left(\frac{1}{X'_d} - \frac{1}{X_d}\right)e^{-t/T'_d} + \frac{1}{X_d}\right]\sin\left(\omega t + \alpha - \frac{\pi}{2}\right) \tag{8.2.1}$$

where E_g is the rms line-to-neutral prefault terminal voltage of the unloaded synchronous machine. Armature resistance is neglected in (8.2.1). Note that at $t = 0$, when the fault occurs, the rms value of $i_{ac}(t)$ in (8.2.1) is

$$I_{ac}(0) = \frac{E_g}{X''_d} = I'' \tag{8.2.2}$$

which is called the rms *subtransient fault current*, I″. The duration of I″ is determined by the time constant T''_d, called the *direct axis short-circuit subtransient time constant*.

At a later time, when t is large compared to T''_d but small compared to the *direct axis short-circuit transient time constant* T'_d, the first exponential term in (8.2.1) has decayed almost to zero but the second exponential has not decayed significantly. The rms ac fault current then equals the rms *transient fault current*, given by

$$I' = \frac{E_g}{X'_d} \tag{8.2.3}$$

When t is much larger than T_d', the rms ac fault current approaches its steady-state value, given by

$$I_{ac}(\infty) = \frac{E_g}{X_d} = I \tag{8.2.4}$$

Since the three-phase no-load voltages are displaced 120° from each other, the three-phase ac fault currents are also displaced 120° from each other. In addition to the ac fault current, each phase has a different dc offset. The maximum dc offset in any one phase, which occurs when $\alpha = 0$ in (8.2.1), is

$$i_{dcmax}(t) = \frac{\sqrt{2}E_g}{X_d''} e^{-t/T_A} = \sqrt{2}\, I'' e^{-t/T_A} \tag{8.2.5}$$

where T_A is called the *armature time constant*. Note that the magnitude of the maximum dc offset depends only on the rms subtransient fault current I''. The above synchronous machine short-circuit currents are summarized in Table 8.2.

Table 8.2
Short-circuit current—unloaded synchronous machine*

COMPONENT \ UNITS	INSTANTANEOUS CURRENT A	rms CURRENT A
Symmetrical (ac)	(8.2.1)	$I_{ac}(t) = E_g\left[\left(\frac{1}{X_d''} - \frac{1}{X_d'}\right)e^{-t/T_d''} + \left(\frac{1}{X_d'} - \frac{1}{X_d}\right)e^{-t/T_d'} + \frac{1}{X_d}\right]$
Subtransient		$I'' = E_g/X_d''$
Transient		$I' = E_g/X_d'$
Steady-state		$I = E_g/X_d$
Maximum dc offset	$i_{dc}(t) = \sqrt{2}\, I'' e^{-t/T_A}$	
Asymmetrical (total)	$i(t) = i_{ac}(t) + i_{dc}(t)$	$I_{rms}(t) = \sqrt{I_{ac}(t)^2 + i_{dc}(t)^2}$ with maximum dc offset: $I_{rms}(t) = \sqrt{I_{ac}(t)^2 + [\sqrt{2}\, I'' e^{-t/T_A}]^2}$

*See Figure 8.2 and (8.2.1)–(8.2.5).

Machine reactances X_d'', X_d', and X_d as well as time constants T_d'', T_d', and T_A are usually provided by synchronous machine manufacturers. They can also be obtained from a three-phase short-circuit test, by analyzing an oscillogram such as that in Figure 8.2 [2]. Typical values of synchronous machine reactances and time constants are given in Table A.1.

Example 8.2 A 500-MVA 20-kV, 60-Hz synchronous generator with reactances $X''_d = 0.15$, $X'_d = 0.24$, $X_d = 1.1$ per unit and time constants $T''_d = 0.035$, $T'_d = 2.0$, $T_A = 0.20$ s, is connected to a circuit breaker. The generator is operating at 5% above rated voltage and at no-load when a bolted three-phase short circuit occurs on the load side of the breaker. The breaker interrupts the fault 3 cycles after fault inception. Determine (a) the subtransient fault current in per-unit and kA rms; (b) maximum dc offset as a function of time; and (c) rms asymmetrical fault current, which the breaker interrupts, assuming maximum dc offset.

Solution

a. The no-load voltage before the fault occurs is $E_g = 1.05$ per unit. From (8.2.2), the subtransient fault current that occurs in each of the three phases is

$$I'' = \frac{1.05}{0.15} = 7.0 \quad \text{per unit}$$

The generator base current is

$$I_{\text{base}} = \frac{S_{\text{rated}}}{\sqrt{3}\,V_{\text{rated}}} = \frac{500}{(\sqrt{3})(20)} = 14.43 \quad \text{kA}$$

The rms subtransient fault current in kA is the per-unit value multiplied by the base current:

$$I'' = (7.0)(14.43) = 101.0 \quad \text{kA}$$

b. From (8.2.5), the maximum dc offset that may occur in any one phase is

$$i_{\text{dcmax}}(t) = \sqrt{2}\,(101.0)\,e^{-t/0.20} = 142.9\,e^{-t/0.20} \quad \text{kA}$$

c. From (8.2.1), the rms ac fault current at $t = 3$ cycles $= 0.05$ s is

$$I_{\text{ac}}(0.05\text{ s}) = 1.05\left[\left(\frac{1}{0.15} - \frac{1}{0.24}\right)e^{-0.05/0.035} + \left(\frac{1}{0.24} - \frac{1}{1.1}\right)e^{-0.05/2.0} + \frac{1}{1.1}\right]$$

$$= 4.920 \quad \text{per unit}$$

$$= (4.920)(14.43) = 71.01 \quad \text{kA}$$

Modifying (8.1.10) to account for the time-varying symmetrical component of fault current, we obtain

$$I_{rms}(0.05) = \sqrt{[I_{ac}(0.05)]^2 + [\sqrt{2}\,I'' e^{-t/T_a}]^2}$$

$$= I_{ac}(0.05)\sqrt{1 + 2\left[\frac{I''}{I_{ac}(0.05)}\right]^2 e^{-2t/T_a}}$$

$$= (71.01)\sqrt{1 + 2\left[\frac{101}{71.01}\right]^2 e^{-2(0.05)/0.20}}$$

$$= (71.01)(1.8585)$$

$$= 132 \quad \text{kA}$$

◁

Section 8.3

Power-System Three-phase Short Circuits

In order to calculate the subtransient fault current for a three-phase short circuit in a power system, we make the following assumptions:

1. Transformers are represented by their leakage reactances. Winding resistances, shunt admittances, and Δ–Y phase shifts are neglected.
2. Transmission lines are represented by their positive-sequence equivalent series reactances. Series resistances and shunt admittances are neglected.
3. Synchronous machines are represented by constant-voltage sources behind subtransient reactances. Armature resistance, saliency, and saturation are neglected.
4. All nonrotating impedance loads are neglected.
5. Induction motors are either neglected (especially for small motors rated less than 50 hp) or represented in the same manner as synchronous machines.

These assumptions are made for simplicity in this text, and in practice they should not be made for all cases. For example, in distribution systems, resistances of primary and secondary distribution lines may in some cases significantly reduce fault current magnitudes.

Figure 8.3 shows a single-line diagram consisting of a synchronous generator feeding a synchronous motor through two transformers and a transmission line. We shall consider a three-phase short circuit at bus 1. The positive-sequence equivalent circuit is shown in Figure 8.4(a), where the voltages E''_g and E''_m are the prefault internal voltages behind the subtransient reactances of the machines, and the closing of switch SW represents the fault. For purposes of calculating the subtransient fault current, E''_g and E''_m are assumed to be constant-voltage sources.

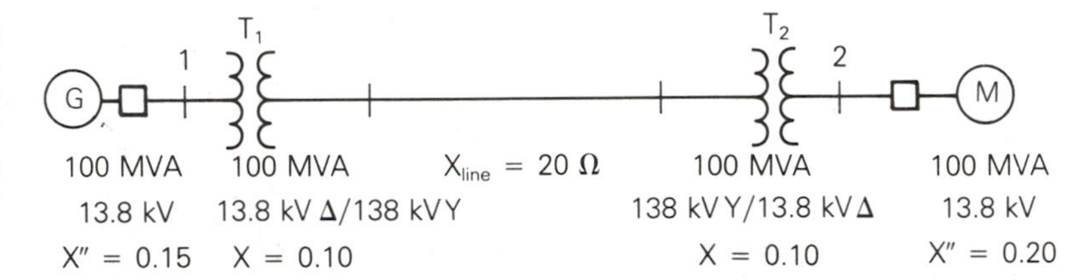

Figure 8.3 Single-line diagram of a synchronous generator feeding a synchronous motor

In Figure 8.4(b) the fault is represented by two opposing voltage sources with equal phasor values V_F. Using superposition, the fault current can then be calculated from the two circuits shown in Figure 8.4(c). However, if V_F equals the prefault voltage at the fault, then the second circuit in Figure 8.4(c) represents the system before the fault occurs. As such, $I''_{F2} = 0$ and V_F, which has no effect, can be removed from the second circuit, as shown in Figure 8.4(d). The subtransient fault current is then determined from the first circuit in Figure 8.4(d), $I''_F = I''_{F1}$. The contribution to the fault from the generator is $I''_g = I''_{g1} + I''_{g2} = I''_{g1} + I_L$, where I_L is the prefault generator current. Similarly, $I''_m = I''_{m1} - I_L$.

Example 8.3 The synchronous generator in Figure 8.3 is operating at rated MVA, 0.95 p.f. lagging and at 5% above rated voltage when a bolted three-phase short circuit occurs at bus 1. Calculate the per-unit values of (a) subtransient fault current, (b) subtransient generator and motor currents, neglecting prefault current, and (c) subtransient generator and motor currents including prefault current.

Solution **a.** Using a 100-MVA base, the base impedance in the zone of the transmission line is

$$Z_{\text{base,line}} = \frac{(138)^2}{100} = 190.44 \quad \Omega$$

and

$$X_{\text{line}} = \frac{20}{190.44} = 0.1050 \quad \text{per unit}$$

The per-unit reactances are shown in Figure 8.4. From the first circuit in Figure 8.4(d), the Thévenin impedance as viewed from the fault is

$$Z_{\text{Th}} = jX_{\text{Th}} = j\frac{(0.15)(0.505)}{(0.15 + 0.505)} = j0.11565 \quad \text{per unit}$$

Figure 8.4

Application of superposition to a power-system three-phase short circuit

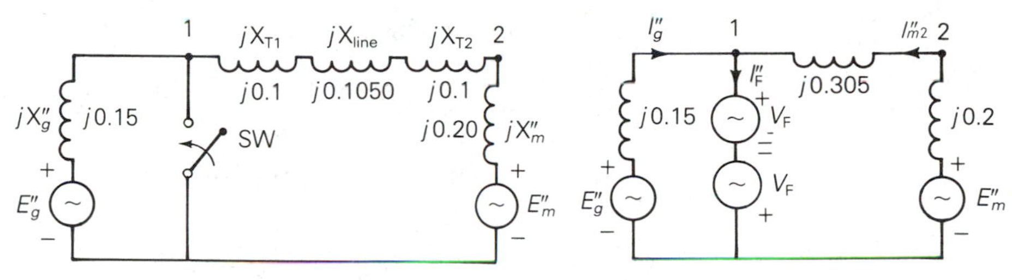

(a) Three-phase short circuit

(b) Short circuit represented by two opposing voltage sources

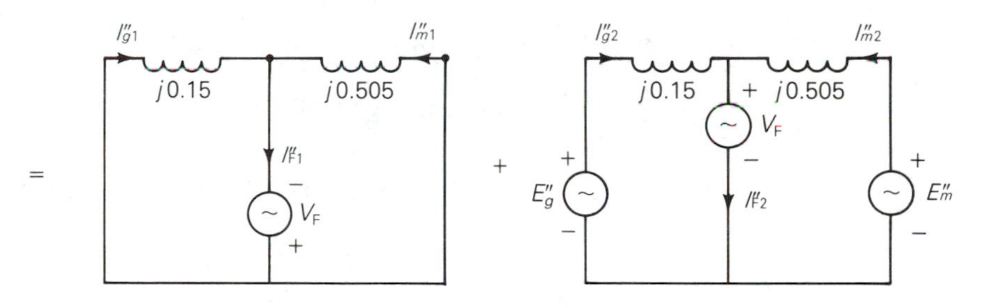

(c) Application of superposition

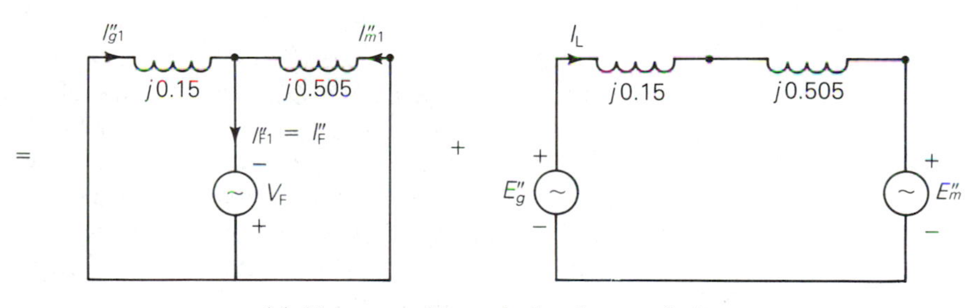

(d) V_F set equal to prefault voltage at fault

and the prefault voltage at the generator terminals is

$$V_F = 1.05\underline{/0^\circ} \quad \text{per unit}$$

The subtransient fault current is then

$$I''_F = \frac{V_F}{Z_{Th}} = \frac{1.05\underline{/0^\circ}}{j0.11565} = -j9.079 \quad \text{per unit}$$

b. Using current division in the first circuit of Figure 8.4(d),

$$I''_{g1} = \left(\frac{0.505}{0.505 + 0.15}\right) I''_F = (0.7710)(-j9.079) = -j7.000 \quad \text{per unit}$$

$$I''_{m1} = \left(\frac{0.15}{0.505 + 0.15}\right) I''_F = (0.2290)(-j9.079) = -j2.079 \quad \text{per unit}$$

c. The generator base current is

$$I_{\text{base, gen}} = \frac{100}{(\sqrt{3})(13.8)} = 4.1837 \quad \text{kA}$$

and the prefault generator current is

$$I_L = \frac{100}{(\sqrt{3})(1.05 \times 13.8)} \underline{/-\cos^{-1} 0.95} = 3.9845\underline{/-18.19^\circ} \quad \text{kA}$$

$$= \frac{3.9845\underline{/-18.19^\circ}}{4.1837} = 0.9524\underline{/-18.19^\circ}$$

$$= 0.9048 - j0.2974 \quad \text{per unit}$$

The subtransient generator and motor currents, including prefault current, are then

$$I_g'' = I_{g1}'' + I_L = -j7.000 + 0.9048 - j0.2974$$

$$= 0.9048 - j7.297 = 7.353\underline{/-82.9^\circ} \quad \text{per unit}$$

$$I_m'' = I_{m1}'' - I_L = -j2.079 - 0.9048 + j0.2974$$

$$= -0.9048 - j1.782 = 1.999\underline{/243.1^\circ} \quad \text{per unit}$$

An alternate method of solving Example 8.3 is to first calculate the internal voltages E_g'' and E_m'' using the prefault load current I_L. Then, instead of using superposition, the fault currents can be resolved directly from the circuit in Figure 8.4(a) (see Problem 8.6). However, in a system with many synchronous machines, the superposition method has the advantage that all machine voltage sources are shorted, and the prefault voltage is the only source required to calculate the fault current. Also, when calculating the contributions to fault current from each branch, prefault currents are usually small, and hence can be neglected. Otherwise, prefault load currents could be obtained from a power-flow program. ◁

Section 8.4

Bus Impedance Matrix

We now extend the results of the previous section to calculate subtransient fault currents for three-phase faults in an N-bus power system. The system is modeled by its positive-sequence network, where lines and transformers are represented by series reactances, and synchronous machines are represented by constant-voltage sources behind subtransient reactances. As before, all resistances, shunt admittances, and nonrotating impedance loads are neglected. For simplicity, we also neglect prefault load currents.

Consider a three-phase short circuit at any bus n. Using the superposition method described in Section 8.3, we analyze two separate circuits. [For example, see Figure 8.4(d).] In the first circuit all machine-voltage sources are short-circuited and the only source is due to the prefault voltage at the fault. Writing nodal equations for the first circuit,

$$\boldsymbol{Y}_{\text{bus}} \boldsymbol{E}^{(1)} = \boldsymbol{I}^{(1)} \tag{8.4.1}$$

where $\boldsymbol{Y}_{\text{bus}}$ is the positive-sequence bus admittance matrix, $\boldsymbol{E}^{(1)}$ is the vector of bus voltages, and $\boldsymbol{I}^{(1)}$ is the vector of current sources. The superscript (1) denotes the first circuit. Solving (8.4.1),

$$\boldsymbol{Z}_{\text{bus}} \boldsymbol{I}^{(1)} = \boldsymbol{E}^{(1)} \tag{8.4.2}$$

where

$$\boldsymbol{Z}_{\text{bus}} = \boldsymbol{Y}_{\text{bus}}^{-1} \tag{8.4.3}$$

$\boldsymbol{Z}_{\text{bus}}$, the inverse of $\boldsymbol{Y}_{\text{bus}}$, is called the positive-sequence *bus impedance matrix*. Both $\boldsymbol{Z}_{\text{bus}}$ and $\boldsymbol{Y}_{\text{bus}}$ are symmetric matrices.

Since the first circuit contains only one source, located at faulted bus n, the current source vector contains only one nonzero component, $I_n^{(1)} = -I''_{\text{F}n}$. Also, the voltage at faulted bus n in the first circuit is $E_n^{(1)} = -V_{\text{F}}$. Rewriting (8.4.2),

$$\begin{bmatrix} Z_{11} & Z_{12} & \cdots & Z_{1n} & \cdots & Z_{1N} \\ Z_{21} & Z_{22} & \cdots & Z_{2n} & \cdots & Z_{2N} \\ \vdots & & & & & \\ Z_{n1} & Z_{n2} & \cdots & Z_{nn} & \cdots & Z_{nN} \\ \vdots & & & & & \\ Z_{N1} & Z_{N2} & \cdots & Z_{Nn} & \cdots & Z_{NN} \end{bmatrix} \begin{bmatrix} 0 \\ 0 \\ \vdots \\ -I''_{\text{F}n} \\ \vdots \\ 0 \end{bmatrix} = \begin{bmatrix} E_1^{(1)} \\ E_2^{(1)} \\ \vdots \\ -V_{\text{F}} \\ \vdots \\ E_N^{(1)} \end{bmatrix} \tag{8.4.4}$$

The minus sign associated with the current source in (8.4.4) indicates that the current injected into bus n is the negative of $I''_{\text{F}n}$, since $I''_{\text{F}n}$ flows away from bus n to the neutral. From (8.4.4), the subtransient fault current is

$$I''_{\text{F}n} = \frac{V_{\text{F}}}{Z_{nn}} \tag{8.4.5}$$

Also from (8.4.4) and (8.4.5), the voltage at any bus k in the first circuit is

$$E_k^{(1)} = Z_{kn}(-I''_{\text{F}n}) = \frac{-Z_{kn}}{Z_{nn}} V_{\text{F}} \tag{8.4.6}$$

The second circuit represents the prefault conditions. Neglecting prefault load current, all voltages throughout the second circuit are equal to the prefault voltage; that is, $E_k^{(2)} = V_F$ for each bus k. Applying superposition,

$$E_k = E_k^{(1)} + E_k^{(2)} = \frac{-Z_{kn}}{Z_{nn}} V_F + V_F = \left(1 - \frac{Z_{kn}}{Z_{nn}}\right) V_F \qquad k = 1,2,\ldots,N \tag{8.4.7}$$

Example 8.4 Faults at bus 1 and 2 in Figure 8.3 are of interest. The prefault voltage is 1.05 per unit and prefault load current is neglected. (a) Determine the 2 × 2 positive-sequence bus impedance matrix. (b) For a bolted three-phase short circuit at bus 1, use $\boldsymbol{Z}_{\text{bus}}$ to calculate the subtransient fault current and the contribution to the fault current from the transmission line. (c) Repeat part (b) for a bolted three-phase short circuit at bus 2.

Solution **a.** The circuit of Figure 8.4(a) is redrawn in Figure 8.5 showing per-unit admittance rather than per-unit impedance values. Neglecting prefault load current, $E_g'' = E_m'' = V_F = 1.05\underline{/0^\circ}$ per unit. From Figure 8.5, the positive-sequence bus admittance matrix is

$$\boldsymbol{Y}_{\text{bus}} = -j\begin{bmatrix} 9.9454 & -3.2787 \\ -3.2787 & 8.2787 \end{bmatrix} \quad \text{per unit}$$

Inverting $\boldsymbol{Y}_{\text{bus}}$,

$$\boldsymbol{Z}_{\text{bus}} = \boldsymbol{Y}_{\text{bus}}^{-1} = +j\begin{bmatrix} 0.11565 & 0.04580 \\ 0.04580 & 0.13893 \end{bmatrix} \quad \text{per unit}$$

b. Using (8.4.5) the subtransient fault current at bus 1 is

$$I_{F1}'' = \frac{V_F}{Z_{11}} = \frac{1.05\underline{/0^\circ}}{j0.11565} = -j9.079 \quad \text{per unit}$$

which agrees with the result in Example 8.3, part (a). The voltages at buses 1 and 2 during the fault are, from (8.4.7),

$$E_1 = \left(1 - \frac{Z_{11}}{Z_{11}}\right) V_F = 0$$

$$E_2 = \left(1 - \frac{Z_{21}}{Z_{11}}\right) V_F = \left(1 - \frac{j0.04580}{j0.11565}\right) 1.05\underline{/0^\circ} = 0.6342\underline{/0^\circ}$$

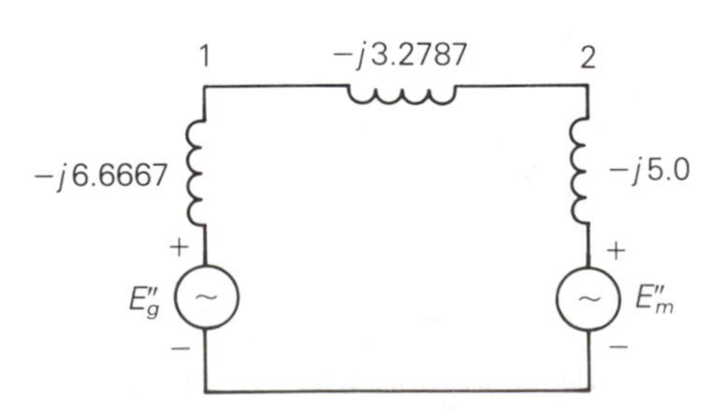

Figure 8.5 Circuit of Figure 8.4(a) showing per-unit admittance values

The current to the fault from the transmission line is obtained from the voltage drop from bus 2 to 1 divided by the impedance of the line and transformers T_1 and T_2:

$$I_{21} = \frac{E_2 - E_1}{j(X_{\text{line}} + X_{T1} + X_{T2})} = \frac{0.6342 - 0}{j0.3050} = -j2.079 \quad \text{per unit}$$

which agrees with the motor current calculated in Example 8.3, part (b), where prefault load current is neglected.

c. Using (8.4.5), the subtransient fault current at bus 2 is

$$I''_{F2} = \frac{V_F}{Z_{22}} = \frac{1.05\underline{/0^\circ}}{j0.13893} = -j7.558 \quad \text{per unit}$$

and from (8.4.7),

$$E_1 = \left(1 - \frac{Z_{12}}{Z_{22}}\right) V_F = \left(1 - \frac{j0.04580}{j0.13893}\right) 1.05\underline{/0^\circ} = 0.7039\underline{/0^\circ}$$

$$E_2 = \left(1 - \frac{Z_{22}}{Z_{22}}\right) V_F = 0$$

The current to the fault from the transmission line is

$$I_{12} = \frac{E_1 - E_2}{j(X_{\text{line}} + X_{T1} + X_{T2})} = \frac{0.7039 - 0}{j0.3050} = -j2.308 \quad \text{per unit} \quad \triangleleft$$

Figure 8.6 shows a bus impedance equivalent circuit that illustrates the short-circuit currents in an N-bus system. This circuit is given the name *rake equivalent* in Neuenswander [5] due to its shape, which is similar to a garden rake.

The diagonal elements $Z_{11}, Z_{22}, \ldots, Z_{NN}$ of the bus impedance matrix,

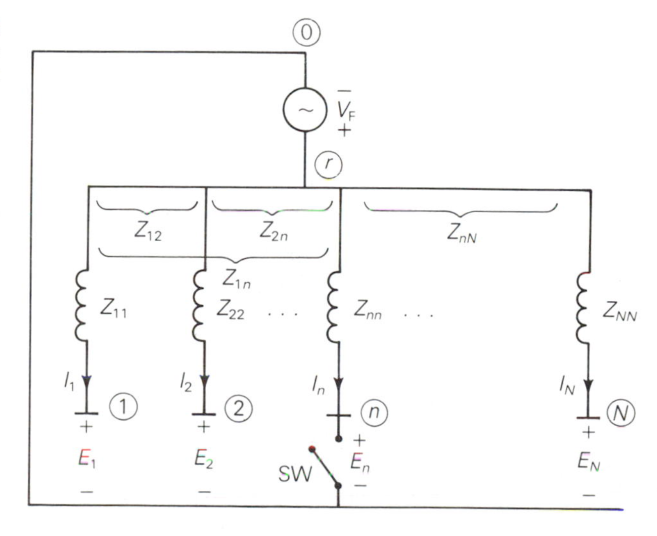

Figure 8.6

Bus impedance equivalent circuit (*rake equivalent*)

which are the *self-impedances*, are shown in Figure 8.6. The off-diagonal elements, or the *mutual impedances*, are indicated by the brackets in the figure.

Neglecting prefault load currents, the internal voltage sources of all synchronous machines are equal both in magnitude and phase. As such, they can be connected, as shown in Figure 8.7, and replaced by one equivalent source V_F from neutral bus 0 to a references bus, denoted r. This equivalent source is also shown in the rake equivalent of Figure 8.6.

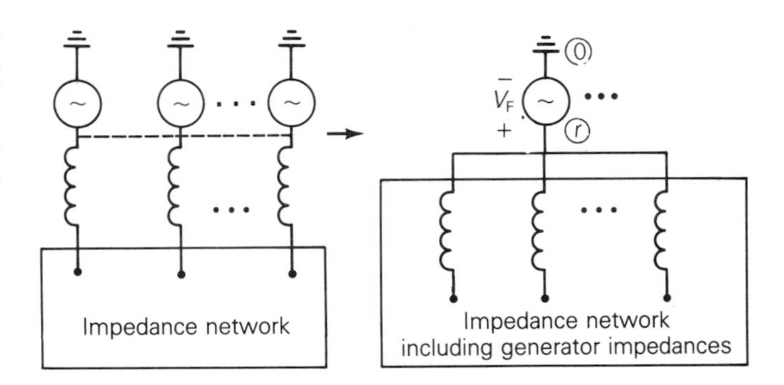

Figure 8.7 Parallel connection of unloaded synchronous machine internal-voltage sources

Using $\boldsymbol{Z}_{\text{bus}}$, the fault currents in Figure 8.6 are given by

$$\begin{bmatrix} Z_{11} & Z_{12} & \cdots & Z_{1n} & \cdots & Z_{1N} \\ Z_{21} & Z_{22} & \cdots & Z_{2n} & \cdots & Z_{2N} \\ \vdots & & & & & \vdots \\ Z_{n1} & Z_{n2} & \cdots & Z_{nn} & \cdots & Z_{nN} \\ \vdots & & & & & \\ Z_{N1} & Z_{N2} & \cdots & Z_{Nn} & \cdots & Z_{NN} \end{bmatrix} \begin{bmatrix} I_1 \\ I_2 \\ \vdots \\ I_n \\ \vdots \\ I_N \end{bmatrix} = \begin{bmatrix} V_F - E_1 \\ V_F - E_2 \\ \vdots \\ V_F - E_n \\ \vdots \\ V_F - E_N \end{bmatrix} \tag{8.4.8}$$

where $I_1, I_2, \ldots$ are the branch currents and $(V_F - E_1)$, $(V_F - E_2)$, ... are the voltages across the branches.

If switch SW in Figure 8.6 is open, all currents are zero and the voltage at each bus with respect to the neutral equals V_F. This corresponds to prefault conditions, neglecting prefault load currents. If switch SW is closed, corresponding to a short circuit at bus n, $E_n = 0$ and all currents except I_n remain zero. The fault current is $I''_{Fn} = I_n = V_F/Z_{nn}$, which agrees with (8.4.5). This fault current also induces a voltage drop $Z_{kn}I_n = (Z_{kn}/Z_{nn})V_F$ across each branch k. The voltage at bus k with respect to the neutral then equals V_F minus this voltage drop, which agrees with (8.4.7).

As shown by Figure 8.6 as well as (8.4.5), subtransient fault currents throughout an N-bus system can be determined from the bus impedance matrix and the prefault voltage. Once $\boldsymbol{Z}_{\text{bus}}$ has been obtained, these fault currents are easily computed.

Section 8.5

Formation of $\boldsymbol{Z}_{\text{bus}}$ One Step at a Time

In Section 8.4 $\boldsymbol{Z}_{\text{bus}}$ is obtained by inverting $\boldsymbol{Y}_{\text{bus}}$. In this section we describe a method of building up $\boldsymbol{Z}_{\text{bus}}$ by adding branches one at a time until the complete network is formed [5,6]. This method can be used to construct a new $\boldsymbol{Z}_{\text{bus}}$ or to modify an existing $\boldsymbol{Z}_{\text{bus}}$. It is easily implemented on a digital computer and is much simpler than inverting $\boldsymbol{Y}_{\text{bus}}$, especially for a large number of buses.

Consider the problem of adding a new branch with impedance Z_b to an existing n bus network whose $n \times n$ bus impedance matrix $\boldsymbol{Z}_{\text{bus(old)}}$ is known. It is desired to construct the new bus impedance matrix $\boldsymbol{Z}_{\text{bus(new)}}$. We recognize the following types of branch additions:

1. Add Z_b from the reference bus r to new bus p.
2. Add Z_b from old bus k to new bus p.
3. Add Z_b from the reference bus r to old bus k.
4. Add Z_b between two old buses k and m.

Since only three-phase short circuits in balanced three-phase systems are under consideration, we work with the positive-sequence network. For simplicity, we assume that there is no mutual impedance between the new branch and any existing branch. Also, all impedances are given in per-unit. The four types of branch additions are now discussed.

Type 1 Add Z_b from the reference bus r to new bus p. See Figure 8.8. The new equation set is:

$$\underbrace{\left[\begin{array}{c|c} \boldsymbol{Z}_{\text{bus(old)}} & \begin{matrix} 0 \\ 0 \\ \vdots \\ 0 \end{matrix} \\ \hline 0 \;\; 0 \;\; \ldots \;\; 0 & Z_b \end{array}\right]}_{\boldsymbol{Z}_{\text{bus(new)}}} \left[\begin{array}{c} I_1 \\ I_2 \\ \vdots \\ I_n \\ \hline I_p \end{array}\right] = \left[\begin{array}{c} V_F - E_1 \\ V_F - E_2 \\ \vdots \\ V_F - E_n \\ \hline V_F - E_p \end{array}\right] \tag{8.5.1}$$

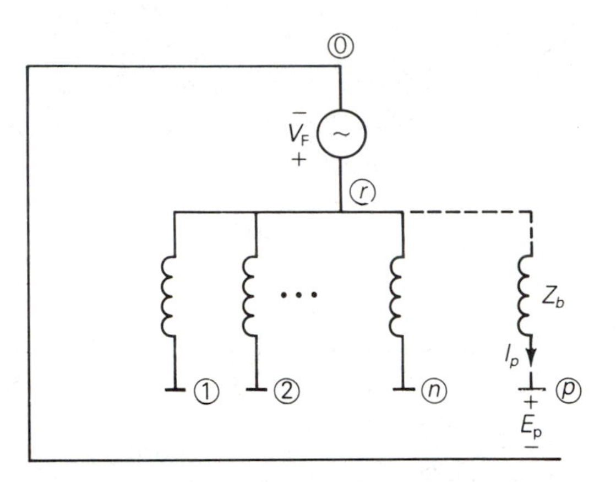

Figure 8.8

Type 1 addition

Since a new bus is added, a new row and column are added to $\boldsymbol{Z}_{\text{bus(old)}}$ to form the $(n + 1) \times (n + 1)$ matrix $\boldsymbol{Z}_{\text{bus(new)}}$. If a fault occurs at new bus p, then $E_p = 0$ and the fault current $I_p = V_F/Z_b$. Therefore, Z_b is the new diagonal element of $\boldsymbol{Z}_{\text{bus(new)}}$. Also, I_p cannot change the existing voltages in the network, since Z_b is not mutually coupled to the existing branches. Therefore, the remaining elements in the new row and column are zero.

Type 2 Add Z_b from old bus k to new bus p. See Figure 8.9. The new equation set is

$$\underbrace{\left[\begin{array}{c|c} & Z_{1k} \\ \boldsymbol{Z}_{\text{bus(old)}} & Z_{2k} \\ & \vdots \\ & Z_{nk} \\ \hline Z_{1k} \quad Z_{2k} \quad \cdots \quad Z_{nk} & Z_{kk}+Z_b \end{array}\right]}_{\boldsymbol{Z}_{\text{bus(new)}}} \left[\begin{array}{c} I_1 \\ I_2 \\ \vdots \\ I_n \\ \hline I_p \end{array}\right] = \left[\begin{array}{c} V_F - E_1 \\ V_F - E_2 \\ \vdots \\ V_F - E_n \\ \hline V_F - E_p \end{array}\right] \tag{8.5.2}$$

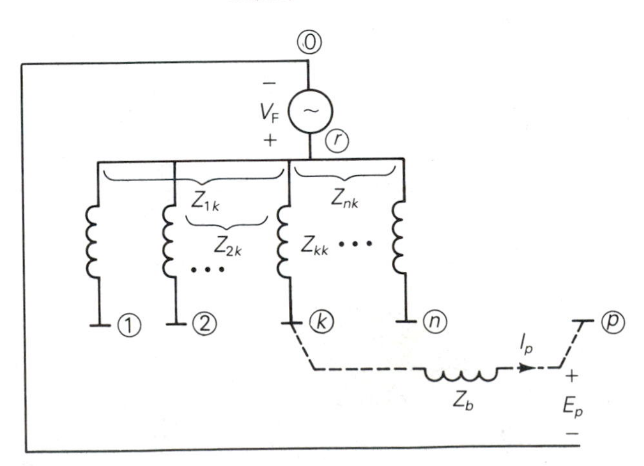

Figure 8.9

Type 2 addition

Again, since a new bus is added, a new row and column are created. If a fault occurs at new bus p, $E_p = 0$ and $I_p = V_F/(Z_{kk} + Z_b)$. Thus $(Z_{kk} + Z_b)$ is the new diagonal element. Also, the current I_p through branch k induces voltage drops $Z_{1k}I_p, Z_{2k}I_p, \ldots, Z_{nk}I_p$ in branches 1, 2, ..., n, respectively. Therefore, the remaining elements of the new column of $\boldsymbol{Z}_{\text{bus(new)}}$ are the elements of the kth column of $\boldsymbol{Z}_{\text{bus(old)}}$. Also, since $\boldsymbol{Z}_{\text{bus(new)}}$ is symmetric, the new row is identical to the new column.

Type 3 Add Z_b from the reference bus r to old bus k. See Figure 8.10. $\boldsymbol{Z}_{\text{bus(new)}}$ is formed in two steps. First, we temporarily create new bus p and connect Z_b from bus k to p, which is a type 2 addition. The equation set is given by (8.5.2). Second, we short-circuit new bus p to the reference. As such, $(V_F - E_P) = 0$ and I_p can be eliminated from (8.5.2). Writing (8.5.2) in partitioned form with $(V_F - E_P) = 0$,

$$\boldsymbol{Z}_{\text{bus(old)}}\, \boldsymbol{I} + \boldsymbol{Z}_{\text{col}} I_p = \boldsymbol{V}_F - \boldsymbol{E} \tag{8.5.3}$$

$$\boldsymbol{Z}_{\text{col}}^{T}\, \boldsymbol{I} + (Z_{kk} + Z_b) I_p = 0 \tag{8.5.4}$$

where

$$\boldsymbol{Z}_{\text{col}} = \begin{bmatrix} Z_{1k} \\ Z_{2k} \\ \vdots \\ Z_{nk} \end{bmatrix} \qquad \boldsymbol{I} = \begin{bmatrix} I_1 \\ I_2 \\ \vdots \\ I_n \end{bmatrix} \qquad (\boldsymbol{V}_F - \boldsymbol{E}) = \begin{bmatrix} V_F - E_1 \\ V_F - E_2 \\ \vdots \\ V_F - E_n \end{bmatrix}$$

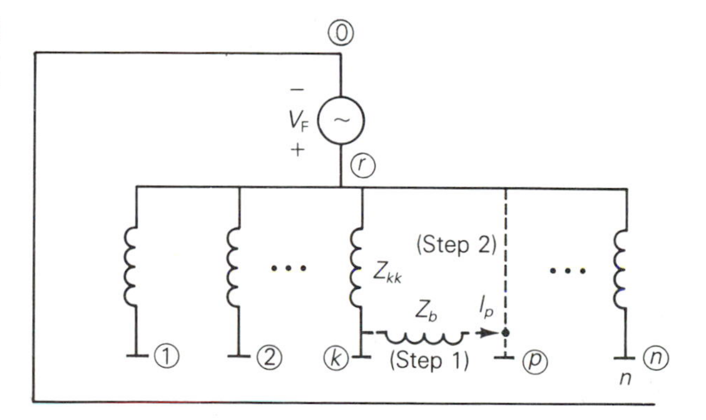

Figure 8.10

Type 3 addition

Solving (8.5.4) for I_p, (8.5.3) becomes

$$\underbrace{\left[\boldsymbol{Z}_{\text{bus(old)}} - \frac{\boldsymbol{Z}_{\text{col}}\, \boldsymbol{Z}_{\text{col}}^{T}}{(Z_{kk} + Z_b)}\right]}_{\boldsymbol{Z}_{\text{bus(new)}}} \boldsymbol{I} = \boldsymbol{V}_F - \boldsymbol{E} \tag{8.5.5}$$

Therefore,

$$\boldsymbol{Z}_{\text{bus(new)}} = \boldsymbol{Z}_{\text{bus(old)}} - \frac{1}{(Z_{kk} + Z_b)} \begin{bmatrix} Z_{1k} \\ Z_{2k} \\ \vdots \\ Z_{nk} \end{bmatrix} [Z_{1k}\ Z_{2k} \ldots Z_{nk}] \tag{8.5.6}$$

Note that $\boldsymbol{Z}_{\text{bus(new)}}$ has the same dimension as the $n \times n$ matrix $\boldsymbol{Z}_{\text{bus(old)}}$.

Type 4 Add Z_b between two old buses k and m. See Figure 8.11. Note that when Z_b is added, the current $(I_k + I_b)$ flows in branch k and $(I_m - I_b)$ flows in branch m. The voltage across branch 1 then becomes

$$\begin{aligned} Z_{11}I_1 + Z_{12}I_2 + \cdots + Z_{1k}(I_k + I_b) + \cdots + Z_{1m}(I_m - I_b) \\ + \cdots + Z_{1n}I_n = (V_F - E_1) \end{aligned} \tag{8.5.7}$$

Similarly,

$$\begin{aligned} Z_{1k}I_1 + Z_{2k}I_2 + \cdots + Z_{kk}(I_k + I_b) + \cdots + Z_{mk}(I_m - I_b) \\ + \cdots + Z_{nk}I_n = (V_F - E_k) \end{aligned} \tag{8.5.8}$$

and

$$\begin{aligned} Z_{1m}I_1 + Z_{2m}I_2 + \cdots + Z_{km}(I_k + I_b) + \cdots + Z_{mm}(I_m - I_b) \\ + \cdots + Z_{nm}I_n = (V_F - E_m) \end{aligned} \tag{8.5.9}$$

Also, the voltage drop across Z_b is:

$$Z_bI_b = E_k - E_m \tag{8.5.10}$$

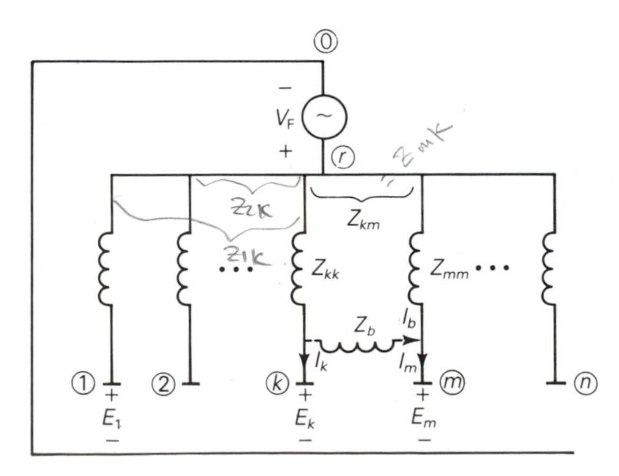

Figure 8.11 Type 4 addition

Using (8.5.8) and (8.5.9) in (8.5.10),

$$Z_b I_b = (Z_{1m} - Z_{1k})I_1 + (Z_{2m} - Z_{2k})I_2 + \cdots + (Z_{km} - Z_{kk})(I_k + I_b) + \cdots + (Z_{mm} - Z_{mk})(I_m - I_b) + \cdots + (Z_{nm} - Z_{nk})I_n \tag{8.5.11}$$

or

$$(Z_{1k} - Z_{1m})I_1 + (Z_{2k} - Z_{2m})I_2 + \cdots + (Z_{kk} - Z_{km})I_k + \cdots + (Z_{mk} - Z_{mm})I_m + \cdots + (Z_{nk} - Z_{nm})I_n + (Z_b + Z_{kk} + Z_{mm} - 2Z_{km})I_b = 0 \tag{8.5.12}$$

Equations (8.5.7)–(8.5.9) and (8.5.12) in matrix format are

$$\left[\begin{array}{c|c} & (Z_{1k} - Z_{1m}) \\ & (Z_{2k} - Z_{2m}) \\ \boldsymbol{Z}_{\text{bus(old)}} & \vdots \\ & (Z_{nk} - Z_{nm}) \\ \hline (Z_{1k} - Z_{1m}) \ldots (Z_{nk} - Z_{nm}) & (Z_b + Z_{kk} + Z_{mm} - 2Z_{km}) \end{array}\right] \left[\begin{array}{c} I_b \\ I_2 \\ \vdots \\ I_n \\ \hline I_b \end{array}\right] = \left[\begin{array}{c} V_F - E_1 \\ V_F - E_2 \\ \vdots \\ V_F - E_n \\ \hline 0 \end{array}\right] \tag{8.5.13}$$

Since the last element in the above voltage vector is zero, I_b can be eliminated from (8.5.13), just as I_p was eliminated from (8.5.3) and (8.5.4). The result is

$$\boldsymbol{Z}_{\text{bus(new)}} = \boldsymbol{Z}_{\text{bus(old)}} - \frac{1}{(Z_b + Z_{kk} + Z_{mm} - 2Z_{km})} \begin{bmatrix} Z_{1k} - Z_{1m} \\ Z_{2k} - Z_{2m} \\ \vdots \\ Z_{nk} - Z_{nm} \end{bmatrix} [(Z_{1k} - Z_{1m}) \ldots (Z_{nk} - Z_{nm})] \tag{8.5.14}$$

When computing $\boldsymbol{Z}_{\text{bus}}$ from scratch or when modifying an existing $\boldsymbol{Z}_{\text{bus}}$, we adopt the following ordering rules:

1. When computing $\boldsymbol{Z}_{\text{bus}}$ from scratch, start with a generator impedance (a type 1 addition) at bus 1 so as to create the reference bus right from the beginning, and also to begin with bus 1. If there is no generator connected to bus 1, temporarily renumber a generator bus as bus 1.
2. When creating new buses, type 2 and 3 additions, proceed in the order of the buses. That is, create bus 2 after bus 1, then bus 3, 4, and so on. If this is not possible with the original bus numbers, then temporarily renumber the buses.

3. Avoid the addition of a branch between two new buses.
4. Subject to the above constraints, branches may be added in arbitrary order.
5. If renumbering of buses is necessary, reorder the elements of $\boldsymbol{Z}_{\text{bus}}$ according to the original bus numbers after $\boldsymbol{Z}_{\text{bus}}$ has been completely constructed.

Example 8.5 Use the one-step-at-a-time method to construct $\boldsymbol{Z}_{\text{bus}}$ for the 2-bus network in Figure 8.3.

Solution We construct $\boldsymbol{Z}_{\text{bus}}$ in the following three steps:

Step 1 Start by adding the generator impedance $jX''_g = j0.15$ per unit from new bus 1 to the reference bus (a type 1 addition):

$$\boldsymbol{Z}_{\text{bus}} = j[0.15] \quad \text{per unit}$$

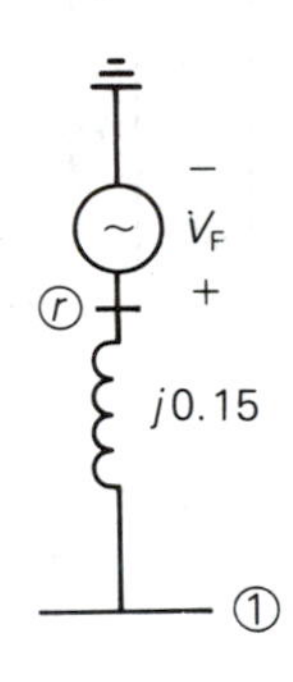

Step 2 Add $j(X_{T1} + X_{\text{line}} + X_{T2}) = j0.305$ per unit from old bus 1 to new bus 2, a type 2 addition:

$$\boldsymbol{Z}_{\text{bus}} = j\left[\begin{array}{c|c} 0.15 & 0.15 \\ \hline 0.15 & (0.305 + 0.15) \end{array}\right] = j\begin{bmatrix} 0.15 & 0.15 \\ 0.15 & 0.455 \end{bmatrix} \quad \text{per unit}$$

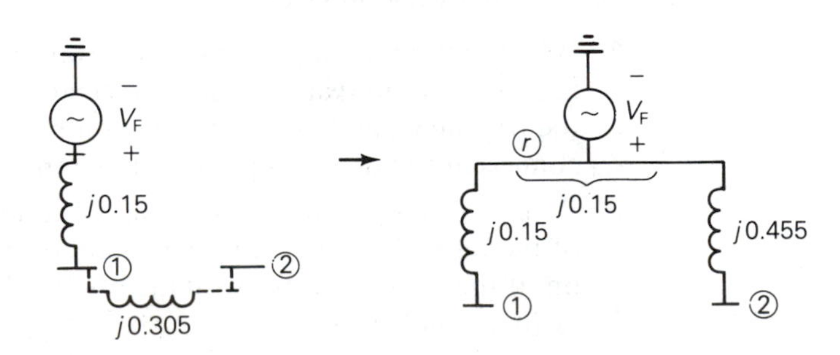

Step 3 Add the synchronous motor impedance $jX''_m = j0.20$ per unit from the reference bus to old bus 2, a type 3 addition:

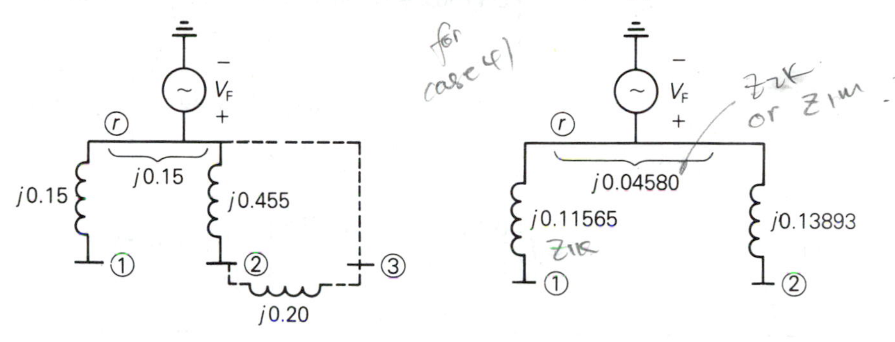

From (8.5.6),

$$Z_{bus} = j\begin{bmatrix} 0.15 & 0.15 \\ 0.15 & 0.455 \end{bmatrix} - \frac{j}{(0.455+0.2)}\begin{bmatrix} 0.15 \\ 0.455 \end{bmatrix}\begin{bmatrix} 0.15 & 0.455 \end{bmatrix}$$

$$= j\left[\begin{array}{c|c} 0.15 - \dfrac{(0.15)^2}{0.655} & 0.15 - \dfrac{(0.15)(0.455)}{0.655} \\ \hline 0.15 - \dfrac{(0.455)(0.15)}{0.655} & 0.455 - \dfrac{(0.455)^2}{0.655} \end{array}\right]$$

$$= j\begin{bmatrix} 0.11565 & 0.04580 \\ 0.04580 & 0.13893 \end{bmatrix} \quad \text{per unit}$$

which is the same as that obtained by inverting Y_{bus} in Example 8.4. Note that bus renumbering is unnecessary for this example. ◁

Example 8.6 Add a new branch with $Z_b = j0.40$ per unit between buses 1 and 2 for the network in Figure 8.3. Construct the new bus impedance matrix assuming Z_b is not mutually coupled to the existing line.

Solution This is a type 4 addition. From (8.5.14) with $k = 1$ and $m = 2$, and using $Z_{bus(old)}$ from the previous example:

$$Z_{bus(new)} = j\begin{bmatrix} 0.11565 & 0.04580 \\ 0.04580 & 0.13893 \end{bmatrix}$$

$$- j\frac{\begin{bmatrix} 0.11565 - 0.04580 \\ 0.04580 - 0.13893 \end{bmatrix}\begin{bmatrix} (0.11565 - 0.04580)(0.04580 - 0.13983) \end{bmatrix}}{(0.4 + 0.11565 + 0.13893 - 2 \times 0.04580)}$$

$$= j\left[\begin{array}{c|c} 0.11565 - \dfrac{(0.06985)^2}{0.56299} & 0.04580 - \dfrac{(0.06985)(-0.09313)}{0.56299} \\ \hline 0.04580 - \dfrac{(-0.09313)(0.06985)}{0.56299} & 0.13893 - \dfrac{(-0.09313)^2}{0.56299} \end{array}\right]$$

$$= j\left[\begin{array}{c|c} 0.10698 & 0.05735 \\ \hline 0.05735 & 0.12352 \end{array}\right] \text{ per unit}$$

Section 8.6

Personal Computer Program: SYMMETRICAL SHORT CIRCUITS

The software package that accompanies this text includes the program "SYMMETRICAL SHORT CIRCUITS," which computes the ac (symmetrical) fault current for a bolted three-phase short circuit at any bus in an N-bus power system. For each fault, the program also computes bus voltages and contributions to the fault current from transmission lines and transformers connected to the fault bus.

Input data for the program include machine, transmission line, and transformer data, as illustrated in Tables 8.3, 8.4, and 8.5, as well as the prefault voltage. When the machine reactance input data consist of direct axis subtransient reactances, the computed ac fault currents are subtransient fault currents. Alternatively, transient or steady-state ac fault currents are computed when these input data consist of direct axis transient or synchronous reactances. Transmission-line reactances are equivalent positive-sequence series reactances X′ for long lines or positive-sequence series reactances X for medium and short lines. All machine, line, and transformer reactances are given in per-unit on a common MVA base. Resistances, shunt admittances, nonrotating impedance loads, and prefault load currents are neglected.

Table 8.3

Synchronous machine data for SYMMETRICAL SHORT CIRCUITS program*

BUS	MACHINE SUBTRANSIENT REACTANCE – X_d''
	per unit
1	0.18
3	0.09

*S_{base} = 400 MVA
V_{base} = 15 kV at buses 1,3
= 345 kV at buses 2,4,5

Table 8.4

Line data for SYMMETRICAL SHORT CIRCUITS program

BUS-TO-BUS	EQUIVALENT POSITIVE-SEQUENCE SERIES REACTANCE
	per unit
2–4	0.4
2–5	0.2
4–5	0.1

Table 8.5

Transformer data for SYMMETRICAL SHORT CIRCUITS program

BUS-TO-BUS	LEAKAGE REACTANCE – X
	per unit
1–5	0.08
3–4	0.04

The program computes the positive-sequence impedance matrix $\boldsymbol{Z}_{\text{bus}}$ using the one-step-at-a-time method described in Section 8.5. The program user can either compute $\boldsymbol{Z}_{\text{bus}}$ from scratch or begin with an existing $\boldsymbol{Z}_{\text{bus}}$ that has already been computed and stored. In both cases, type 1, 2, 3, and 4 modifications to $\boldsymbol{Z}_{\text{bus}}$ can be made. The program computes $\boldsymbol{Z}_{\text{bus}}$ in accordance with the ordering rules given in Section 8.5.

After $\boldsymbol{Z}_{\text{bus}}$ is computed, (8.4.5) and (8.4.7) are employed to compute the fault current and the voltages at each bus during the fault for a three-phase fault at bus 1. Contributions to this fault current from each line or transformer branch connected to the fault bus are also computed by dividing the voltage across the branch by the branch impedance. These computations are then repeated for a fault at bus 2, then bus 3, and so on to bus N.

For a fault at bus 1, 2, . . . , N, output data consist of: the fault current, contributions to the fault current from each branch connected to the fault bus, and bus voltages. $\boldsymbol{Z}_{\text{bus}}$ can also be included in the output, if desired.

Example 8.7 Consider the 5-bus power system whose single-line diagram is shown in Figure 7.2. Machine, line, and transformer data are given in Tables 8.3, 8.4, and 8.5. The prefault voltage is 1.05 per unit. Run the SYMMETRICAL SHORT CIRCUITS program for this system.

Solution The program is run with the input data from Tables 8.3, 8.4, and 8.5 and with a prefault voltage of 1.05 per unit. The 5×5 $\boldsymbol{Z}_{\text{bus}}$ matrix computed by the program is shown in Table 8.6, and the fault currents and bus voltages are given in Table 8.7. Note that these fault currents are subtransient fault currents, since the machine reactance input data consist of direct axis subtransient reactances.

Table 8.6

Z_{bus} for Example 8.7

$$j\begin{bmatrix} 0.11189 & 0.07081 & 0.03405 & 0.04919 & 0.08162 \\ 0.07081 & 0.22781 & 0.05459 & 0.07886 & 0.10228 \\ 0.03405 & 0.05459 & 0.07297 & 0.06541 & 0.04919 \\ 0.04919 & 0.07886 & 0.06541 & 0.09447 & 0.07105 \\ 0.08162 & 0.10228 & 0.04919 & 0.07105 & 0.11790 \end{bmatrix}$$

Table 8.7

Fault currents and bus voltages for Example 8.7

FAULT BUS	FAULT CURRENT	CONTRIBUTIONS TO FAULT CURRENT		
		GEN LINE OR TRSF	BUS-TO-BUS	CURRENT
	PER UNIT			PER UNIT
1	9.384			
		G 1	GRND– 1	5.833
		T 1	5 – 1	3.551
2	4.609			
		L 1	4 – 2	1.716
		L 2	5 – 2	2.893
3	14.389			
		G 2	GRND– 3	11.667
		T 2	4 – 3	2.722
4	11.114			
		L 1	2 – 4	0.434
		L 3	5 – 4	2.603
		T 2	3 – 4	8.077
5	8.906			
		L 2	2 – 5	0.695
		L 3	4 – 5	4.172
		T 1	1 – 5	4.038

VF = 1.05	PER-UNIT BUS VOLTAGES DURING THE FAULT				
FAULT BUS	BUS1	BUS2	BUS3	BUS4	BUS5
1	0.0000	0.3855	0.7304	0.5884	0.2841
2	0.7236	0.0000	0.7984	0.6865	0.5786
3	0.5600	0.2644	0.0000	0.1089	0.3422
4	0.5033	0.1736	0.3231	0.0000	0.2603
5	0.3231	0.1391	0.6119	0.4172	0.0000

◁

Example 8.8 For the power system given in Example 8.7, run the SYMMETRICAL SHORT CIRCUITS program with an additional line installed between buses 2 and 4. This line, whose reactance is 0.3 per unit, is not mutually coupled to any other line.

Solution An additional line with reactance X = 0.3 per unit is added from bus 2 to bus 4 and the program is rerun. $\boldsymbol{Z}_{bus}$ along with the fault currents and bus voltages are shown in Tables 8.8 and 8.9.

Table 8.8

Z_{bus} for Example 8.8

$$
j\begin{bmatrix}
0.11089 & 0.06388 & 0.03456 & 0.04992 & 0.08017 \\
0.06388 & 0.18005 & 0.05806 & 0.08387 & 0.09227 \\
0.03456 & 0.05806 & 0.07272 & 0.06504 & 0.04992 \\
0.04992 & 0.08387 & 0.06504 & 0.09395 & 0.07210 \\
0.08017 & 0.09227 & 0.04992 & 0.07210 & 0.11580
\end{bmatrix}
$$

Table 8.9

Fault currents and bus voltages for Example 8.8

FAULT BUS	FAULT CURRENT	CONTRIBUTIONS TO FAULT CURRENT		
		GEN LINE OR TRSF	BUS-TO-BUS	CURRENT
	PER UNIT			PER UNIT
1	9.469			
		G 1	GRND– 1	5.833
		T 1	5 – 1	3.636
2	5.832			
		L 1	4 – 2	1.402
		L 2	5 – 2	2.560
		L 4	4 – 2	1.870
3	14.439			
		G 2	GRND– 3	11.667
		T 2	4 – 3	2.772
4	11.176			
		L 1	2 – 4	0.282
		L 3	5 – 4	2.442
		L 4	2 – 4	0.376
		T 2	3 – 4	8.077
5	9.067			
		L 2	2 – 5	1.067
		L 3	4 – 5	3.962
		T 1	1 – 5	4.038

VF = 1.05	PER-UNIT BUS VOLTAGES DURING THE FAULT				
FAULT BUS	BUS1	BUS2	BUS3	BUS4	BUS5
1	0.0000	0.4451	0.7228	0.5773	0.2909
2	0.6775	0.0000	0.7114	0.5609	0.5119
3	0.5510	0.2117	0.0000	0.1109	0.3293
4	0.4921	0.1127	0.3231	0.0000	0.2442
5	0.3231	0.2134	0.5974	0.3962	0.0000

Section 8.7

Circuit Breaker and Fuse Selection

The computer programs SYMMETRICAL SHORT CIRCUITS (Section 8.6) and SHORT CIRCUITS (Section 9.6) may be utilized in power-system design to select, set, and coordinate protective equipment such as circuit breakers, fuses, relays, and instrument transformers. In this section we discuss basic principles of circuit breaker and fuse selection.

ac Circuit Breakers

A *circuit breaker* is a mechanical switch capable of interrupting fault currents and of reclosing. When circuit-breaker contacts separate while carrying current, an arc forms. The breaker is designed to extinguish the arc by elongating and cooling it. The fact that ac arc current naturally passes through zero twice during its 60-Hz cycle aids the arc extinction process.

Circuit breakers are classified as *power* circuit breakers when they are intended for service in ac circuits above 1500 V, and *low-voltage* circuit breakers in ac circuits up to 1500 V. There are different types of circuit breakers depending on the medium—air, oil, SF_6 gas, or vacuum—in which the arc is elongated. Also, the arc can be elongated either by a magnetic force or by a blast of air.

Some circuit breakers are equipped with a high-speed automatic reclosing capability. Since most faults are temporary and self-clearing, reclosing is based on the idea that if a circuit is deenergized for a short time, it is likely that whatever caused the fault has disintegrated and the ionized arc in the fault has dissipated.

When reclosing breakers are employed in EHV systems, standard practice is to reclose only once, approximately 15 to 50 cycles (depending on operating voltage) after the breaker interrupts the fault. If the fault persists and the EHV breaker recloses into it, the breaker reinterrupts the fault current and then "locks out," requiring operator resetting. Multiple-shot reclosing in EHV systems is not standard practice because transient stability (Chapter 12) may be compromised. However, for distribution systems (2.4–46 kV) where customer outages are of concern, standard reclosers are equipped for two or more reclosures.

For low-voltage applications, molded case circuit breakers with dual trip capability are available. There is a magnetic instantaneous trip for large fault currents above a specified threshold, and a thermal trip with time delay for smaller fault currents.

Modern circuit-breaker standards are based on symmetrical interrupting current. It is usually necessary to calculate only symmetrical fault current at a system location, and then select a breaker with a symmetrical interrupting capability equal to or above the calculated current. The breaker has the additional capability to interrupt the asymmetrical (or total) fault current if the dc offset is not too large.

Recall from Section 8.1 that the maximum asymmetry factor $K(\tau = 0)$ is $\sqrt{3}$, which occurs at fault inception ($\tau = 0$). After fault inception, the dc fault current decays exponentially with time constant $T = (L/R) = (X/\omega R)$, and the asymmetry factor decreases. Power circuit breakers with a 2-cycle rated interruption time are designed for an asymmetrical interrupting capability up to 1.4 times their symmetrical interrupting capability, whereas slower circuit breakers have a lower asymmetrical interrupting capability.

A simplified method for breaker selection is called the "E/X simplified method" [1,7]. The maximum symmetrical short-circuit current at the system location in question is calculated from the prefault voltage and system reactance characteristics, using, for example, computer programs similar to those described in Sections 8.6 and 9.6. Resistances, shunt admittances, nonrotating impedance loads, and prefault load currents are neglected. Then, if the X/R ratio at the system location is less than 15, a breaker with a symmetrical interrupting capability equal to or above the calculated current at the given operating voltage is satisfactory. However, if X/R is greater than 15, the dc offset may not have decayed to a sufficiently low value. In this case, a method for correcting the calculated fault current to account for dc and ac time constants as well as breaker speed can be used [10]. If X/R is unknown, the calculated fault current should not be greater than 80% of the breaker interrupting capability.

When selecting circuit breakers for generators, two cycle breakers are employed in practice, and the subtransient fault current is calculated; therefore, subtransient machine reactances X_d'' are used in fault calculations. For synchronous motors, subtransient reactances X_d'' or transient reactances X_d' are used, depending on breaker speed. Also, induction motors can momentarily contribute to fault current. Large induction motors are usually modeled as sources in series with X_d'' or X_d', depending on breaker speed. Smaller induction motors (below 50 hp) are often neglected entirely.

Table 8.10 shows a schedule of preferred ratings for outdoor power circuit breakers. Some of the more important ratings shown are described next.

Voltage ratings

Rated maximum voltage: Designated the maximum rms line-to-line operating voltage. The breaker should be used in systems with an operating voltage less than or equal to this rating.

Rated low frequency withstand voltage: The maximum 60-Hz rms line-to-line voltage that the circuit breaker can withstand without insulation damage.

Rated impulse withstand voltage: The maximum crest voltage of a voltage pulse with standard rise and delay times that the breaker insulation can withstand.

Rated voltage range factor K: The range of voltage for which the symmetrical interrupting capability times the operating voltage is constant.

Current ratings

Rated continuous current: The maximum 60-Hz rms current that the breaker can carry continuously while it is in the closed position without overheating.

Rated short-circuit current: The maximum rms symmetrical current that the breaker can safely interrupt at rated maximum voltage.

Rated momentary current: The maximum rms asymmetrical current that the breaker can withstand while in the closed position without damage. Rated momentary current for standard breakers is 1.6 times the symmetrical interrupting capability.

Rated interrupting time: The time in cycles on a 60-Hz basis from the instant the trip coil is energized to the instant the fault current is cleared.

Rated interrupting MVA: For a three-phase circuit breaker, this is $\sqrt{3}$ times the rated maximum voltage in kV times the rated short-circuit current in kA. It is more common to work with current and voltage ratings than with MVA rating.

As an example, the symmetrical interrupting capability of the 69-kV class breaker listed in Table 8.10 is plotted versus operating voltage in Figure 8.12. As shown, the symmetrical interrupting capability increases from its rated short-circuit current $I = 19\,\text{kA}$ at rated maximum voltage $V_{max} = 72.5\,\text{kV}$ up to $I_{max} = KI = (1.21)(19) = 23\,\text{kA}$ at an operating voltage $V_{min} = V_{max}/K = 72.5/1.21 = 60\,\text{kV}$. At operating voltages V between V_{min} and V_{max}, the symmetrical interrupting capability is $I \times V_{max}/V = 1378/V\,\text{kA}$. At operating voltages below V_{min}, the symmetrical interrupting capability remains at $I_{max} = 23\,\text{kA}$.

Figure 8.12

Symmetrical interrupting capability of a 69-kV class breaker

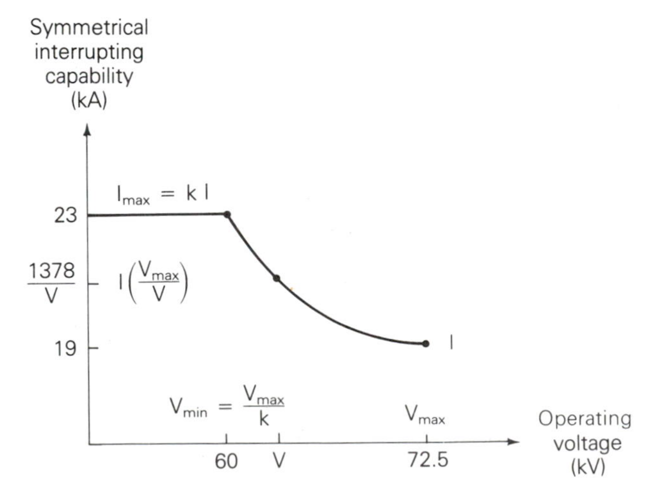

Breakers of the 115-kV class and higher have a voltage range factor $K = 1.0$; that is, their symmetrical interrupting current capability remains constant.

Example 8.9 The calculated symmetrical fault current is 17 kA at a three-phase bus where the operating voltage is 64 kV. The X/R ratio at the bus is unknown. Select a circuit breaker from Table 8.10 for this bus.

Solution The 69-kV-class breaker has a symmetrical interrupting capability $I\left(\frac{V_{max}}{V}\right) = 19\left(\frac{72.5}{64}\right) = 21.5\,\text{kA}$ at the operating voltage $V = 64\,\text{kV}$. The calculated symmetrical fault current, 17 kA, is less than 80% of this capability (less than $0.80 \times 21.5 = 17.2\,\text{kA}$), which is a requirement when X/R is unknown. Therefore, we select the 69-kV-class breaker from Table 8.10. ◁

Fuses

Figure 8.13(a) shows a cutaway view of a fuse, which is one of the simplest overcurrent devices. The fuse consists of a metal "fusible" link or links encapsulated in a tube, packed in filler material, and connected to contact terminals. Silver is a typical link metal, and sand is a typical filler material.

Figure 8.13

Typical fuse

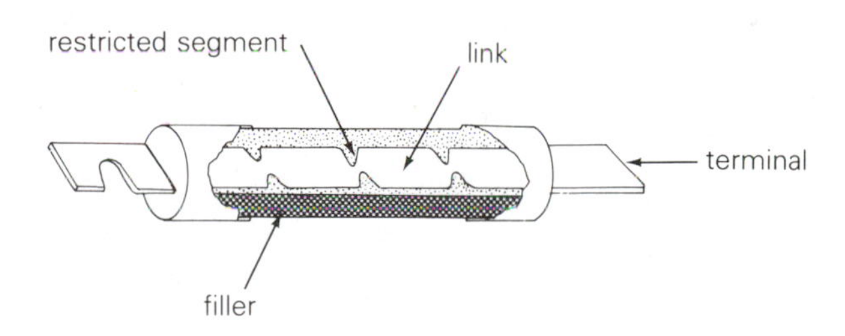

(a) Cutaway view

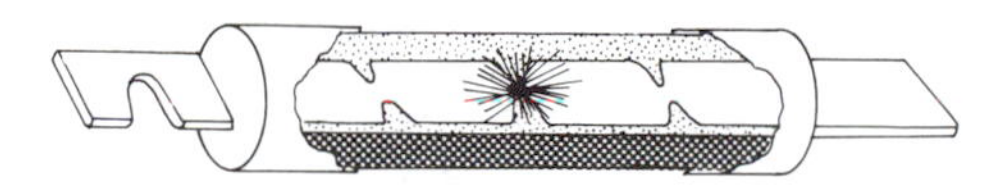

(b) The link melts and an arc is established under sustained overload current

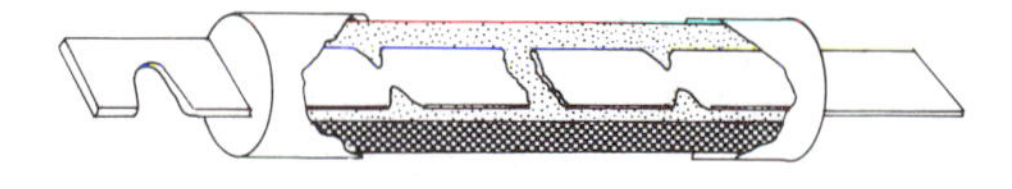

(c) The "open" link after clearing the overload current.

IDENTIFICATION		RATED VALUES					
		VOLTAGE		INSULATION LEVEL		CURRENT	
				RATED WITHSTAND TEST VOLTAGE			
NOMINAL VOLTAGE CLASS kV, rms	NOMINAL 3 PHASE MVA CLASS	RATED MAX VOLTAGE kV, rms	RATED VOLTAGE RANGE FACTOR *K*	LOW FREQUENCY kV, rms	IMPULSE kV, CREST	RATED CONTINU-OUS CURRENT AT 60 Hz AMPERES, rms	RATED SHORT-CIRCUIT CURRENT (AT RATED MAX kV) kA, rms
COL 1	COL 2	COL 3	COL 4	COL 5	COL 6	COL 7	COL 8
14.4	250	15.5	2.67			600	8.9
14.4	500	15.5	1.29			1200	18
23	500	25.8	2.15			1200	11
34.5	1500	38	1.65			1200	22
46	1500	48.3	1.21			1200	17
69	2500	72.5	1.21			1200	19
115		121	1.0			1200	20
115		121	1.0			1600	40
115		121	1.0			2000	40
115		121	1.0			2000	63
115		121	1.0			3000	40
115		121	1.0			3000	63
138		145	1.0			1200	20
138	Not	145	1.0			1600	40
138		145	1.0			2000	40
138		145	1.0			2000	63
138		145	1.0			2000	80
138	Applica-	145	1.0			3000	40
138		145	1.0			3000	63
138		145	1.0			3000	80
161	ble	169	1.0			1200	16
161		169	1.0			1600	31.5
161		169	1.0			2000	40
161		169	1.0			2000	50
230		242	1.0			1600	31.5
230		242	1.0			2000	31.5
230		242	1.0			3000	31.5
230		242	1.0			2000	40
230		242	1.0			3000	40
230		242	1.0			3000	63
345		362	1.0			2000	40
345		362	1.0			3000	40
500		550	1.0			2000	40
500		550	1.0			3000	40
700		765	1.0			2000	40
700		765	1.0			3000	40

Table 8.10 Preferred ratings for outdoor circuit breakers (symmetrical current basis of rating) [10]

RATED VALUES		RELATED REQUIRED CAPABILITIES			
			CURRENT VALUES		
			MAX SYMMETRICAL INTERRUPTING CAPABILITY	3 SECOND SHORT-TIME CURRENT CARRYING CAPABILITY	
			K TIMES RATED SHORT-CIRCUIT CURRENT		
RATED INTERRUPTING TIME CYCLES	RATED PERMISSIBLE TRIPPING DELAY SECONDS	RATED MAX VOLTAGE DIVIDED BY *K* kV, rms	kA, rms	kA, rms	CLOSING AND LATCHING CAPABILITY 1.6*K* TIMES RATED SHORT-CIRCUIT CURRENT kA, rms
COL 9	COL 10	COL 11	COL 12	COL 13	COL 14
5	2	5.8	24	24	38
5	2	12	23	23	37
5	2	12	24	24	38
5	2	23	36	36	58
5	2	40	21	21	33
5	2	60	23	23	37
3	1	121	20	20	32
3	1	121	40	40	64
3	1	121	40	40	64
3	1	121	63	63	101
3	1	121	40	40	64
3	1	121	63	63	101
3	1	145	20	20	32
3	1	145	40	40	64
3	1	145	40	40	64
3	1	145	63	63	101
3	1	145	80	80	128
3	1	145	40	40	64
3	1	145	63	63	101
3	1	145	80	80	128
3	1	169	16	16	26
3	1	169	31.5	31.5	50
3	1	169	40	40	64
3	1	169	50	50	80
3	1	242	31.5	31.5	50
3	1	242	31.5	31.5	50
3	1	242	31.5	31.5	50
3	1	242	40	40	64
3	1	242	40	40	64
3	1	242	63	63	101
3	1	362	40	40	64
3	1	362	40	40	64
2	1	550	40	40	64
2	1	550	40	40	64
2	1	765	40	40	64
2	1	765	40	40	64

During normal operation, when the fuse is operating below its continuous current rating, the electrical resistance of the link is so low that it simply acts as a conductor. If an overload current from one to about six times its continuous current rating occurs and persists for more than a short interval of time, the temperature of the link eventually reaches a level that causes a restricted segment of the link to melt. As shown in Figure 8.13(b), a gap is then formed and an electric arc is established. As the arc causes the link metal to burn back, the gap width increases. The resistance of the arc eventually reaches such a high level that the arc cannot be sustained and it is extinguished, as in Figure 8.13(c). The current flow within the fuse is then completely cut off.

If the fuse is subjected to fault currents higher than about six times its continuous current rating, several restricted segments melt simultaneously, resulting in rapid arc suppression and fault clearing. Arc suppression is accelerated by the filler material in the fuse.

Many modern fuses are current limiting. As shown in Figure 8.14, a current-limiting fuse has such a high speed of response that it cuts off a high fault current in less than a half cycle—before it can build up to its full peak value. By limiting fault currents, these fuses permit the use of motors, transformers, conductors, and bus structures that could not otherwise withstand the destructive forces of high fault currents.

Figure 8.14

Operation of a current-limiting fuse

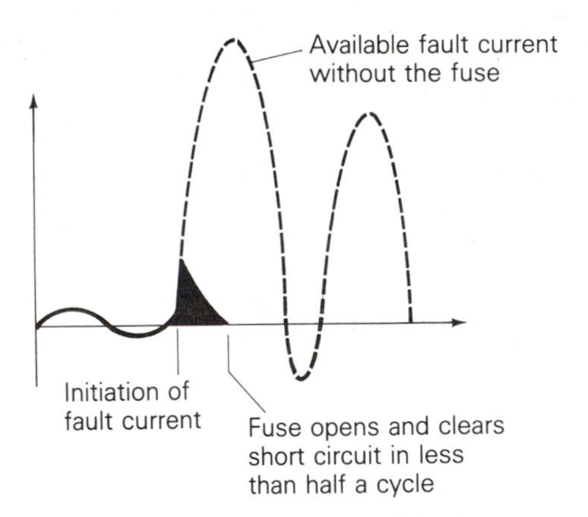

Fuse specification is normally based on the following four factors.

1. Voltage rating. This rms voltage determines the ability of a fuse to suppress the internal arc that occurs after the fuse link melts. A blown fuse should be able to withstand its voltage rating. Most low-voltage power distribution fuses have 250- or 600-V ratings. Ratings of medium voltage fuses range from 2.4 to 34.5 kV.
2. Continuous current rating. The fuse should carry this rms current indefinitely, without melting and clearing.
3. Interrupting current rating. This is the largest rms asymmetrical current that the fuse can safely interrupt. Most modern, low-voltage

current-limiting fuses have a 200-kA interrupting rating. Standard interrupting ratings for medium-voltage current-limiting fuses include 65, 80, and 100 kA.

4. Time response. The melting and clearing time of a fuse depends on the magnitude of the overcurrent or fault current, and is usually specified by a "time–current" curve. Figure 8.15 shows the time–current curve of a 15.5-kV, 100-A (continuous) current-limiting fuse. As shown, the fuse link melts within 2 s and clears within 5 s for a 500-A current. For a 5-kA current, the fuse link melts in less than 0.01 s and clears within 0.015 s.

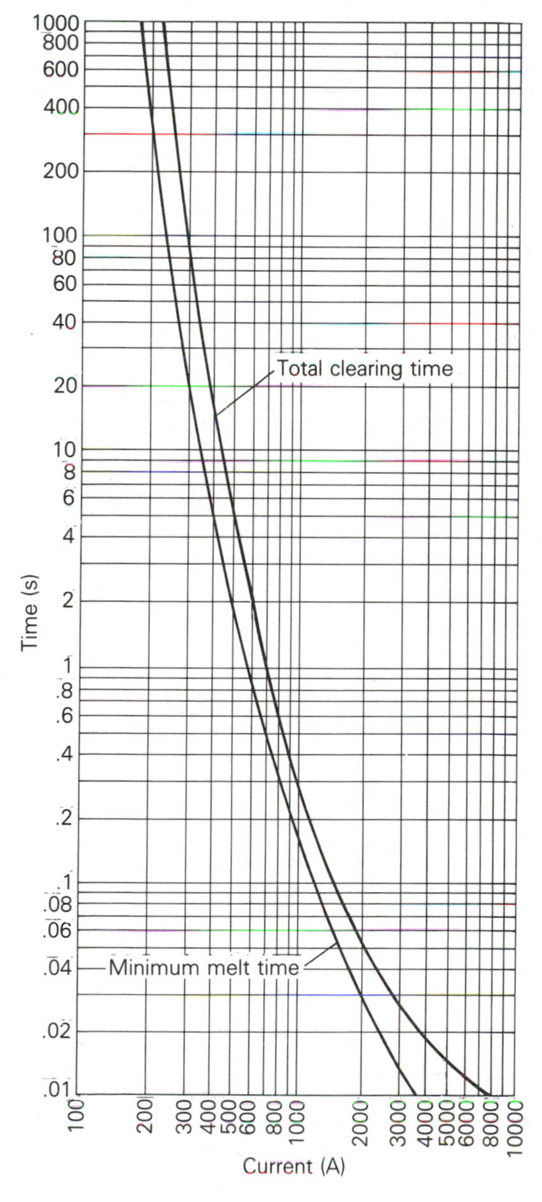

Figure 8.15

Time–current curves for a 15.5-kV, 100-A current-limiting fuse

It is usually a simple matter to coordinate fuses in a power circuit such that only the fuse closest to the fault opens the circuit. In a radial circuit, fuses with larger continuous current ratings are located closer to the source, such that the fuse closest to the fault clears before other, upstream fuses melt.

Fuses are inexpensive, fast operating, easily coordinated, and reliable, and they do not require protective relays or instrument transformers. Their chief disadvantage is that the fuse or the fuse link must be manually replaced after it melts. They are basically one-shot devices that are, for example, incapable of high-speed reclosing.

Problems

Section 8.1

8.1 In the circuit of Figure 8.1, $V = 277$ volts, $L = 2\,\text{mH}$, $R = 0.4\,\Omega$, and $\omega = 2\pi 60\,\text{rad/s}$. Determine (a) the rms symmetrical fault current; (b) the rms asymmetrical fault current at the instant the switch closes, assuming maximum dc offset; (c) the rms asymmetrical fault current 5 cycles after the switch closes, assuming maximum dc offset; (d) the dc offset as a function of time if the switch closes when the instantaneous source voltage is 300 volts.

8.2 Repeat Example 8.1 with $V = 4\,\text{kV}$, $X = 2\,\Omega$, and $R = 1\,\Omega$.

Section 8.2

8.3 A 1000-MVA 20-kV, 60-Hz three-phase generator is connected through a 1000-MVA 20-kV Δ/345-kV Y transformer to a 345-kV circuit breaker and a 345-kV transmission line. The generator reactances are $X_d'' = 0.17$, $X_d' = 0.30$, and $X_d = 1.5$ per unit, and its time constants are $T_d'' = 0.05$, $T_d' = 1.0$, and $T_A = 0.10\,\text{s}$. The transformer series reactance is 0.10 per unit; transformer losses and exciting current are neglected. A three-phase short-circuit occurs on the line side of the circuit breaker when the generator is operated at rated terminal voltage and at no-load. The breaker interrupts the fault 3 cycles after fault inception. Determine (a) the subtransient current through the breaker in per-unit and in kA rms, (b) the rms asymmetrical fault current the breaker interrupts, assuming maximum dc offset. Neglect the effect of the transformer on the time constants.

8.4 For Problem 8.3, determine (a) the instantaneous symmetrical fault current in kA in phase a of the generator as a function of time, assuming maximum dc offset occurs in this generator phase; and (b) the maximum dc offset current in kA as a function of time that can occur in any one generator phase.

Section 8.3

8.5 Recalculate the subtransient current through the breaker in Problem 8.3 if the generator is initially delivering rated MVA at 0.80 p.f. lagging and at rated terminal voltage.

8.6 Solve Example 8.3, parts (a) and (c) without using the superposition principle. First calculate the internal machine voltages E_g'' and E_m'', using the prefault load current. Then determine the subtransient fault, generator, and motor currents directly from Figure 8.4(a). Compare your answers with those of Example 8.3.

8.7 Equipment ratings for the four-bus power system shown in Figure 8.16 are as follows:

Generator G1: 500 MVA, 13.8 kV, X″ = 0.20 per unit
Generator G2: 750 MVA, 18 kV, X″ = 0.18 per unit
Generator G3: 1000 MVA, 20 kV, X″ = 0.17 per unit
Transformer T1: 500 MVA, 13.8 Δ/500 Y kV, X = 0.12 per unit
Transformer T2: 750 MVA, 18 Δ/500 Y kV, X = 0.10 per unit
Transformer T3: 1000 MVA, 20 Δ/500 Y kV, X = 0.10 per unit
Each 500-kV line: $X_1 = 50\,\Omega$

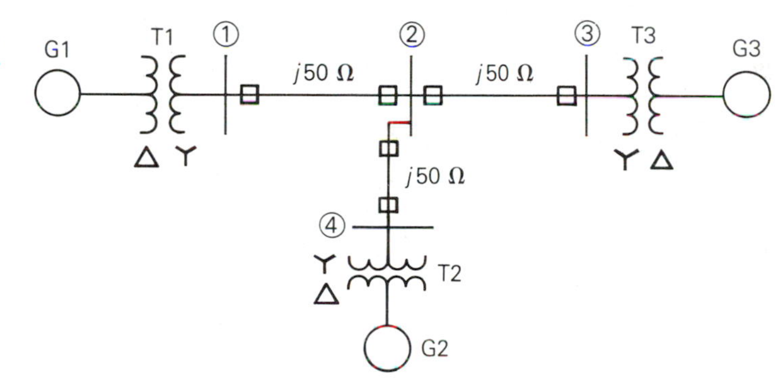

Figure 8.16 Problem 8.7

A three-phase short circuit occurs at bus 1, where the prefault voltage is 525 kV. Prefault load current is neglected. Draw the positive-sequence reactance diagram in per-unit on a 1000-MVA, 20-kV base in the zone of generator G3. Determine (a) the Thévenin reactance in per-unit at the fault, (b) the subtransient fault current in per-unit and in kA rms, (c) contributions to the fault current from generator G1 and from line 1–2.

Section 8.4

8.8 The bus impedance matrix for a three-bus power system is

$$\boldsymbol{Z}_{\text{bus}} = j \begin{bmatrix} 0.4 & 0.1 & 0.3 \\ 0.1 & 0.8 & 0.5 \\ 0.3 & 0.5 & 1.2 \end{bmatrix} \quad \text{per unit}$$

where subtransient reactances were used to compute $\boldsymbol{Z}_{\text{bus}}$. Prefault voltage is 1.0 per unit and prefault current is neglected. (a) Draw the bus impedance matrix equivalent circuit (rake equivalent). Identify the per-unit self and mutual impedances as well as the prefault voltage in the circuit. (b) A three-phase short circuit occurs at bus 2. Determine the subtransient fault current and the voltages at buses 1, 2, and 3 during the fault.

8.9 Determine $\boldsymbol{Y}_{\text{bus}}$ in per-unit for the circuit in Problem 8.7. Then invert $\boldsymbol{Y}_{\text{bus}}$ to obtain $\boldsymbol{Z}_{\text{bus}}$.

Section 8.5

8.10 For the three-bus power system whose $\boldsymbol{Z}_{\text{bus}}$ is given in Problem 8.8, a new impedance $Z_b = j0.7$ per unit is added from old bus 2 to new bus 4. The new impedance element is not mutually coupled to any of the existing impedance elements. Determine the new bus impedance matrix.

8.11 Repeat Problem 8.10 if Z_b is added from old bus 2 to the reference bus.

8.12 Repeat Problem 8.10 if Z_b is added from old bus 2 to old bus 3.

8.13 Repeat Problem 8.10 if Z_b is added from new bus 4 to the reference bus.

8.14 Determine $\boldsymbol{Z}_{\text{bus}}$ for the circuit given in Problem 8.7, using the one-step-at-a-time method.

Section 8.6

8.15 Using the SYMMETRICAL SHORT CIRCUITS program for the circuit given in Problem 8.7, compute: (a) $\boldsymbol{Z}_{\text{bus}}$ and (b) the fault current and bus voltages during the fault for a three-phase short circuit at each bus.

8.16 Rework Problems 8.10, 8.11, 8.12, and 8.13 using the SYMMETRICAL SHORT CIRCUITS program. Refer to SAMPLE RUN 7.2 in the software manual to construct Z_{bus}.

8.17 Run the SYMMETRICAL SHORT CIRCUITS program for the power system given in Example 8.7 with a new line installed between buses 4 and 5. This line, whose series impedance is $Z_b = j0.4$ per unit, is not mutually coupled to any other line.

8.18 Starting with the $\boldsymbol{Z}_{\text{bus}}$ matrix computed in Problem 8.17, remove the new line by adding an impedance $Z_b = -j0.4$ per unit between buses 4 and 5. Verify that the computed $\boldsymbol{Z}_{\text{bus}}$ is the same as that shown in Table 8.6.

Section 8.7

8.19 A three-phase circuit breaker has a 15.5-kV rated maximum voltage, 9.0-kA rated short-circuit current, and a 2.67-rated voltage range factor. (a) Determine the symmetrical interrupting current at 10-kV and at 5-kV operating voltages. (b) Can this breaker be safely installed at a three-phase bus where the symmetrical fault current is 10 kA, the operating voltage is 13.8 kV, and the (X/R) ratio is 12?

References

1. Westinghouse Electric Corporation, *Electrical Transmission and Distribution Reference Book*, 4th ed. (East Pittsburgh, Pa., 1964).
2. E. W. Kimbark, *Power System Stability, Synchronous Machines*, Vol. 3 (New York: John Wiley and Sons, 1956).
3. A. E. Fitzgerald, C. Kingsley, and S. Umans, *Electric Machinery*, 4th ed. (New York: McGraw-Hill, 1985).
4. M. S. Sarma, *Electric Machines* (Dubuque, Iowa: Wm. C. Brown, 1985).
5. J. R. Neuenswander, *Modern Power Systems* (New York: Intext Educational Publishers, 1971).
6. H. E. Brown, *Solution of Large Networks by Matrix Methods* (New York: John Wiley and Sons, 1975).
7. G. N. Lester, "High Voltage Circuit Breaker Standards in the USA—Past, Present and Future," *IEEE Transactions PAS*, Vol. PAS—93 (1974): pp. 590–600.
8. W. D. Stevenson, Jr., *Elements of Power System Analysis*, 4th ed. (New York: McGraw-Hill, 1982).
9. C. A. Gross, *Power System Analysis* (New York: John Wiley and Sons, 1979).
10. *Application Guide for AC High-Voltage Circuit Breakers Rated on a Symmetrical Current Basis*, ANSI C 37.010 (New York: American National Standards Institute, 1972).

CHAPTER 9

Unsymmetrical Faults

Short circuits occur in three-phase power systems as follows, in order of frequency of occurrence: single line-to-ground, line-to-line, double line-to-ground and balanced three-phase faults. The path of the fault current may have either zero impedance, which is called a *bolted* short circuit, or nonzero impedance. Other types of faults include one-conductor-open and two-conductors-open, which can occur when conductors break or when one or two phases of a circuit breaker inadvertently open.

Although the three-phase short circuit occurs the least, it was considered first, in Chapter 8, because of its simplicity. When a balanced three-phase fault occurs in a balanced three-phase system, there is only positive-sequence fault current; the zero-, positive-, and negative-sequence networks are completely uncoupled.

When an unsymmetrical fault occurs in an otherwise balanced system, the sequence networks are interconnected only at the fault location. As such, the computation of fault currents is greatly simplified by the use of sequence networks. Readers may wish to briefly review Chapter 3, Symmetrical Components, before proceeding with unsymmetrical faults.

As in the case of balanced three-phase faults, unsymmetrical faults have two components of fault current, an ac or symmetrical component, including subtransient, transient, and steady-state currents, and a dc component. The simplified E/X method for breaker selection described in Section 8.7 is also applicable to unsymmetrical faults. The dc offset current need not be considered unless it is too large—for example, when the X/R ratio is too large.

We begin this chapter by using the per-unit zero-, positive-, and negative-sequence networks to represent a three-phase system. Also, certain assumptions are made to simplify fault current calculations, and the balanced three-phase fault is briefly reviewed. Single line-to-ground, line-to-line, and double line-to-ground faults are presented in Sections 9.2, 9.3, and 9.4. The use of the positive-sequence bus impedance matrix for three-phase fault calculations in Section 8.4 is extended in Section 9.5 to unsymmetrical fault calculations by considering a bus impedance matrix for each sequence network. In Section 9.6 a personal computer program, SHORT CIRCUITS, which is based on the use of bus impedance matrices, is presented; it computes symmetrical fault currents for both three-phase and unsymmetrical faults. This program may be used in power-system design to select, set, and coordinate protective equipment.

Section 9.1

System Representation

A three-phase power system is represented by its sequence networks in this chapter. The zero-, positive-, and negative-sequence networks of system components—generators, motors, transformers, and transmission lines—as developed in Chapters 3, 4, and 5 can be used to construct system zero-, positive-, and negative-sequence networks. We make the following assumptions:

1. The power system operates under balanced steady-state conditions before the fault occurs. Thus the zero-, positive-, and negative-sequence networks are uncoupled before the fault occurs. During unsymmetrical faults they are interconnected only at the fault location.
2. Prefault load current is neglected. Because of this, the positive-sequence internal voltages of all machines are equal to the prefault

voltage V_F. Therefore, the prefault voltage at each bus in the positive-sequence network equals V_F.

3. Transformer winding resistances and shunt admittances are neglected.
4. Transmission-line series resistances and shunt admittances are neglected.
5. Synchronous machine armature resistance, saliency, and saturation are neglected.
6. All nonrotating impedance loads are neglected.
7. Induction motors are either neglected (especially for motors rated 50hp or less) or represented in the same manner as synchronous machines.

Note that these assumptions are made for simplicity in this text, and in practice should not be made for all cases. For example, in primary and secondary distribution systems, prefault currents may be in some cases comparable to short-circuit currents, and in other cases line resistances may significantly reduce fault currents.

Although fault currents as well as contributions to fault currents on the fault side of Δ–Y transformers are not affected by Δ–Y phase shifts, contributions to the fault from the other side of such transformers are affected by Δ–Y phase shifts for unsymmetrical faults. Therefore, we include Δ–Y phase-shift effects in this chapter.

We consider faults at the general three-phase bus shown in Figure 9.1. Terminals *abc*, denoted the *fault terminals*, are brought out in order to make external connections that represent faults. Before a fault occurs, the currents I_a, I_b, and I_c are zero.

Figure 9.1 General three-phase bus

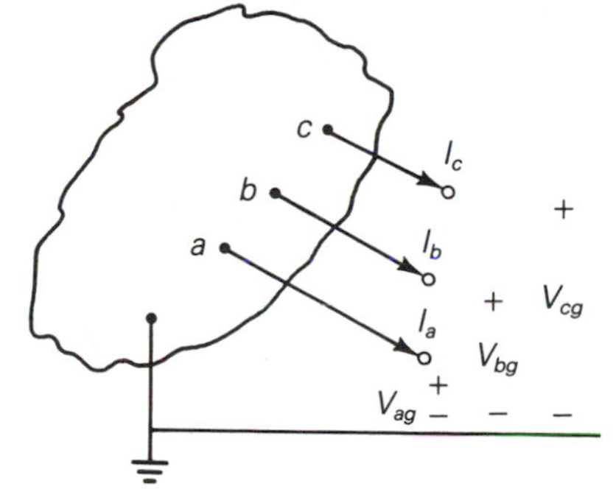

Figure 9.2(a) shows general sequence networks as viewed from the fault terminals. Since the prefault system is balanced, these zero-, positive-, and negative-sequence networks are uncoupled. Also, the sequence components of the fault currents, I_0, I_1, and I_2 are zero before a fault occurs. The general

sequence networks in Figure 9.2(a) are reduced to their Thévenin equivalents as viewed from the fault terminals in Figure 9.2(b). Each sequence network has a Thévenin equivalent impedance. Also, the positive-sequence network has a Thévenin equivalent voltage source, which equals the prefault voltage V_F.

Figure 9.2 Sequence networks at a general three-phase bus in a balanced system

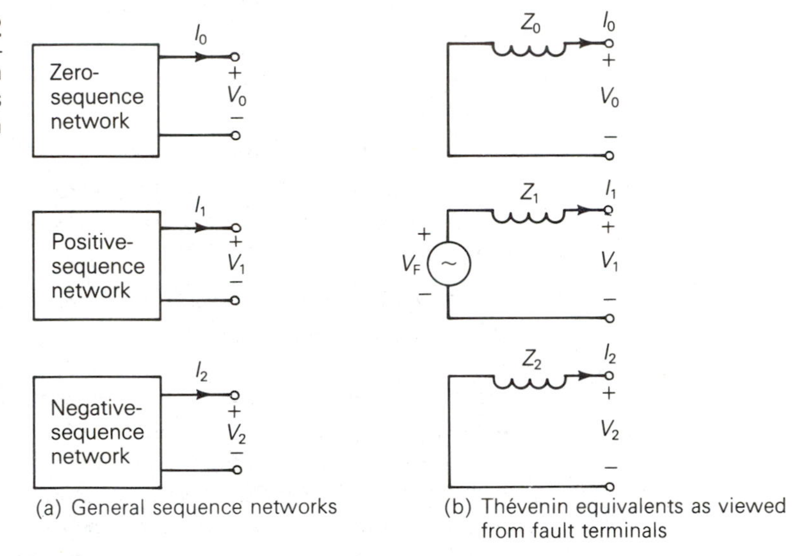

Example 9.1 A single-line diagram of the power system considered in Example 8.3 is shown in Figure 9.3, where negative- and zero-sequence reactances are also given. The neutrals of the generator and Δ–Y transformers are solidly grounded. The motor neutral is grounded through a reactance $X_n = 0.05$ per unit on the motor base. (a) Draw the per-unit zero-, positive-, and negative-sequence networks on a 100-MVA, 13.8-kV base in the zone of the generator. (b) Reduce the sequence networks to their Thévenin equivalents, as viewed from bus 2. Prefault voltage is $V_F = 1.05\underline{/0°}$ per unit. Prefault load current and Δ–Y transformer phase shift are neglected.

Figure 9.3 Single-line diagram for Example 9.1

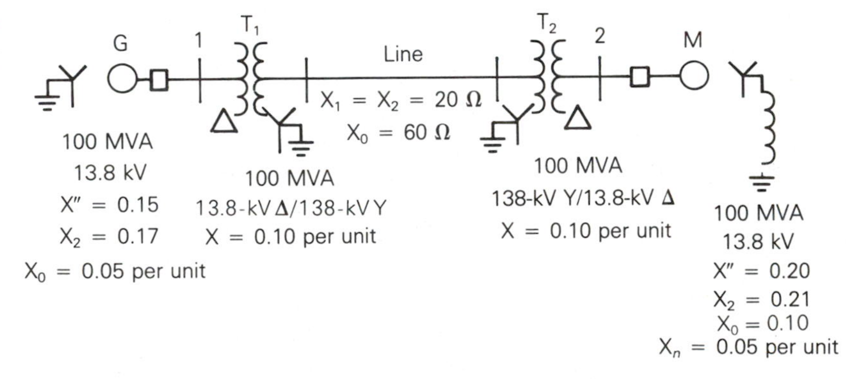

Solution **a.** The sequence networks are shown in Figure 9.4. The positive-sequence network is the same as that shown in Figure 8.4(a). The negative-sequence

network is similar to the positive-sequence network, except that there are no sources, and negative-sequence machine reactances are shown. Δ–Y phase shifts are omitted from the positive- and negative-sequence networks for this example. In the zero-sequence network the zero-sequence generator, motor, and transmission-line reactances are shown. Since the motor neutral is grounded through a neutral reactance X_n, $3X_n$ is included in the zero-sequence motor circuit. Also, the zero-sequence Δ–Y transformer models are taken from Figure 4.17.

Figure 9.4

Sequence networks for Example 9.1

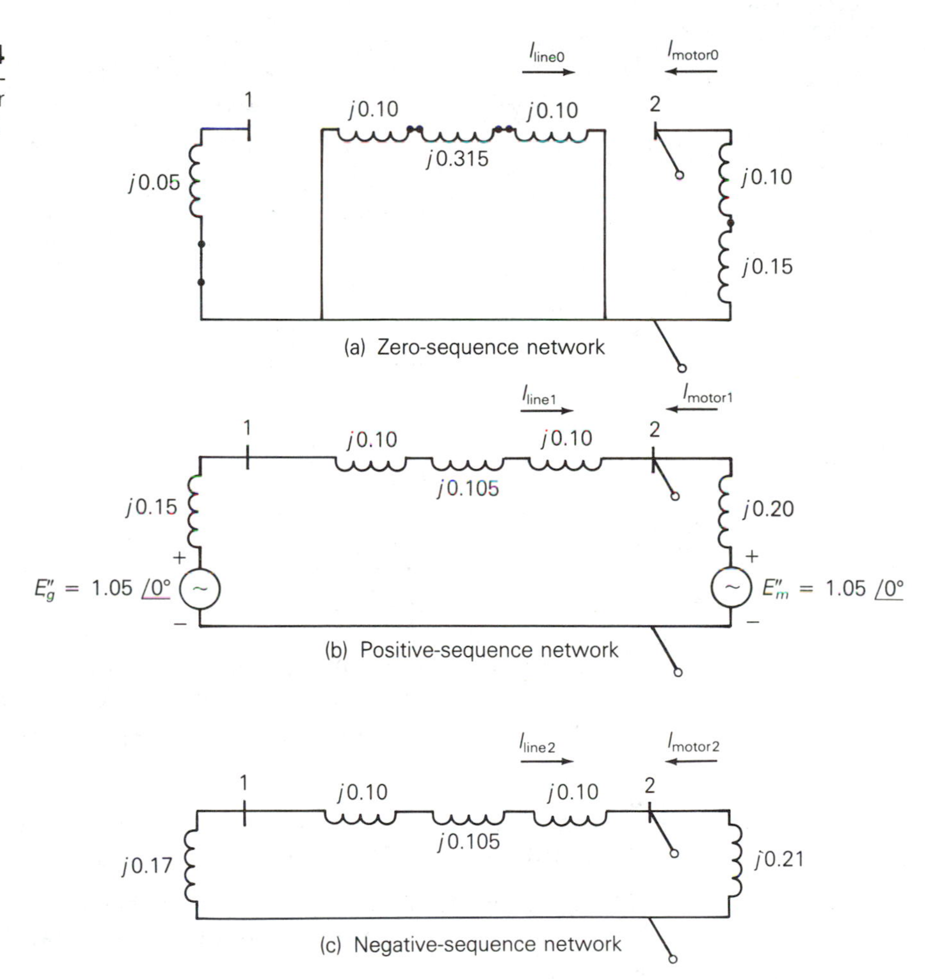

b. Figure 9.5 shows the sequence networks reduced to their Thévenin equivalents, as viewed from bus 2. For the positive-sequence equivalent, the Thévenin voltage source is the prefault voltage $V_F = 1.05\underline{/0°}$ per unit. From Figure 9.4, the positive-sequence Thévenin impedance at bus 2 is the motor impedance $j0.20$, as seen to the right of bus 2, in parallel with $j(0.15 + 0.10 + 0.105 + 0.10) = j0.455$, as seen to the left; the parallel combination is $j0.20//j0.455 = j0.13893$ per unit. Similarly, the negative-sequence Thévenin impedance is $j0.21//j(0.17 + 0.10 + 0.105 + 0.10) = j0.21//j0.475 = j0.14562$ per unit. In the zero-sequence network of Figure 9.4, the Thévenin impedance at bus 2 consists only of $j(0.10 + 0.15) = j0.25$ per unit, as seen to the right of bus 2; due to the Δ connection of transformer T_2, the zero-sequence network looking to the left of bus 2 is open. ◁

Figure 9.5

Thévenin equivalents of sequence networks for Example 9.1

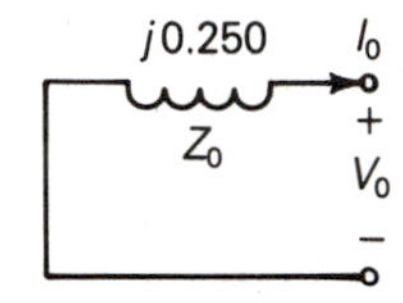

(a) Zero-sequence network

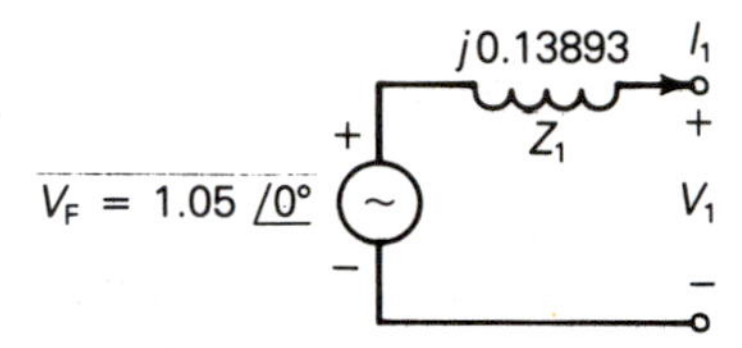

(b) Positive-sequence network

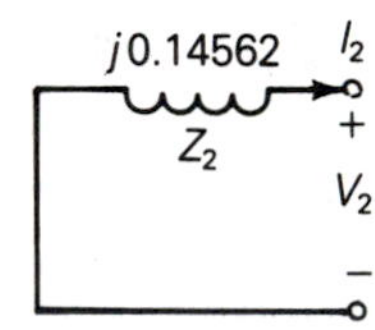

(c) Negative-sequence network

Recall that for three-phase faults, as considered in Chapter 8, the fault currents are balanced and have only a positive-sequence component. Therefore, we work only with the positive-sequence network when calculating three-phase fault currents.

Example 9.2 Calculate the per-unit subtransient fault currents in phases a, b, and c for a bolted three-phase-to-ground short circuit at bus 2 in Example 9.1.

Solution The terminals of the positive-sequence network in Figure 9.5(b) are shorted, as shown in Figure 9.6. The positive-sequence fault current is

$$I_1 = \frac{V_F}{Z_1} = \frac{1.05\underline{/0^\circ}}{j0.13893} = -j7.558 \quad \text{per unit}$$

which is the same result as obtained in part (c) of Example 8.4. Note that since subtransient machine reactances are used in Figures 9.4 to 9.6, the current calculated above is the positive-sequence subtransient fault current at bus 2. Also, the zero-sequence current I_0 and negative-sequence current I_2 are both zero. Therefore, the subtransient fault currents in each phase are, from (3.1.16),

$$\begin{bmatrix} I''_a \\ I''_b \\ I''_c \end{bmatrix} = \begin{bmatrix} 1 & 1 & 1 \\ 1 & a^2 & a \\ 1 & a & a^2 \end{bmatrix} \begin{bmatrix} 0 \\ -j7.558 \\ 0 \end{bmatrix} = \begin{bmatrix} 7.558\underline{/-90^\circ} \\ 7.558\underline{/150^\circ} \\ 7.558\underline{/30^\circ} \end{bmatrix} \quad \text{per unit}$$

Figure 9.6

Example 9.2: Bolted three-phase-to-ground fault at bus 2

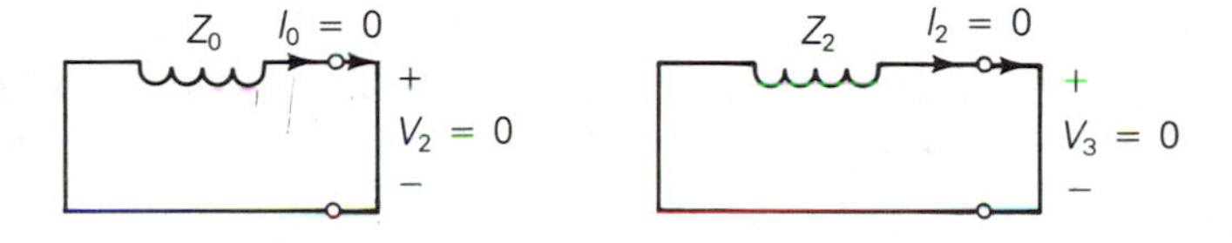

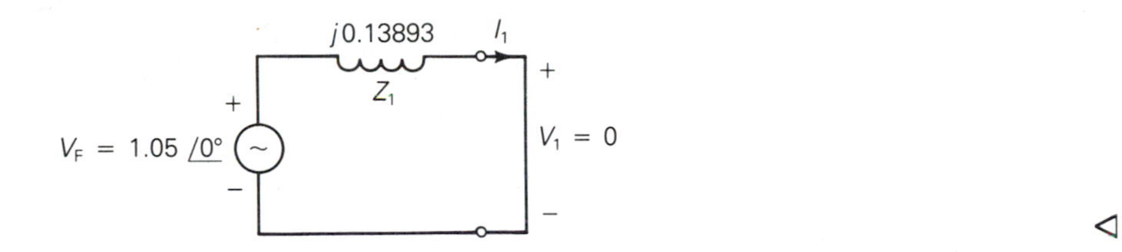

◁

The sequence components of the line-to-ground voltages at the fault terminals are, from Figure 9.2(b),

$$\begin{bmatrix} V_0 \\ V_1 \\ V_2 \end{bmatrix} = \begin{bmatrix} 0 \\ V_F \\ 0 \end{bmatrix} - \begin{bmatrix} Z_0 & 0 & 0 \\ 0 & Z_1 & 0 \\ 0 & 0 & Z_2 \end{bmatrix} \begin{bmatrix} I_0 \\ I_1 \\ I_2 \end{bmatrix} \tag{9.1.1}$$

During a bolted three-phase fault, the sequence fault currents are $I_0 = I_2 = 0$ and $I_1 = V_F/Z_1$; therefore, from (9.1.1), the sequence fault voltages are $V_0 = V_1 = V_2 = 0$, which must be true since $V_{ag} = V_{bg} = V_{cg} = 0$. However, fault voltages need not be zero during unsymmetrical faults, which are considered next.

Section 9.2

Single Line-to-Ground Fault

Consider a single line-to-ground fault from phase a to ground at the general three-phase bus shown in Figure 9.7(a). For generality, we include a fault impedance Z_F. In the case of a bolted fault, $Z_F = 0$, whereas for an arcing fault, Z_F is the arc impedance. In the case of a transmission-line insulator flashover, Z_F includes the total fault impedance between the line and ground, including the impedances of the arc and the transmission tower, as well as the tower footing if there are no neutral wires.

Figure 9.7 Single line-to-ground fault

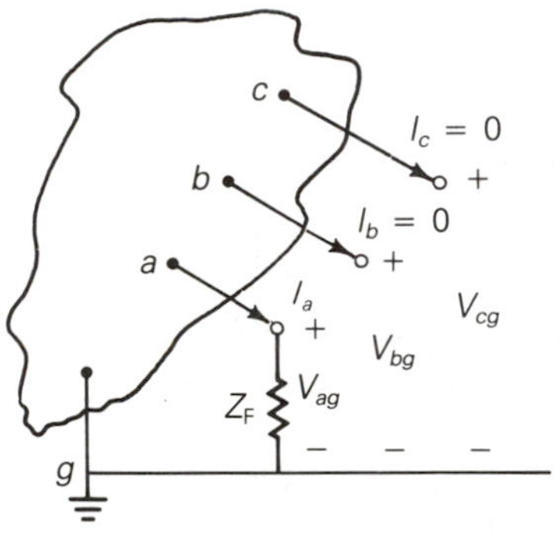

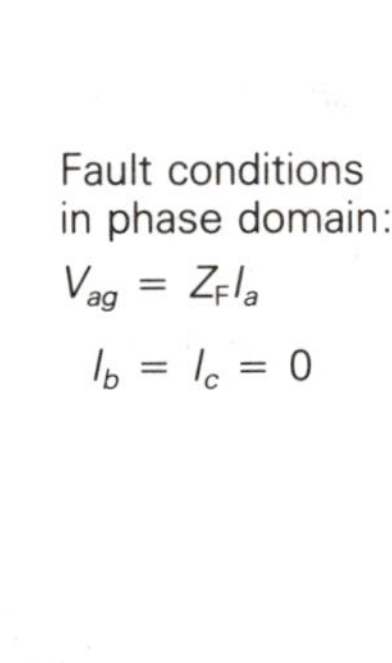

(a) General three-phase bus

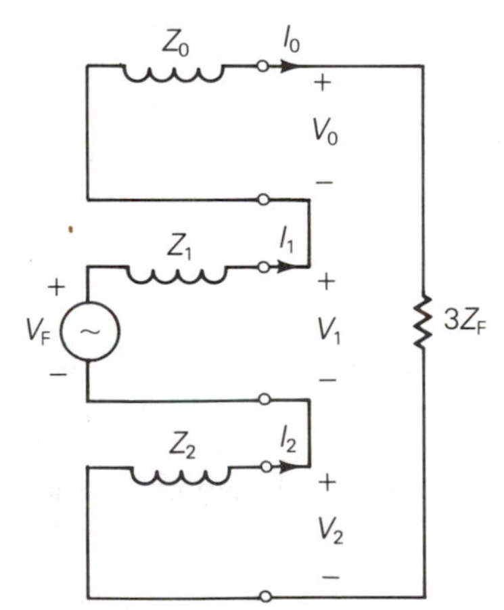

Fault conditions in sequence domain:

$I_0 = I_1 = I_2$

$(V_0 + V_1 + V_2) = 3Z_F I_1$

(b) Interconnected sequence networks

The relations to be derived here apply only to a single line-to-ground fault on phase a. However, since any of the three phases can be arbitrarily labeled phase a, we do not consider single line-to-ground faults on other phases.

From Figure 9.7(a):

$$\left.\begin{array}{l}\text{Fault conditions in phase domain}\\ \text{Single line-to-ground fault}\end{array}\right\}\quad \begin{array}{ll} I_b = I_c = 0 & (9.2.1)\\ V_{ag} = Z_F I_a & (9.2.2)\end{array}$$

We now transform (9.2.1) and (9.2.2) to the sequence domain. Using (9.2.1) in (3.1.19),

$$\begin{bmatrix} I_0 \\ I_1 \\ I_2 \end{bmatrix} = \frac{1}{3}\begin{bmatrix} 1 & 1 & 1 \\ 1 & a & a^2 \\ 1 & a^2 & a \end{bmatrix}\begin{bmatrix} I_a \\ 0 \\ 0 \end{bmatrix} = \frac{1}{3}\begin{bmatrix} I_a \\ I_a \\ I_a \end{bmatrix} \tag{9.2.3}$$

Also, using (3.1.3) and (3.1.20) in (9.2.2),

$$(V_0 + V_1 + V_2) = Z_F(I_0 + I_1 + I_2) \tag{9.2.4}$$

From (9.2.3) and (9.2.4):

$$\left.\begin{array}{l}\text{Fault conditions in sequence domain}\\ \text{Single line-to-ground fault}\end{array}\right\}\quad \begin{array}{ll} I_0 = I_1 = I_2 & (9.2.5)\\ (V_0 + V_1 + V_2) = (3Z_F)I_1 & (9.2.6)\end{array}$$

Equations (9.2.5) and (9.2.6) can be satisfied by interconnecting the sequence networks in series at the fault terminals through the impedance $(3Z_F)$, as shown in Figure 9.7(b). From this figure, the sequence components of the fault currents are:

$$I_0 = I_1 = I_2 = \frac{V_F}{Z_0 + Z_1 + Z_2 + (3Z_F)} \tag{9.2.7}$$

Transforming (9.2.7) to the phase domain via (3.1.20),

$$I_a = I_0 + I_1 + I_2 = 3I_1 = \frac{3V_F}{Z_0 + Z_1 + Z_2 + (3Z_F)} \tag{9.2.8}$$

Note also from (3.1.21) and (3.1.22),

$$I_b = (I_0 + a^2 I_1 + aI_2) = (1 + a^2 + a)I_1 = 0 \tag{9.2.9}$$

$$I_c = (I_0 + aI_1 + a^2 I_2) = (1 + a + a^2)I_1 = 0 \tag{9.2.10}$$

These are obvious, since the single line-to-ground fault is on phase a, not phase b or c.

The sequence components of the line-to-ground voltages at the fault are determined from (9.1.1). The line-to-ground voltages at the fault can then be obtained by transforming the sequence voltages to the phase domain.

Example 9.3 Calculate the subtransient fault current in per-unit and in kA for a bolted single line-to-ground short circuit from phase a to ground at bus 2 in Example 9.1. Also calculate the per-unit line-to-ground voltages at faulted bus 2.

Solution The zero-, positive-, and negative-sequence networks in Figure 9.5 are connected in series at the fault terminals, as shown in Figure 9.8. Since the short circuit is bolted, $Z_F = 0$. From (9.2.7), the sequence currents are:

$$I_0 = I_1 = I_2 = \frac{1.05\underline{/0^\circ}}{j(0.25 + 0.13893 + 0.14562)}$$

$$= \frac{1.05}{j0.53455} = -j1.96427 \quad \text{per unit}$$

From (9.2.8), the subtransient fault current is

$$I_a'' = 3(-j1.96427) = -j5.8928 \quad \text{per unit}$$

The base current at bus 2 is $100/(13.8\sqrt{3}) = 4.1837$kA. Therefore,

$$I_a'' = (-j5.8928)(4.1837) = 24.65\underline{/-90^\circ} \quad \text{kA}$$

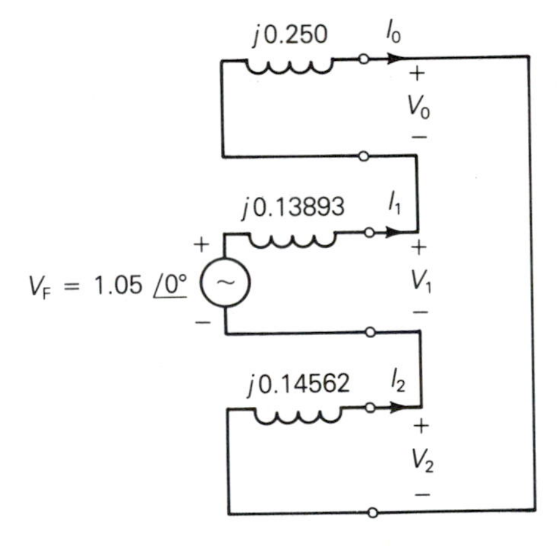

Figure 9.8 Example 9.3: Single line-to-ground fault at bus 2

From (9.1.1), the sequence components of the voltages at the fault are

$$\begin{bmatrix} V_0 \\ V_1 \\ V_2 \end{bmatrix} = \begin{bmatrix} 0 \\ 1.05\underline{/0^\circ} \\ 0 \end{bmatrix} - \begin{bmatrix} j0.25 & 0 & 0 \\ 0 & j0.13893 & 0 \\ 0 & 0 & j0.14562 \end{bmatrix} \begin{bmatrix} -j1.96427 \\ -j1.96427 \\ -j1.96427 \end{bmatrix}$$

$$= \begin{bmatrix} -0.49107 \\ 0.77710 \\ -0.28604 \end{bmatrix} \quad \text{per unit}$$

Transforming to the phase domain, the line-to-ground voltages at faulted bus 2 are

$$\begin{bmatrix} V_{ag} \\ V_{bg} \\ V_{cg} \end{bmatrix} = \begin{bmatrix} 1 & 1 & 1 \\ 1 & a^2 & a \\ 1 & a & a^2 \end{bmatrix} \begin{bmatrix} -0.49107 \\ 0.77710 \\ -0.28604 \end{bmatrix} = \begin{bmatrix} 0 \\ 1.179\underline{/231.3^\circ} \\ 1.179\underline{/128.7^\circ} \end{bmatrix} \quad \text{per unit}$$

Note that $V_{ag} = 0$, as specified by the fault conditions. Also $I''_b = I''_c = 0$. ◁

Section 9.3

Line-to-Line Fault

Consider a line-to-line fault from phase b to c, shown in Figure 9.9(a). Again, we include a fault impedance Z_F for generality. From Figure 9.9(a):

Fault conditions in phase domain / Line-to-line fault:

$$I_a = 0 \tag{9.3.1}$$

$$I_c = -I_b \tag{9.3.2}$$

$$V_{bg} - V_{cg} = Z_F I_b \tag{9.3.3}$$

Figure 9.9

Line-to-line fault

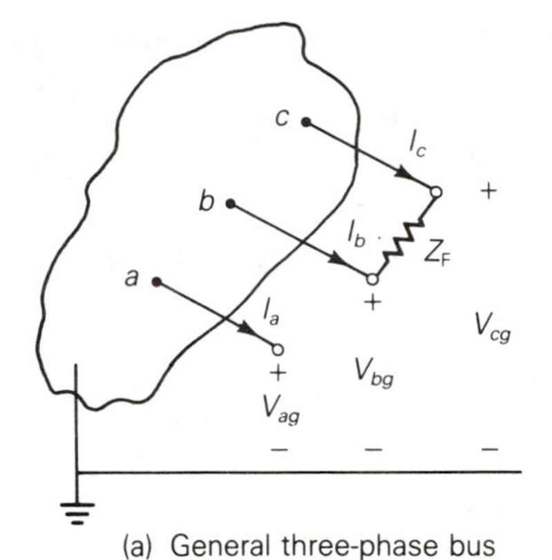

(a) General three-phase bus

Fault conditions in phase domain:

$I_a = 0$

$I_c = -I_b$

$(V_{bg} - V_{cg}) = Z_F I_b$

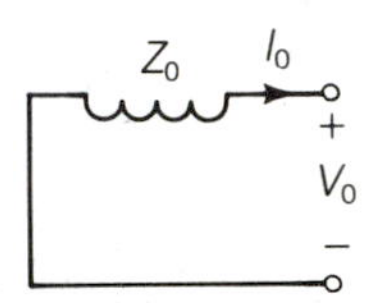

Fault conditions in sequence domain:

$I_0 = 0$

$I_2 = -I_1$

$(V_1 - V_2) = Z_F I_1$

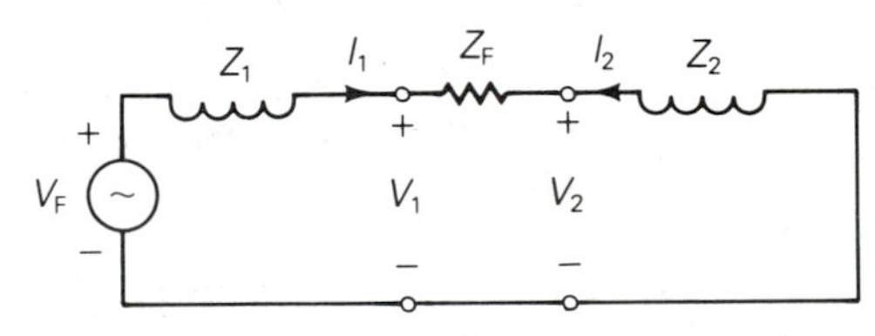

(b) Interconnected sequence networks

We transform (9.3.1)–(9.3.3) to the sequence domain. Using (9.3.1) and (9.3.2) in (3.1.19),

$$\begin{bmatrix} I_0 \\ I_1 \\ I_2 \end{bmatrix} = \frac{1}{3}\begin{bmatrix} 1 & 1 & 1 \\ 1 & a & a^2 \\ 1 & a^2 & a \end{bmatrix}\begin{bmatrix} 0 \\ I_b \\ -I_b \end{bmatrix} = \begin{bmatrix} 0 \\ \frac{1}{3}(a - a^2)I_b \\ \frac{1}{3}(a^2 - a)I_b \end{bmatrix} \tag{9.3.4}$$

Using (3.1.4), (3.1.5), and (3.1.21) in (9.3.3),

$$(V_0 + a^2V_1 + aV_2) - (V_0 + aV_1 + a^2V_2) = Z_F(I_0 + a^2I_1 + aI_2) \tag{9.3.5}$$

Noting from (9.3.4) that $I_0 = 0$ and $I_2 = -I_1$, (9.3.5) simplifies to

$$(a^2 - a)V_1 - (a^2 - a)V_2 = Z_F(a^2 - a)I_1$$

or

$$V_1 - V_2 = Z_F I_1 \tag{9.3.6}$$

Therefore, from (9.3.4) and (9.3.6):

Fault conditions in sequence domain
Line-to-line fault

$$I_0 = 0 \tag{9.3.7}$$

$$I_2 = -I_1 \tag{9.3.8}$$

$$V_1 - V_2 = Z_F I_1 \tag{9.3.9}$$

Equations (9.3.7)–(9.3.9) are satisfied by connecting the positive- and negative-sequence networks in parallel at the fault terminals through the fault impedance Z_F, as shown in Figure 9.9(b). From this figure, the fault currents are:

$$I_1 = -I_2 = \frac{V_F}{(Z_1 + Z_2 + Z_F)} \qquad I_0 = 0 \tag{9.3.10}$$

Transforming (9.3.10) to the phase domain and using the identity $(a^2 - a) = -j\sqrt{3}$, the fault current in phase b is

$$\begin{aligned} I_b &= I_0 + a^2 I_1 + a I_2 = (a^2 - a) I_1 \\ &= -j\sqrt{3} I_1 = \frac{-j\sqrt{3} V_F}{(Z_1 + Z_2 + Z_F)} \end{aligned} \tag{9.3.11}$$

Note also from (3.1.20) and (3.1.22) that

$$I_a = I_0 + I_1 + I_2 = 0 \tag{9.3.12}$$

and

$$I_c = I_0 + a I_1 + a^2 I_2 = (a - a^2) I_1 = -I_b \tag{9.3.13}$$

which verify the fault conditions given by (9.3.1) and (9.3.2). The sequence components of the line-to-ground voltages at the fault are given by (9.1.1).

Example 9.4 Calculate the subtransient fault current in per-unit and in kA for a bolted line-to-line fault from phase b to c at bus 2 in Example 9.1.

Solution The positive- and negative-sequence networks in Figure 9.5 are connected in parallel at the fault terminals, as shown in Figure 9.10. From (9.3.10) with $Z_F = 0$, the sequence fault currents are

$$I_1 = -I_2 = \frac{1.05\underline{/0^\circ}}{j(0.13893 + 0.14562)} = 3.690\underline{/-90^\circ}$$

$$I_0 = 0$$

From (9.3.11), the subtransient fault current in phase b is

$$I_b'' = (-j\sqrt{3})(3.690\underline{/-90^\circ}) = -6.391 = 6.391\underline{/180^\circ} \quad \text{per unit}$$

Using 4.1837 kA as the base current at bus 2,

$$I_b'' = (6.391\underline{/180^\circ})(4.1837) = 26.74\underline{/180^\circ} \quad \text{kA}$$

Figure 9.10

Example 9.4: Line-to-line fault at bus 2

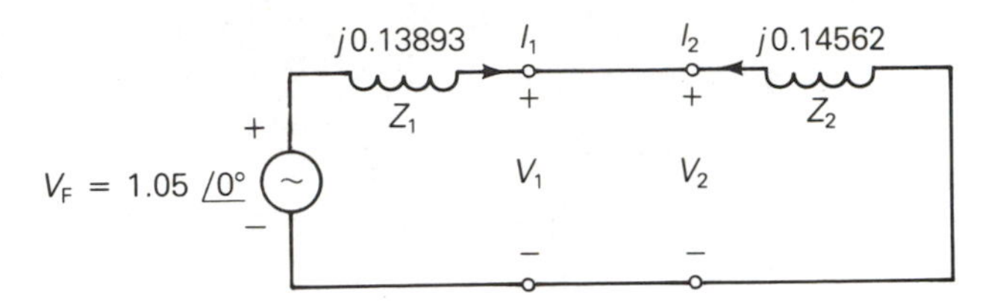

Also, from (9.3.12) and (9.3.13),

$$I''_a = 0 \qquad I''_c = 26.74\underline{/0^\circ} \quad \text{kA}$$

◁

Section 9.4

Double Line-to-Ground Fault

A double line-to-ground fault from phase b to phase c through fault impedance Z_F to ground is shown in Figure 9.11(a). From this figure:

Fault conditions in the phase domain
Double line-to-ground fault

$$I_a = 0 \tag{9.4.1}$$

$$V_{cg} = V_{bg} \tag{9.4.2}$$

$$V_{bg} = Z_F(I_b + I_c) \tag{9.4.3}$$

Figure 9.11

Double line-to-ground fault

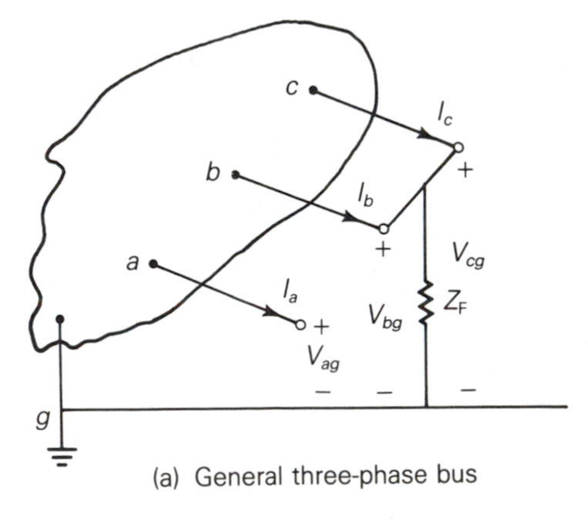

Fault conditions in phase domain:

$I_a = 0$

$V_{bg} = V_{cg} = Z_F(I_b + I_c)$

(a) General three-phase bus

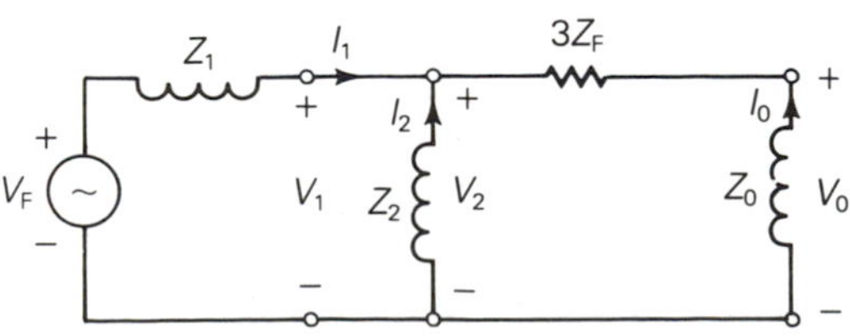

Fault conditions in sequence domain:

$I_0 + I_1 + I_2 = 0$

$V_0 - V_1 = (3Z_F)I_0$

$V_1 = V_2$

(b) Interconnected sequence networks

Transforming (9.4.1) to the sequence domain via (3.1.20),

$$I_0 + I_1 + I_2 = 0 \tag{9.4.4}$$

Also, using (3.1.4) and (3.1.5) in (9.4.2),

$$(V_0 + aV_1 + a^2V_2) = (V_0 + a^2V_1 + aV_2)$$

Simplifying:

$$(a^2 - a)V_2 = (a^2 - a)V_1$$

or

$$V_2 = V_1 \tag{9.4.5}$$

Now, using (3.1.4), (3.1.21), and (3.1.22) in (9.4.3),

$$(V_0 + a^2V_1 + aV_2) = Z_F(I_0 + a^2I_1 + aI_2 + I_0 + aI_1 + a^2I_2) \tag{9.4.6}$$

Using (9.4.5) and the identity $a^2 + a = -1$ in (9.4.6),

$$(V_0 - V_1) = Z_F(2I_0 - I_1 - I_2) \tag{9.4.7}$$

From (9.4.4), $I_0 = -(I_1 + I_2)$; therefore, (9.4.7) becomes

$$V_0 - V_1 = (3Z_F)I_0 \tag{9.4.8}$$

From (9.4.4), (9.4.5), and (9.4.8), we summarize:

Fault conditions in the sequence domain
Double line-to-ground fault

$$I_0 + I_1 + I_2 = 0 \tag{9.4.9}$$

$$V_2 = V_1 \tag{9.4.10}$$

$$V_0 - V_1 = (3Z_F)I_0 \tag{9.4.11}$$

Equations (9.4.9)–(9.4.11) are satisfied by connecting the zero-, positive-, and negative-sequence networks in parallel at the fault terminal; additionally, $(3Z_F)$ is included in series with the zero-sequence network. This connection is shown in Figure 9.11(b). From this figure the positive-sequence fault current is

$$I_1 = \frac{V_F}{Z_1 + [Z_2//(Z_0 + 3Z_F)]} = \frac{V_F}{Z_1 + \left[\dfrac{Z_2(Z_0 + 3Z_F)}{Z_2 + Z_0 + 3Z_F}\right]} \tag{9.4.12}$$

Using current division in Figure 9.11(b), the negative- and zero-sequence fault currents are

$$I_2 = (-I_1)\left(\frac{Z_0 + 3Z_F}{Z_0 + 3Z_F + Z_2}\right) \tag{9.4.13}$$

$$I_0 = (-I_1)\left(\frac{Z_2}{Z_0 + 3Z_F + Z_2}\right) \tag{9.4.14}$$

These sequence fault currents can be transformed to the phase domain via (3.1.16). Also, the sequence components of the line-to-ground voltages at the fault are given by (9.1.1).

Example 9.5 Calculate (a) the subtransient fault current in each phase, (b) neutral fault current, and (c) contributions to the fault current from the motor and from the transmission line, for a bolted double line-to-ground fault from phase b to c to ground at bus 2 in Example 9.1. Neglect the Δ–Y transformer phase shifts.

Solution **a.** The zero-, positive-, and negative-sequence networks in Figure 9.5 are connected in parallel at the fault terminals in Figure 9.12. From (9.4.12) with $Z_F = 0$,

$$I_1 = \frac{1.05\underline{/0^\circ}}{j\left[0.13893 + \dfrac{(0.14562)(0.25)}{0.14562 + 0.25}\right]} = \frac{1.05\underline{/0^\circ}}{j0.23095}$$

$$= -j4.5464 \quad \text{per unit}$$

From (9.4.13) and (9.4.14),

$$I_2 = (+j4.5464)\left(\frac{0.25}{0.25 + 0.14562}\right) = j2.8730 \quad \text{per unit}$$

$$I_0 = (+j4.5464)\left(\frac{0.14562}{0.25 + 0.14562}\right) = j1.6734 \quad \text{per unit}$$

Transforming to the phase domain, the subtransient fault currents are:

$$\begin{bmatrix} I_a'' \\ I_b'' \\ I_c'' \end{bmatrix} = \begin{bmatrix} 1 & 1 & 1 \\ 1 & a^2 & a \\ 1 & a & a^2 \end{bmatrix} \begin{bmatrix} +j1.6734 \\ -j4.5464 \\ +j2.8730 \end{bmatrix} = \begin{bmatrix} 0 \\ 6.8983\underline{/158.66^\circ} \\ 6.8983\underline{/21.34^\circ} \end{bmatrix} \quad \text{per unit}$$

Using the base current of 4.1837 kA at bus 2,

$$\begin{bmatrix} I_a'' \\ I_b'' \\ I_c'' \end{bmatrix} = \begin{bmatrix} 0 \\ 6.8983\underline{/158.66^\circ} \\ 6.8983\underline{/21.34^\circ} \end{bmatrix} (4.1837) = \begin{bmatrix} 0 \\ 28.86\underline{/158.66^\circ} \\ 28.86\underline{/21.34^\circ} \end{bmatrix} \quad \text{kA}$$

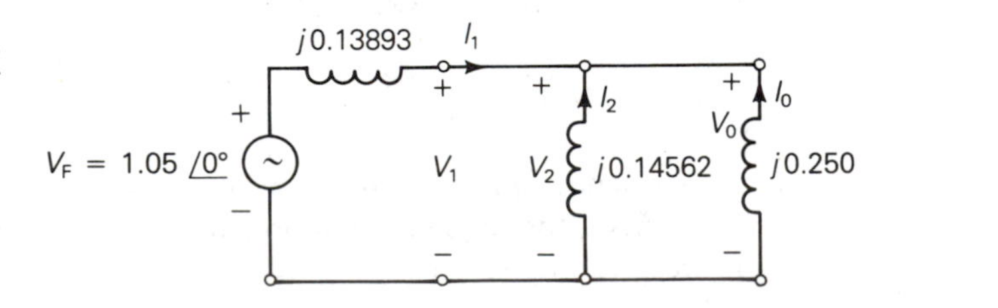

Figure 9.12 Example 9.5: Double line-to-ground fault at bus 2

b. The neutral fault current is

$$I_n = (I_b'' + I_c'') = 3I_0 = j5.0202 \quad \text{per unit}$$
$$= (j5.0202)(4.1837) = 21.00\underline{/90^\circ} \quad \text{kA}$$

c. Neglecting Δ–Y transformer phase shifts, the contributions to the fault current from the motor and transmission line can be obtained from Figure 9.4. From the zero-sequence network, Figure 9.4(a), the contribution to the zero-sequence fault current from the line is zero, due to the transformer connection. That is,

$$I_{\text{line } 0} = 0$$
$$I_{\text{motor } 0} = I_0 = j1.6734 \quad \text{per unit}$$

From the positive-sequence network, Figure 9.4(b), the positive terminals of the internal machine voltages can be connected, since $E_g'' = E_m''$. Then, by current division,

$$I_{\text{line } 1} = \frac{X_m''}{X_m'' + (X_g'' + X_{T1} + X_{\text{line } 1} + X_{T2})} I_1$$
$$= \frac{0.20}{0.20 + (0.455)}(-j4.5464) = -j1.3882 \quad \text{per unit}$$
$$I_{\text{motor } 1} = \frac{0.455}{0.20 + 0.455}(-j4.5464) = -j3.1582 \quad \text{per unit}$$

From the negative-sequence network, Figure 9.4(c), using current division,

$$I_{\text{line } 2} = \frac{0.21}{0.21 + 0.475}(j2.8730) = j0.8808 \quad \text{per unit}$$
$$I_{\text{motor } 2} = \frac{0.475}{0.21 + 0.475}(j2.8730) = j1.9922 \quad \text{per unit}$$

Transforming to the phase domain with base currents of 0.41837 kA for the line and 4.1837 kA for the motor,

$$\begin{bmatrix} I''_{\text{line}\,a} \\ I''_{\text{line}\,b} \\ I''_{\text{line}\,c} \end{bmatrix} = \begin{bmatrix} 1 & 1 & 1 \\ 1 & a^2 & a \\ 1 & a & a^2 \end{bmatrix} \begin{bmatrix} 0 \\ -j1.3882 \\ j0.8808 \end{bmatrix} = \begin{bmatrix} 0.5074\underline{/-90^\circ} \\ 1.9813\underline{/172.643^\circ} \\ 1.9813\underline{/7.357^\circ} \end{bmatrix} \quad \text{per unit}$$

$$= \begin{bmatrix} 0.2123\underline{/-90^\circ} \\ 0.8289\underline{/172.643^\circ} \\ 0.8289\underline{/7.357^\circ} \end{bmatrix} \quad \text{kA}$$

$$\begin{bmatrix} I''_{\text{motor}\,a} \\ I''_{\text{motor}\,b} \\ I''_{\text{motor}\,c} \end{bmatrix} = \begin{bmatrix} 1 & 1 & 1 \\ 1 & a^2 & a \\ 1 & a & a^2 \end{bmatrix} \begin{bmatrix} j1.6734 \\ -j3.1582 \\ j1.9922 \end{bmatrix} = \begin{bmatrix} 0.5074\underline{/90^\circ} \\ 4.9986\underline{/153.17^\circ} \\ 4.9986\underline{/26.83^\circ} \end{bmatrix} \quad \text{per unit}$$

$$= \begin{bmatrix} 2.123\underline{/90^\circ} \\ 20.91\underline{/153.17^\circ} \\ 20.91\underline{/26.83^\circ} \end{bmatrix} \quad \text{kA}$$

◁

Example 9.6 Rework Example 9.5, with the Δ–Y transformer phase shifts included. Assume American standard phase shift.

Solution The sequence networks of Figure 9.4 are redrawn in Figure 9.13 with ideal phase-shifting transformers representing Δ–Y phase shifts. In accordance with the American standard, positive-sequence quantities on the high-voltage side of the transformers lead their corresponding quantities on the low-voltage side by 30°. Also, the negative-sequence phase shifts are the reverse of the positive-sequence phase shifts.

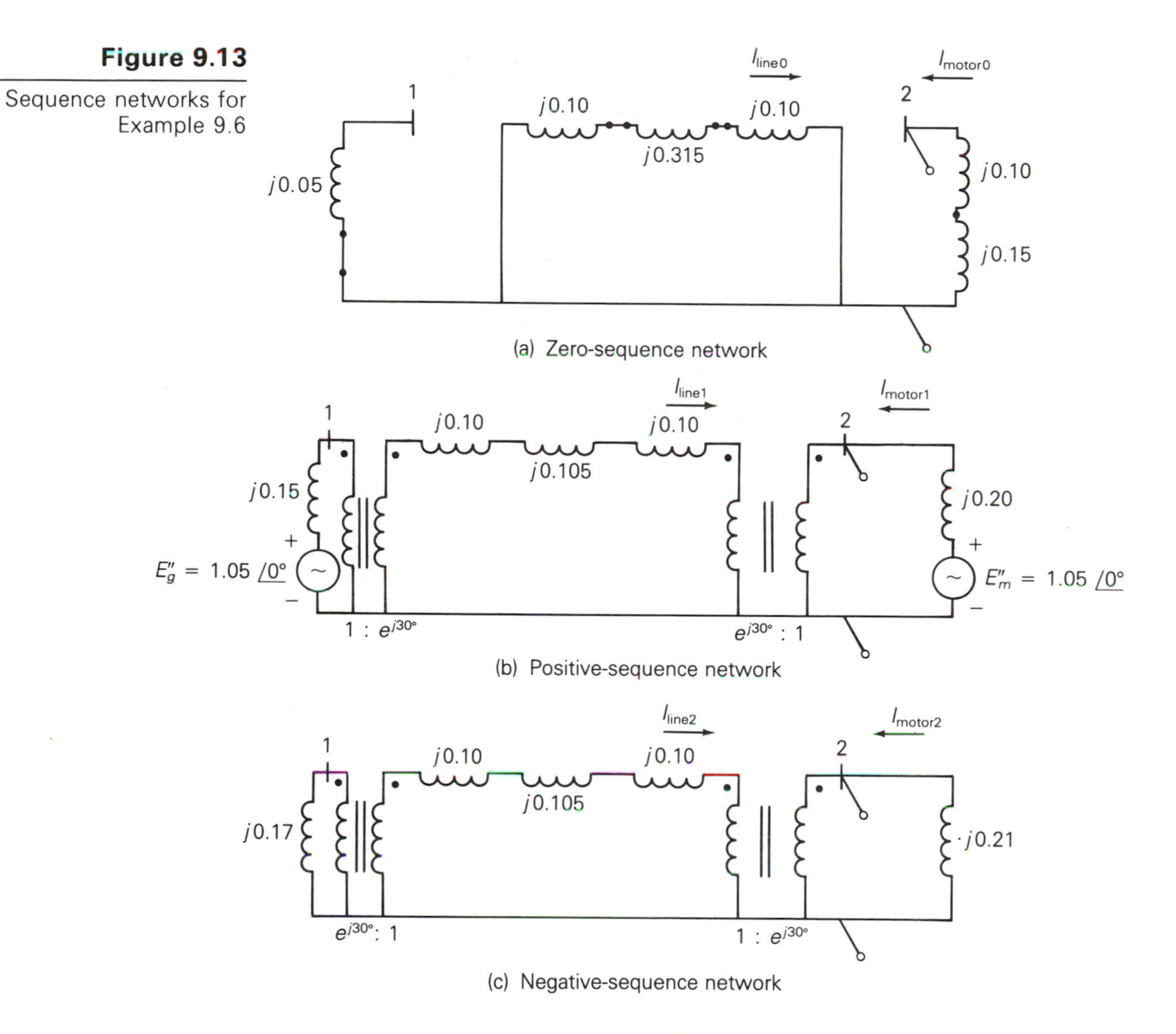

Figure 9.13 Sequence networks for Example 9.6

a. Recall from Section 4.1 and (4.1.26) that per-unit impedance is unchanged when it is referred from one side of an ideal phase-shifting transformer to the other. Accordingly, the Thévenin equivalents of the sequence networks in Figure 9.13, as viewed from fault bus 2, are the same as those given in Figure 9.5. Therefore, the sequence components as well as the phase components of the fault currents are the same as those given in Example 9.5(a).

b. The neutral fault current is the same as that given in Example 9.5(b).

c. The zero-sequence network, Figure 9.13(a), is the same as that given in Figure 9.4(a). Therefore, the contributions to the zero-sequence fault current from the line and motor are the same as those given in Example 9.5(c).

$$I_{\text{line}\,0} = 0 \qquad I_{\text{motor}\,0} = I_0 = j1.6734 \quad \text{per unit}$$

The contribution to the positive-sequence fault current from the line in Figure 9.13(b) leads that in Figure 9.4(b) by 30°. That is,

$$I_{\text{line }1} = (-j1.3882)(1\underline{/30^\circ}) = 1.3882\underline{/-60^\circ} \quad \text{per unit}$$

$$I_{\text{motor }1} = -j3.1582 \quad \text{per unit}$$

Similarly, the contribution to the negative-sequence fault current from the line in Figure 9.13(c) lags that in Figure 9.4(c) by 30°. That is,

$$I_{\text{line }2} = (j0.8808)(1\underline{/-30^\circ}) = 0.8808\underline{/60^\circ} \quad \text{per unit}$$

$$I_{\text{motor }2} = j1.9922 \quad \text{per unit}$$

Thus the sequence currents as well as the phase currents from the motor are the same as those given in Example 9.5(c). Also, the sequence currents from the line have the same magnitudes as those given in Example 9.5(c), but the positive- and negative-sequence line currents are shifted by $+30^\circ$ and -30°, respectively. Transforming the line currents to the phase domain:

$$\begin{bmatrix} I''_{\text{line}\,a} \\ I''_{\text{line}\,b} \\ I''_{\text{line}\,c} \end{bmatrix} = \begin{bmatrix} 1 & 1 & 1 \\ 1 & a^2 & a \\ 1 & a^2 & a^2 \end{bmatrix} \begin{bmatrix} 0 \\ 1.3882\underline{/-60^\circ} \\ 0.8808\underline{/60^\circ} \end{bmatrix}$$

$$= \begin{bmatrix} 1.2166\underline{/-21.17^\circ} \\ 2.2690\underline{/180^\circ} \\ 1.2166\underline{/21.17^\circ} \end{bmatrix} \text{per unit} = \begin{bmatrix} 0.5090\underline{/-21.17^\circ} \\ 0.9492\underline{/180^\circ} \\ 0.5090\underline{/21.17^\circ} \end{bmatrix} \quad \text{kA}$$

In conclusion, Δ–Y transformer phase shifts have no effect on the fault currents and no effect on the contribution to the fault currents on the fault side of the Δ–Y transformers. However, on the other side of the Δ–Y transformers, the positive- and negative-sequence components of the contributions to the fault currents are shifted by $\pm 30^\circ$, which affects both the magnitude as well as the angle of the phase components of these fault contributions for unsymmetrical faults. ◁

Figure 9.14 summarizes the sequence network connections for both the balanced three-phase fault and the unsymmetrical faults that we have considered. Sequence network connections for two additional faults, one-conductor-open and two-conductors-open, are also shown in Figure 9.14 and are left as an exercise for the reader to verify (see Problems 9.10 and 9.11).

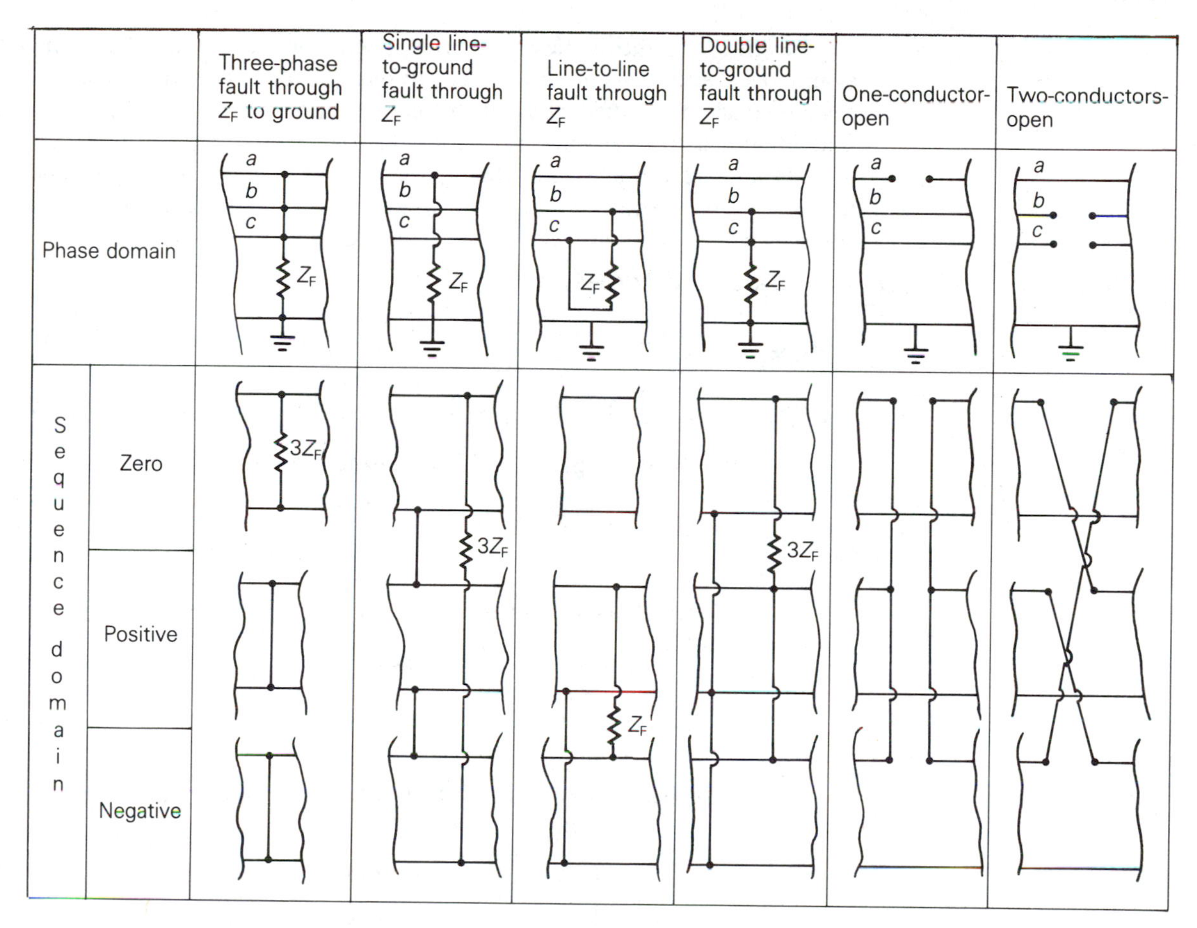

Figure 9.14 Summary of faults

Section 9.5

Sequence Bus Impedance Matrices

We use the positive-sequence bus impedance matrix in Section 8.4 for calculating currents and voltages during balanced three-phase faults. This method is extended here to unsymmetrical faults by representing each sequence network as a bus impedance equivalent circuit (or as a *rake* equivalent). A bus impedance matrix can be computed for each sequence network using the one-step-at-a-time method given in Section 8.5. For simplicity, resistances, shunt admittances, nonrotating impedance loads, and prefault load currents are neglected.

Figure 9.15 shows the connection of sequence rake equivalents for both symmetrical and unsymmetrical faults at bus n of an N-bus three-phase power system. Each bus impedance element has an additional subscript, 0, 1, or 2, that identifies the sequence rake equivalent in which it is located. Mutual impedances are not shown in the figure. The prefault voltage V_F is included in the positive-sequence rake equivalent. From the figure the sequence components of the fault current for each type of fault at bus n are:

Balanced three-phase fault:

$$I_{n-1} = \frac{V_F}{Z_{nn-1}} \tag{9.5.1}$$

$$I_{n-0} = I_{n-2} = 0 \tag{9.5.2}$$

Single line-to-ground fault (phase *a* to ground):

$$I_{n-0} = I_{n-1} = I_{n-2} = \frac{V_F}{Z_{nn-0} + Z_{nn-1} + Z_{nn-2} + 3Z_F} \tag{9.5.3}$$

Line-to-line fault (phase *b* to *c*):

$$I_{n-1} = -I_{n-2} = \frac{V_F}{Z_{nn-1} + Z_{nn-2} + Z_F} \tag{9.5.4}$$

$$I_{n-0} = 0 \tag{9.5.5}$$

Double line-to-ground fault (phase *b* to *c* to ground):

$$I_{n-1} = \frac{V_F}{Z_{nn-1} + \left[\dfrac{Z_{nn-2}(Z_{nn-0} + 3Z_F)}{Z_{nn-2} + Z_{nn-0} + 3Z_F}\right]} \tag{9.5.6}$$

$$I_{n-2} = (-I_{n-1})\left(\frac{Z_{nn-0} + 3Z_F}{Z_{nn-0} + 3Z_F + Z_{nn-2}}\right) \tag{9.5.7}$$

$$I_{n-0} = (-I_{n-1})\left(\frac{Z_{nn-2}}{Z_{nn-0} + 3Z_F + Z_{nn-2}}\right) \tag{9.5.8}$$

Also from Figure 9.15, the sequence components of the line-to-ground voltages at any bus k during a fault at bus n are:

$$\begin{bmatrix} V_{k-0} \\ V_{k-1} \\ V_{k-2} \end{bmatrix} = \begin{bmatrix} 0 \\ V_F \\ 0 \end{bmatrix} - \begin{bmatrix} Z_{kn-0} & 0 & 0 \\ 0 & Z_{kn-1} & 0 \\ 0 & 0 & Z_{kn-2} \end{bmatrix} \begin{bmatrix} I_{n-0} \\ I_{n-1} \\ I_{n-2} \end{bmatrix} \tag{9.5.9}$$

If bus k is on the unfaulted side of a Δ–Y transformer, then the phase angles of V_{k-1} and V_{k-2} in (9.5.9) are modified to account for Δ–Y phase shifts. Also, the above sequence fault currents and sequence voltages can be transformed to the phase domain via (3.1.16) and (3.1.9).

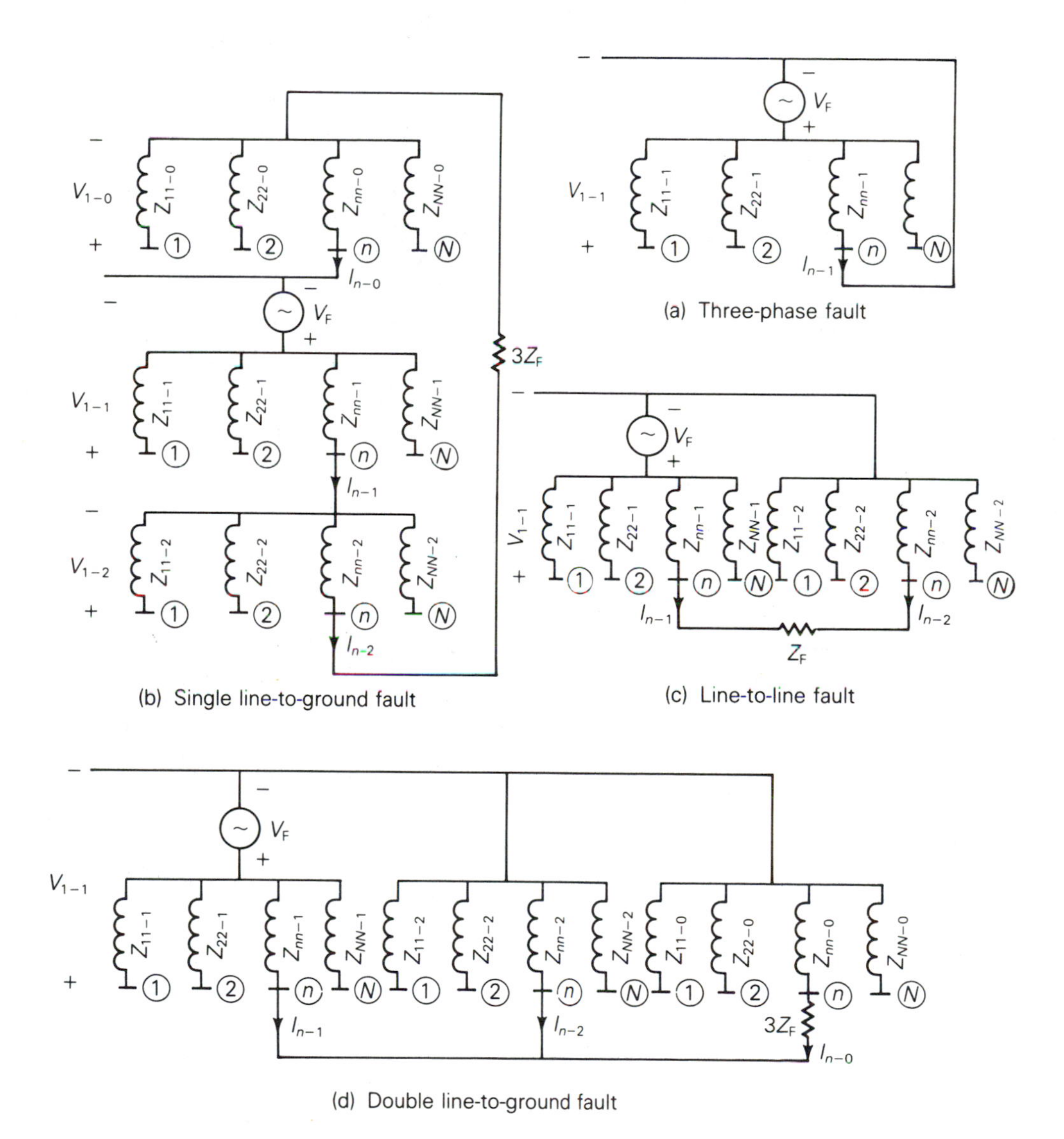

Figure 9.15 Connection of rake equivalent sequence networks for three-phase system faults (mutual impedances not shown)

Example 9.7 Faults at buses 1 and 2 for the three-phase power system given in Example 9.1 are of interest. The prefault voltage is 1.05 per unit. Prefault load current is neglected. **(a)** Determine the per-unit zero-, positive-, and negative-sequence bus impedance matrices using the one-step-at-a-time method given in Section 8.5. Find the subtransient fault current in per-unit for a bolted single line-to-ground fault current from phase a to ground **(b)** at bus 1 and **(c)** at bus 2. Find the per-unit line-to-ground voltages at **(d)** bus 1 and **(e)** bus 2 during the single line-to-ground fault at bus 1.

Solution **a.** Referring to Figure 9.4(a), the zero-sequence bus impedance matrix is determined in two steps as follows:

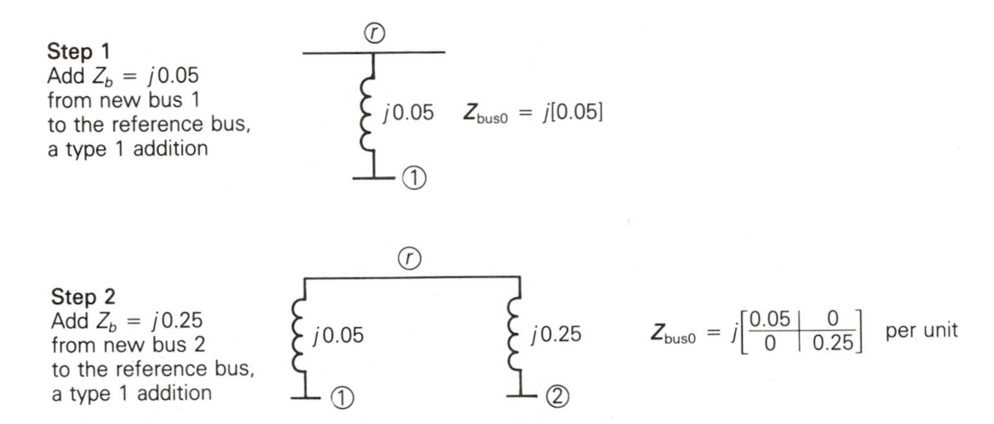

Note that the transformer leakage reactances and the zero-sequence transmission-line reactance in Figure 9.4(a) have no effect on $\boldsymbol{Z}_{\text{bus}\,0}$. The transformer Δ connections block the flow of zero-sequence current from the transformers to bus 1 and 2.

The positive-sequence bus impedance matrix, determined in Example 8.5, is

$$\boldsymbol{Z}_{\text{bus}\,1} = j\left[\begin{array}{c|c} 0.11565 & 0.04580 \\ \hline 0.04580 & 0.13893 \end{array}\right] \quad \text{per unit}$$

Referring to Figure 9.4(c), the negative-sequence bus impedance matrix is determined in three steps, as follows:

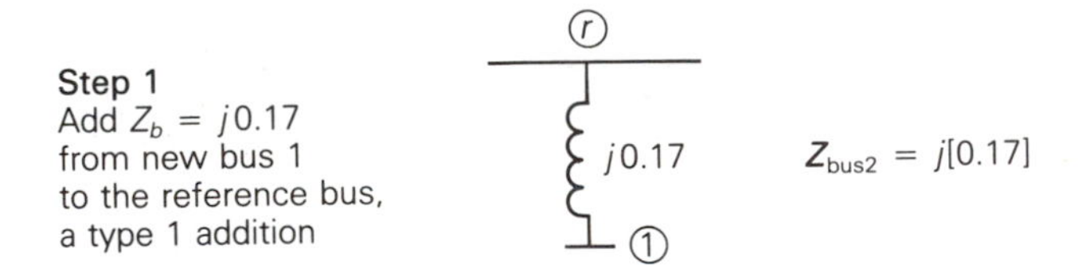

Step 2
Add $Z_b = j(0.10 + 0.105 + 0.10) = j0.305$ from old bus 1 to new bus 2, a type 2 addition

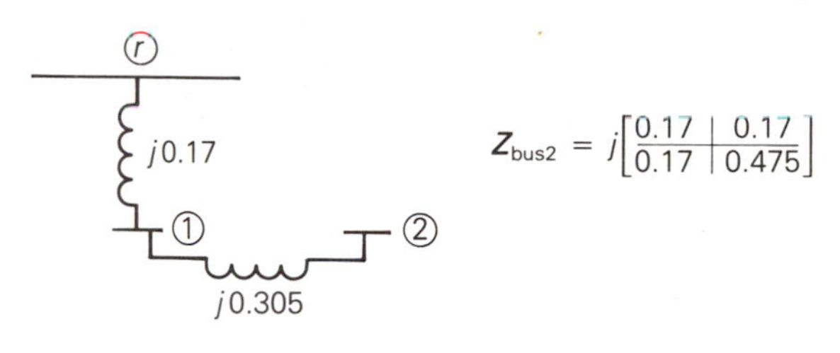

$$Z_{\text{bus2}} = j\left[\begin{array}{c|c} 0.17 & 0.17 \\ \hline 0.17 & 0.475 \end{array}\right]$$

Step 3
Add $Z_b = j0.21$ from old bus 2 to the reference bus, a type 3 addition

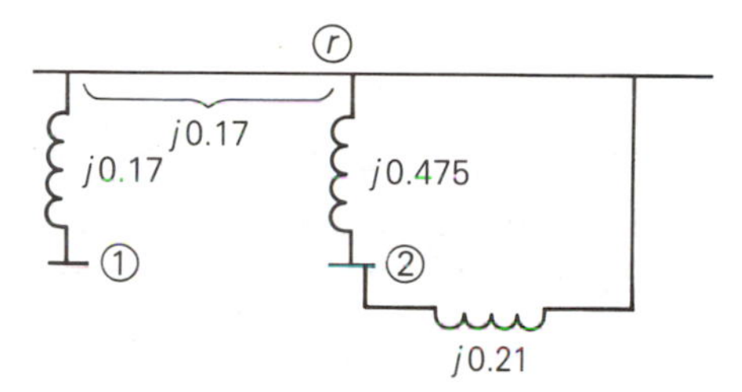

$$Z_{\text{bus}\,2} = j\left\{\begin{bmatrix} 0.17 & 0.17 \\ 0.17 & 0.475 \end{bmatrix} - \frac{1}{(0.475 + 0.21)}\begin{bmatrix} 0.17 \\ 0.475 \end{bmatrix}[0.17 \quad 0.475]\right\}$$

$$= j\left[\begin{array}{c|c} 0.12781 & 0.05212 \\ \hline 0.05212 & 0.14562 \end{array}\right] \quad \text{per unit}$$

b. From (9.5.3), with $n = 1$ and $Z_F = 0$, the sequence fault currents are

$$I_{1-0} = I_{1-1} = I_{1-2} = \frac{V_F}{Z_{11-0} + Z_{11-1} + Z_{11-2}}$$

$$= \frac{1.05\underline{/0^\circ}}{j(0.05 + 0.11565 + 0.12781)} = \frac{1.05}{j0.29346} = -j3.578 \quad \text{per unit}$$

The subtransient fault currents at bus 1 are, from (3.1.16),

$$\begin{bmatrix} I''_{1a} \\ I''_{1b} \\ I''_{1c} \end{bmatrix} = \begin{bmatrix} 1 & 1 & 1 \\ 1 & a^2 & a \\ 1 & a & a^2 \end{bmatrix} \begin{bmatrix} -j3.578 \\ -j3.578 \\ -j3.578 \end{bmatrix} = \begin{bmatrix} -j10.73 \\ 0 \\ 0 \end{bmatrix} \quad \text{per unit}$$

c. Again from (9.5.3), with $n = 2$ and $Z_F = 0$,

$$I_{2-0} = I_{2-1} = I_{2-2} = \frac{V_F}{Z_{22-0} + Z_{22-1} + Z_{22-2}}$$

$$= \frac{1.05\underline{/0^\circ}}{j(0.25 + 0.13893 + 0.14562)} = \frac{1.05}{j0.53455}$$

$$= -j1.96427 \quad \text{per unit}$$

and

$$\begin{bmatrix} I''_{2a} \\ I''_{2b} \\ I''_{2c} \end{bmatrix} = \begin{bmatrix} 1 & 1 & 1 \\ 1 & a^2 & a \\ 1 & a & a^2 \end{bmatrix} \begin{bmatrix} -j1.96427 \\ -j1.96427 \\ -j1.96427 \end{bmatrix} = \begin{bmatrix} -j5.8928 \\ 0 \\ 0 \end{bmatrix}$$

This is the same result as obtained in Example 9.3.

d. The sequence components of the line-to-ground voltages at bus 1 during the fault at bus 1 are, from (9.5.9), with $k = 1$ and $n = 1$,

$$\begin{bmatrix} V_{1-0} \\ V_{1-1} \\ V_{1-2} \end{bmatrix} = \begin{bmatrix} 0 \\ 1.05\underline{/0^\circ} \\ 0 \end{bmatrix} - \begin{bmatrix} j0.05 & 0 & 0 \\ 0 & j0.11565 & 0 \\ 0 & 0 & j0.12781 \end{bmatrix} \begin{bmatrix} -j3.578 \\ -j3.578 \\ -j3.578 \end{bmatrix}$$

$$= \begin{bmatrix} -0.1789 \\ 0.6362 \\ -0.4573 \end{bmatrix} \text{ per unit}$$

and the line-to-ground voltages at bus 1 during the fault at bus 1 are

$$\begin{bmatrix} V_{1-ag} \\ V_{1-bg} \\ V_{1-cg} \end{bmatrix} = \begin{bmatrix} 1 & 1 & 1 \\ 1 & a^2 & a \\ 1 & a & a^2 \end{bmatrix} \begin{bmatrix} -0.1789 \\ +0.6362 \\ -0.4573 \end{bmatrix} = \begin{bmatrix} 0 \\ 0.9843\underline{/254.2^\circ} \\ 0.9843\underline{/105.8^\circ} \end{bmatrix} \text{ per unit}$$

e. The sequence components of the line-to-ground voltages at bus 2 during the fault at bus 1 are, from (9.5.9), with $k = 2$ and $n = 1$,

$$\begin{bmatrix} V_{2-0} \\ V_{2-1} \\ V_{2-2} \end{bmatrix} = \begin{bmatrix} 0 \\ 1.05\underline{/0^\circ} \\ 0 \end{bmatrix} - \begin{bmatrix} 0 & 0 & 0 \\ 0 & j0.04580 & 0 \\ 0 & 0 & j0.05212 \end{bmatrix} \begin{bmatrix} -j3.578 \\ -j3.578 \\ -j3.578 \end{bmatrix}$$

$$= \begin{bmatrix} 0 \\ 0.8861 \\ -0.18649 \end{bmatrix} \text{ per unit}$$

Note that since both bus 1 and 2 are on the low-voltage side of the Δ–Y transformers in Figure 9.3, there is no shift in the phase angles of these sequence voltages. From the above, the line-to-ground voltages at bus 2 during the fault at bus 1 are

$$\begin{bmatrix} V_{2-ag} \\ V_{2-bg} \\ V_{2-cg} \end{bmatrix} = \begin{bmatrix} 1 & 1 & 1 \\ 1 & a^2 & a \\ 1 & a & a^2 \end{bmatrix} \begin{bmatrix} 0 \\ 0.8861 \\ -0.18649 \end{bmatrix} = \begin{bmatrix} 0.70 \\ 0.9926\underline{/249.4^\circ} \\ 0.9926\underline{/110.6^\circ} \end{bmatrix} \text{ per unit}$$

◁

Section 9.6

Personal Computer Program: SHORT CIRCUITS

The software package that accompanies this text includes the program "SHORT CIRCUITS," which computes the symmetrical fault current for each of the following faults at any bus in an N-bus power system: balanced three-phase fault, single line-to-ground fault, line-to-line fault, or double line-to-ground fault. For each fault, the program also computes bus voltages and contributions to the fault current from transmission lines and transformers connected to the fault bus.

Input data for the program include machine, transmission-line, and transformer data, as illustrated in Tables 9.1, 9.2, and 9.3 as well as the prefault voltage V_F and fault impedance Z_F. When the machine positive-sequence reactance input data consist of direct axis subtransient reactances, the computed symmetrical fault currents are subtransient fault currents. Alternatively, transient or steady-state fault currents are computed when this input data consist of direct axis transient or synchronous reactances. Transmission-line positive- and zero-sequence series reactances are those of the equivalent π circuits for long lines or of the nominal π circuit for medium or short lines. Also, recall that the negative-sequence transmission-line reactance equals the positive-sequence transmission-line reactance. All machine, line, and transformer reactances are given in per-unit on a common MVA base. Resistances, shunt admittances, nonrotating impedance loads, and prefault load currents are neglected.

Table 9.1

Synchronous machine data for Example 9.8

BUS	X_0	$X_1 = X''_d$	X_2	NEUTRAL REACTANCE X_n
	per unit	per unit	per unit	per unit
1	0.05	0.18	0.18	0
3	0.02	0.09	0.09	0.01

Table 9.2

Line data for Example 9.8

BUS-TO-BUS	X_0	X_1
	per unit	per unit
2–4	1.2	0.4
2–5	0.6	0.2
4–5	0.3	0.1

Table 9.3

Transformer data for Example 9.8

LOW-VOLTAGE (connection) bus	HIGH-VOLTAGE (connection) bus	LEAKAGE REACTANCE	NEUTRAL REACTANCE
		per unit	per unit
1 (Δ)	5 (Y)	0.08	0
3 (Δ)	4 (Y)	0.04	0

$S_{base} = 400$ MVA

$V_{base} = \begin{cases} 15 \text{ kV at buses 1, 3} \\ 345 \text{ kV at buses 2, 4, 5} \end{cases}$

The program computes the zero-, positive-, and negative-sequence bus impedance matrices $\boldsymbol{Z}_{\text{bus}\,0}$, $\boldsymbol{Z}_{\text{bus}\,1}$, and $\boldsymbol{Z}_{\text{bus}\,2}$, using the one-step-at-a-time method described in Section 8.5. The program user can either compute these impedance matrices from scratch or begin with existing matrices that have already been computed and stored. In both cases, type 1, 2, 3, and 4 modifications, as described in Section 8.5, can be made. The program computes the matrices in accordance with the ordering rules given in Section 8.5.

After $\boldsymbol{Z}_{\text{bus}\,0}$, $\boldsymbol{Z}_{\text{bus}\,1}$, and $\boldsymbol{Z}_{\text{bus}\,2}$ are computed, (9.5.1)–(9.5.9) are used to compute the sequence fault currents and the sequence voltages at each bus during a fault at bus 1 for the fault type selected by the program user (for example, three-phase fault, or single line-to-ground fault, and so on). Contributions to the sequence fault currents from each line or transformer branch connected to the fault bus are computed by dividing the sequence voltage across the branch by the branch sequence impedance. The phase angles of positive- and negative-sequence voltages are also modified to account for Δ–Y transformer phase shifts. The sequence currents and sequence voltages are then transformed to the phase domain via (3.1.16) and (3.1.9). All these computations are then repeated for a fault at bus 2, then bus 3, and so on to bus N.

For large power systems with hundreds or thousands of buses, computer storage requirements can be reduced by about one-third if $\boldsymbol{Z}_{\text{bus}\,2}$ is set equal to $\boldsymbol{Z}_{\text{bus}\,1}$. This is a good approximation when calculating subtransient fault currents because the negative-sequence reactance of a synchronous machine is approximately equal to its direct axis subtransient reactance.

Output data for the fault type and fault impedance selected by the user consist of: the fault current in each phase, contributions to the fault current from each branch connected to the fault bus for each phase, and the line-to-ground voltages at each bus—for a fault at bus 1, then bus 2, and so on to bus N. $\boldsymbol{Z}_{\text{bus}\,0}$, $\boldsymbol{Z}_{\text{bus}\,1}$, and $\boldsymbol{Z}_{\text{bus}\,2}$ can also be included in the output if desired.

Example 9.8 Consider the five-bus power system whose single-line diagram is shown in Figure 7.2. Machine, line, and transformer data are given in Tables 9.1, 9.2, and 9.3. Note that the neutrals of both transformers and generator 1 are solidly grounded, as indicated by a neutral reactance of zero for these equipments. However, a neutral reactance = 0.01 per unit is connected to the generator 2 neutral. The prefault voltage is 1.05 per unit. Run the SHORT CIRCUITS program for a bolted single line-to-ground fault at bus 1, then bus 2, and so on to bus 5.

Solution The program was run with the input data from Tables 9.1, 9.2, and 9.3, with $V_F = 1.05$, and with $Z_F = 0$ for a single line-to-ground fault. The 5×5 bus impedance matrices $\boldsymbol{Z}_{\text{bus}\,0}$, $\boldsymbol{Z}_{\text{bus}\,1}$, and $\boldsymbol{Z}_{\text{bus}\,2}$ are shown in Table 9.4 and the fault currents and bus voltages are given in Tables 9.5 and 9.6. Note that these fault currents are subtransient currents, since the machine positive-sequence reactance input consists of direct axis subtransient reactances.

Table 9.4

$\boldsymbol{Z}_{\text{bus}}$ impedance matrices for Example 9.8

$Z_F = 0 + j0$

$$\boldsymbol{Z}_{\text{bus0}} = j\begin{bmatrix} 0.05000 & 0.00000 & 0.00000 & 0.00000 & 0.00000 \\ 0.00000 & 0.43576 & 0.00000 & 0.01758 & 0.04485 \\ 0.00000 & 0.00000 & 0.05000 & 0.00000 & 0.00000 \\ 0.00000 & 0.01758 & 0.00000 & 0.03576 & 0.00848 \\ 0.00000 & 0.04485 & 0.00000 & 0.00848 & 0.06303 \end{bmatrix}$$

$$\boldsymbol{Z}_{\text{bus1}} = j\begin{bmatrix} 0.11189 & 0.07081 & 0.03405 & 0.04919 & 0.08162 \\ 0.07081 & 0.22781 & 0.05459 & 0.07886 & 0.10228 \\ 0.03405 & 0.05459 & 0.07297 & 0.06541 & 0.04919 \\ 0.04919 & 0.07886 & 0.06541 & 0.09447 & 0.07105 \\ 0.08162 & 0.10228 & 0.04919 & 0.07105 & 0.11790 \end{bmatrix}$$

$$\boldsymbol{Z}_{\text{bus2}} = j\begin{bmatrix} 0.11189 & 0.07081 & 0.03405 & 0.04919 & 0.08162 \\ 0.07081 & 0.22781 & 0.05459 & 0.07886 & 0.10228 \\ 0.03405 & 0.05459 & 0.07297 & 0.06541 & 0.04919 \\ 0.04919 & 0.07886 & 0.06541 & 0.09447 & 0.07105 \\ 0.08162 & 0.10228 & 0.04919 & 0.07105 & 0.11790 \end{bmatrix}$$

FAULT BUS	SINGLE LINE-TO-GROUND FAULT CURRENT (PHASE A) per unit/degrees	CONTRIBUTIONS TO FAULT CURRENT				
		GEN LINE OR TRSF	BUS-TO-BUS	CURRENT PHASE A per unit/degrees	PHASE B	PHASE C
1	11.505/−90.00	G 1	GRND–1	8.6032/ −90.00	1.4511/ −90.00	1.4511/ −90.00
		T 1	5–1	2.9023/ −90.00	1.4511/ 90.00	1.4511/ 90.00
2	3.534/−90.00	L 1	4–2	1.2878/ −90.00	0.0281/ 90.00	0.0281/ 90.00
		L 2	5–2	2.2461/ −90.00	0.0281/ −90.00	0.0281/ −90.00
3	16.076/−90.00	G 2	GRND–3	14.0483/ −90.00	1.0138/ −90.00	1.0138/ −90.00
		T 2	4–3	2.0276/ −90.00	1.0138/ 90.00	1.0138/ 90.00
4	14.018/−90.00	L 1	2–4	0.4356/ −90.00	0.1116/ 90.00	0.1116/ 90.00
		L 3	5–4	2.6138/ −90.00	0.6697/ 90.00	0.6697/ 90.00
		T 2	3–4	10.9688/ −90.00	0.7813/ −90.00	0.7813/ −90.00
5	10.541/−90.00	L 2	2–5	0.6552/ −90.00	0.1679/ 90.00	0.1679/ 90.00
		L 3	4–5	3.9310/ −90.00	1.0072/ 90.00	1.0072/ 90.00
		T 1	1–5	5.9551/ −90.00	1.1751/ −90.00	1.1751/. −90.00

Table 9.5 Fault currents for Example 9.8

$V_F = 1.05/0$		BUS VOLTAGES DURING FAULT		
FAULT BUS	BUS #	PHASE A	PHASE B per unit/degrees	PHASE C
1	1	0.0000/0.00	0.9537/252.45	0.9537/107.55
	2	0.6843/50.10	1.0500/−90.00	0.6843/129.90
	3	0.7888/0.00	0.9912/246.55	0.9912/113.45
	4	0.7842/42.02	1.0500/−90.00	0.7842/137.98
	5	0.6406/55.03	1.0500/−90.00	0.6406/124.97
2	1	0.9277/−34.47	0.9277/214.47	1.0500/90.00
	2	0.0000/0.00	1.1915/229.74	1.1915/130.26
	3	0.9552/−33.34	0.9552/213.34	1.0500/90.00
	4	0.8435/0.00	1.0158/243.53	1.0158/116.47
	5	0.7562/0.00	1.0179/243.30	1.0179/116.70
3	1	0.6850/0.00	0.9717/249.36	0.9717/110.64
	2	0.6616/52.52	1.0500/−90.00	0.6616/127.48
	3	0.0000/0.00	0.9942/246.16	0.9942/113.84
	4	0.6058/60.07	1.0500/−90.00	0.6058/119.93
	5	0.6933/49.22	1.0500/−90.00	0.6933/130.78
4	1	0.7328/−45.76	0.7328/225.76	1.0500/90.00
	2	0.2309/0.00	0.9401/255.30	0.9401/104.70
	3	0.6481/−54.10	0.6481/234.10	1.0500/90.00
	4	0.0000/0.00	0.9432/254.59	0.9432/105.41
	5	0.3463/0.00	0.9386/255.65	0.9386/104.35
5	1	0.6677/−51.84	0.6677/231.84	1.0500/90.00
	2	0.1736/0.00	0.9651/250.43	0.9651/109.57
	3	0.8048/−40.72	0.8048/220.72	1.0500/90.00
	4	0.5209/0.00	0.9592/251.45	0.9592/108.55
	5	0.0000/0.00	0.9681/249.93	0.9681/110.07

Table 9.6 Bus voltages for Example 9.8 ◁

Problems

Section 9.1

9.1 The single-line diagram of a three-phase power system is shown in Figure 9.16. Equipment ratings are given as follows:

Synchronous generators:

G_1	1000 MVA	15 kV	$X''_d = X_2 = 0.15$, $X_0 = 0.07$ per unit
G_2	1000 MVA	15 kV	$X''_d = X_2 = 0.25$, $X_0 = 0.10$ per unit
G_3	500 MVA	13.8 kV	$X''_d = X_2 = 0.20$, $X_0 = 0.05$ per unit
G_4	750 MVA	13.8 kV	$X''_d = 0.30$, $X_2 = 0.40$, $X_0 = 0.10$ per unit

Transformers:

T_1	1000 MVA	15 kV Δ/500 kV Y	X = 0.10 per unit
T_2	1000 MVA	15 kV Δ/500 kV Y	X = 0.10 per unit
T_3	500 MVA	15 kV Y/500 kV Y	X = 0.12 per unit
T_4	750 MVA	15 kV Y/500 kV Y	X = 0.11 per unit

Transmission lines:

1–2	500 kV	$X_1 = 50\,\Omega$, $X_0 = 150\,\Omega$
1–3	500 kV	$X_1 = 40\,\Omega$, $X_0 = 100\,\Omega$
2–3	500 kV	$X_1 = 40\,\Omega$, $X_0 = 100\,\Omega$

The inductor connected to generator 3 neutral has a reactance of 0.05 per unit using generator 3 ratings as a base. Draw the zero-, positive-, and negative-sequence reactance diagrams using a 1000-MVA, 500-kV base in the zone of line 1–2. Neglect the Δ–Y transformer phase shifts.

Figure 9.16 Problem 9.1

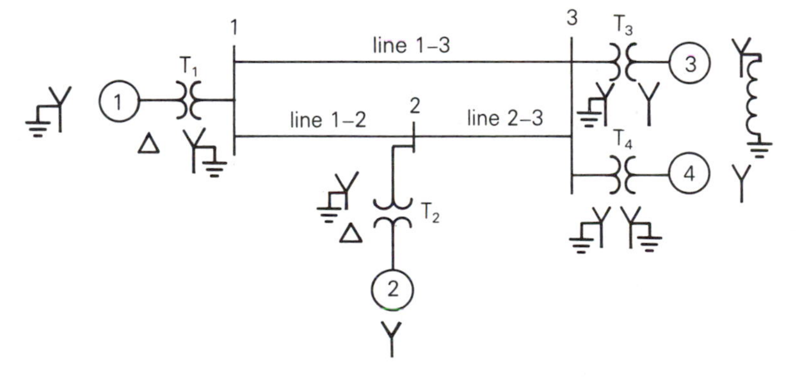

9.2 Faults at bus n in Problem 9.1 are of interest (the instructor selects $n = 1$, 2, or 3). Determine the Thévenin equivalent of each sequence network as viewed from the fault bus. Prefault voltage is 1.0 per unit. Prefault load currents and Δ–Y transformer phase shifts are neglected.
Hint: Use the Y–Δ conversion in Figure 2.21.

9.3 Determine the subtransient fault current in per-unit and in kA during a bolted three-phase fault at the fault bus selected in Problem 9.2.

Section 9.2–9.4

9.4 Determine the subtransient fault current in per-unit and in kA, as well as the per-unit line-to-ground voltages at the fault bus for a bolted single line-to-ground fault at the fault bus selected in Problem 9.2.

9.5 Repeat Problem 9.4 for a single line-to-ground arcing fault with arc impedance $Z_F = 15 + j0\,\Omega$.

9.6 Repeat Problem 9.4 for a bolted line-to-line fault.

9.7 Repeat Problem 9.4 for a bolted double line-to-ground fault.

9.8 Repeat Problems 9.1 and 9.4 including Δ–Y transformer phase shifts. Assume American standard phase shift. Also calculate the sequence components and phase components of the contribution to the fault current from generator n ($n = 1, 2,$ or 3 as specified by the instructor in Problem 9.2).

9.9 A 500-MVA, 13.8-kV synchronous generator with $X_d'' = X_2 = 0.20$ and $X_0 = 0.05$ per unit is connected to a 500-MVA, 13.8-kV Δ/500-kV Y transformer with 0.10 per-unit leakage reactance. The generator and transformer neutrals are solidly grounded. The generator is operated at no-load and rated voltage, and the high-voltage side of the transformer is disconnected from the power system. Compare the subtransient fault currents for the following bolted faults at the transformer high-voltage terminals: three-phase fault, single line-to-ground fault, line-to-line fault, and double line-to-ground fault.

9.10 As shown in Figure 9.17(a), two three-phase buses abc and $a'b'c'$ are interconnected by short circuits between phases b and b' and between c and c', with an open circuit between phases a and a'. The fault conditions in the phase domain are $I_a = I_{a'} = 0$ and $V_{bb'} = V_{cc'} = 0$. Determine the fault conditions in the sequence domain and verify the interconnection of the sequence networks as shown in Figure 9.14 for this one-conductor-open fault.

Figure 9.17

Problems 9.10 and 9.11: open conductor faults

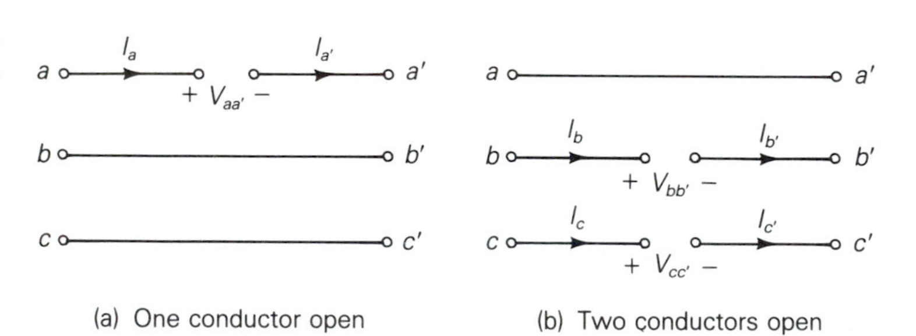

9.11 Repeat Problem 9.10 for the two-conductors-open fault shown in Figure 9.17(b). The fault conditions in the phase domain are

$$I_b = I_{b'} = I_c = I_{c'} = 0 \qquad \text{and} \qquad V_{aa'} = 0$$

Section 9.5

9.12 The zero-, positive-, and negative-sequence bus impedance matrices for a two-bus three-phase power system are

$$\mathbf{Z}_{\text{bus}\,0} = j\left[\begin{array}{c|c} 0.10 & 0 \\ \hline 0 & 0.10 \end{array}\right] \quad \text{per unit}$$

$$\mathbf{Z}_{\text{bus}\,1} = \mathbf{Z}_{\text{bus}\,2} = j\left[\begin{array}{c|c} 0.20 & 0.10 \\ \hline 0.10 & 0.30 \end{array}\right] \quad \text{per unit}$$

Determine the per-unit fault current and per-unit voltage at bus 2 for a bolted three-phase fault at bus 1. The prefault voltage is 1.0 per unit.

9.13 Repeat Problem 9.12 for a bolted single line-to-ground fault at bus 1.

9.14 Repeat Problem 9.12 for a bolted line-to-line fault at bus 1.

9.15 Repeat Problem 9.12 for a bolted double line-to-ground fault at bus 1.

9.16 Use the one-step-at-a-time method to find the 3×3 per-unit zero-, positive-, and negative-sequence bus impedance matrices for the power system given in Problem 9.1. Use a base of 1000 MVA and 500 kV in the zone of line 1–2.

9.17 Using the bus impedance matrices determined in Problem 9.16, verify the fault currents for the faults given in Problems 9.2 through 9.7.

Section 9.6

9.18 Rework Problem 9.16 using the SHORT CIRCUITS program.

9.19 Rework Problem 9.17 using the SHORT CIRCUITS program.

9.20 Run the SHORT CIRCUITS program for the power system given in Example 9.8 for a bolted double line-to-ground fault at bus 1, then bus 2, and so on to bus 5. Prefault voltage is 1.05 per unit. Compare these fault currents with the single line-to-ground faults computed in Example 9.8.

9.21 Rework Problem 9.20 with a new line installed between buses 4 and 5. This line has a positive-sequence series impedance $Z_{b1} = j0.4$ per unit and a zero-sequence series impedance $Z_{b0} = j1.2$ per unit. The new line is not mutually coupled to any other line. Are the fault currents larger or smaller than those of Problem 9.20?

References

1. Westinghouse Electric Corporation, *Electrical Transmission and Distribution Reference Book*, 4th ed. (East Pittsburgh, Pa., 1964).
2. Westinghouse Electric Corporation, *Applied Protective Relaying* (Newark, N.J., 1976).
3. P. M. Anderson, *Analysis of Faulted Power Systems* (Ames: Iowa State University Press, 1973).
4. J. R. Neuenswander, *Modern Power Systems* (New York: Intext Educational Publishers, 1971).
5. H. E. Brown, *Solution of Large Networks by Matrix Methods* (New York: John Wiley and Sons, 1975).
6. W. D. Stevenson, Jr., *Elements of Power System Analysis*, 4th ed. (New York: McGraw-Hill, 1982).
7. C. A. Gross, *Power System Analysis* (New York: John Wiley and Sons, 1979).

CHAPTER 10

Power-System Controls

Automatic control systems are used extensively in power systems. Local controls are employed at turbine-generator units and at selected voltage-controlled buses. Central controls are employed at area control centers.

Figure 10.1 shows two basic controls of a steam turbine-generator: the voltage regulator and turbine-governor. The voltage regulator adjusts the power output of the generator exciter in order to control the magnitude of generator terminal voltage V_t. When a reference voltage V_{ref} is raised (or lowered), the output voltage V_r of the regulator increases (or decreases) the exciter voltage E_{fd} applied to the generator field winding, which in turn acts to increase (or decrease) V_t. Also a potential transformer and rectifier monitor V_t, which is used as a feedback signal in the voltage regulator. If V_t decreases, the voltage regulator increases V_r to increase E_{fd}, which in turn acts to increase V_t.

The turbine-governor shown in Figure 10.1 adjusts the steam valve position to control the mechanical power output p_m of the turbine. When a reference power level p_{ref} is raised (or lowered), the governor moves the steam valve in the open (or close) direction to increase (or decrease) p_m. The governor also monitors rotor speed ω_m, which is used as a feedback signal to control the balance between p_m and the electrical power output p_e of the generator. Neglecting losses, if p_m is greater than p_e, ω_m increases, and the governor moves the steam valve in the close direction to reduce p_m. Similarly, if p_m is less than p_e, ω_m decreases, and the governor moves the valve in the open direction.

In addition to voltage regulators at generator buses, equipment is used to control voltage magnitudes at other selected buses. Tap-changing transformers, switched capacitor banks, and static var systems can be automatically regulated for rapid voltage control.

Central controls also play an important role in modern power systems. Today's systems are composed of interconnected areas, where each area has its own control center. There are many advantages to interconnections. For example, interconnected areas can share their reserve power to handle anticipated load peaks and unanticipated generator outages. Interconnected areas can also tolerate larger load changes with smaller frequency deviations than an isolated area.

Figure 10.2 shows how a typical area meets its daily load cycle. The base load is carried by base-loaded generators running at 100% of their rating for 24 hours. Nuclear units and large fossil-fuel units are typically base-loaded. The variable part of the load is

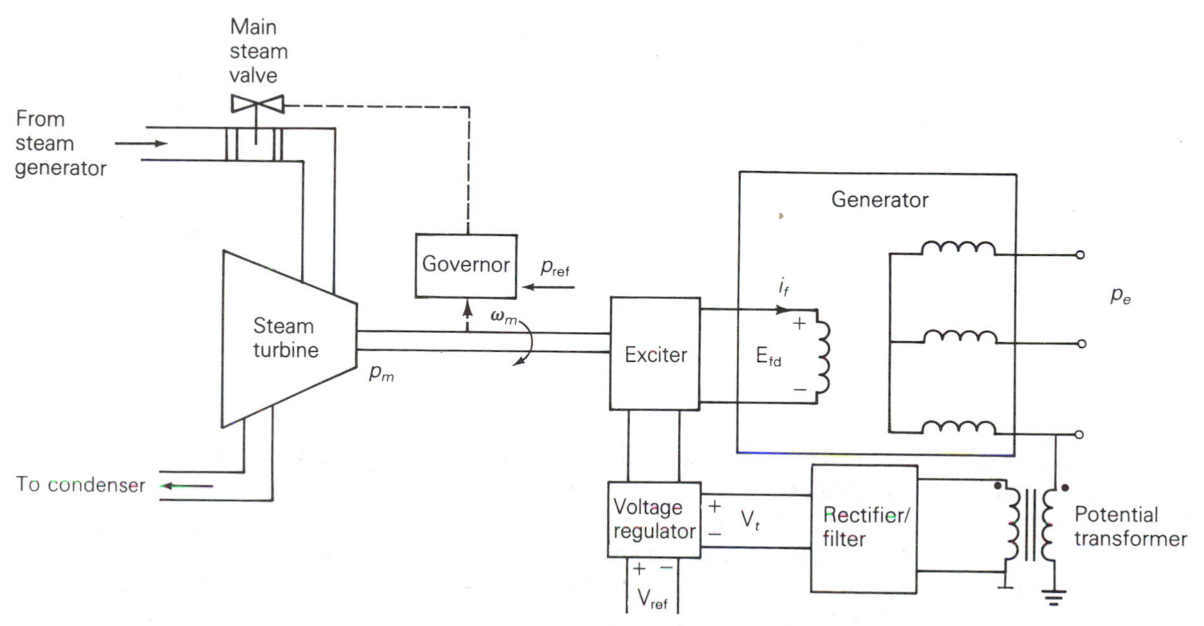

Figure 10.1 Voltage regulator and turbine-governor controls for a steam turbine-generator

carried by units that are controlled from the central control center. Medium-sized fossil-fuel units and hydro units are used for control. During peak load hours, smaller, less efficient units such as gas-turbine or diesel-generating units are employed. In addition, generators operating at partial output (with *spinning reserve*) and stand-by generators provide a reserve margin.

The central control center monitors information including area frequency, generating unit outputs, and tie-line power flows to interconnected areas. This information is used by automatic *load-frequency control* (LFC) in order to maintain area frequency at its scheduled value (60 Hz) and net tie-line power flow out of the area at its scheduled value. Raise and lower reference power signals are dispatched to the turbine-governors of controlled units.

Operating costs vary widely among controlled units. Larger units tend to be more efficient, but the varying cost of different fuels such as coal, oil, and gas is an important factor. *Economic dispatch* determines the megawatt outputs of the controlled units that minimize the total operating cost for a given load demand. Economic dispatch is coordinated with LFC such that reference power signals dispatched to controlled units move the units toward their economic loadings and satisfy LFC objectives.

In this chapter we investigate automatic controls employed in power systems under normal operation. Sections 10.1 and 10.2 describe the operation of the two generator controls: voltage and turbine-governor. Load-frequency control is discussed in Section 10.3 and economic dispatch is presented in Section 10.4.

Figure 10.2

Daily load cycle

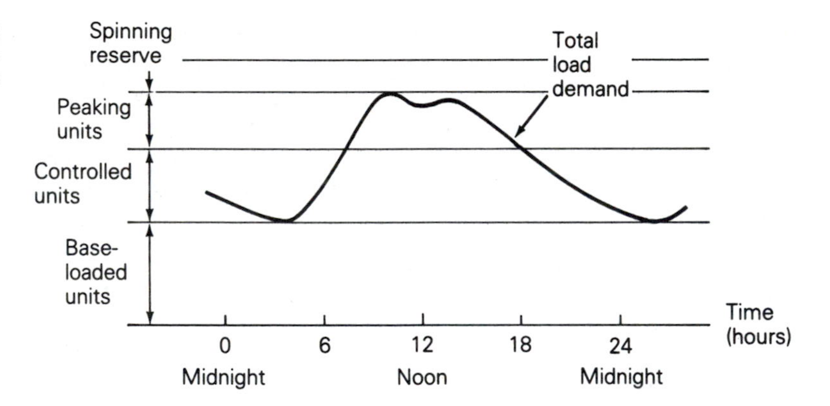

Section 10.1

Generator-Voltage Control

The *exciter* delivers dc power to the field winding on the rotor of a synchronous generator. For older generators, the exciter consists of a dc generator driven by the rotor. The dc power is transferred to the rotor via slip rings and brushes. For newer generators, *static* or *brushless* exciters are often employed.

For static exciters, ac power is obtained directly from the generator terminals or a nearby station service bus. The ac power is then rectified via thyristors and transferred to the rotor of the synchronous generator via slip rings and brushes.

For brushless exciters, ac power is obtained from an "inverted" synchronous generator whose three-phase armature windings are located on the main generator rotor and whose field winding is located on the stator. The ac power from the armature windings is rectified via diodes mounted on the rotor and is transferred directly to the field winding. For this design, slip rings and brushes are eliminated.

Block diagrams of several standard types of generator-voltage control systems have been developed by the IEEE Working Group on Exciters [1]. A simplified block diagram of generator-voltage control, similar to those given in [1], is shown in Figure 10.3. Nonlinearities due to exciter saturation and limits on exciter output are not shown in this figure.

Figure 10.3

Simplified block diagram—generator-voltage control

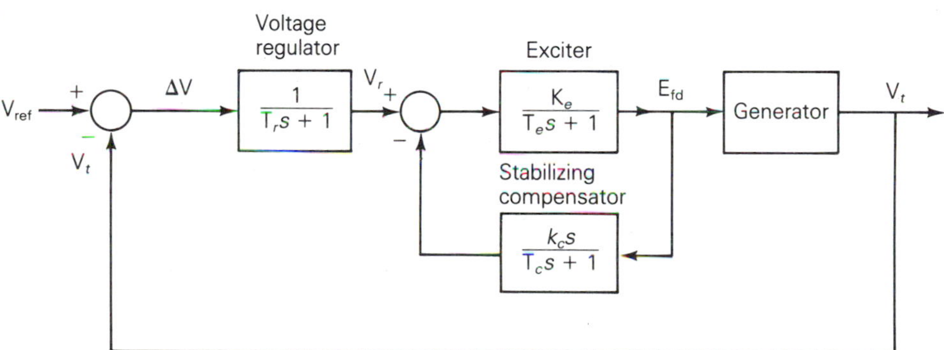

The generator terminal voltage V_t in Figure 10.3 is compared with a voltage reference V_{ref} to obtain a voltage error signal ΔV, which in turn is applied to the voltage regulator. The $1/(T_r s + 1)$ block accounts for voltage-regulator time delay, where s is the Laplace operator and T_r is the voltage-regulator time constant. Note that if a unit step is applied to a $1/(T_r s + 1)$ block, the output rises exponentially to unity with time constant T_r.

Neglecting the stabilizing compensator in Figure 10.3, the output V_r of the voltage regulator is applied to the exciter, which is represented by a $K_e/(T_e s + 1)$ block. The output of this exciter block is the field voltage E_{fd}, which is applied to the generator field winding and acts to adjust the generator terminal voltage. The generator block, which relates the effect of changes in E_{fd} to V_t, can be developed from synchronous-machine equations [2].

The stabilizing compensator shown in Figure 10.3 is used to improve the dynamic response of the exciter by reducing excessive overshoot. The compensator is represented by a $(K_c s)/(T_c s + 1)$ block, which provides a filtered first derivative. The input to this block is the exciter voltage E_{fd}, and the output is a stabilizing feedback signal that is subtracted from the regulator voltage V_r.

Block diagrams such as those shown in Figure 10.3 are used for computer representation of generator-voltage control in transient stability computer programs (see Chapter 12). In practice, high-gain, fast-responding exciters provide large, rapid increases in field voltage E_{fd} during short circuits at the generator terminals in order to improve transient stability after fault clearing. Equations represented in the block diagram can be used to compute the transient response of generator-voltage control.

Section 10.2

Turbine-Governor Control

Turbine-generator units operating in a power system contain stored kinetic energy due to their rotating masses. If the system load suddenly increases, stored kinetic energy is released to initially supply the load increase. Also, the electrical torque T_e of each turbine-generating unit increases to supply the load increase, while the mechanical torque T_m of the turbine initially remains constant. From Newton's second law, $J\alpha = T_m - T_e$, the acceleration α is therefore negative. That is, each turbine-generator decelerates and the rotor speed drops as kinetic energy is released to supply the load increase. The electrical frequency of each generator, which is proportional to rotor speed for synchronous machines, also drops.

From this, we conclude that either rotor speed or generator frequency indicates a balance or imbalance of generator electrical torque T_e and turbine mechanical torque T_m. If speed or frequency is decreasing, then T_e is greater than T_m (neglecting generator losses). Similarly, if speed or frequency is increasing, T_e is less than T_m. Accordingly, generator frequency is an appropriate control signal for governing the mechanical output power of the turbine.

The steady-state frequency–power relation for turbine-governor control is

$$\Delta p_m = \Delta p_{\text{ref}} - \frac{1}{R}\Delta f \tag{10.2.1}$$

where Δf is the change in frequency, Δp_m is the change in turbine mechanical power output, and Δp_{ref} is the change in a reference power setting. R is called the *regulation constant*. The equation is plotted in Figure 10.4 as a family of curves, with Δp_{ref} as a parameter. Note that when Δp_{ref} is fixed, Δp_m is directly proportional to the drop in frequency.

Figure 10.4 illustrates a steady-state frequency–power relation. When an electrical load change occurs, the turbine-generator rotor accelerates or decelerates, and frequency undergoes a transient disturbance. Under normal operating conditions, the rotor acceleration eventually becomes zero, and the frequency reaches a new steady-state, shown in the figure.

The regulation constant R in (10.2.1) is the negative of the slope of the

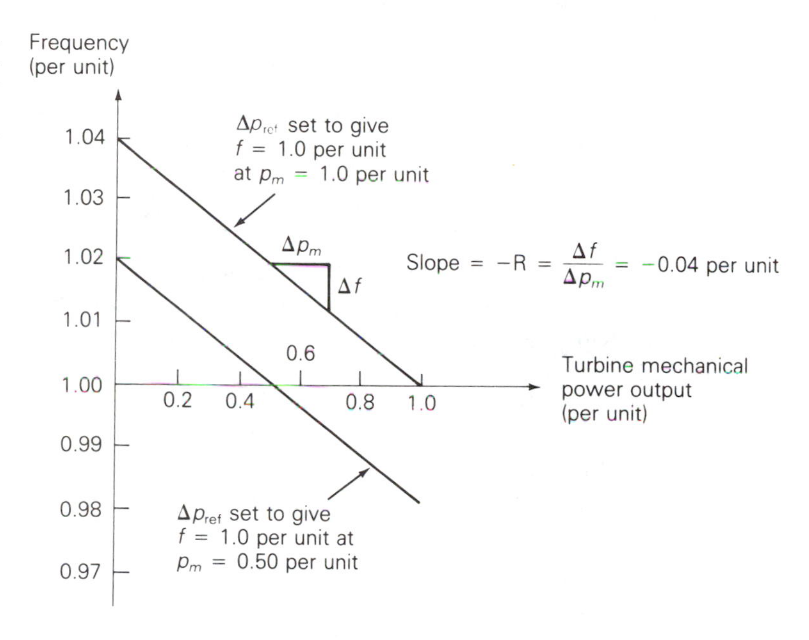

Figure 10.4 Steady-state frequency–power relation for a turbine-governor

Δf versus Δp_m curves shown in Figure 10.4. The units of R are Hz/MW when Δf is in Hz and Δp_m is in MW. When Δf and Δp_m are given in per-unit, however, R is also in per-unit.

Example 10.1 A 500-MVA, 60-Hz turbine-generator has a regulation constant R = 0.05 per unit based on its own rating. If the generator frequency increases by 0.01 Hz in steady-state, what is the decrease in turbine mechanical power output? Assume a fixed reference power setting.

Solution The per-unit change in frequency is

$$\Delta f_{\text{p.u.}} = \frac{\Delta f}{f_{\text{base}}} = \frac{0.01}{60} = 1.6667 \times 10^{-4} \quad \text{per unit}$$

Then, from (10.2.1), with $\Delta p_{\text{ref}} = 0$,

$$\Delta p_{m\text{p.u.}} = \left(\frac{-1}{0.05}\right)(1.6667 \times 10^{-4}) = -3.3333 \times 10^{-4} \quad \text{per unit}$$

$$\Delta p_m = (\Delta p_{m\text{p.u.}})\text{S}_{\text{base}} = (-3.3333 \times 10^{-4})(500) = -1.6667 \quad \text{MW}$$

The turbine mechanical power output decreases by 1.67 MW. ◁

The steady-state frequency–power relation for one area of an interconnected power system can be determined by summing (10.2.1) for each turbine-generating unit in the area. Noting that Δf is the same for each unit,

$$\begin{aligned}\Delta p_m &= \Delta p_{m1} + \Delta p_{m2} + \Delta p_{m3} + \cdots \\ &= (\Delta p_{\text{ref1}} + \Delta p_{\text{ref2}} + \cdots) - \left(\frac{1}{R_1} + \frac{1}{R_2} + \cdots\right)\Delta f \\ &= \Delta p_{\text{ref}} - \left(\frac{1}{R_1} + \frac{1}{R_2} + \cdots\right)\Delta f \end{aligned} \tag{10.2.2}$$

where Δp_m is the total change in turbine mechanical powers and Δp_{ref} is the total change in reference power settings within the area. We define the *area frequency response characteristic* β as

$$\beta = \left(\frac{1}{R_1} + \frac{1}{R_2} \cdots\right) \tag{10.2.3}$$

Using (10.2.3) in (10.2.2),

$$\Delta p_m = \Delta p_{\text{ref}} - \beta \Delta f \tag{10.2.4}$$

Equation (10.2.4) is the area steady-state frequency–power relation. The units of β are MW/Hz when Δf is in Hz and Δp_m is in MW. β can also be given in per-unit. In practice, β is somewhat higher than that given by (10.2.3) due to system losses and the frequency dependence of loads.

A standard figure for the regulation constant is R = 0.05 per unit. When all turbine-generating units have the same per-unit value of R based on their own ratings, then each unit shares total power changes in proportion to its own ratings. This desirable feature is illustrated by the following example.

Example 10.2

One area of an interconnected 60-Hz power system has three turbine-generator units rated 1000, 750, and 500 MVA, respectively. The regulation constant of each unit is R = 0.05 per unit based on its own rating. Each unit is initially operating at one-half of its own rating, when the system load suddenly increases by 200 MW. Determine (a) the per-unit area frequency response characteristic β on a 1000 MVA system base; (b) the steady-state drop in area frequency, and (c) the increase in turbine mechanical power output of each unit. Assume that the reference power setting of each turbine-generator remains constant. Neglect losses and the dependence of load on frequency.

Solution

a. The regulation constants are converted to per-unit on the system base using

$$R_{\text{p.u.new}} = R_{\text{p.u.old}} \frac{S_{\text{base(new)}}}{S_{\text{base(old)}}}$$

We obtain

$$R_{1\text{p.u.new}} = R_{1\text{p.u.old}} = 0.05$$

$$R_{2\text{p.u.new}} = (0.05)\left(\frac{1000}{750}\right) = 0.06667$$

$$R_{3\text{p.u.new}} = (0.05)\left(\frac{1000}{500}\right) = 0.10 \quad \text{per unit}$$

Using (10.2.3),

$$\beta = \frac{1}{R_1} + \frac{1}{R_2} + \frac{1}{R_3} = \frac{1}{0.05} + \frac{1}{0.06667} + \frac{1}{0.10} = 45.0 \quad \text{per unit}$$

b. Neglecting losses and dependence of load on frequency, the steady-state increase in total turbine mechanical power equals the load increase, 200 MW or 0.20 per unit. Using (10.2.4) with $\Delta p_{ref} = 0$,

$$\Delta f = \left(\frac{-1}{\beta}\right)\Delta p_m = \left(\frac{-1}{45}\right)(0.20) = -4.444 \times 10^{-3} \quad \text{per unit}$$

$$= (-4.444 \times 10^{-3})(60) = -0.2667 \quad \text{Hz}$$

The steady-state frequency drop is 0.2667 Hz.

c. From (10.2.1), using $\Delta f = -4.444 \times 10^{-3}$ per unit,

$$\Delta p_{m1} = \left(\frac{-1}{0.05}\right)(-4.444 \times 10^{-3}) = 0.08888 \quad \text{per unit}$$

$$= 88.88 \quad \text{MW}$$

$$\Delta p_{m2} = \left(\frac{-1}{0.06667}\right)(-4.444 \times 10^{-3}) = 0.06666 \quad \text{per unit}$$

$$= 66.66 \quad \text{MW}$$

$$\Delta p_{m3} = \left(\frac{-1}{0.10}\right)(-4.444 \times 10^{-3}) = 0.04444 \quad \text{per unit}$$

$$= 44.44 \quad \text{MW}$$

◁

Note that unit 1, whose MVA rating is $33\frac{1}{3}\%$ larger than that of unit 2 and 100% larger than that of unit 3, picks up $33\frac{1}{3}\%$ more load than unit 2 and 100% more load than unit 3. That is, each unit shares the total load change in proportion to its own rating.

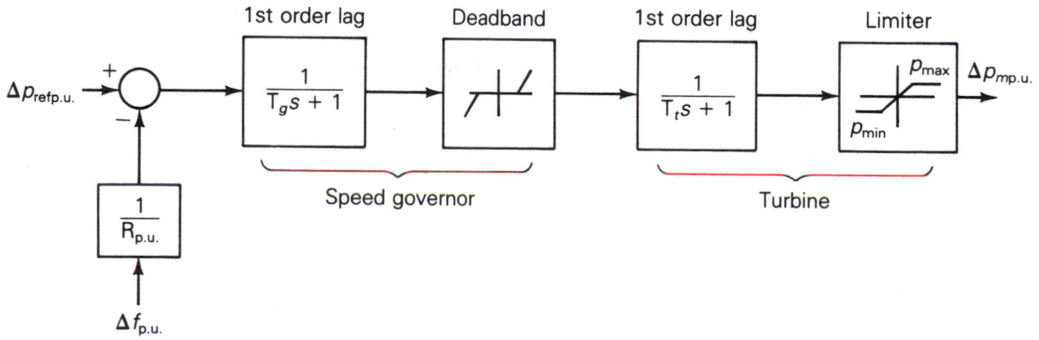

Figure 10.5 Turbine-governor block diagram

Figure 10.5 shows a block diagram of a nonreheat steam turbine-governor, which includes nonlinearities and time delays that were not included in (10.2.1). The deadband block in this figure accounts for the fact that speed governors do not respond to changes in frequency or to reference power settings that are smaller than a specified value. The limiter block accounts for the fact that turbines have minimum and maximum outputs. The $1/(\mathrm{T}s + 1)$ blocks account for time delays, where s is the Laplace operator and T is a time constant. Typical values are $\mathrm{T}_g = 0.10$ and $\mathrm{T}_t = 1.0$ seconds. Block diagrams for steam turbine-governors with reheat and hydro turbine-governors are also available [3].

Section 10.3

Load-Frequency Control

As shown in Section 10.2, turbine-governor control eliminates rotor accelerations and decelerations following load changes during normal operation. However, there is a steady-state frequency error Δf when the change in turbine-governor reference setting Δp_{ref} is zero. One of the objectives of load-frequency control (LFC), therefore, is to return Δf to zero.

In a power system consisting of interconnected areas, each area agrees to export or import a scheduled amount of power through transmission-line interconnections, or tie-lines, to its neighboring areas. Thus a second LFC objective is to have each area absorb its own load changes during normal operation. This objective is achieved by maintaining the net tie-line power flow out of each area at its scheduled value.

The following summarizes the two basic LFC objectives for an interconnected power system:

1. Following a load change, each area should assist in returning the steady-state frequency error Δf to zero.
2. Each area should maintain the net tie-line power flow out of the area at its scheduled value, in order for the area to absorb its own load changes.

The following control strategy developed by N. Cohn [4] meets these LFC objectives. We first define the *area control error* (ACE) as follows:

$$\begin{aligned} \mathrm{ACE} &= (p_{\mathrm{tie}} - p_{\mathrm{tie,sched}}) + \mathrm{B}_f(f - 60) \\ &= \Delta p_{\mathrm{tie}} + \mathrm{B}_f \Delta f \qquad (10.3.1) \end{aligned}$$

where Δp_{tie} is the deviation in net tie-line power flow out of the area from its scheduled value $p_{\mathrm{tie,sched}}$, and Δf is the deviation of area frequency from its scheduled value (60 Hz). Thus the ACE for each area consists of a linear combination of tie-line error Δp_{tie} and frequency error Δf. The constant B_f is called a *frequency bias constant.*

The change in reference power setting $\Delta p_{\text{ref}i}$ of each turbine-governor operating under LFC is proportional to the integral of the area control error. That is,

$$\Delta p_{\text{ref}i} = -K_i \int \text{ACE}\, dt \tag{10.3.2}$$

Each area monitors its own tie-line power flows and frequency at the area control center. The ACE given by (10.3.1) is computed and a percentage of the ACE is allocated to each controlled turbine-generator unit. Raise or lower commands are dispatched to the turbine-governors at discrete time intervals of two or more seconds in order to adjust the reference power settings. As the commands accumulate, the integral action in (10.3.2) is achieved.

The constant K_i in (10.3.2) is an integrator gain. The minus sign in (10.3.2) indicates that if either the net tie-line power flow out of the area or the area frequency is low—that is, if the ACE is negative—then the area should increase its generation.

When a load change occurs in any area, a new steady-state operation can be obtained only after the power output of every turbine-generating unit in the interconnected system reaches a constant value. This occurs only when all reference power settings are zero, which in turn occurs only when the ACE of every area is zero. Furthermore, the ACE is zero in every area only when both Δp_{tie} and Δf are zero. Therefore, in steady-state, both LFC objectives are satisfied.

Example 10.3

As shown in Figure 10.6, a 60-Hz power system consists of two interconnected areas. Area 1 has 2000 MW of total generation and an area frequency response characteristic $\beta_1 = 700$ MW/Hz. Area 2 has 4000 MW of total generation and $\beta_2 = 1400$ MW/Hz. Each area is initially generating one-half of its total generation, at $\Delta p_{\text{tie}1} = \Delta p_{\text{tie}2} = 0$ and at 60 Hz when the load in area 1 suddenly increases by 100 MW. Determine the steady-state frequency error Δf and the steady-state tie-line error Δp_{tie} of each area for the following two cases: (a) without LFC, and (b) with LFC given by (10.3.1) and (10.3.2). Neglect losses and the dependence of load on frequency.

Figure 10.6

Example 10.3

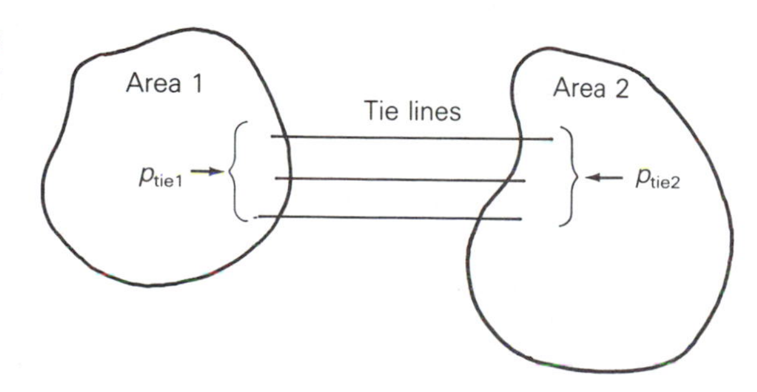

Solution **a.** Since the two areas are interconnected, the steady-state frequency error Δf is the same for both areas. Adding (10.2.4) for each area,

$$(\Delta p_{m1} + \Delta p_{m2}) = (\Delta p_{\text{ref}1} + \Delta p_{\text{ref}2}) - (\beta_1 + \beta_2)\Delta f$$

Neglecting losses and the dependence of load on frequency, the steady-state increase in total mechanical power of both areas equals the load increase, 100 MW. Also, without LFC, $\Delta p_{\text{ref}1}$ and $\Delta p_{\text{ref}2}$ are both zero. The above equation then becomes

$$100 = -(\beta_1 + \beta_2)\Delta f = -(700 + 1400)\Delta f$$

$$\Delta f = -100/2100 = -0.0476 \quad \text{Hz}$$

Next, using (10.2.4) for each area, with $\Delta p_{\text{ref}} = 0$,

$$\Delta p_{m1} = -\beta_1 \Delta f = -(700)(-0.0476) = 33.33 \quad \text{MW}$$

$$\Delta p_{m2} = -\beta_2 \Delta f = -(1400)(-0.0476) = 66.67 \quad \text{MW}$$

In response to the 100-MW load increase in area 1, area 1 picks up 33.33 MW and area 2 picks up 66.67 MW of generation. The 66.67-MW increase in area 2 generation is transferred to area 1 through the tie-lines. Therefore, the change in net tie-line power flow out of each area is

$$\Delta p_{\text{tie}2} = +66.67 \quad \text{MW}$$

$$\Delta p_{\text{tie}1} = -66.67 \quad \text{MW}$$

b. From (10.3.1), the area control error for each area is

$$\text{ACE}_1 = \Delta p_{\text{tie}1} + \text{B}_1 \Delta f_1$$

$$\text{ACE}_2 = \Delta p_{\text{tie}2} + \text{B}_2 \Delta f_2$$

Neglecting losses, the sum of the net tie-line flows must be zero; that is, $\Delta p_{\text{tie}1} + \Delta p_{\text{tie}2} = 0$ or $\Delta p_{\text{tie}2} = -\Delta p_{\text{tie}1}$. Also, in steady-state $\Delta f_1 = \Delta f_2 = \Delta f$. Using these relations in the above equations,

$$\text{ACE}_1 = \Delta p_{\text{tie}1} + \text{B}_1 \Delta f$$

$$\text{ACE}_2 = -\Delta p_{\text{tie}1} + \text{B}_2 \Delta f$$

In steady-state, $\text{ACE}_1 = \text{ACE}_2 = 0$; otherwise, the LFC given by (10.3.2) would be changing the reference power settings of turbine-governors on LFC. Adding the above two equations,

$$\text{ACE}_1 + \text{ACE}_2 = 0 = (\text{B}_1 + \text{B}_2)\Delta f$$

Therefore, $\Delta f = 0$ and $\Delta p_{\text{tie}1} = \Delta p_{\text{tie}2} = 0$. That is, in steady-state the frequency error is returned to zero, area 1 picks up its own 100-MW load increase, and area 2 returns to its original operating condition—that is, the condition before the load increase occurred.

We note that turbine-governor controls act almost instantaneously, subject only to the time delays shown in Figure 10.5. However, LFC acts more slowly. LFC raise and lower signals are dispatched from the area control center to turbine-governors at discrete-time intervals of 2 or more seconds. Also, it takes time for the raise or lower signals to accumulate. Thus case (a) represents the first action. Turbine-governors in both areas rapidly respond to the load increase in area 1 in order to stabilize the frequency drop. Case (b) represents the second action. As LFC signals are dispatched to turbine-governors, Δf and Δp_{tie} are slowly returned to zero. ◁

The choice of the B_f and K_i constants in (10.3.1) and (10.3.2) affects the transient response to load changes—for example, the speed and stability of the response. The frequency bias B_f should be high enough such that each area adequately contributes to frequency control. Cohn [4] has shown that choosing B_f equal to the area frequency response characteristic, $B_f = \beta$, gives satisfactory performance of the interconnected system. The integrator gain K_i should not be too high, otherwise instability may result. Also, the time interval at which LFC signals are dispatched, two or more seconds, should be long enough so that LFC does not attempt to follow random or spurious load changes. A detailed investigation of the effect of B_f, K_i and LFC time interval on the transient response of LFC and turbine-governor controls is beyond the scope of this text.

Two additional LFC objectives are to return the integral of frequency error and the integral of net tie-line error to zero in steady-state. By meeting these objectives, LFC controls both the time of clocks that are driven by 60-Hz motors and energy transfers out of each area. These two objectives are achieved by making temporary changes in the frequency schedule and tie-line schedule in (10.3.1).

Finally, note that LFC maintains control during normal changes in load and frequency—that is, changes that are not too large. During emergencies, when large imbalances between generation and load occur, LFC is bypassed and other, emergency controls are applied.

Section 10.4

Economic Dispatch

Section 10.3 describes how LFC adjusts the reference power settings of turbine-governors in an area to control frequency and net tie-line power flow out of the area. This section describes how the real power output of each controlled generating unit in an area is selected to meet a given load and to minimize the total operating costs in the area. This is the *economic dispatch* problem [5].

We begin this section by considering an area with only fossil-fuel generating units, with no constraints on maximum and minimum generator outputs, and with no transmission losses. The economic dispatch problem is first solved for this idealized case. Then we include inequality constraints on generator outputs; then we include transmission losses. Next we discuss the coordination of economic dispatch with LFC. Finally, we briefly discuss the dispatch of other types of units including nuclear, pumped-storage hydro, and hydro units.

Fossil-Fuel Units, No Inequality Constraints, No Transmission Losses

Figure 10.7 shows the operating cost C_i of a fossil-fuel generating unit versus real power output P_i. Fuel cost is the major portion of the variable cost of operation, although other variable costs, such as maintenance, could have been included in the figure. Fixed costs, such as the capital cost of installing the unit, are not included. Only those costs that are a function of unit power output—that is, those costs that can be controlled by operating strategy—enter into the economic dispatch formulation.

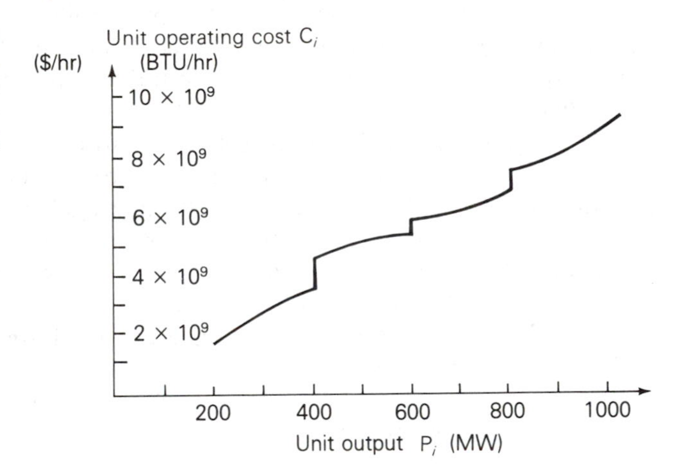

Figure 10.7

Unit operating cost versus real power output—fossil-fuel generating unit

In practice, C_i is constructed of piecewise continuous functions valid for ranges of output P_i, based on empirical data. The discontinuities in Figure 10.7 may be due to the firing of equipment such as additional boilers or condensers as power output is increased. It is often convenient to express C_i in terms of BTU/hr, which is relatively constant over the lifetime of the unit, rather than $/hr, which can change monthly or daily. C_i can be converted to $/hr by multiplying the fuel input in BTU/hr by the cost of fuel in $/BTU.

Figure 10.8 shows the unit incremental operating cost dC_i/dP_i versus unit output P_i, which is the slope or derivative of the C_i versus P_i curve in Figure 10.7. When C_i consists of only fuel costs, dC_i/dP_i is the ratio of the fuel energy input in BTU to energy output in kWhr, which is called *heat rate*. Note that the reciprocal of the heat rate, which is the ratio of output energy to input energy, gives a measure of fuel efficiency for the unit. For the unit shown in Figure 10.8, maximum efficiency occurs at $P_i = 600$ MW, where the heat rate is 9000 BTU/kWhr. The efficiency at this output is

$$\text{percent efficiency} = \left(\frac{1}{9000}\frac{\text{kWhr}}{\text{BTU}}\right)\left(3413\frac{\text{BTU}}{\text{kWhr}}\right) \times 100 = 37.92\%$$

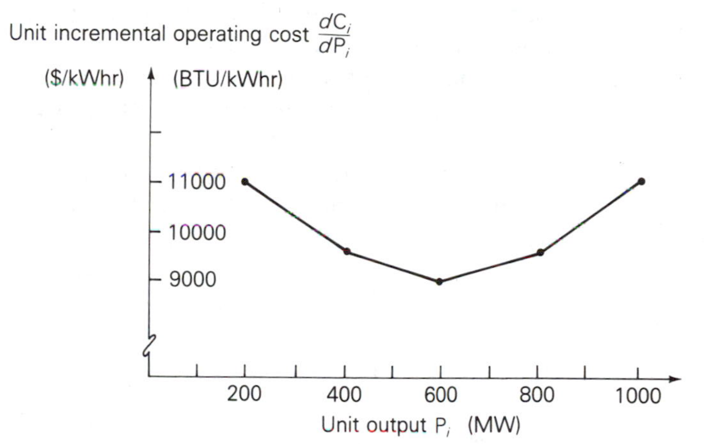

Figure 10.8

Unit incremental operating cost versus real power output—fossil-fuel generating unit

The dC_i/dP_i curve in Figure 10.8 is also represented by piecewise continuous functions valid for ranges of output P_i. For analytical work, the actual curves are often approximated by straight lines. The ratio dC_i/dP_i can also be converted to $/kWhr by multiplying the heat rate in BTU/kWhr by the cost of fuel in $/BTU.

For the area of an interconnected power system consisting of N units operating on economic dispatch, the total variable cost C_T of operating these units is

$$C_T = \sum_{i=i}^{N} C_i$$

$$= C_1(P_1) + C_2(P_2) + \cdots + C_N(P_N) \quad \$/\text{hr} \tag{10.4.1}$$

where C_i, expressed in $/hr, includes fuel cost as well as any other variable costs of unit i. Let P_T equal the total load demand in the area. Neglecting transmission losses,

$$P_1 + P_2 + \cdots + P_N = P_T \tag{10.4.2}$$

Due to relatively slow changes in load demand, P_T may be considered constant for periods of 2 to 10 minutes. The economic dispatch problem can be stated as follows:

Find the values of unit outputs $P_1, P_2, \cdots, P_N$ that minimize C_T given by (10.4.1), subject to the equality constraint given by (10.4.2).

A criterion for the solution to this problem is: All units on economic dispatch should operate at equal incremental operating cost. That is,

$$\frac{dC_1}{dP_1} = \frac{dC_2}{dP_2} = \cdots = \frac{dC_N}{dP_N} \tag{10.4.3}$$

An intuitive explanation of this criterion is the following. Suppose one unit is operating at a higher incremental operating cost than the other units. If the output power of that unit is reduced and transferred to units with lower incremental operating costs, then the total operating cost C_T decreases. That is, reducing the output of the unit with the *higher* incremental cost results in a *greater cost decrease* than the cost increase of adding that same output reduction to units with lower incremental costs. Therefore, all units must operate at the same incremental operating cost (the economic dispatch criterion).

A mathematical solution to the economic dispatch problem can also be given. The minimum value of C_T occurs when the total differential dC_T is zero. That is,

$$dC_T = \frac{\partial C_T}{\partial P_1} dP_1 + \frac{\partial C_T}{\partial P_2} dP_2 + \cdots + \frac{\partial C_T}{\partial P_N} dP_N = 0 \tag{10.4.4}$$

Using (10.4.1), (10.4.4) becomes

$$dC_T = \frac{dC_1}{dP_1} dP_1 + \frac{dC_2}{dP_2} dP_2 + \cdots + \frac{dC_N}{dP_N} dP_N = 0 \tag{10.4.5}$$

Also, assuming P_T is constant, the differential of (10.4.2) is

$$dP_1 + dP_2 + \cdots + dP_N = 0 \tag{10.4.6}$$

Multiplying (10.4.6) by λ and subtracting the resulting equation from (10.4.5),

$$\left(\frac{dC_1}{dP_1} - \lambda\right) dP_1 + \left(\frac{dC_2}{dP_2} - \lambda\right) dP_2 + \cdots + \left(\frac{dC_N}{dP_N} - \lambda\right) dP_N = 0 \tag{10.4.7}$$

Equation (10.4.7) is satisfied when each term in parentheses equals zero. That is,

$$\frac{dC_1}{dP_1} = \frac{dC_2}{dP_2} = \cdots = \frac{dC_N}{dP_N} = \lambda \tag{10.4.8}$$

Therefore, all units have the same incremental operating cost, denoted here by λ, in order to minimize the total operating cost C_T.

Example 10.4

An area of an interconnected power system has two fossil-fuel units operating on economic dispatch. The variable operating costs of these units are given by

$$C_1 = 10P_1 + 8 \times 10^{-3} P_1^2 \quad \$/\text{hr}$$

$$C_2 = 8P_2 + 9 \times 10^{-3} P_2^2 \quad \$/\text{hr}$$

where P_1 and P_2 are in megawatts. Determine the power output of each unit, the incremental operating cost, and the total operating cost C_T that minimizes C_T as the total load demand P_T varies from 500 to 1500 MW. Generating unit inequality constraints and transmission losses are neglected.

Solution The incremental operating costs of the units are

$$\frac{dC_1}{dP_1} = 10 + 16 \times 10^{-3}P_1 \quad \$/\text{MWhr}$$

$$\frac{dC_2}{dP_2} = 8 + 18 \times 10^{-3}P_2 \quad \$/\text{MWhr}$$

Using (10.4.8), the minimum total operating cost occurs when

$$\frac{dC_1}{dP_1} = 10 + 16 \times 10^{-3}P_1 = \frac{dC_2}{dP_2} = 8 + 18 \times 10^{-3}P_2$$

Using $P_2 = P_T - P_1$, the preceding equation becomes

$$10 + 16 \times 10^{-3}P_1 = 8 + 18 \times 10^{-3}(P_T - P_1)$$

Solving for P_1,

$$P_1 = \frac{18 \times 10^{-3}P_T - 2}{34 \times 10^{-3}} = 0.5294P_T - 58.82 \quad \text{MW}$$

Also, the incremental operating cost when C_T is minimized is

$$\frac{dC_2}{dP_2} = \frac{dC_1}{dP_1} = 10 + 16 \times 10^{-3}P_1 = 10 + 16 \times 10^{-3}(0.5294P_T - 58.82)$$

$$= 9.0589 + 8.4704 \times 10^{-3}P_T \quad \$/\text{MWhr}$$

and the minimum total operating cost is

$$C_T = C_1 + C_2 = (10P_1 + 8 \times 10^{-3}P_1^2) + (8P_2 + 9 \times 10^{-3}P_2^2) \quad \$/\text{hr}$$

The economic dispatch solution is shown in Table 10.1 for values of P_T from 500 to 1500 MW.

Table 10.1

Economic dispatch solution for Example 10.4

P_T MW	P_1 MW	P_2 MW	dC_1/dP_1 \$/MWhr	C_T \$/hr
500	206	294	13.29	5529
600	259	341	14.14	6901
700	312	388	14.99	8358
800	365	435	15.84	9899
900	418	482	16.68	11525
1000	471	529	17.53	13235
1100	524	576	18.38	15030
1200	576	624	19.22	16910
1300	629	671	20.07	18875
1400	682	718	20.92	20924
1500	735	765	21.76	23058

◁

Effect of Inequality Constraints

Each generating unit must not operate above its rating or below some minimum value. That is,

$$P_{i\min} < P_i < P_{i\max} \qquad i = 1, 2, \ldots, N \tag{10.4.9}$$

Other inequality constraints may also be included in the economic dispatch problem. For example, some unit outputs may be restricted so that certain transmission lines or other equipments are not overloaded. Also, under adverse weather conditions, generation at some units may be limited to reduce emissions.

When inequality constraints are included, we modify the economic dispatch solution as follows. If one or more units reach their limit values, then these units are held at their limits, and the remaining units operate at equal incremental operating cost λ. The incremental operating cost of the area equals the common λ for the units that are not at their limits.

Example 10.5 Rework Example 10.4 if the units are subject to the following inequality constraints:

$$100 \leqq P_1 \leqq 600 \quad \text{MW}$$

$$400 \leqq P_2 \leqq 1000 \quad \text{MW}$$

Solution At light loads, unit 2 operates at its lower limit of 400 MW, where its incremental operating cost is $dC_2/dP_2 = 15.2$ \$/MWhr. Additional load comes from unit 1 until $dC_1/dP_1 = 15.2$ \$/MWhr, or

$$\frac{dC_1}{dP_1} = 10 + 16 \times 10^{-3} P_1 = 15.2$$

$$P_1 = 325 \quad \text{MW}$$

For P_T less than 725 MW, where P_1 is less than 325 MW, the incremental operating cost of the area is determined by unit 1 alone.

At heavy loads, unit 1 operates at its upper limit of 600 MW, where its incremental operating cost is $dC_1/dP_1 = 19.60$ \$/MWhr. Additional load comes from unit 2 for all values of dC_2/dP_2 greater than 19.60 \$/MWhr. At $dC_2/dP_2 = 19.60$ \$/MWhr,

$$\frac{dC_2}{dP_2} = 8 + 18 \times 10^{-3} P_2 = 19.60$$

$$P_2 = 644 \quad \text{MW}$$

For P_T greater than 1244 MW, where P_2 is greater than 644 MW, the incremental operating cost of the area is determined by unit 2 alone.

For $725 < P_T < 1244$ MW, neither unit has reached a limit value, and the economic dispatch solution is the same as that given in Table 10.1.

The solution to this example is summarized in Table 10.2 for values of P_T from 500 to 1500 MW.

Table 10.2

Economic dispatch solution for Example 10.5

P_T MW	P_1 MW	P_2 MW	dC/dP \$/MWhr	C_T \$/hr
500	100	400	$\frac{dC_1}{dP_1}$ 11.60	5720
600	200	400	13.20	6960
700	300	400	14.80	8360
725	325	400	15.20	8735
800	365	435	15.84	9899
900	418	482	16.68	11525
1000	471	529	17.53	13235
1100	524	576	18.38	15030
1200	576	624	19.22	16910
1244	600	644	$\frac{dC_2}{dP_2}$ 19.60	17765
1300	600	700	20.60	18890
1400	600	800	22.40	21040
1500	600	900	24.20	23370

◁

Effect of Transmission Losses

Although one unit may be very efficient with a low incremental operating cost, it may also be located far from the load center. The transmission losses associated with this unit may be so high that the economic dispatch solution requires the unit to decrease its output, while other units with higher incremental operating costs but lower transmission losses increase their outputs.

When transmission losses are included in the economic dispatch problem, (10.4.2) becomes

$$P_1 + P_2 + \cdots + P_N - P_L = P_T \tag{10.4.10}$$

where P_T is the total load demand and P_L is the total transmission loss in the area. In general, P_L is not constant, but depends on the unit outputs $P_1, P_2, \ldots, P_N$. The total differential of (10.4.10) is

$$(dP_1 + dP_2 + \cdots + dP_N) - \left(\frac{\partial P_L}{\partial P_1} dP_1 + \frac{\partial P_L}{\partial P_2} dP_2 + \cdots + \frac{\partial P_L}{\partial P_N} dP_N\right) = 0 \tag{10.4.11}$$

Multiplying (10.4.11) by λ and subtracting the resulting equation from (10.4.5),

$$\left(\frac{dC_1}{dP_1} + \lambda\frac{\partial P_L}{\partial P_1} - \lambda\right) dP_1 + \left(\frac{dC_2}{dP_2} + \lambda\frac{\partial P_L}{\partial P_2} - \lambda\right) dP_2 + \cdots + \left(\frac{dC_N}{dP_N} + \lambda\frac{\partial P_L}{\partial P_N} - \lambda\right) dP_N = 0 \tag{10.4.12}$$

Equation (10.4.12) is satisfied when each term in parentheses equals zero. That is,

$$\frac{dC_i}{dP_i} + \lambda\frac{\partial P_L}{\partial P_i} - \lambda = 0$$

or

$$\lambda = \frac{dC_i}{dP_i}(L_i) = \frac{dC_i}{dP_i}\left(\frac{1}{1 - \frac{\partial P_L}{\partial P_i}}\right) \qquad i = 1, 2, \ldots, N \tag{10.4.13}$$

Equation (10.4.13) gives the economic dispatch criterion, including transmission losses. Each unit that is not at a limit value operates such that its incremental operating cost dC_i/dP_i multiplied by the *penalty factor* L_i is the same. Note that when transmission losses are negligible, $\partial P_L/\partial P_i = 0$, $L_i = 1$, and (10.4.13) reduces to (10.4.8).

Example 10.6 Total transmission losses for the power-system area given in Example 10.5 are given by

$$P_L = 1.5 \times 10^{-4} P_1^2 + 2 \times 10^{-5} P_1 P_2 + 3 \times 10^{-5} P_2^2 \quad \text{MW}$$

where P_1 and P_2 are given in megawatts. Determine the output of each unit, total transmission losses, total load demand, and total operating cost C_T when the area $\lambda = 16.00$ \$/MWhr.

Solution Using the incremental operating costs from Example 10.4 in (10.4.13),

$$\frac{dC_1}{dP_1}\left(\frac{1}{1 - \frac{\partial P_L}{\partial P_1}}\right) = \frac{10 + 16 \times 10^{-3} P_1}{1 - (3 \times 10^{-4} P_1 + 2 \times 10^{-5} P_2)} = 16.00$$

$$\frac{dC_2}{dP_2}\left(\frac{1}{1 - \frac{\partial P_L}{\partial P_2}}\right) = \frac{8 + 18 \times 10^{-3} P_2}{1 - (6 \times 10^{-5} P_2 + 2 \times 10^{-5} P_1)} = 16.00$$

Rearranging the above two equations,

$$20.8 \times 10^{-3} P_1 + 32 \times 10^{-5} P_2 = 6.00$$

$$32 \times 10^{-5} P_1 + 18.96 \times 10^{-3} P_2 = 8.00$$

Solving,

$$P_1 = 282 \quad \text{MW} \qquad P_2 = 417 \quad \text{MW}$$

Using the equation for total transmission losses,

$$P_L = 1.5 \times 10^{-4}(282)^2 + 2 \times 10^{-5}(282)(417) + 3 \times 10^{-5}(417)^2$$

$$= 19.5 \quad \text{MW}$$

From (10.4.10), the total load demand is

$$P_T = P_1 + P_2 - P_L = 282 + 417 - 19.5 = 679.5 \quad \text{MW}$$

Also, using the cost formulas given in Example 10.4, the total operating cost is

$$C_T = C_1 + C_2 = 10(282) + 8 \times 10^{-3}(282)^2 + 8(417) + 9 \times 10^{-3}(417)^2$$
$$= 8357 \quad \$/\text{hr}$$

Note that when transmission losses are included, λ given by (10.4.13) is no longer the incremental operating cost of the area. Instead, λ is the unit incremental operating cost dC_i/dP_i multiplied by the unit penalty factor L_i.

◁

In Example 10.6, total transmission losses are expressed as a quadratic function of unit output powers. For an area with N units, this formula generalizes to

$$P_L = \sum_{i=1}^{N} \sum_{j=1}^{N} P_i B_{ij} P_j \tag{10.4.14}$$

where the B_{ij} terms are called *loss coefficients* or B *coefficients*. The B coefficients are not truly constant, but vary with unit loadings. However, the B coefficients are often assumed constant in practice since the calculation of $\partial P_L/\partial P_i$ is thereby simplified. Using (10.4.14),

$$\frac{\partial P_L}{\partial P_i} = 2 \sum_{j=1}^{N} B_{ij} P_j \tag{10.4.15}$$

This equation can be used to compute the penalty factor L_i in (10.4.13).

Various methods of evaluating B coefficients from power-flow studies are available [6]. In practice, more than one set of B coefficients may be used during the daily load cycle.

When the unit incremental cost curves are linear, an analytic solution to the economic dispatch problem is possible, as illustrated by Examples 10.4–10.6. However, in practice, the incremental cost curves are nonlinear and contain discontinuities. In this case, an iterative solution by digital computer can be obtained. Given the load demand P_T, the unit incremental cost curves, generator limits, and B coefficients, such an iterative solution can be obtained by the following nine steps. Assume that the incremental cost curves are stored in tabular form, such that a unique value of P_i can be read for each dC_i/dP_i.

Step 1 Set iteration index $m = 1$.

Step 2 Estimate mth value of λ.

Step 3 Skip this step for all $m > 1$. Determine initial unit outputs $P_i (i = 1, 2, \ldots, N)$. Use $dC_i/dP_i = \lambda$ and read P_i from each incremental operating cost table. Transmission losses are neglected here.

Step 4 Compute $\partial P_L/\partial P_i$ from (10.4.15) ($i = 1, 2, \ldots, N$).

Step 5 Compute dC_i/dP_i from (10.4.13) ($i = 1, 2, \ldots, N$).

Step 6 Determine updated values of unit output P_i ($i = 1, 2, \ldots, N$). Read P_i from each incremental operating cost table. If P_i exceeds a limit value, set P_i to the limit value.

Step 7 Compare P_i determined in Step 6 with the previous value ($i = 1, 2, \ldots, N$). If the change in each unit output is less than a specified tolerance ε_1, go to Step 8. Otherwise, return to Step 4.

Step 8 Compute P_L from (10.4.14).

Step 9 If $\left|\left(\sum_{i=1}^{N} P_i\right) - P_L - P_T\right|$ is less than a specified tolerance ε_2, stop. Otherwise set $m = m + 1$ and return to Step 2.

Instead of having their values stored in tabular form for this procedure, the incremental cost curves could instead be represented by nonlinear functions such as polynomials. Then, in Step 3 and Step 5 each unit output P_i would be computed from the nonlinear functions instead of being read from a table. Note that this procedure assumes that the total load demand P_T is constant. In practice, this economic dispatch program is executed every few minutes with updated values of P_T.

Coordination of Economic Dispatch with LFC

Both the load-frequency control (LFC) and economic dispatch objectives are achieved by adjusting the reference power settings of turbine-governors on control. Figure 10.9 shows an *automatic generation control* strategy for achieving both objectives in a coordinated manner. As shown, the area control error (ACE) is first computed, and a share K_{1i} ACE is allocated to each unit. Second, the deviation of total actual generation from total desired generation is computed, and a share $K_{2i} \sum (P_{iD} - P_i)$ is allocated to unit i. Third, the deviation of actual generation from desired generation of unit i is computed, and $(P_{iD} - P_i)$ is allocated to unit i. An error signal formed from these three components and multiplied by a control gain K_{3i} determines the raise or lower signals that are sent to the turbine-governor of each unit i on control.

In practice, raise or lower signals are dispatched to the units at discrete time intervals of 2 to 10 seconds. The desired outputs P_{iD} of units on control, determined from the economic dispatch program, are updated at slower intervals, typically every 2 to 10 minutes.

Other Types of Units

The economic dispatch criterion has been derived for a power-system area consisting of fossil-fuel generating units. In practice, however, an area has a mix of different types of units including fossil-fuel, nuclear, pumped-storage hydro, hydro, and other types.

Although the fixed costs of a nuclear unit may be high, their operating costs are low due to inexpensive nuclear fuel. As such, nuclear units are

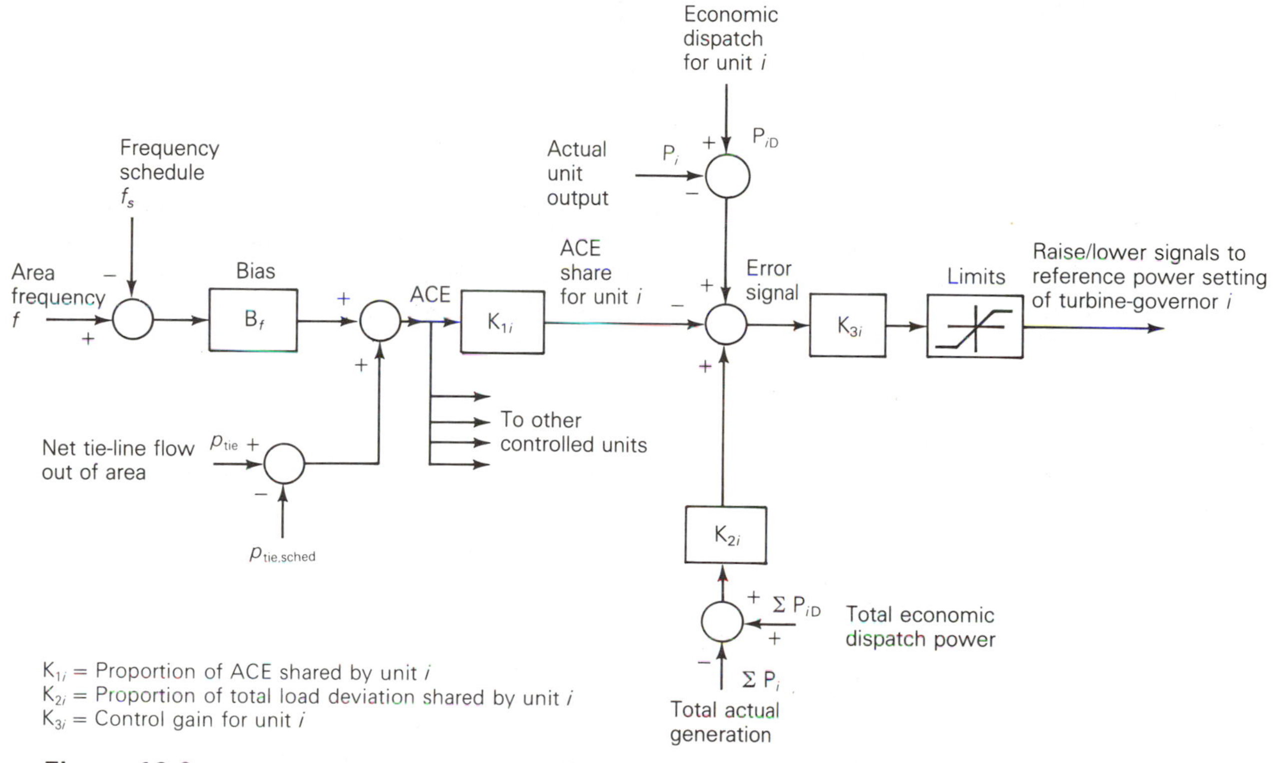

Figure 10.9 Automatic generation control [15]

normally base-loaded at their rated outputs. That is, the reference power settings of turbine-governors for nuclear units are held constant at rated output; therefore, these units do not participate in LFC or economic dispatch.

Pumped-storage hydro is a form of energy storage. During off-peak hours these units are operated as synchronous motors to pump water to a higher elevation. Then during peak-load hours the water is released and the units are operated as synchronous generators to supply power. As such, pumped-storage hydro units are used for light-load build-up and peak-load shaving. Economic operation of the area is improved by pumping during off-peak hours when the area λ is low, and by generating during peak-load hours when λ is high. Techniques are available for incorporating pumped-storage hydro units into economic dispatch of fossil-fuel units [7].

In an area consisting of hydro plants located along a river, the objective is to maximize the energy generated over the yearly water cycle rather than to minimize total operating costs. Reservoirs are used to store water during high-water or light-load periods, although some water may have to be released through spillways. Also, there are constraints on water levels due to river transportation, irrigation, or fishing requirements. Optimal strategies are available for coordinating outputs of plants along a river [8]. Economic dispatch strategies for mixed fossil-fuel/hydro systems are also available [9, 10].

Techniques are also available for including reactive power flows in the economic dispatch formulation, whereby both active and reactive powers are selected to minimize total operating costs. In particular, reactive injections from generators, switched capacitor banks, and static var systems, along with transformer tap settings, can be selected to minimize transmission-line losses [13, 14]. However, electric utility companies usually control reactive power locally. That is, the reactive power output of each generator is selected to control the generator terminal voltage, and the reactive power output of each capacitor bank or static var system located at a power system bus is selected to control the voltage magnitude at that bus. In this way, the reactive power flows on transmission lines are low, and the need for central dispatch of reactive power is eliminated.

Problems

Section 3.2

10.1 An area of an interconnected 60-Hz power system has three turbine-generator units rated 100, 200, and 500 MVA. The regulation constants of the units are 0.04, 0.05, and 0.06 per unit, respectively, based on their ratings. Each unit is initially operating at one-half its own rating when the load suddenly decreases by 50 MW. Determine (a) the unit area frequency response characteristic β on a 100-MVA base, (b) the steady-state increase in area frequency, and (c) the MW decrease in mechanical power output of each turbine. Assume that the reference power setting of each turbine-governor remains constant. Neglect losses and the dependence of load on frequency.

10.2 Each unit in Problem 10.1 is initially operating at one-half its own rating when the frequency increases by 0.002 per unit. Determine the MW decrease of each unit. The reference power setting of each turbine-governor is fixed. Neglect losses and the dependence of load on frequency.

Section 10.3

10.3 A 60-Hz power system consists of two interconnected areas. Area 1 has 1000 MW of generation and an area frequency response characteristic $\beta_1 = 400$ MW/Hz. Area 2 has 2000 MW of generation and $\beta_2 = 800$ MW/Hz. Each area is initially operating at one-half its total generation, at $\Delta p_{\text{tie}1} = \Delta p_{\text{tie}2} = 0$ and at 60 Hz, when the load in area 1 suddenly increases by 300 MW. Determine the steady-state frequency error and the steady-state tie-line error Δp_{tie} of each area. Assume that the reference power settings of all turbine-governors are fixed. That is, LFC is not employed in any area. Neglect losses and the dependence of load on frequency.

10.4 Repeat Problem 10.3 if LFC is employed in area 2 alone. The area 2 frequency bias coefficient is set at $B_{f2} = \beta_2 = 800$ MW/Hz. Assume that LFC in area 1 is inoperative due to a computer failure.

10.5 Repeat Problem 10.3 if LFC is employed in both areas. The frequency bias coefficients are $B_{f1} = \beta_1 = 400$ MW/Hz and $B_{f2} = \beta_2 = 800$ MW/Hz.

10.6 Repeat Problem 10.4 if there is a third area, with 4000 MW of generation and $\beta_3 = 1500$ MW/Hz. The load in area 1 increases by 300 MW. LFC is employed in area 2 alone. All three areas are interconnected.

Section 10.4

10.7 An area of an interconnected power system has two fossil-fuel units operating on economic dispatch. The variable operating costs of these units are given by

$$C_1 = \begin{cases} 2P_1 + 0.02P_1^2 & \text{for } 0 < P_1 \le 100 \quad \text{MW} \\ 6P_1 \dfrac{\$}{\text{hr}} & \text{for } P_1 > 100 \quad \text{MW} \end{cases}$$

$$C_2 = 0.03P_2^2 \quad \frac{\$}{\text{hr}}$$

where P_1 and P_2 are in megawatts. Determine the power output of each unit, the incremental operating cost, and the total cost C_T that minimizes C_T as the load demand P_T varies from 200 to 700 MW. Generating-unit inequalities and transmission losses are neglected.

10.8 Rework Problem 10.7 if the units are subject to the following inequality constraints:

$$100 \le P_1 \le 500 \quad \text{MW}$$

$$50 \le P_2 \le 300 \quad \text{MW}$$

10.9 Rework Problem 10.8 if total transmission losses for the power system are given by

$$P_L = 2 \times 10^{-4}P_1^2 + 1 \times 10^{-4}P_2^2 \quad \text{MW}$$

10.10 Expand the summations in (10.4.14) for $N = 2$, and verify the formula for $\partial P_L/\partial P_i$ given by (10.4.15). Assume $B_{ij} = B_{ji}$.

References

1. IEEE Committee Report, "Computer Representation of Excitation Systems," *IEEE Transactions PAS*, Vol. PAS-87 (June 1968): pp. 1460–1464.
2. M. S. Sarma, *Electric Machines* (Dubuque, Iowa: William C. Brown, 1985).
3. IEEE Committee Report, "Dynamic Models for Steam and Hydro Turbines in Power System Studies," *IEEE Transactions PAS*, Vol. PAS-92, No. 6 (November/December 1973): pp. 1904–1915.
4. N. Cohn, *Control of Generation and Power Flow on Interconnected Systems* (New York: John Wiley and Sons, 1971).
5. L. K. Kirchmayer, *Economic Operation of Power Systems* (New York: John Wiley and Sons, 1958).
6. L. K. Kirchmayer, and G. W. Stagg, "Evaluation of Methods of Coordinating Incremental Fuel Costs and Incremental Transmission Losses," *Transactions AIEE*, Vol. 71, part III (1952): pp. 513–520.
7. G. H. McDaniel, and A. F. Gabrielle, "Dispatching Pumped Storage Hydro," *IEEE Transmission PAS*, Vol. PAS-85 (May 1966): pp. 465–471.
8. E. B. Dahlin, and E. Kindingstad, "Adaptive Digital River Flow Predictor for Power Dispatch," *IEEE Transactions PAS*, Vol. PAS-83 (April 1964): pp. 320–327.
9. L. K. Kirchmayer, *Economic Control of Interconnected Systems* (New York: John Wiley and Sons, 1959).
10. J. H. Drake, et al., "Optimum Operation of a Hydrothermal System," *Transactions AIEE* (*Power Apparatus and Systems*), Vol. 62 (August 1962): pp. 242–250.

11. W. D. Stevenson, Jr., *Elements of Power System Analysis*, 4th ed. (New York: McGraw-Hill, 1982).

12. O. I. Elgerd, *Electric Energy Systems Theory, An Introduction* (New York: McGraw-Hill, 1982).

13. H. H. Happ, "Optimal Power Dispatch—A Comprehensive Survey," *IEEE Transactions PAS*, Vol. PAS-96, No. 3 (May/June 1977): pp. 841–850.

14. J. Peschon, et al., "Optimum Control of Reactive Power Flow", *IEEE Transactions PAS*, Vol. PAS-87, No. 1 (January 1968): pp. 40–49.

15. G. L. Kusic, *Computer-Aided Power Systems Analysis* (New Jersey: Prentice-Hall, 1986).

CHAPTER 11

Transmission Lines: Transient Operation

Transient overvoltages caused by lightning strikes to transmission lines and by switching operations are of fundamental importance in selecting equipment insulation levels and surge protection devices. We must, therefore, understand the nature of transmission-line transients.

During our study of the steady-state performance of transmission lines in Chapter 6, the line constants R, L, G, and C were recognized as distributed rather than lumped constants. When a line with distributed constants is subjected to a disturbance such as a lightning strike or a switching operation, voltage and current waves arise and travel along the line at a velocity near the speed of light. When these waves arrive at the line terminals, reflected voltage and current waves arise and travel back down the line, superimposed on the initial waves.

Because of line losses, traveling waves are attenuated and essentially die out after a few reflections. Also, the series inductances of transformer windings effectively block the disturbances, thereby preventing them from entering generator windings. However, due to the reinforcing action of several reflected waves, it is possible for voltage to build up to a level that could cause transformer insulation or line insulation to arc over and suffer damage.

Circuit breakers, which can operate within 50 ms, are too slow to protect against lightning or switching surges. Lightning surges can rise to peak levels within a few microseconds and switching surges within a few hundred microseconds—fast enough to destroy insulation before a circuit breaker could open. However, protective devices are available. Called surge arresters, these can be used to protect equipment insulation against transient overvoltages. These devices limit voltage to a ceiling level and absorb the energy from lightning and switching surges.

We begin this chapter with a discussion of traveling waves on single-phase lossless lines (Section 11.1). Boundary conditions are given in Section 11.2, and the Bewley lattice diagram for organizing reflections is presented in Section 11.3. Discrete-time models of single-phase lines and of lumped RLC elements are derived in Section 11.4, and the effects of line losses and multiconductor lines are discussed in Sections 11.5 and 11.6. In Section 11.7 power-system overvoltages including lightning surges, switching surges, and power-frequency overvoltages are discussed, followed by an introduction to insulation coordination in Section 11.8. Finally, the personal computer program TRANSMISSION-LINE TRANSIENTS, suitable for power-system analysis and design with respect to insulation coordination, is described in Section 11.9.

Section 11.1

Traveling Waves on Single-phase Lossless Lines

We first consider a single-phase two-wire lossless transmission line. Figure 11.1 shows a line section of length Δx meters. If the line has a loop inductance L H/m and a line-to-line capacitance C F/m, then the line section has a series inductance $L\,\Delta x$ H and shunt capacitance $C\,\Delta x$ F, as shown. In Chapter 6, the direction of line position x was selected to be from the receiving end ($x = 0$) to the sending end ($x = l$); this selection was unimportant, since the variable

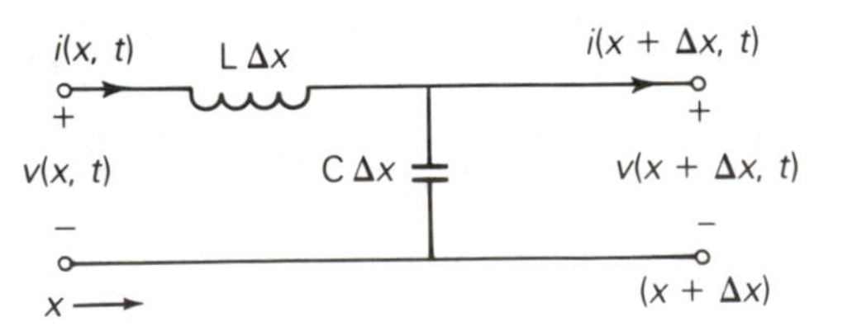

Figure 11.1 Single-phase two-wire lossless line section of length Δx

x was subsequently eliminated when relating the steady-state sending-end quantities V_s and I_s to the receiving-end quantities V_R and I_R. Here, however, we are interested in voltages and current waveforms traveling along the line. Therefore, we select the direction of increasing x as being from the sending end ($x = 0$) toward the receiving end ($x = l$).

Writing a KVL and KCL equation for the circuit in Figure 11.1,

$$v(x + \Delta x, t) - v(x, t) = -\mathrm{L}\,\Delta x \frac{\partial i(x, t)}{\partial t} \tag{11.1.1}$$

$$i(x + \Delta x, t) - i(x, t) = -\mathrm{C}\,\Delta x \frac{\partial v(x, t)}{\partial t} \tag{11.1.2}$$

Dividing (11.1.1) and (11.1.2) by Δx and taking the limit as $\Delta x \to 0$, we obtain

$$\frac{\partial v(x, t)}{\partial x} = -\mathrm{L}\frac{\partial i(x, t)}{\partial t} \tag{11.1.3}$$

$$\frac{\partial i(x, t)}{\partial x} = -\mathrm{C}\frac{\partial v(x, t)}{\partial t} \tag{11.1.4}$$

We use partial derivatives here because $v(x, t)$ and $i(x, t)$ are differentiated with respect to both position x and time t. Also, the negative signs in (11.1.3) and (11.1.4) are due to the reference direction for x. For example, with a positive value of $\partial i/\partial t$ in Figure 11.1, $v(x, t)$ decreases as x increases.

Taking the Laplace transform of (11.1.3) and (11.1.4),

$$\frac{d\mathrm{V}(x, s)}{dx} = -s\mathrm{L}\mathrm{I}(x, s) \tag{11.1.5}$$

$$\frac{d\mathrm{I}(x, s)}{dx} = -s\mathrm{C}\mathrm{V}(x, s) \tag{11.1.6}$$

where zero initial conditions are assumed. $\mathrm{V}(x, s)$ and $\mathrm{I}(x, s)$ are the Laplace transforms of $v(x, t)$ and $i(x, t)$. Also, ordinary rather than partial derivatives are used since the derivatives are now with respect to only one variable, x.

Next we differentiate (11.1.5) with respect to x and use (11.1.6), in order to eliminate $\mathrm{I}(x, s)$:

$$\frac{d^2\mathrm{V}(x, s)}{dx^2} = -s\mathrm{L}\frac{d\mathrm{I}(x, s)}{dx} = s^2\mathrm{LC}\mathrm{V}(x, s)$$

or

$$\frac{d^2\mathrm{V}(x, s)}{dx^2} - s^2\mathrm{LC}\mathrm{V}(x, s) = 0 \tag{11.1.7}$$

Similarly, (11.1.6) can be differentiated in order to obtain

$$\frac{d^2\mathrm{I}(x, s)}{dx^2} - s^2\mathrm{LC}\mathrm{I}(x, s) = 0 \tag{11.1.8}$$

Equation (11.1.7) is a linear, second-order homogeneous differential equation. By inspection, its solution is

$$\mathrm{V}(x, s) = \mathrm{V}^{+}(s)e^{-sx/v} + \mathrm{V}^{-}(s)e^{+sx/v} \tag{11.1.9}$$

where

$$v = \frac{1}{\sqrt{\mathrm{LC}}} \quad \text{m/s} \tag{11.1.10}$$

Similarly, the solution to (11.1.8) is

$$\mathrm{I}(x, s) = \mathrm{I}^{+}(s)e^{-sx/v} + \mathrm{I}^{-}(s)e^{+sx/v} \tag{11.1.11}$$

The reader can quickly verify that these solutions satisfy (11.1.7) and (11.1.8). The "constants" $\mathrm{V}^{+}(s)$, $\mathrm{V}^{-}(s)$, $\mathrm{I}^{+}(s)$, and $\mathrm{I}^{-}(s)$, which in general are functions of s but are independent of x, can be evaluated from the boundary conditions at the sending and receiving ends of the line. The superscripts + and − refer to waves traveling in the positive x and negative x directions, soon to be explained.

Taking the inverse Laplace transform of (11.1.9) and (11.1.11), and recalling the time shift properly, $\mathscr{L}[f(t-\tau)] = \mathrm{F}(s)e^{-s\tau}$, we obtain

$$v(x, t) = v^{+}\left(t - \frac{x}{v}\right) + v^{-}\left(t + \frac{x}{v}\right) \tag{11.1.12}$$

$$i(x, t) = i^{+}\left(t - \frac{x}{v}\right) + i^{-}\left(t + \frac{x}{v}\right) \tag{11.1.13}$$

where the functions $v^{+}(\;)$, $v^{-}(\;)$, $i^{+}(\;)$, and $i^{-}(\;)$, can be evaluated from the boundary conditions.

We now show that $v^{+}(t - x/v)$ represents a voltage wave traveling in the positive x direction with velocity $v = 1/\sqrt{\mathrm{LC}}$ m/s. Consider any wave $f^{+}(u)$, where $u = t - x/v$. Suppose that this wave begins at $u = u_0$, as shown in Figure 11.2(a). At time $t = t_1$, the wavefront is at $u_0 = (t_1 - x_1/v)$, or at $x_1 = v(t_1 - u_0)$. At a later time, t_2, the wavefront is at $u_0 = (t_2 - x_2/v)$ or at $x_2 = v(t_2 - u_0)$. As shown in Figure 11.2(b), the wavefront has moved in the positive x direction a distance $(x_2 - x_1) = v(t_2 - t_1)$ during time $(t_2 - t_1)$. The velocity is therefore $(x_2 - x_1)/(t_2 - t_1) = v$.

Figure 11.2

The function $f^{+}(u)$, where $u = \left(t - \frac{x}{v}\right)$

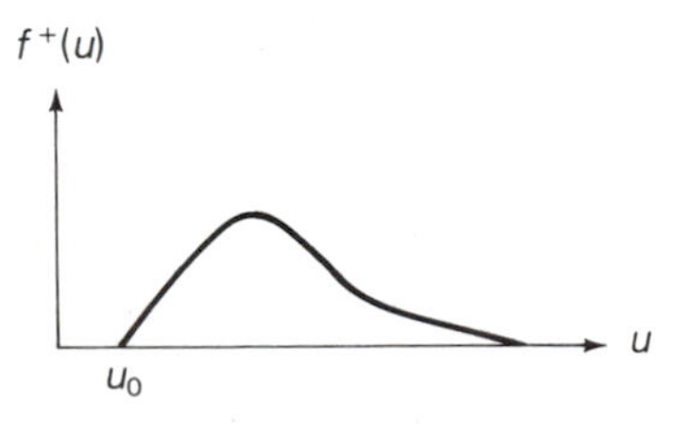

(a) $f^{+}(u)$ versus u

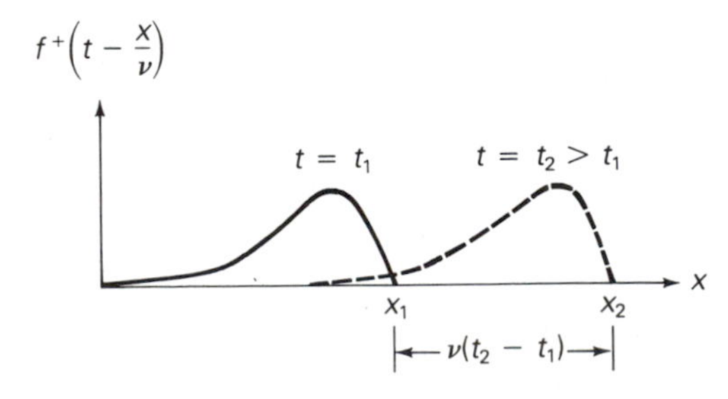

(b) $f^{+}\left(t - \frac{x}{v}\right)$ versus x

Similarly, $i^+(t - x/v)$ represents a current wave traveling in the positive x direction with velocity v. We call $v^+(t - x/v)$ and $i^+(t - x/v)$ the *forward* traveling voltage and current waves. It can be shown analogously that $v^-(t + x/v)$ and $i^-(t + x/v)$ travel in the negative x direction with velocity v. We call $v^-(t + x/v)$ and $i^-(t + x/v)$ the *backward* traveling voltage and current waves.

Recall from (6.4.16) that for a lossless line $f\lambda = 1/\sqrt{LC}$. It is now evident that the term $1/\sqrt{LC}$ in this equation is v, the velocity of propagation of voltage and current waves along the lossless line. Also, recall from Chapter 5 that L is proportional to μ and C is proportional to ε. For overhead lines, $v = 1/\sqrt{LC}$ is approximately equal to $1/\sqrt{\mu\varepsilon} = 1/\sqrt{\mu_0\varepsilon_0} = 3 \times 10^8$ m/s, the speed of light in free space. For cables, the relative permitivity $\varepsilon/\varepsilon_0$ may be 3 to 5 or even higher, resulting in a value of v lower than that for overhead lines.

We next evaluate the terms $I^+(s)$ and $I^-(s)$. Using (11.1.9) and (11.1.10) in (11.1.6),

$$\frac{s}{v}\left[-I^+(s)e^{-sx/v} + I^-(s)e^{+sx/v}\right] = -sC\left[V^+(s)e^{-sx/v} + V^-(s)e^{+sx/v}\right]$$

Equating the coefficients of $e^{-sx/v}$ on both sides of this equation,

$$I^+(s) = (vC)\,V^+(s) = \frac{V^+(s)}{\sqrt{\dfrac{L}{C}}} = \frac{V^+(s)}{Z_c} \tag{11.1.14}$$

where

$$Z_c = \sqrt{\frac{L}{C}} \quad \Omega \tag{11.1.15}$$

Similarly, equating the coefficients of $e^{+sx/v}$,

$$I^-(s) = \frac{-V^-(s)}{Z_c} \tag{11.1.16}$$

Thus we can rewrite (11.1.11) and (11.1.13) as

$$I(x, s) = \frac{1}{Z_c}\left[V^+(s)e^{-sx/v} - V^-(s)e^{+sx/v}\right] \tag{11.1.17}$$

$$i(x, t) = \frac{1}{Z_c}\left[v^+\left(t - \frac{x}{v}\right) - v^-\left(t + \frac{x}{v}\right)\right] \tag{11.1.18}$$

Recall from (6.4.3) that $Z_c = \sqrt{L/C}$ is the characteristic impedance (also called surge impedance) of a lossless line.

Section 11.2

Boundary Conditions for Single-phase Lossless Lines

Figure 11.3 shows a single-phase two-wire lossless line terminated by an impedance $Z_R(s)$ at the receiving end and a source with Thévenin voltage $E_G(s)$ and with Thévenin impedance $Z_G(s)$ at the sending end. $V(x, s)$ and $I(x, s)$ are the Laplace transforms of the voltage and current at position x. The line has length l, surge impedance $Z_c = \sqrt{L/C}$, and velocity $v = 1/\sqrt{LC}$. We assume that the line is initially unenergized.

Figure 11.3

Single-phase two-wire lossless line with source and load terminations

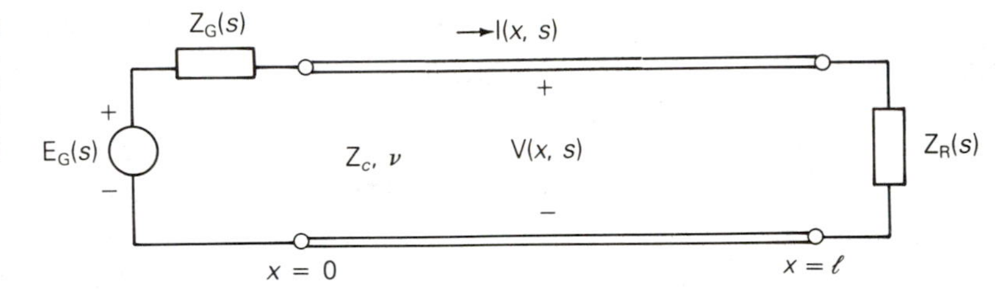

From Figure 11.3, the boundary condition at the receiving end is

$$V(l, s) = Z_R(s)\, I(l, s) \tag{11.2.1}$$

Using (11.1.9) and (11.1.17) in (11.2.1),

$$V^+(s)e^{-sl/v} + V^-(s)e^{+sl/v} = \frac{Z_R(s)}{Z_c}\left[V^+(s)e^{-sl/v} - V^-(s)e^{+sl/v}\right]$$

Solving for $V^-(l, s)$

$$V^-(l, s) = \Gamma_R(s)V^+(s)e^{-2s\tau} \tag{11.2.2}$$

where

$$\Gamma_R(s) = \frac{\dfrac{Z_R(s)}{Z_c} - 1}{\dfrac{Z_R(s)}{Z_c} + 1} \quad \text{per unit} \tag{11.2.3}$$

$$\tau = \frac{l}{v} \quad \text{seconds} \tag{11.2.4}$$

$\Gamma_R(s)$ is called the *receiving-end voltage reflection coefficient*. Also, τ, called the *transit time* of the line, is the time it takes a wave to travel the length of the line.

Using (11.2.2) in (11.1.9) and (11.1.17),

$$V(x, s) = V^+(s)[e^{-sx/v} + \Gamma_R(s)e^{s[(x/v) - 2\tau]}] \tag{11.2.5}$$

$$I(x, s) = \frac{V^+(s)}{Z_c}[e^{-sx/v} - \Gamma_R(s)e^{s[(x/v) - 2\tau]}] \tag{11.2.6}$$

From Figure 11.3 the boundary condition at the sending end is

$$V(0, s) = E_G(s) - Z_G(s)I(0, s) \tag{11.2.7}$$

Using (11.2.5) and (11.2.6) in (11.2.7),

$$V^+(s)\left[1 + \Gamma_R(s)e^{-2s\tau}\right] = E_G(s) - \left[\frac{Z_G(s)}{Z_c}\right]V^+(s)\left[1 - \Gamma_R(s)e^{-2s\tau}\right]$$

Solving for $V^+(s)$,

$$V^+(s)\left\{\left[\frac{Z_G(s)}{Z_c} + 1\right] - \Gamma_R(s)e^{-2s\tau}\left[\frac{Z_G(s)}{Z_c} - 1\right]\right\} = E_G(s)$$

$$V^+(s)\left[\frac{Z_G(s)}{Z_c} + 1\right]\left\{1 - \Gamma_R(s)\Gamma_S(s)e^{-2s\tau}\right\} = E_G(s)$$

or

$$V^+(s) = E_G(s)\left[\frac{Z_c}{Z_G(s) + Z_c}\right]\left[\frac{1}{1 - \Gamma_R(s)\Gamma_S(s)e^{-2s\tau}}\right] \tag{11.2.8}$$

where

$$\Gamma_S(s) = \frac{\dfrac{Z_G(s)}{Z_c} - 1}{\dfrac{Z_G(s)}{Z_c} + 1} \tag{11.2.9}$$

$\Gamma_S(s)$ is called the *sending-end voltage reflection coefficient*. Using (11.2.9) in (11.2.5) and (11.2.6), the complete solution is

$$V(x, s) = E_G(s)\left[\frac{Z_c}{Z_G(s) + Z_c}\right]\left[\frac{e^{-sx/v} + \Gamma_R(s)e^{s[(x/v) - 2\tau]}}{1 - \Gamma_R(s)\Gamma_S(s)e^{-2s\tau}}\right] \tag{11.2.10}$$

$$I(x, s) = \left[\frac{E_G(s)}{Z_G(s) + Z_c}\right]\left[\frac{e^{-sx/v} - \Gamma_R(s)e^{s[(x/v) - 2\tau]}}{1 - \Gamma_R(s)\Gamma_S(s)e^{-2s\tau}}\right] \tag{11.2.11}$$

where

$$\Gamma_R(s) = \frac{\dfrac{Z_R(s)}{Z_c} - 1}{\dfrac{Z_R(s)}{Z_c} + 1} \quad \text{per unit} \qquad \Gamma_S(s) = \frac{\dfrac{Z_G(s)}{Z_c} - 1}{\dfrac{Z_G(s)}{Z_c} + 1} \quad \text{per unit} \tag{11.2.12}$$

$$Z_c = \sqrt{\frac{L}{C}} \quad \Omega \qquad v = \frac{1}{\sqrt{LC}} \quad \text{m/s} \qquad \tau = \frac{l}{v} \quad \text{s} \tag{11.2.13}$$

The following four examples illustrate this general solution. All four examples refer to the line shown in Figure 11.3, which has length l, velocity v, characteristic impedance Z_c, and is initially unenergized.

Example 11.1 Let $Z_R = Z_c$ and $Z_G = 0$. The source voltage is a step, $e_G(t) = Eu_{-1}(t)$. (a) Determine $v(x, t)$ and $i(x, t)$. (b) Plot the voltage and current versus time t at the center of the line and at the receiving end.

Solution **a.** From (11.2.12) with $Z_R = Z_c$ and $Z_G = 0$,

$$\Gamma_R(s) = \frac{1-1}{1+1} = 0 \qquad \Gamma_S(s) = \frac{0-1}{0+1} = -1$$

The Laplace transform of the source voltage is $E_G(s) = E/s$. Then, from (11.2.10) and (11.2.11),

$$V(x, s) = \left(\frac{E}{s}\right)(1)(e^{-sx/v}) = \frac{Ee^{-sx/v}}{s}$$

$$I(x, s) = \frac{(E/Z_c)}{s}e^{-sx/v}$$

Taking the inverse Laplace transform,

$$v(x, t) = Eu_{-1}\left(t - \frac{x}{v}\right)$$

$$i(x, t) = \frac{E}{Z_c}u_{-1}\left(t - \frac{x}{v}\right)$$

b. At the center of the line, where $x = l/2$,

$$v\left(\frac{l}{2}, t\right) = Eu_{-1}\left(t - \frac{\tau}{2}\right) \qquad i\left(\frac{l}{2}, t\right) = \frac{E}{Z_c}u_{-1}\left(t - \frac{\tau}{2}\right)$$

At the receiving end, where $x = l$,

$$v(l, t) = Eu_{-1}(t - \tau) \qquad i(l, t) = \frac{E}{Z_c}u_{-1}(t - \tau)$$

These waves, plotted in Figure 11.4, can be explained as follows. At $t = 0$ the ideal step voltage of E volts, applied to the sending end, encounters Z_c, the characteristic impedance of the line. Therefore, a forward traveling step voltage wave of E volts is initiated at the sending end. Also, since the ratio of the forward traveling voltage to current is Z_c, a forward traveling step current wave of (E/Z_c) amperes is initiated. These waves travel in the positive x direction, arriving at the center of the line at $t = \tau/2$, and at the end of the line at $t = \tau$. The receiving-end load is *matched* to the line; that is, $Z_R = Z_c$. For a matched load, $\Gamma_R = 0$, and therefore no backward traveling waves are initiated. In steady-state, the line with matched load is energized at E volts with current E/Z_c amperes.

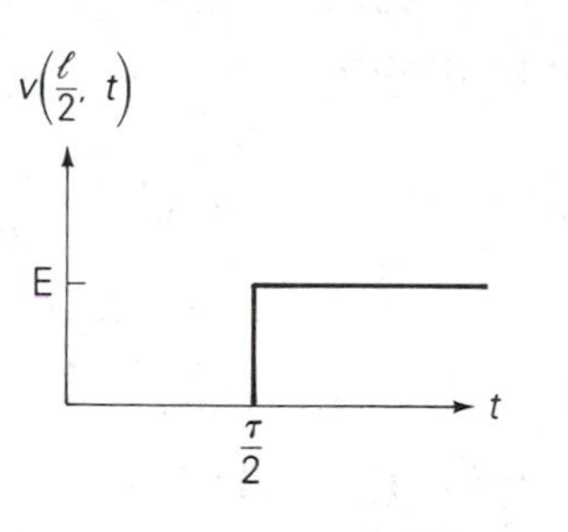

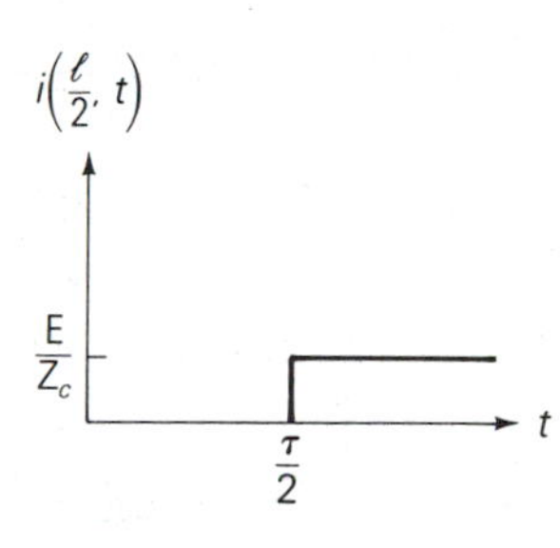

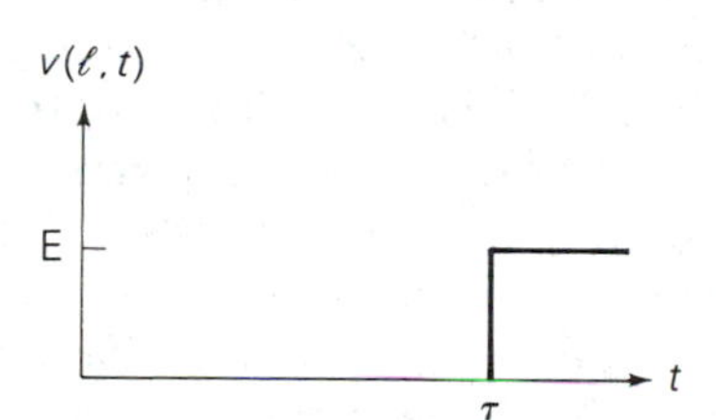

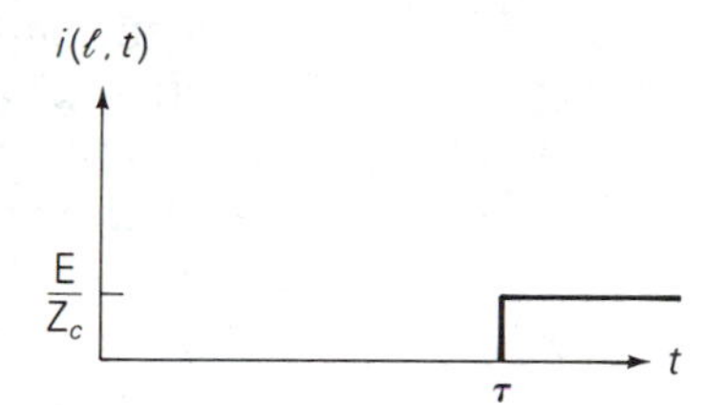

Figure 11.4 Voltage and current waveforms for Example 11.1

◁

Example 11.2 The receiving end is open. The source voltage at the sending end is a step $e_G(t) = Eu_{-1}(t)$, with $Z_G(s) = Z_c$. (a) Determine $v(x,t)$ and $i(x,t)$. (b) Plot the voltage and current versus time t at the center of the line.

Solution **a.** From (11.2.12),

$$\Gamma_R(s) = \lim_{Z_R \to \infty} \frac{\dfrac{Z_R}{Z_c} - 1}{\dfrac{Z_R}{Z_c} + 1} = 1 \qquad \Gamma_S(s) = \frac{1-1}{1+1} = 0$$

The Laplace transform of the source voltage is $E_G(s) = E/s$. Then, from (11.2.10) and (11.2.11),

$$V(x,s) = \frac{E}{s}\left(\frac{1}{2}\right)\left[e^{-sx/v} + e^{s[(x/v)-2\tau]}\right]$$

$$I(x,s) = \frac{E}{s}\left(\frac{1}{2Z_c}\right)\left[e^{-sx/v} - e^{s[(x/v)-2\tau]}\right]$$

Taking the inverse Laplace transform,

$$v(x,t) = \frac{E}{2}u_{-1}\left(t - \frac{x}{v}\right) + \frac{E}{2}u_{-1}\left(t + \frac{x}{v} - 2\tau\right)$$

$$i(x,t) = \frac{E}{2Z_c}u_{-1}\left(t - \frac{x}{v}\right) - \frac{E}{2Z_c}u_{-1}\left(t + \frac{x}{v} - 2\tau\right)$$

b. At the center of the line, where $x = l/2$,

$$v\left(\frac{l}{2}, t\right) = \frac{E}{2}u_{-1}\left(t - \frac{\tau}{2}\right) + \frac{E}{2}u_{-1}\left(t - \frac{3\tau}{2}\right)$$

$$i\left(\frac{l}{2}, t\right) = \frac{E}{2Z_c}u_{-1}\left(t - \frac{\tau}{2}\right) - \frac{E}{2Z_c}u_{-1}\left(t - \frac{3\tau}{2}\right)$$

These waves are plotted in Figure 11.5. At $t = 0$ the step voltage source of E volts encounters the source impedance $Z_G = Z_c$ in series with the characteristic impedance of the line, Z_c. Using voltage division, the sending-end voltage at $t = 0$ is E/2. Therefore, a forward traveling step voltage wave of E/2 volts and a forward traveling step current wave of $E/(2Z_c)$ amperes are initiated at the sending end. These waves arrive at the center of the line at $t = \tau/2$. Also, with $\Gamma_R = 1$, the backward traveling voltage wave equals the forward traveling voltage wave, and the backward traveling current wave is the negative of the forward traveling current wave. These backward traveling waves, which are initiated at the receiving end at $t = \tau$ when the forward traveling waves arrive there, arrive at the center of the line at $t = 3\tau/2$ and are superimposed on the forward traveling waves. No additional forward or backward traveling waves are initiated because the source impedance is matched to the line; that is, $\Gamma_S(s) = 0$. In steady-state the line, which is open at the receiving end, is energized at E volts with zero current.

Figure 11.5

Voltage and current waveforms for Example 11.2

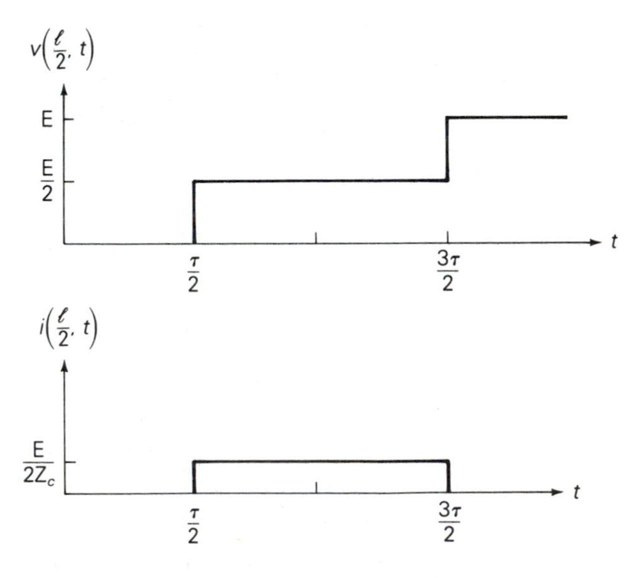

◁

Example 11.3 The receiving end is terminated by a capacitor with C_R farads, which is initially unenergized. The source voltage at the sending end is a unit step $e_G(t) = Eu_{-1}(t)$, with $Z_G = Z_c$. Determine and plot $v(x, t)$ versus time t at the sending end of the line.

Solution From (11.2.12) with $Z_R = \dfrac{1}{sC_R}$ and $Z_G = Z_c$,

$$\Gamma_R(s) = \frac{\dfrac{1}{sC_RZ_c} - 1}{\dfrac{1}{sC_RZ_c} + 1} = \frac{-s + \dfrac{1}{Z_cC_R}}{s + \dfrac{1}{Z_cC_R}}$$

$$\Gamma_S(s) = \frac{1-1}{1+1} = 0$$

Then, from (11.2.10), with $E_G(s) = E/s$,

$$V(x, s) = \frac{E}{s}\left(\frac{1}{2}\right)\left[e^{-sx/v} + \left(\frac{-s + \dfrac{1}{Z_cC_R}}{s + \dfrac{1}{Z_cC_R}}\right)e^{s[(x/v) - 2\tau]}\right]$$

$$= \frac{E}{2}\left[\frac{e^{-sx/v}}{s} + \frac{1}{s}\left(\frac{-s + \dfrac{1}{Z_cC_R}}{s + \dfrac{1}{Z_cC_R}}\right)e^{s[(x/v) - 2\tau]}\right]$$

Using partial fraction expansion of the second term above,

$$V(x, s) = \frac{E}{2}\left[\frac{e^{-sx/v}}{s} + \left(\frac{1}{s} - \frac{2}{s + \dfrac{1}{Z_cC_R}}\right)e^{s[(x/v) - 2\tau]}\right]$$

The inverse Laplace transform is

$$v(x, t) = \frac{E}{2}u_{-1}\left(t - \frac{x}{v}\right) + \frac{E}{2}[1 - 2e^{(-1/Z_cC_R)(t + x/v - 2\tau)}]\,u_{-1}\left(t + \frac{x}{v} - 2\tau\right)$$

At the sending end, where $x = 0$,

$$v(0, t) = \frac{E}{2}u_{-1}(t) + \frac{E}{2}[1 - 2e^{(-1/Z_cC_R)(t - 2\tau)}]\,u_{-1}(t - 2\tau)$$

$v(0, t)$ is plotted in Figure 11.6. As in Example 11.2, a forward traveling step voltage wave of E/2 volts is initiated at the sending end at $t = 0$. At $t = \tau$, when the forward traveling wave arrives at the receiving end, a backward traveling wave is initiated. The backward traveling voltage wave, an exponential with initial value $-E/2$, steady-state value $+E/2$, and time constant Z_cC_R, arrives at the sending end at $t = 2\tau$, where it is superimposed on the forward traveling wave. No additional waves are initiated, since the source impedance is matched to the line. In steady-state, the line and the capacitor at the receiving end are energized at E volts with zero current.

The capacitor at the receiving end can also be viewed as a short circuit

Figure 11.6

Voltage waveform for Example 11.3

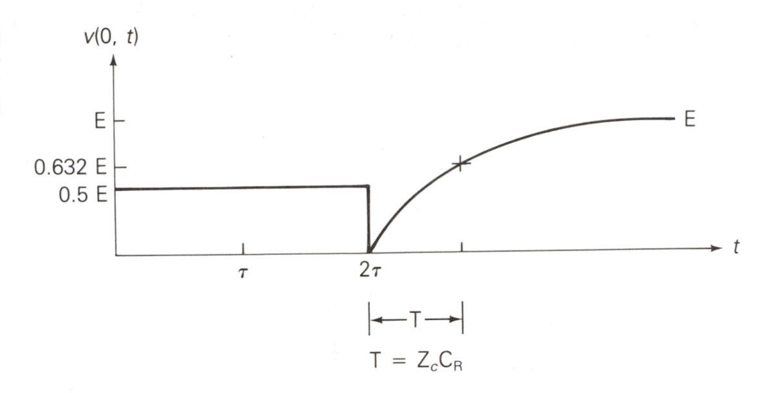

at the instant $t = \tau$, when the forward traveling wave arrives at the receiving end. For a short circuit at the receiving end, $\Gamma_R = -1$, and therefore the backward traveling voltage wavefront is $-E/2$, the negative of the forward traveling wave. However, in steady-state the capacitor is an open circuit, for which $\Gamma_R = +1$, and the steady-state backward traveling voltage wave equals the forward traveling voltage wave. ◁

Example 11.4 At the receiving end, $Z_R = Z_c/3$. At the sending end, $e_G(t) = Eu_{-1}(t)$ and $Z_G = 2Z_c$. Determine and plot the voltage versus time at the center of the line.

Solution From (11.2.12),

$$\Gamma_R = \frac{\frac{1}{3} - 1}{\frac{1}{3} + 1} = -\frac{1}{2} \qquad \Gamma_S = \frac{2-1}{2+1} = \frac{1}{3}$$

From (11.2.10), with $E_G(s) = E/s$,

$$V(x, s) = \frac{E}{s}\left(\frac{1}{3}\right)\frac{\left[e^{-sx/v} - \frac{1}{2}e^{s[(x/v) - 2\tau]}\right]}{1 + \left(\frac{1}{6}e^{-2s\tau}\right)}$$

The preceding equation can be rewritten using the following geometric series:

$$\frac{1}{1+y} = 1 - y + y^2 - y^3 + y^4 - \cdots$$

with $y = \frac{1}{6}e^{-2s\tau}$,

$$V(x, s) = \frac{E}{3s}\left[e^{-sx/v} - \frac{1}{2}e^{s[(x/v) - 2\tau]}\right]\left[1 - \frac{1}{6}e^{-2s\tau} + \frac{1}{36}e^{-4s\tau} - \frac{1}{216}e^{-6s\tau} + \cdots\right]$$

Multiplying the terms within the brackets,

$$V(x, s) = \frac{E}{3s}\left[e^{-sx/v} - \frac{1}{2}e^{s[(x/v) - 2\tau]} - \frac{1}{6}e^{-s[(x/v) + 2\tau]} + \frac{1}{12}e^{s[(x/v) - 4\tau]} + \frac{1}{36}e^{-s[(x/v) + 4\tau]} - \frac{1}{72}e^{s[(x/v) - 6\tau]} + \cdots\right]$$

Taking the inverse Laplace transform,

$$v(x, t) = \frac{E}{3}\left[u_{-1}\left(t - \frac{x}{v}\right) - \frac{1}{2}u_{-1}\left(t + \frac{x}{v} - 2\tau\right) - \frac{1}{6}u_{-1}\left(t - \frac{x}{v} - 2\tau\right)\right.$$
$$+ \frac{1}{12}u_{-1}\left(t + \frac{x}{v} - 4\tau\right) + \frac{1}{36}u_{-1}\left(t - \frac{x}{v} - 4\tau\right)$$
$$\left. - \frac{1}{72}u_{-1}\left(t + \frac{x}{v} - 6\tau\right)\cdots\right]$$

At the center of the line, where $x = l/2$,

$$v\left(\frac{l}{2}, t\right) = \frac{E}{3}\left[u_{-1}\left(t - \frac{\tau}{2}\right) - \frac{1}{2}u_{-1}\left(t - \frac{3\tau}{2}\right) - \frac{1}{6}u_{-1}\left(t - \frac{5\tau}{2}\right)\right.$$
$$+ \frac{1}{12}u_{-1}\left(t - \frac{7\tau}{2}\right) + \frac{1}{36}u_{-1}\left(t - \frac{9\tau}{2}\right)$$
$$\left. - \frac{1}{72}u_{-1}\left(t - \frac{11\tau}{2}\right)\cdots\right]$$

$v(l/2, t)$ is plotted in Figure 11.7(a). Since neither the source nor the load is matched to the line, the voltage at any point along the line consists of an infinite series of forward and backward traveling waves. At the center of the line, the first forward traveling wave arrives at $t = \tau/2$; then a backward traveling wave arrives at $3\tau/2$, another forward traveling wave arrives at $5\tau/2$, another backward traveling wave at $7\tau/2$, and so on.

Figure 11.7

Example 11.4

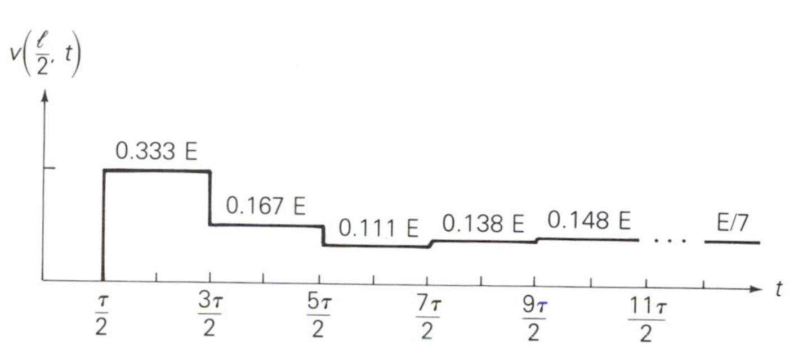

(a) Voltage waveform

$Z_G = 2Z_c$ $\quad i_{ss}(x)$

E $\quad v_{ss}(x)$ $\quad Z_R = \frac{Z_c}{3}$

(b) Steady-state solution

The steady-state voltage can be evaluated from the final value theorem. That is,

$$v_{ss}(x) = \lim_{t\to\infty} v(x,t) = \lim_{s\to 0} s\mathrm{V}(x,s)$$

$$= \lim_{s\to 0}\left\{ s\left(\frac{\mathrm{E}}{s}\right)\left(\frac{1}{3}\right)\frac{\left[e^{-sx/v} - \frac{1}{2}e^{s[(x/v)-2\tau]}\right]}{1+\frac{1}{6}e^{-2s\tau}}\right\}$$

$$= \mathrm{E}\left(\frac{1}{3}\right)\left(\frac{1-\frac{1}{2}}{1+\frac{1}{6}}\right) = \frac{\mathrm{E}}{7}$$

The steady-state solution can also be evaluated from the circuit in Figure 11.7(b). Since there is no steady-state voltage drop across the lossless line when a dc source is applied, the line can be eliminated, leaving only the source and load. The steady-state voltage is then, by voltage division,

$$v_{ss}(x) = \mathrm{E}\left(\frac{Z_R}{Z_R + Z_G}\right) = \mathrm{E}\left(\frac{\frac{1}{3}}{\frac{1}{3}+2}\right) = \frac{\mathrm{E}}{7}$$

◁

Section 11.3

Bewley Lattice Diagram

A lattice diagram developed by L. V. Bewley [2] conveniently organizes the reflections that occur during transmission-line transients. For the Bewley lattice diagram, shown in Figure 11.8, the vertical scale represents time and is scaled in units of τ, the transient time of the line. The horizontal scale represents line position x, and the diagonal lines represent traveling waves. Each reflection is determined by multiplying the incident wave arriving at an end by the reflection coefficient at that end. The voltage $v(x,t)$ at any point x and t on the diagram is determined by adding all the terms directly above that point.

Example 11.5 For the line and terminations given in Example 11.4, draw the lattice diagram and plot $v(l/3, t)$ versus time t.

Solution The lattice diagram is shown in Figure 11.8. At $t = 0$, the source voltage encounters the source impedance and the line characteristic impedance, and the first forward traveling wave is determined by voltage division:

$$\mathrm{V}_1(s) = \mathrm{E}_G(s)\left[\frac{Z_c}{Z_c + Z_G}\right] = \frac{\mathrm{E}}{s}\left[\frac{1}{1+2}\right] = \frac{\mathrm{E}}{3s}$$

which is a step with magnitude (E/3) volts. The next traveling wave, a backward one, is $\mathrm{V}_2(s) = \Gamma_R(s)\mathrm{V}_1(s) = (-\frac{1}{2})\mathrm{V}_1(s) = -\mathrm{E}/(6s)$, and the next wave, a forward one, is $\mathrm{V}_3(s) = \Gamma_S(s)\mathrm{V}_2(s) = (\frac{1}{3})\mathrm{V}_2(s) = -\mathrm{E}/(18s)$. Subsequent waves are calculated in a similar manner.

Figure 11.8

Bewley lattice diagram for Example 11.5

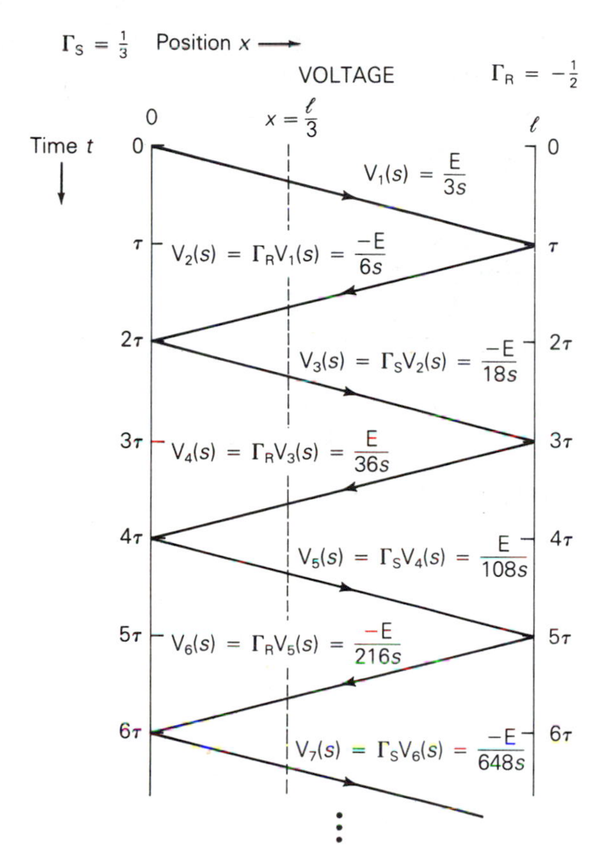

The voltage at $x = l/3$ is determined by drawing a vertical line at $x = l/3$ on the lattice diagram, shown dashed in Figure 11.8. Starting at the top of the dashed line, where $t = 0$, and moving down, each voltage wave is added at the time it intersects the dashed line. The first wave v_1 arrives at $t = \tau/3$, the second v_2 arrives at $5\tau/3$, v_3 at $7\tau/3$, and so on. $v(l/3, t)$ is plotted in Figure 11.9.

Figure 11.9

Voltage waveform for Example 11.5

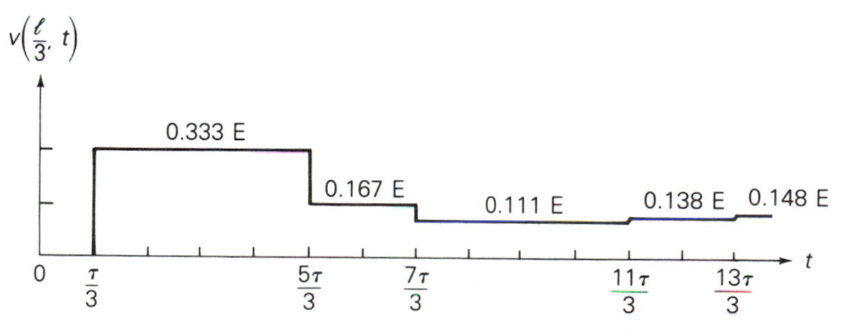

◁

Figure 11.10 shows a forward traveling voltage wave V_A^+ arriving at the junction of two lossless lines A and B with characteristic impedances Z_A and Z_B, respectively. This could be, for example, the junction of an overhead line and a cable. When V_A^+ arrives at the junction, both a reflection V_A^- on line A and a refraction V_B^+ on line B will occur. Writing a KVL and KCL equation at the junction,

$$V_A^+ + V_A^- = V_B^+ \tag{11.3.1}$$

$$I_A^+ + I_A^- = I_B^+ \tag{11.3.2}$$

Figure 11.10

Junction of two single-phase lossless lines

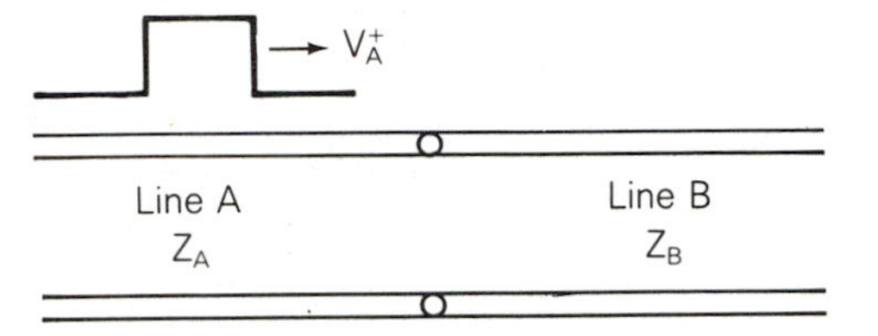

Recall that $I_A^+ = V_A^+/Z_A$, $I_A^- = -V_A^-/Z_A$, and $I_B^+ = V_B^+/Z_B$. Using these relations in (11.3.2),

$$\frac{V_A^+}{Z_A} - \frac{V_A^-}{Z_A} = \frac{V_B^+}{Z_B} \tag{11.3.3}$$

Solving (11.3.1) and (11.3.3) for V_A^- and V_B^+ in terms of V_A^+ yields

$$V_A^- = \Gamma_{AA} V_A^+ \tag{11.3.4}$$

where

$$\Gamma_{AA} = \frac{\dfrac{Z_B}{Z_A} - 1}{\dfrac{Z_B}{Z_A} + 1} \tag{11.3.5}$$

and

$$V_B^+ = \Gamma_{BA} V_A \tag{11.3.6}$$

where

$$\Gamma_{BA} = \frac{2\left(\dfrac{Z_B}{Z_A}\right)}{\dfrac{Z_B}{Z_A} + 1} \tag{11.3.7}$$

Note that Γ_{AA}, given by (11.3.5), is similar to Γ_R, given by (11.2.12), except that Z_B replaces Z_R. Thus for waves arriving at the junction from line A, the "load" at the receiving end of line A is the characteristic impedance of line B.

Example 11.6

As shown in Figure 11.10, a single-phase lossless overhead line with $Z_A = 400\,\Omega$, $v_A = 3 \times 10^8$ m/s, and $l_A = 30$ km is connected to a single-phase lossless cable with $Z_B = 100\,\Omega$, $v_B = 2 \times 10^8$ m/s, and $l_B = 20$ km. At the sending end of line A, $e_g(t) = Eu_{-1}(t)$ and $Z_G = Z_A$. At the receiving end of line B, $Z_R = 2Z_B = 200\ \Omega$. Draw the lattice diagram for $0 \le t \le 0.6$ ms and plot the voltage at the junction versus time. The line and cable are initially unenergized.

Solution From (11.2.13),

$$\tau_A = \frac{30 \times 10^3}{3 \times 10^8} = 0.1 \times 10^{-3} \quad \text{s} \qquad \tau_B = \frac{20 \times 10^3}{2 \times 10^8} = 0.1 \times 10^{-3} \quad \text{s}$$

From (11.2.12), with $Z_G = Z_A$ and $Z_R = 2Z_B$,

$$\Gamma_S = \frac{1-1}{1+1} = 0 \qquad \Gamma_R = \frac{2-1}{2+1} = \frac{1}{3}$$

From (11.3.5) and (11.3.6), the reflection and refraction coefficients for waves arriving at the junction from line A are

$$\Gamma_{AA} = \frac{\frac{100}{400} - 1}{\frac{100}{400} + 1} = \frac{-3}{5} \qquad \Gamma_{BA} = \frac{2\frac{100}{400}}{\frac{100}{400} + 1} = \frac{2}{5} \Bigg\} \text{ from line A}$$

Reversing A and B, the reflection and refraction coefficients for waves returning to the junction from line B are

$$\Gamma_{BB} = \frac{\frac{400}{100} - 1}{\frac{400}{100} + 1} = \frac{3}{5} \qquad \Gamma_{AB} = \frac{2\frac{400}{100}}{\frac{400}{100} + 1} = \frac{8}{5} \Bigg\} \text{ from line B}$$

The lattice diagram is shown in Figure 11.11. Using voltage division, the first forward traveling voltage wave is

$$V_1(s) = E_G(s)\left(\frac{Z_A}{Z_A + Z_G}\right) = \frac{E}{s}\left(\frac{1}{2}\right) = \frac{E}{2s}$$

When v_1 arrives at the junction, a reflected wave v_2 and refracted wave v_3 are initiated. Using the reflection and refraction coefficients for line A,

$$V_2(s) = \Gamma_{AA}V_1(s) = \left(\frac{-3}{5}\right)\left(\frac{E}{2s}\right) = \frac{-3E}{10s}$$

$$V_3(s) = \Gamma_{BA}V_1(s) = \left(\frac{2}{5}\right)\left(\frac{E}{2s}\right) = \frac{E}{5s}$$

When v_2 arrives at the receiving end of line B, a reflected wave $V_4(s) = \Gamma_R V_3(s) = \frac{1}{3}(E/5s) = (E/15s)$ is initiated. When v_4 arrives at the junction, reflected wave v_5 and refracted wave v_6 are initiated. Using the reflection and refraction coefficients for line B,

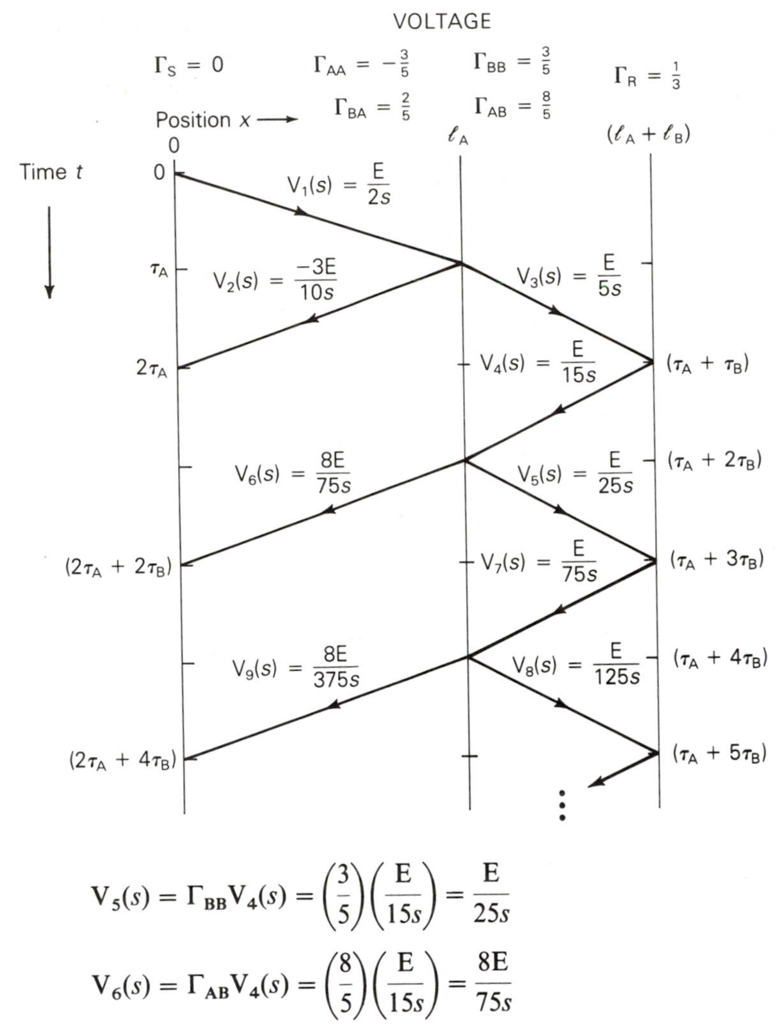

Figure 11.11

Lattice diagram for Example 11.6

$$V_5(s) = \Gamma_{BB}V_4(s) = \left(\frac{3}{5}\right)\left(\frac{E}{15s}\right) = \frac{E}{25s}$$

$$V_6(s) = \Gamma_{AB}V_4(s) = \left(\frac{8}{5}\right)\left(\frac{E}{15s}\right) = \frac{8E}{75s}$$

Subsequent reflections and refractions are calculated in a similar manner.

The voltage at the junction is determined by starting at $x = l_A$ at the top

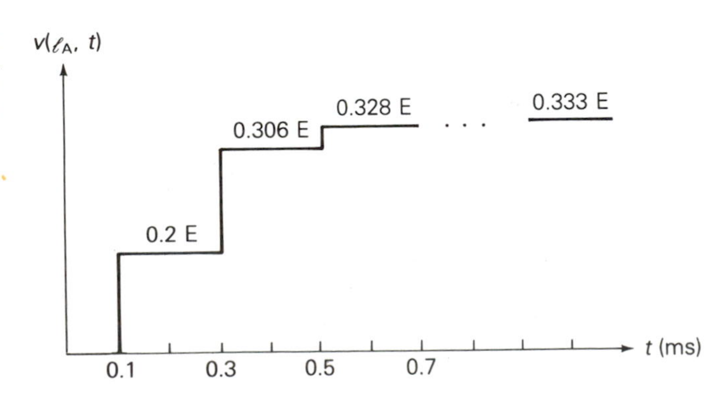

Figure 11.12

Junction voltage for Example 11.6

of the lattice diagram, where $t = 0$. Then, moving down the lattice diagram, voltage waves either just to the left or just to the right of the junction are added when they occur. For example, looking just to the right of the junction at $x = l_A^+$, the voltage wave v_3, a step of magnitude E/5 volts occurs at $t = \tau_A$. Then at $t = (\tau_A + 2\tau_B)$, two waves v_4 and v_5, which are steps of magnitude E/15 and E/25, are added to v_3. $v(l_A, t)$ is plotted in Figure 11.12.

The steady-state voltage is determined by removing the lossless lines and calculating the steady-state voltage across the receiving-end load:

$$v_{ss}(x) = E\left(\frac{Z_R}{Z_R + Z_G}\right) = E\left(\frac{200}{200 + 400}\right) = \frac{E}{3} \qquad \triangleleft$$

The preceding analysis can be extended to the junction of more than two lossless lines, as shown in Figure 11.13. Writing a KVL and KCL equation at the junction for a voltage V_A^+ arriving at the junction from line A,

$$V_A^+ + V_A^- = V_B^+ = V_C^+ = V_D^+ = \cdots \tag{11.3.8}$$

$$I_A^+ + I_A^- = I_B^+ + I_C^+ + I_D^+ + \cdots \tag{11.3.9}$$

Using $I_A^+ = V_A^+/Z_A$, $I_A^- = -V_A^-/Z_A$, $I_B^+ = V_B^+/Z_B$, and so on in (11.3.9),

$$\frac{V_A^+}{Z_A} - \frac{V_A^-}{Z_A} = \frac{V_B^+}{Z_B} + \frac{V_C^+}{Z_C} + \frac{V_D^+}{Z_D} + \cdots \tag{11.3.10}$$

Figure 11.13

Junction of lossless lines A, B, C, D, and so on

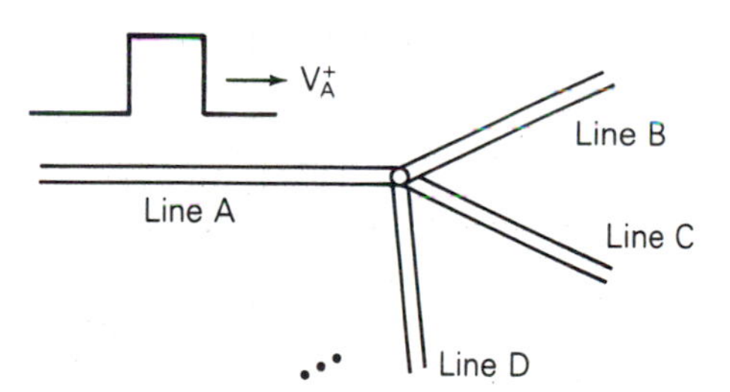

Equations (11.3.8) and (11.3.10) can be solved for V_A^-, V_B^+, V_C^+, V_D^+, and so on in terms of V_A^+. (See Problem 11.11.)

Section 11.4

Discrete-Time Models of Single-phase Lossless Lines and Lumped RLC Elements

Our objective in this section is to develop discrete-time models of single-phase lossless lines and of lumped RLC elements suitable for computer calculation of transmission-line transients at discrete-time intervals $t = \Delta t$, $2\Delta t$, $3\Delta t$, and so on. The discrete-time models are presented as equivalent circuits consisting of lumped resistors and current sources. The current sources in the models represent the past history of the circuit—that is, the history at times $t - \Delta t$, $t - 2\Delta t$, and so on. After interconnecting the equivalent circuits of all the components in any given circuit, nodal equations can then be written for each discrete time. Discrete-time models, first developed by L. Bergeron [3], are presented first.

Single-phase Lossless Line

From the general solution of a single-phase lossless line, given by (11.1.12) and (11.1.18), we obtain

$$v(x, t) + Z_c i(x, t) = 2v^{+}\left(t - \frac{x}{v}\right) \tag{11.4.1}$$

$$v(x, t) - Z_c i(x, t) = 2v^{-}\left(t + \frac{x}{v}\right) \tag{11.4.2}$$

In (11.4.1), the left side $(v + Z_c i)$ remains constant when the argument $(t - x/v)$ is constant. Therefore, to a fictitious observer traveling at velocity v in the positive x direction along the line, $(v + Z_c i)$ remains constant. If τ is the transit time from terminal k to terminal m of the line, the value of $(v + Z_c i)$ observed at time $(t - \tau)$ at terminal k must equal the value at time t at terminal m. That is,

$$v_k(t - \tau) + Z_c i_k(t - \tau) = v_m(t) + Z_c i_m(t) \tag{11.4.3}$$

where k and m denote terminals k and m, as shown in Figure 11.14(a).

Similarly, $(v - Z_c i)$ in (11.4.2) remains constant when $(t + x/v)$ is constant. To a fictitious observer traveling at velocity v in the negative x direction, $(v - Z_c i)$ remains constant. Therefore, the value of $(v - Z_c i)$ at time $(t - \tau)$ at terminal m must equal the value at time t at terminal k. That is,

$$v_m(t - \tau) - Z_c i_m(t - \tau) = v_k(t) - Z_c i_k(t) \tag{11.4.4}$$

Equation (11.4.3) is rewritten as

$$i_m(t) = I_m(t - \tau) - \frac{1}{Z_c} v_m(t) \tag{11.4.5}$$

where

$$I_m(t - \tau) = i_k(t - \tau) + \frac{1}{Z_c} v_k(t - \tau) \tag{11.4.6}$$

Similarly, (11.4.4) is rewritten as

$$i_k(t) = I_k(t - \tau) + \frac{1}{Z_c} v_k(t) \tag{11.4.7}$$

where

$$I_k(t-\tau) = i_m(t-\tau) - \frac{1}{Z_c} v_m(t-\tau) \tag{11.4.8}$$

Also, using (11.4.7) in (11.4.6),

$$I_m(t-\tau) = I_k(t-2\tau) + \frac{2}{Z_c} v_k(t-\tau) \tag{11.4.9}$$

and using (11.4.5) in (11.4.8),

$$I_k(t-\tau) = I_m(t-2\tau) - \frac{2}{Z_c} v_m(t-\tau) \tag{11.4.10}$$

Figure 11.14

Single-phase two-wire lossless line

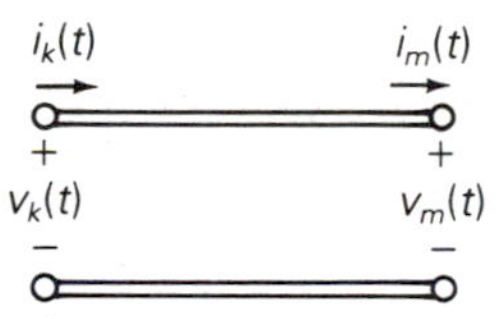

(a) Terminal variables

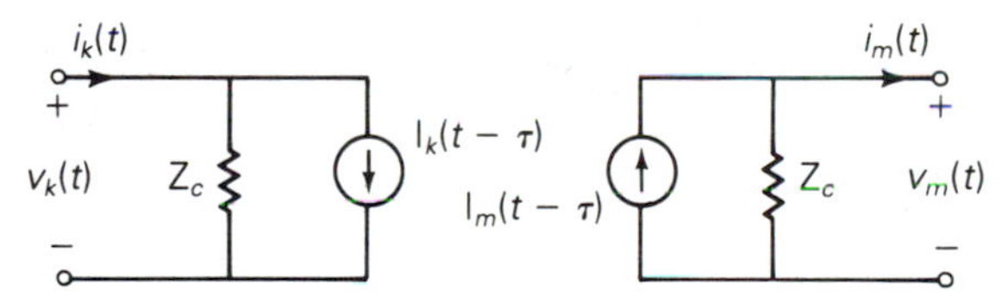

(b) Discrete-time equivalent circuit

Equations (11.4.5) and (11.4.7) are represented by the circuit shown in Figure 11.14(b). The current sources $I_m(t-\tau)$ and $I_k(t-\tau)$ shown in this figure, which are given by (11.4.9) and (11.4.10), represent the past history of the transmission line.

Note that in Figure 11.14(b) terminals k and m are not directly connected. The conditions at one terminal are "felt" indirectly at the other terminal after a delay of τ seconds.

Lumped Inductance

As shown in Figure 11.15(a) for a constant lumped inductance L,

$$v(t) = L\frac{di(t)}{dt} \tag{11.4.11}$$

Integrating this equation from time $(t - \Delta t)$ to t,

$$\int_{t-\Delta t}^{t} di(t) = \frac{1}{L}\int_{t-\Delta t}^{t} v(t)\,dt \tag{11.4.12}$$

Using the trapezoidal rule of integration,

$$i(t) - i(t-\Delta t) = \left(\frac{1}{L}\right)\left(\frac{\Delta t}{2}\right)[v(t) + v(t-\Delta t)]$$

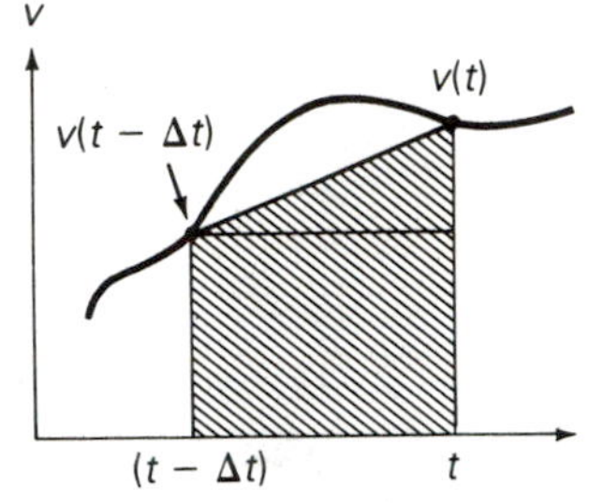

Trapezoidal Integration Rule

Rearranging:

$$i(t) = \frac{v(t)}{(2L/\Delta t)} + \left[i(t-\Delta t) + \frac{v(t-\Delta t)}{(2L/\Delta t)}\right]$$

or

$$i(t) = \frac{v(t)}{(2L/\Delta t)} + I_L(t-\Delta t) \tag{11.4.13}$$

where

$$I_L(t-\Delta t) = i(t-\Delta t) + \frac{v(t-\Delta t)}{(2L/\Delta t)} = I_L(t-2\Delta t) + \frac{v(t-\Delta t)}{(L/\Delta t)} \tag{11.4.14}$$

Figure 11.15

Lumped inductance

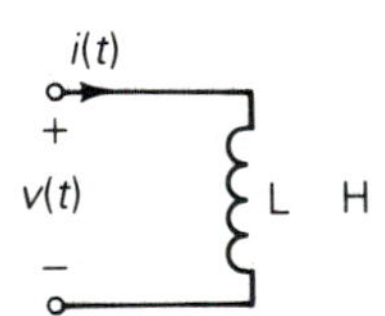

(a) Continuous time circuit

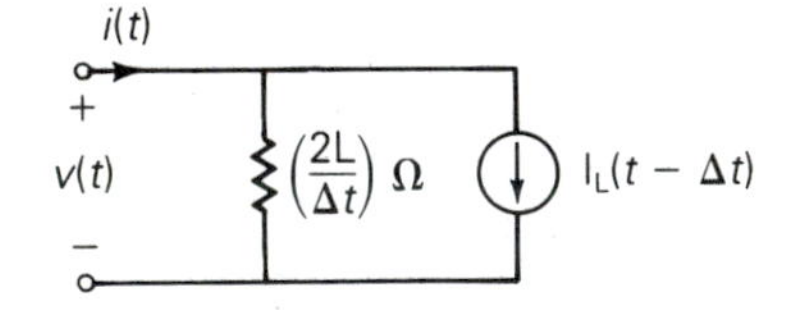

(b) Discrete-time circuit

Equations (11.4.13) and (11.4.14) are represented by the circuit shown in Figure 11.15(b). As shown, the inductor is replaced by a resistor with resistance $(2L/\Delta t)$ Ω. A current source $I_L(t-\Delta t)$ given by (11.4.14) is also included. $I_L(t-\Delta t)$ represents the past history of the inductor. Note that the trapezoidal rule introduces an error of the order $(\Delta t)^3$.

Lumped Capacitance

As shown in Figure 11.16(a) for a constant lumped capacitance C,

$$i(t) = C\frac{dv(t)}{dt} \tag{11.4.15}$$

Figure 11.16

Lumped capacitance

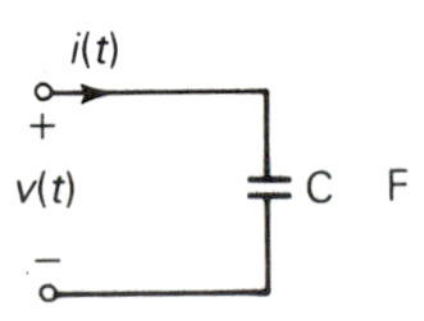

(a) Continuous time circuit

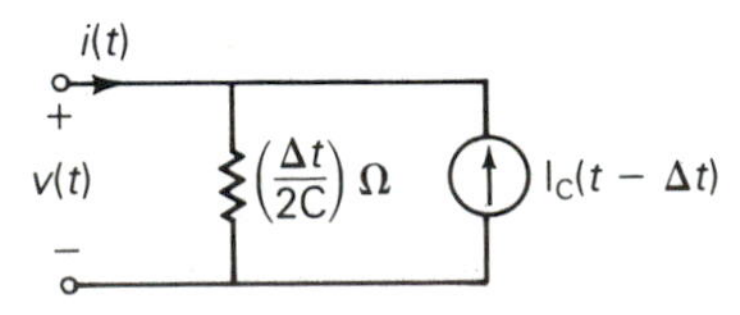

(b) Discrete-time circuit

Integrating from time $(t - \Delta t)$ to t,

$$\int_{t-\Delta t}^{t} dv(t) = \frac{1}{C}\int_{t-\Delta t}^{t} i(t)\,dt \tag{11.4.16}$$

Using the trapezoidal rule of integration,

$$v(t) - v(t - \Delta t) = \frac{1}{C}\left(\frac{\Delta t}{2}\right)[i(t) + i(t - \Delta t)]$$

Rearranging,

$$i(t) = \frac{v(t)}{(\Delta t/2C)} - I_C(t - \Delta t) \tag{11.4.17}$$

where

$$I_C(t - \Delta t) = i(t - \Delta t) + \frac{v(t - \Delta t)}{(\Delta t/2C)} = -I_C(t - 2\Delta t) + \frac{v(t - \Delta t)}{(\Delta t/4C)} \tag{11.4.18}$$

Equations (11.4.17) and (11.4.18) are represented by the circuit in Figure 11.16(b). The capacitor is replaced by a resistor with resistance $(\Delta t/2C)\,\Omega$. A current source $I_C(t - \Delta t)$, which represents the capacitor's past history, is also included.

Lumped Resistance

The discrete model of a constant lumped resistance R, as shown in Figure 11.17, is the same as the continuous model. That is,

$$v(t) = Ri(t) \tag{11.4.19}$$

Figure 11.17

Lumped resistance

(a) Continuous time circuit (b) Discrete-time circuit

Nodal Equations

A circuit consisting of single-phase lossless transmission lines and constant lumped RLC elements can be replaced by the equivalent circuits given in Figures 11.14(b), 11.15(b), 11.16(b), and 11.17(b). Then, writing nodal equations, the result is a set of linear algebraic equations that determine the bus voltages at each instant t.

Example 11.7 For the circuit given in Example 11.3, replace the circuit elements by their discrete-time equivalent circuits and write the nodal equations that determine the sending-end and receiving-end voltages. Then, using a digital computer, compute the sending-end and receiving-end voltages for $0 \leqq t \leqq 9$ ms. For numerical calculations, assume $E = 100$ V, $Z_c = 400\ \Omega$, $C_R = 5\ \mu F$, $\tau = 1.0$ ms, and $\Delta t = 0.1$ ms.

Solution The discrete model is shown in Figure 11.18, where $v_k(t)$ represents the sending-end voltage $v(0, t)$ and $v_m(t)$ represents the receiving-end voltage $v(l, t)$. Also, the sending-end voltage source $e_G(t)$ in series with Z_G is converted to an equivalent current source in parallel with Z_G. Writing nodal equations for this circuit,

$$\left[\begin{array}{c|c} \left(\frac{1}{400} + \frac{1}{400}\right) & 0 \\ \hline 0 & \left(\frac{1}{400} + \frac{1}{10}\right) \end{array}\right] \begin{bmatrix} v_k(t) \\ v_m(t) \end{bmatrix} = \begin{bmatrix} \frac{1}{4} - I_k(t - 1.0) \\ I_m(t - 1.0) + I_C(t - 0.1) \end{bmatrix}$$

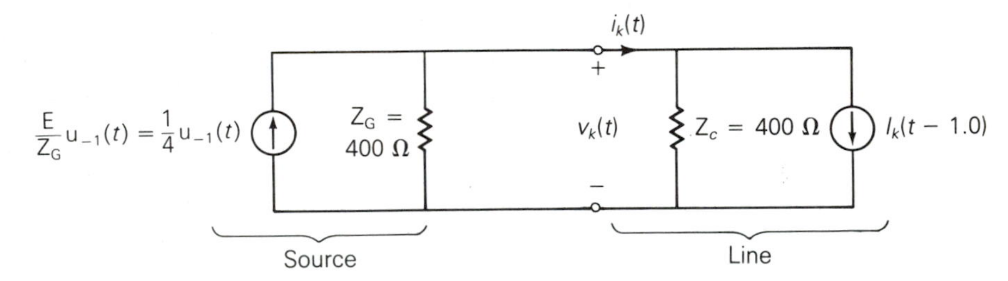

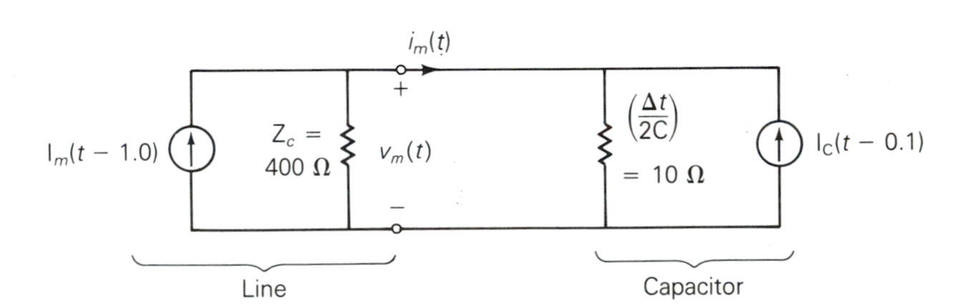

Figure 11.18 Discrete-time equivalent circuit for Example 11.7

Solving,

$$v_k(t) = 200[\tfrac{1}{4} - I_k(t - 1.0)] \tag{a}$$

$$v_m(t) = 9.75610[I_m(t - 1.0) + I_C(t - 0.1)] \tag{b}$$

The current sources in these equations are, from (11.4.9), (11.4.10), and (11.4.18), with the argument $(t - \tau)$ replaced by t,

$$I_m(t) = I_k(t - 1.0) + \tfrac{2}{400}v_k(t) \tag{c}$$

$$I_k(t) = I_m(t - 1.0) - \tfrac{2}{400} v_m(t) \qquad \text{(d)}$$

$$I_C(t) = -I_C(t - 0.1) + \tfrac{1}{5} v_m(t) \qquad \text{(e)}$$

Equations (a) through (e) above are in a form suitable for digital computer solution. A scheme for iteratively computing v_k and v_m is as follows, starting at $t = 0$:

1. Compute $v_k(t)$ and $v_m(t)$ from equations (a) and (b).
2. Compute $I_m(t)$, $I_k(t)$, and $I_C(t)$ from equations (c), (d), and (e). Store $I_m(t)$ and $I_k(t)$.
3. Change t to $(t + \Delta t) = (t + 0.1)$ and return to (1) above.

Note that since the transmission line and capacitor are unenergized for time t less than zero, the current sources $I_m(\ )$, $I_k(\ )$, and $I_C(\ )$ are zero whenever their arguments ( ) are negative. Note also from equations (a) through (e) that it is necessary to store the past ten values of $I_m(\ )$ and $I_k(\ )$.

A personal computer program written in BASIC that executes the above scheme and the computational results are shown in Figure 11.19. The plotted sending-end voltage $v_k(t)$ can be compared with the results of Example 11.3. ◁

Example 11.7 can be generalized to compute bus voltages at discrete-time intervals for an arbitrary number of buses, single-phase lossless lines, and lumped RLC elements. When current sources instead of voltage sources are employed, the unknowns are all bus voltages, for which nodal equations $\mathbf{YV} = \mathbf{I}$ can be written at each discrete-time instant. Also, the dependent current sources in $\mathbf{I}$ are written in terms of bus voltages and current sources at prior times. For computational convenience, the time interval Δt can be chosen constant so that the bus admittance matrix $\mathbf{Y}$ is a constant real symmetric matrix as long as the RLC elements are constant.

OUTPUT

Time ms	VK Volts	VM Volts
0.00	50.00	0.00
0.20	50.00	0.00
0.40	50.00	0.00
0.60	50.00	0.00
0.80	50.00	0.00
1.00	50.00	2.44
1.20	50.00	11.73
1.40	50.00	20.13
1.60	50.00	27.73
1.80	50.00	34.61
2.00	2.44	40.83
2.20	11.73	46.46
2.40	20.13	51.56
2.60	27.73	56.17
2.80	34.61	60.34
3.00	40.83	64.12
3.20	46.46	67.53
3.40	51.56	70.62
3.60	56.17	73.42
3.80	60.34	75.95
4.00	64.12	78.24
4.20	67.53	80.31
4.40	70.62	82.18
4.60	73.42	83.88
4.80	75.95	85.41
5.00	78.24	86.80
5.20	80.31	88.06
5.40	82.18	89.20
5.60	83.88	90.22
5.80	85.41	91.15
6.00	86.80	92.00
6.20	88.06	92.76
6.40	89.20	93.45
6.60	90.22	94.07
6.80	91.15	94.64
7.00	92.00	95.15
7.20	92.76	95.61
7.40	93.45	96.03
7.60	94.07	96.40
7.80	94.64	96.75
8.00	95.15	97.06
8.20	95.61	97.34
8.40	96.03	97.59
8.60	96.40	97.82
8.80	96.75	98.03
9.00	97.06	98.22

COMPUTER PROGRAM LISTING

```
10  REM EXAMPLE 11.7
20  LPRINT " TIME     VK      VM  "
30  LPRINT " ms      Volts   Volts  "
40  IC = 0
50  T = 0
60  KPRINT = 2
65  REM T IS TIME. KPRINT DETERMINES THE PRINTOUT INTERVAL.
70  REM LINES 110 TO 210 COMPUTE EQS(a)-(e) FOR THE FIRST
80  REM TEN TIME STEPS (A TOTAL OF ONE ms) DURING WHICH
90  REM THE CURRENT SOURCES ON THE RIGHT HAND SIDE
100 REM OF THE EQUATIONS ARE ZERO. TEN VALUES OF
105 REM CURRENT SOURCES IK(J) AND IM(J) ARE STORED.
110 FOR J = 1 TO 10
120 VK = 200/4
130 VM = 9.7561*IC
140 IM(J) = (2/400)*VK
150 IK(J) = (-2/400)*VM
160 IC = -IC + (1/5)*VM
170 Z = (J - 1)/KPRINT
180 M = INT(Z)
190 IF M = Z THEN LPRINT USING "*** **  "; T,VK,VM
200 T = T + .1
210 NEXT J
220 REM LINES 250 TO 420 COMPUTE EQS(a)-(e) FOR TIME T
230 REM EQUAL TO AND GREATER THAN 1.0 ms. THE PAST TEN
240 REM VALUES OF IK(J) AND IM(J) ARE STORED.
250 FOR J = 1 TO 10
260 REM LINE 270 IS EQ(a).
270 VK = 200*((1/4) -IK(J))
280 REM LINE 290 IS EQ(b).
290 VM = 9.7561*(IM(J) + IC)
300 REM LINE 310 IS EQ(e).
310 IC = -IC + (1/5)*VM
320 REM LINES 330-360 ARE EQS (c) and (d).
330 C1 = IK(J) + (2/400)*VK
340 C2 = IM(J) - (2/400)*VM
350 IM(J) = C1
360 IK(J) = C2
370 Z = (J - 1)/KPRINT
380 M = INT(Z)
390 IF M = Z THEN LPRINT USING "*** **  "; T,VK,VM
400 T = T + .1
410 NEXT J
420 IF T < 9.0 THEN GOTO 250
430 STOP
```

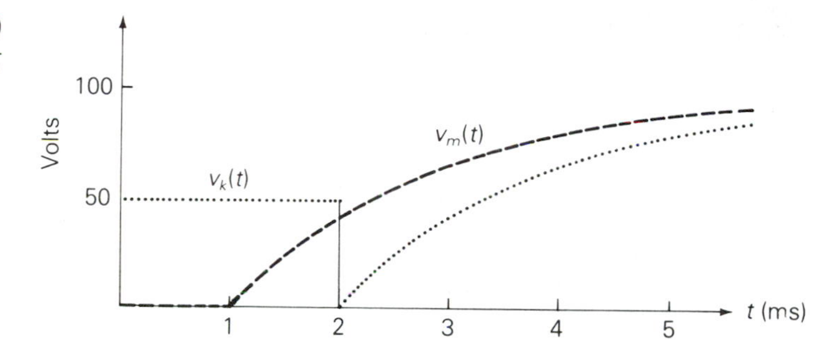

Figure 11.19

Example 11.7

Section 11.5

Lossy Lines

Transmission-line series resistance or shunt conductance causes the following:

1. Attenuation
2. Distortion
3. Power losses

These effects are briefly discussed as follows.

Attenuation When constant series resistance R Ω/m and shunt conductance G S/m are included in the circuit of Figure 11.1 for a single-phase two-wire line, (11.1.3) and (11.1.4) become

$$\frac{\partial v(x,t)}{\partial x} = -\mathrm{R}i(x,t) - \mathrm{L}\frac{\partial i(x,t)}{\partial t} \tag{11.5.1}$$

$$\frac{\partial i(x,t)}{\partial x} = -\mathrm{G}v(x,t) - \mathrm{C}\frac{\partial v(x,t)}{\partial t} \tag{11.5.2}$$

Taking the Laplace transform of these, equations analogous to (11.1.7) and (11.1.8) are

$$\frac{d^2\mathrm{V}(x,s)}{dx^2} - \gamma^2(s)\mathrm{V}(x,s) = 0 \tag{11.5.3}$$

$$\frac{d^2\mathrm{I}(x,s)}{dx^2} - \gamma^2(s)\mathrm{I}(x,s) = 0 \tag{11.5.4}$$

where $\gamma(s) = \sqrt{(\mathrm{R} + s\mathrm{L})(\mathrm{G} + s\mathrm{C})}$ (11.5.5)

The solution to these equations is

$$V(x, s) = V^+(s)e^{-\gamma(s)x} + V^-(s)e^{+\gamma(s)x} \tag{11.5.6}$$

$$I(x, s) = I^+(s)e^{-\gamma(s)x} + I^-(s)e^{+\gamma(s)x} \tag{11.5.7}$$

In general, it is impossible to obtain a closed form expression for $v(x, t)$ and $i(x, t)$, which are the inverse Laplace transforms of these equations. However, for the special case of a *distortionless* line, which has the property $R/L = G/C$, the inverse Laplace transform can be obtained as follows. Rewrite (11.5.5) as

$$\gamma(s) = \sqrt{LC[(s + \delta)^2 - \sigma^2]} \tag{11.5.8}$$

where

$$\delta = \frac{1}{2}\left(\frac{R}{L} + \frac{G}{C}\right) \tag{11.5.9}$$

$$\sigma = \frac{1}{2}\left(\frac{R}{L} - \frac{G}{C}\right) \tag{11.5.10}$$

For a distortionless line, $\sigma = 0$, $\delta = R/L$, and (11.5.6) and (11.5.7) become

$$V(x, s) = V^+(s)e^{-\sqrt{LC}[s + (R/L)]x} + V^-(s)e^{+\sqrt{LC}[s + (R/L)]x} \tag{11.5.11}$$

$$I(x, s) = I^+(s)e^{-\sqrt{LC}[s + (R/L)]x} + I^-(s)e^{+\sqrt{LC}[s + (R/L)]x} \tag{11.5.12}$$

Using $v = 1/\sqrt{LC}$ and $\sqrt{LC}(R/L) = \sqrt{RG} = \alpha$ for the distortionless line, the inverse transform of these equations is

$$v(x, t) = e^{-\alpha x}v^+\left(t - \frac{x}{v}\right) + e^{+\alpha x}v^-\left(t + \frac{x}{v}\right) \tag{11.5.13}$$

$$i(x, t) = e^{-\alpha x}i^+\left(t - \frac{x}{v}\right) + e^{\alpha x}i^-\left(t + \frac{x}{v}\right) \tag{11.5.14}$$

These voltage and current waves consist of forward and backward traveling waves similar to (11.1.12) and (11.1.13) for a lossless line. However, for the lossy distortionless line, the waves are attenuated versus x due to the $e^{\pm\alpha x}$ terms. Note that the attenuation term $\alpha = \sqrt{RG}$ is constant. Also, the attenuated waves travel at constant velocity $v = 1/\sqrt{LC}$. Therefore, waves traveling along the distortionless line do not change their shape; only their magnitudes are attenuated.

Distortion For sinusoidal-steady-state waves, the propagation constant $\gamma(j\omega)$ is, from (11.5.5), with $s = j\omega$

$$\gamma(j\omega) = \sqrt{(R + j\omega L)(G + j\omega C)} = \alpha + j\beta \tag{11.5.15}$$

For a lossless line, $R = G = 0$; therefore, $\alpha = 0$, $\beta = \omega\sqrt{LC}$, and the phase velocity $v = \omega/\beta = 1/\sqrt{LC}$ is constant. Thus sinusoidal waves of all frequencies travel at constant velocity v without attenuation along a lossless line.

For a distortionless line (R/L) = (G/C), and $\gamma(j\omega)$ can be rewritten, using (11.5.8)–(11.5.10), as

$$\gamma(j\omega) = \sqrt{LC\left(j\omega + \frac{R}{L}\right)^2} = \sqrt{LC}\left(j\omega + \frac{R}{L}\right)$$

$$= \sqrt{RG} + j\frac{\omega}{v} = \alpha + j\beta \tag{11.5.16}$$

Since $\alpha = \sqrt{RG}$ and $v = 1/\sqrt{LC}$ are constant, sinusoidal waves of all frequencies travel along the distortionless line at constant velocity with constant attenuation—that is, without distortion.

It can also be shown that for frequencies above 1 MHz, practical transmission lines with typical constants R, L, G, and C tend to be distortionless. Above 1 MHz, α and β can be approximated by

$$\alpha \simeq \frac{R}{2}\sqrt{\frac{C}{L}} + \frac{G}{2}\sqrt{\frac{L}{C}} \tag{11.5.17}$$

$$\beta \simeq \omega\sqrt{LC} = \frac{\omega}{v} \tag{11.5.18}$$

Therefore, sinusoidal waves with frequencies above 1 MHz travel along a practical line undistorted at constant velocity $v = 1/\sqrt{LC}$, with attenuation α given by (11.5.17).

At frequencies below 1 MHz these approximations do not hold, and lines are generally not distortionless. For typical transmission and distribution lines, (R/L) is much greater than (G/C) by a factor of 1000 or so. Therefore, the condition (R/L) = (G/C) for a distortionless line does not hold.

Figure 11.20 shows the effect of distortion and attenuation of voltage surges based on experiments on a 132-kV overhead transmission line [4]. The shapes of the surges at three points along the line are shown. Note how distortion reduces the front of the wave and builds up the tail as it travels along the line.

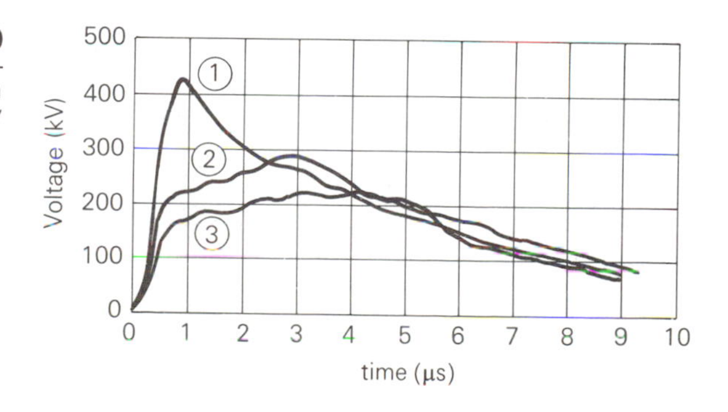

Figure 11.20

Distortion and attenuation of surges on a 132-kV overhead line [4]

Power Losses Power losses are associated with series resistance R and shunt conductance G. When a current I flows along a line, I^2R losses occur, and when a voltage V appears across the conductors, V^2G losses occur. V^2G losses are primarily due to insulator leakage and corona for overhead lines, and to dielectric losses for cables. For practical lines operating at rated voltage and rated current, I^2R losses are much greater than V^2G losses.

As discussed above, the analysis of transients on single-phase two-wire lossy lines with constant parameters R, L, G, and C is complicated. The analysis becomes more complicated when skin effect is included, which means that R is not constant but frequency-dependent. Additional complications arise for a single-phase line consisting of one conductor with earth return, where Carson [5] has shown that both series resistance and inductance are frequency-dependent.

In view of these complications, the solution of transients on lossy lines is best handled via digital computation techniques. A single-phase line of length l can be approximated by a lossless line with half the total resistance $(Rl/2)$ Ω lumped in series with the line at both ends. For improved accuracy, the line can be divided into various line sections, and each section can be approximated by a lossless line section, with a series resistance lumped at both ends. Simulations have shown that accuracy does not significantly improve with more than two line sections.

Discrete-time equivalent circuits of a single-phase lossless line, Figure 11.14, together with a constant lumped resistance, Figure 11.17, can be used to approximate a lossy line section, as shown in Figure 11.21. Also, digital techniques for modeling frequency-dependent line parameters [6, 7] are available but are not discussed here.

Figure 11.21

Approximate model of a lossy line segment

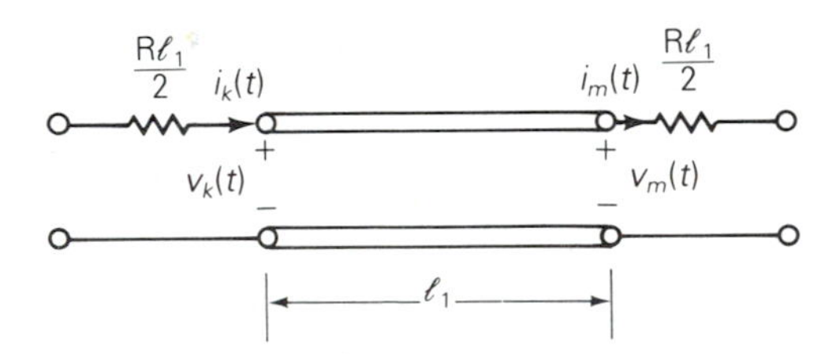

(a) Lossless line segment of length ℓ_1 with lumped line resistance

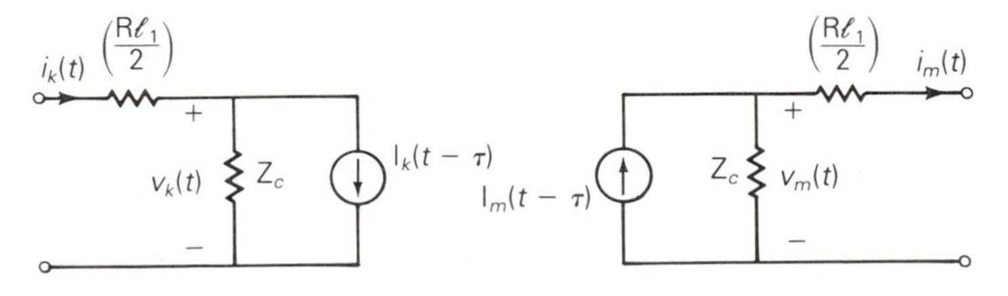

(b) Discrete-time model

Section 11.6

Multiconductor Lines

Up to now we have considered transients on single-phase two-wire lines. For a transmission line with n conductors above a ground plane, waves travel in n "modes," where each mode has its own wave velocity and its own surge impedance. In this section we illustrate "modal analysis" for a relatively simple three-phase line [8].

Given a three-phase, lossless, completely transposed line consisting of three conductors above a perfectly conducting ground plane, the transmission-line equations are

$$\frac{d\mathbf{V}(x,s)}{dx} = -s\mathbf{L}\mathbf{I}(x,s) \tag{11.6.1}$$

$$\frac{d\mathbf{I}(x,s)}{dx} = -s\mathbf{C}\mathbf{V}(x,s) \tag{11.6.2}$$

where

$$\mathbf{V}(x,s) = \begin{bmatrix} V_{ag}(x,s) \\ V_{bg}(x,s) \\ V_{cg}(x,s) \end{bmatrix} \qquad \mathbf{I}(x,s) = \begin{bmatrix} I_a(x,s) \\ I_b(x,s) \\ I_c(x,s) \end{bmatrix} \tag{11.6.3}$$

Equations (11.6.1) and (11.6.2) are identical to (11.1.5) and (11.1.6) except that scalar quantities are replaced by vector quantities. $\mathbf{V}(x,s)$ is the vector of line-to-ground voltages and $\mathbf{I}(x,s)$ is the vector of line currents. For a completely transposed line, the 3×3 inductance matrix $\mathbf{L}$ and capacitance matrix $\mathbf{C}$ are given by

$$\mathbf{L} = \begin{bmatrix} L_s & L_m & L_m \\ L_m & L_s & L_m \\ L_m & L_m & L_s \end{bmatrix} \quad \text{H/m} \tag{11.6.4}$$

$$\mathbf{C} = \begin{bmatrix} C_s & C_m & C_m \\ C_m & C_s & C_m \\ C_m & C_m & C_s \end{bmatrix} \quad \text{F/m} \tag{11.6.5}$$

For any given line configuration, $\mathbf{L}$ and $\mathbf{C}$ can be computed from the equations given in Sections 5.7 and 5.11. Note that L_s, L_m, and C_s are positive, whereas C_m is negative.

We now transform the phase quantities to modal quantities. First we define

$$\begin{bmatrix} V_{ag}(x,s) \\ V_{bg}(x,s) \\ V_{cg}(x,s) \end{bmatrix} = \mathbf{T}_V \begin{bmatrix} V^0(x,s) \\ V^+(x,s) \\ V^-(x,s) \end{bmatrix} \tag{11.6.6}$$

$$\begin{bmatrix} I_a(x,s) \\ I_b(x,s) \\ I_c(x,s) \end{bmatrix} = \mathbf{T}_I \begin{bmatrix} I^0(x,s) \\ I^+(x,s) \\ I^-(x,s) \end{bmatrix} \tag{11.6.7}$$

$V^0(x,s)$, $V^+(x,s)$, and $V^-(x,s)$ are denoted *zero-mode*, *positive-mode*, and *negative-mode* voltages, respectively. Similarly, $I^0(x,s)$, $I^+(x,s)$, and $I^-(x,s)$ are *zero-*, *positive-*, and *negative-mode* currents. $\mathbf{T}_V$ and $\mathbf{T}_I$ are 3×3 constant transformation matrices, soon to be specified. Denoting $\mathbf{V}_m(x,s)$ and $\mathbf{I}_m(x,s)$ as the modal voltage and modal current vectors,

$$\mathbf{V}(x,s) = \mathbf{T}_V\mathbf{V}_m(x,s) \tag{11.6.8}$$

$$\mathbf{I}(x,s) = \mathbf{T}_I\mathbf{I}_m(x,s) \tag{11.6.9}$$

Using (11.6.8) and (11.6.9) in (11.6.1),

$$\mathbf{T}_V \frac{d\mathbf{V}_m(x,s)}{dx} = -s\mathbf{L}\mathbf{T}_I\mathbf{I}_m(x,s)$$

or

$$\frac{d\mathbf{V}_m(x,s)}{dx} = -s(\mathbf{T}_v^{-1}\mathbf{L}\mathbf{T}_I)\mathbf{I}_m(x,s) \tag{11.6.10}$$

Similarly, using (11.6.8) and (11.6.9) in (11.6.2),

$$\frac{d\mathbf{I}_m(x,s)}{dx} = -s(\mathbf{T}_I^{-1}\mathbf{C}\mathbf{T}_v)\mathbf{V}_m(x,s) \tag{11.6.11}$$

The objective of the modal transformation is to diagonalize the matrix products within the parentheses of (11.6.10) and (11.6.11), thereby decoupling these vector equations. For a three-phase completely transposed line, $\mathbf{T}_V$ and $\mathbf{T}_I$ are given by

$$\mathbf{T}_V = \mathbf{T}_I = \begin{bmatrix} 1 & 1 & 1 \\ 1 & -2 & 1 \\ 1 & 1 & -2 \end{bmatrix} \tag{11.6.12}$$

Also, the inverse transformation matrices are

$$\mathbf{T}_V^{-1} = \mathbf{T}_I^{-1} = \frac{1}{3}\begin{bmatrix} 1 & 1 & 1 \\ 1 & -1 & 0 \\ 1 & 0 & -1 \end{bmatrix} \tag{11.6.13}$$

Substituting (11.6.12), (11.6.13), (11.6.4), and (11.6.5) into (11.6.10) and (11.6.11) yields

$$\frac{d}{dx}\begin{bmatrix} V^0(x,s) \\ V^+(x,s) \\ V^-(x,s) \end{bmatrix} = \begin{bmatrix} -s(L_s+2L_m) & 0 & 0 \\ 0 & -s(L_s-L_m) & 0 \\ 0 & 0 & -s(L_s-L_m) \end{bmatrix}\begin{bmatrix} I^0(x,s) \\ I^+(x,s) \\ I^-(x,s) \end{bmatrix} \tag{11.6.14}$$

$$\frac{d}{dx}\begin{bmatrix} I^0(x,s) \\ I^+(x,s) \\ I^-(x,s) \end{bmatrix} = \begin{bmatrix} -s(C_s+2C_m) & 0 & 0 \\ 0 & -s(C_s-C_m) & 0 \\ 0 & 0 & -s(C_s-C_m) \end{bmatrix}\begin{bmatrix} V^0(x,s) \\ V^+(x,s) \\ V^-(x,s) \end{bmatrix} \tag{11.6.15}$$

From (11.6.14) and (11.6.15), the zero-mode equations are

$$\frac{dV^0(x,s)}{dx} = -s(L_s + 2L_m)I^0(x,s) \tag{11.6.16}$$

$$\frac{dI^0(x,s)}{dx} = -s(C_s + 2C_m)V^0(x,s) \tag{11.6.17}$$

These equations are identical in form to those of a two-wire lossless line, (11.1.5) and (11.1.6). By analogy, the zero-mode waves travel at velocity

$$v^0 = \frac{1}{\sqrt{(L_s + 2L_m)(C_s + 2C_m)}} \quad \text{m/s} \tag{11.6.18}$$

and the zero-mode surge impedance is

$$Z_c^0 = \sqrt{\frac{L_s + 2L_m}{C_s + 2C_m}} \quad \Omega \tag{11.6.19}$$

Similarly, the positive- and negative-mode velocities and surge impedances are

$$v^{+} = v^{-} = \frac{1}{\sqrt{(L_s - L_m)(C_s - C_m)}} \quad \text{m/s} \tag{11.6.20}$$

$$Z_c^{+} = Z_c^{-} = \sqrt{\frac{L_s - L_m}{C_s - C_m}} \quad \Omega \tag{11.6.21}$$

These equations can be extended to more than three conductors—for example, to a three-phase line with shield wires or to a double-circuit three-phase line. Although the details are more complicated, the modal transformation is straightforward. There is one mode for each conductor, and each mode has its own wave velocity and its own surge impedance.

The solution of transients on multiconductor lines is best handled via digital computer methods, and such programs are available [9, 10]. Digital techniques are also available to model the following effects:

1. Nonlinear and time-varying RLC elements [8]
2. Lossy lines with frequency-dependent line parameters [6, 7, 12]

Section 11.7

Power-System Overvoltages

Overvoltages encountered by power-system equipment are of three types:

Lightning surges
Switching surges
Power frequency (50 or 60 Hz) overvoltages

Lightning

Lightning is the greatest single cause of overhead transmission and distribution line outages. Data obtained over a 14-year period from electric utility companies in the United States and Canada and covering 25,000 miles of transmission show that lightning accounted for about 26% of outages on 230-kV circuits and about 65% of outages on 345-kV circuits [13]. A similar study in Britain, also over a 14-year period, covering 50,000 faults on distribution lines shows that lightning accounted for 47% of outages on circuits up to and including 33 kV [14].

The electrical phenomena that occur within clouds leading to a lightning strike are complex and not totally understood. Several theories [15, 16, 17] generally agree, however, that charge separation occurs within clouds. Wilson

[15] postulates that falling raindrops attract negative charges and therefore leave behind masses of positively charged air. The falling raindrops bring the negative charge to the bottom of the cloud, and upward air drafts carry the positively charged air to the top of the cloud, as shown in Figure 11.22. As a result of charge separation within the cloud, positive charges are induced on the earth directly under the negative charges at the bottom of the cloud. The electric field lines shown in Figure 11.22 originate from the positive charges and terminate at the negative charges.

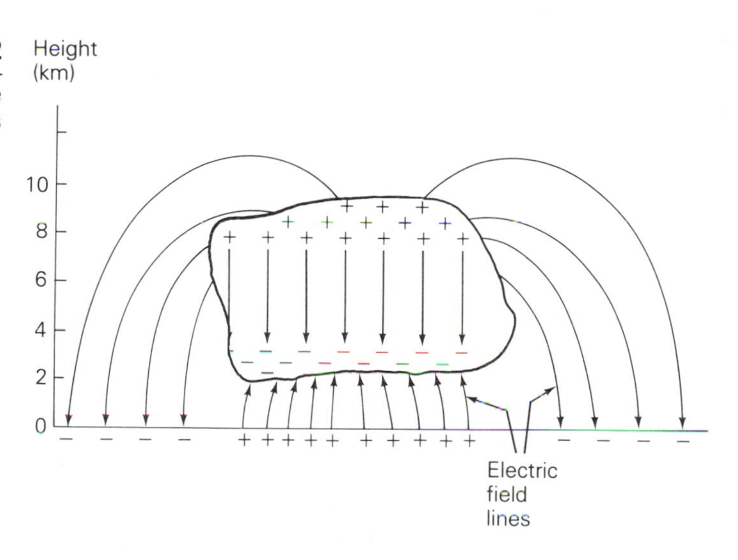

Figure 11.22 Postulation of charge separation within clouds [16]

When voltage gradients reach the breakdown strength of the humid air within the cloud, typically 5 to 15 kV/cm, an ionized path or *leader* moves from the cloud toward the earth. The leader progresses somewhat randomly along an irregular path, in steps. These leader steps, about 50 m long, move at a velocity of about 10^5 m/s. As a result of the opposite charge distribution under the cloud, another ionized path or *streamer* may rise to meet the leader. When the leader and streamer meet, a lightning discharge occurs, which neutralizes the charges.

The current involved in a lightning stroke typically rises to a peak value within 1 to 10 μs, and then diminishes to one-half the peak within 20 to 100 μs. The distribution of peak currents is shown in Figure 11.23 [20]. This curve represents the percentage of strokes that exceed a given peak current. For example, 50% of all strokes have a peak current greater than 45 kA. In extreme cases, the peak current can exceed 200 kA. Also, test results indicate that approximately 90% of all strokes are negative.

It has also been shown that what appears to the eye as a single flash of lightning is often the cumulative effect of many strokes. A typical flash consists of typically 3 to 5, and occasionally as many as 40, strokes, at intervals of 50 ms.

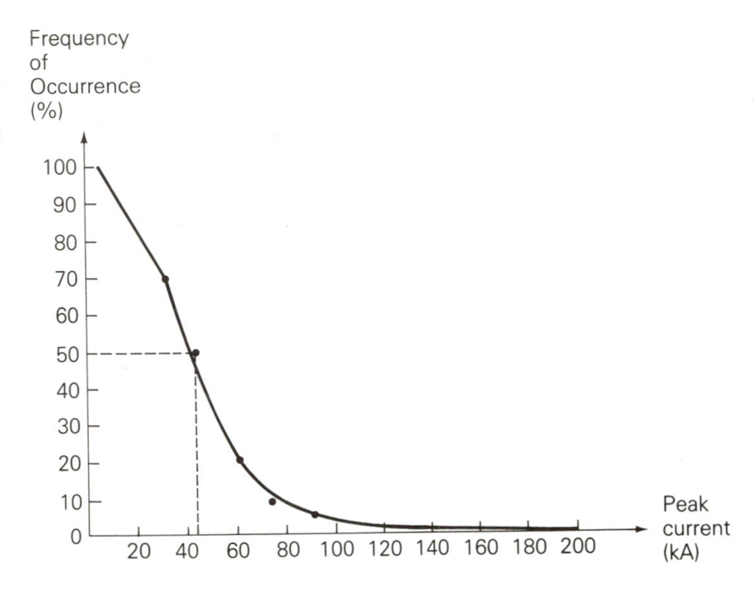

Figure 11.23

Frequency of occurrence of lightning currents that exceed a given peak value [20]

The frequency of occurrence of thunderstorms in a given geographical location is of interest to transmission-line designers. Figure 11.24 is an isokeraunic map prepared by the U.S. Weather Bureau, which shows the average number of thunderstorm days in the contiguous United States [18]. As shown, the greatest thunderstorm activity occurs in Florida, with 70 to 100 storm days per year, and in New Mexico, with a maximum of 70. On the West Coast, the average number of thunderstorm days is as low as 3 to 10 per year.

A typical transmission-line design goal is to have an average of less than 0.50 lightning outages per year per 100 miles of transmission. For a given overhead line with a specified voltage rating, the following factors affect this design goal:

1. Tower height
2. Number and location of shield wires
3. Number of standard insulator discs per phase wire
4. Tower impedance and tower-to-ground impedance

It is well known that lightning strikes tall objects. Thus shorter, H-frame structures are less susceptible to lightning strokes than taller, lattice towers. Also, shorter span lengths with more towers per kilometer can reduce the number of strikes.

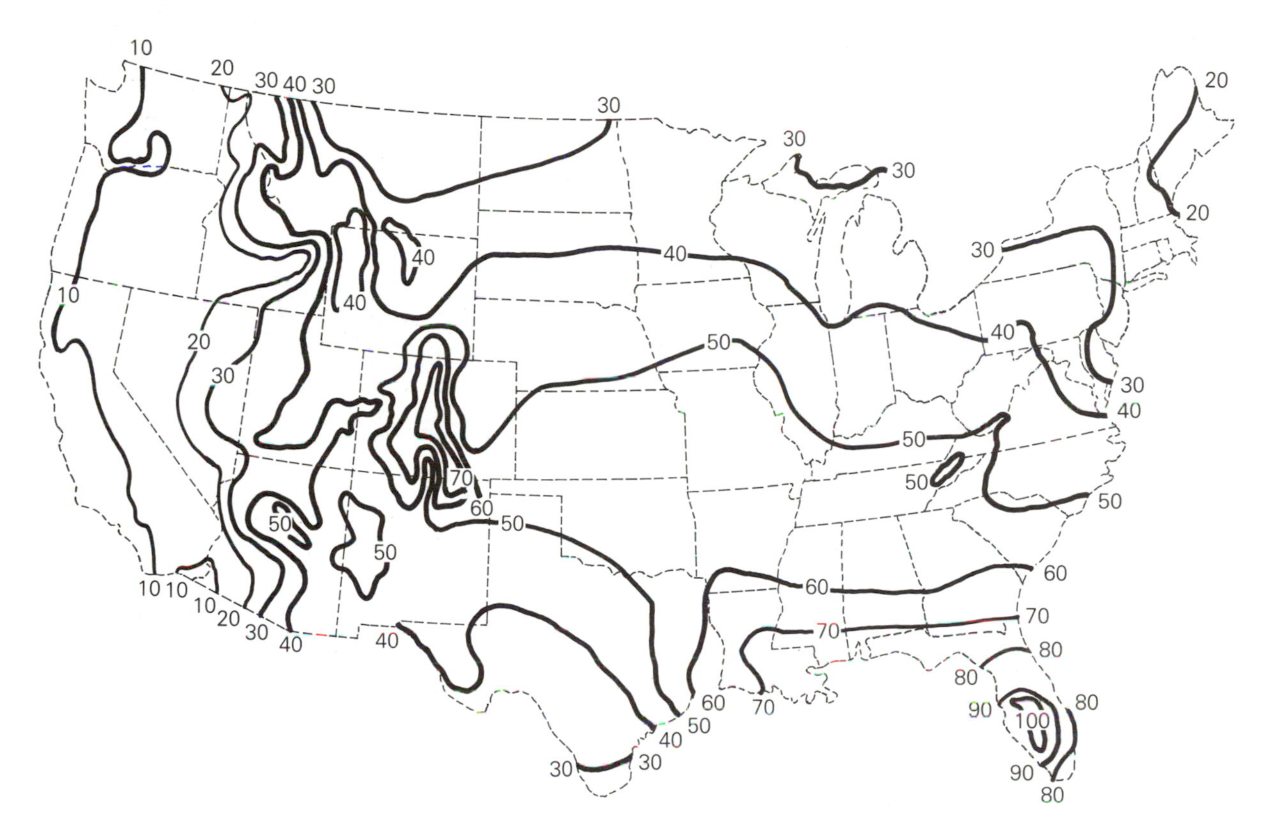

Figure 11.24 Isokeraunic map showing average thunderstorm days per year within the continental United States [18]

Shield wires installed above phase conductors can effectively shield the phase conductors from direct lightning strokes. Figure 11.25 illustrates the effect of shield wires. Experience has shown that the chance of a direct hit to phase conductors located within $\pm 30°$ arcs beneath the shield wires is reduced by a factor of 1000 [18]. Some lightning strokes are, therefore, expected to hit these overhead shield wires. When this occurs, traveling voltage and current waves propagate in both directions along the shield wire that is hit. When a wave arrives at a tower, a reflected wave returns toward the point where the lightning hit, and two refracted waves occur. One refracted wave moves along the shield wire into the next span. And since the shield wire is electrically connected to the tower, the other refracted wave moves down the tower, its energy being harmlessly diverted to ground.

However, if the tower impedance or tower-to-ground impedance is too high, IZ voltages that are produced could exceed the breakdown strength of the insulator discs that hold the phase wires. The number of insulator discs per string (see Table 5.1) is selected to avoid insulator flashover. Also, tower impedances and tower footing resistances are designed to be as low as possible.

Figure 11.25

Effect of shield wires

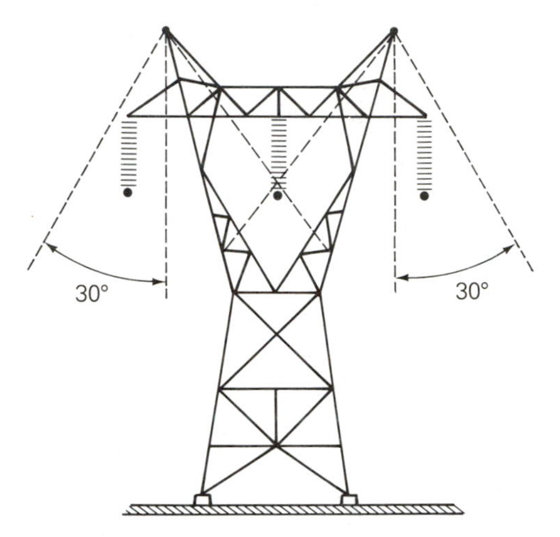

If the inherent tower construction does not give a naturally low resistance to ground, driven ground rods can be employed. Sometimes buried conductors running under the line (called *counterpoise*) are employed.

Switching Surges

The magnitudes of overvoltages due to lightning surges are not significantly affected by the power-system voltage. On the other hand, overvoltages due to switching surges are directly proportional to system voltage. Consequently, lightning surges are less important for EHV transmission above 345 kV and for UHV transmission, which has improved insulation. Switching surges become the limiting factor in insulation coordination for system voltages above 345 kV.

One of the simplest and largest overvoltages can occur when an open-circuited line is energized, as shown in Figure 11.26. Assume that the circuit breaker closes at the instant the sinusoidal source voltage has a peak value $\sqrt{2}\,V$. Assuming zero source impedance, a forward traveling voltage wave of magnitude $\sqrt{2}\,V$ occurs. When this wave arrives at the open-circuited receiving end, where $\Gamma_R = +1$, the reflected voltage wave superimposed on the forward wave results in a maximum voltage of $2\sqrt{2}\,V = 2.83\,V$. Even higher voltages can occur when a line is reclosed after momentary interruption.

Figure 11.26

Energizing an open-circuited line

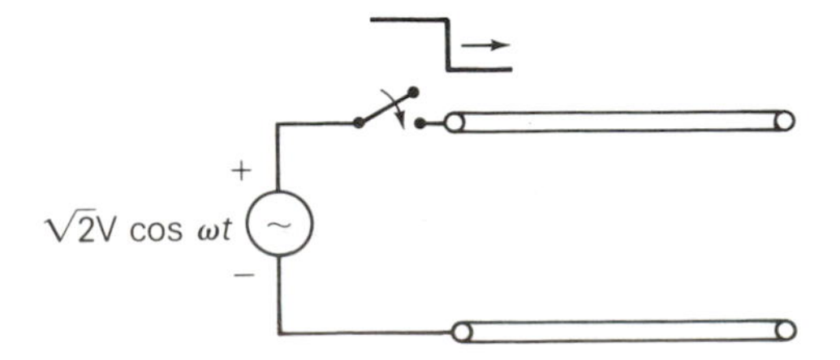

In order to reduce overvoltages due to line energizing or reclosing, resistors are almost always preinserted in circuit breakers at 345 kV and above. Resistors ranging from 200 to 800 Ω are preinserted when EHV circuit breakers are closed, and subsequently bypassed. When a circuit breaker closes, the source voltage divides across the preinserted resistors and the line, thereby reducing the initial line voltage. When the resistors are shorted out, a new transient is initiated, but the maximum line voltage can be substantially reduced by careful design.

Dangerous overvoltages can also occur during a single line-to-ground fault on one phase of a transmission line. When such a fault occurs, a voltage equal and opposite to that on the faulted phase occurs at the instant of fault inception. Traveling waves are initiated on both the faulted phase and, due to capacitive coupling, the unfaulted phases. At the line ends, reflections are produced and are superimposed on the normal operating voltages of the unfaulted phases. Kimbark and Legate [19] show that a line-to-ground fault can create an overvoltage on an unfaulted phase as high as 2.1 times the peak line-to-neutral voltage of the three-phase line.

Power Frequency Overvoltages

Sustained overvoltages at the fundamental power frequency (60 Hz in the United States) or at higher harmonic frequencies (such as 120 Hz, 180 Hz, and so on) occur due to load rejection, to ferroresonance, or to permanent faults. These overvoltages are normally of long duration, seconds to minutes, and are weakly damped.

Section 11.8

Insulation Coordination

Insulation coordination is the process of correlating electric equipment insulation strength with protective device characteristics so that the equipment is protected against expected overvoltages. The selection of equipment insulation strength and the protected voltage level provided by protective devices depends on engineering judgment and cost.

As shown by the top curve in Figure 11.27, equipment insulation strength is a function of time. Equipment insulation can generally withstand high transient overvoltages only if they are of sufficiently short duration. However, determination of insulation strength is somewhat complicated. During repeated tests with identical voltage waveforms under identical conditions, equipment insulation may fail one test and withstand another.

For purposes of insulation testing, a standard impulse voltage wave, as

Figure 11.27

Equipment insulation strength

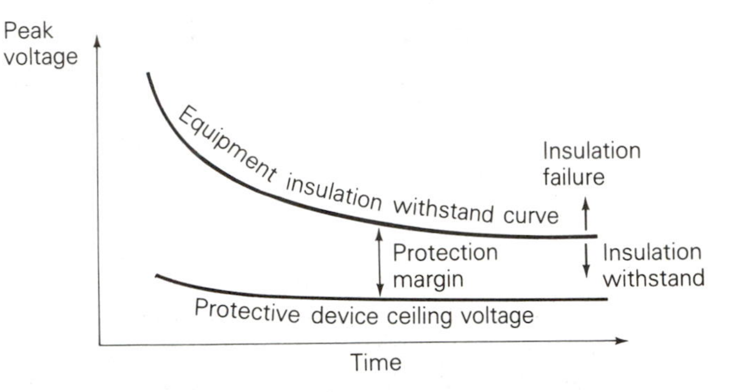

shown in Figure 11.28, is defined. The impulse wave shape is specified by giving the time T_1 in microseconds for the voltage to reach its peak value and the time T_2 for the voltage to decay to one-half its peak. One standard wave is a 1.2×50 wave, which rises to a peak value at $T_1 = 1.2\,\mu s$ and decays to one-half its peak at $T_2 = 50\,\mu s$.

Figure 11.28

Standard impulse voltage waveform

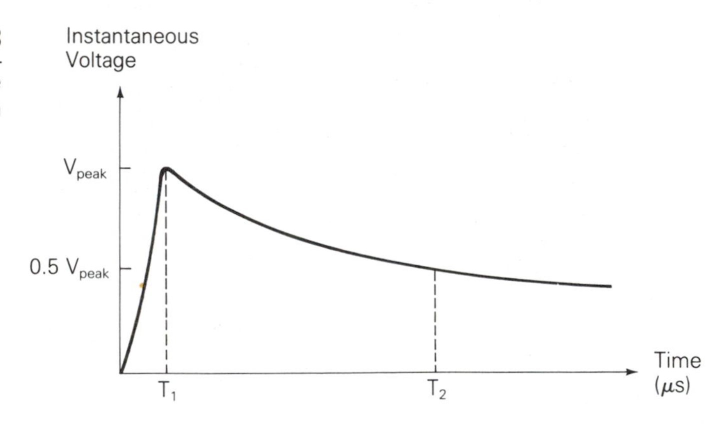

Basic insulation level or BIL is defined as the peak value of the standard impulse voltage wave in Figure 11.28. Standard BILs adopted by the IEEE are shown in Table 11.1. Equipment conforming to these BILs must be capable of withstanding repeated applications of the standard waveform of positive or negative polarity without insulation failure. Also, these standard BILs apply to equipment regardless of how it is grounded. For nominal system voltages 115 kV and above, solidly grounded equipment with the reduced BILs shown in the table have been used.

BILs are often expressed in per-unit, where the base voltage is the maximum value of nominal line-to-ground system voltage. Consider for example a 345-kV system, for which the maximum value of nominal line-to-ground voltage is $\sqrt{2}(345/\sqrt{3}) = 281.7\,\text{kV}$. The 1550-kV standard BIL shown in Table 11.1 is then $(1550/281.7) = 5.5$ per unit.

Note that overhead-transmission-line insulation, which is external insulation, is usually self-restoring. When a transmission-line insulator string flashes over, a short circuit occurs. After circuit breakers open to deenergize the line, the insulation of the string usually recovers, and the line can be rapidly reenergized. However, transformer insulation, which is internal, is not self-restoring. When transformer insulation fails, the transformer must be removed for repair or replaced.

To protect equipment such as a transformer against overvoltages higher than its BIL, a protective device, such as that shown in Figure 11.29, is employed. Such protective devices are generally connected in parallel with the equipment from each phase to ground. As shown in Figure 11.27, the function of the protective device is to maintain its voltage at a ceiling voltage below the BIL of the equipment it protects. The difference between the equipment breakdown voltage and the protective device ceiling voltage is the *protection margin*.

Figure 11.29

Single-line diagram of equipment and protective device

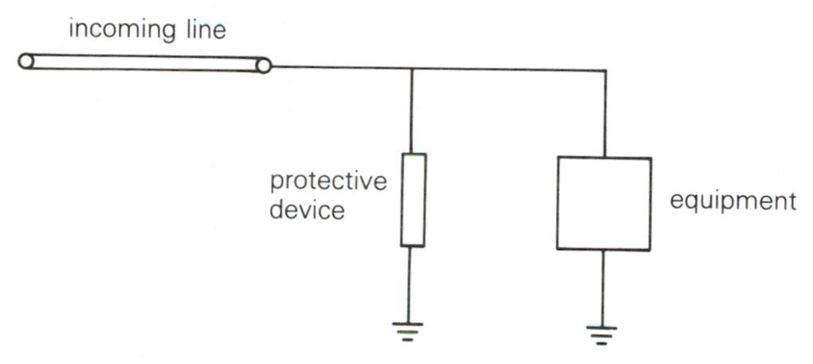

Table 11.1

Standard and reduced basic insulation levels [18]

NOMINAL SYSTEM VOLTAGE	STANDARD BIL	REDUCED BIL*
kVrms	kV	kV
1.2	45	
2.5	60	
5.0	75	
8.7	95	
15	110	
23	150	
34.5	200	
46	250	
69	350	
92	450	
115	550	450
138	650	550
161	750	650
196	900	750
230	1050	825– 900
287	1300	1000–1100
345	1550	1175–1300
500		1300–1800
765		1675–2300

*For solidly grounded systems

These BILs are based on 1.2 × 50 μs voltage waveforms. They apply to internal (or non-self-restoring) insulation such as transformer insulation, as well as to external (or self-restoring) insulation, such as transmission-line insulation, on a statistical basis.

VOLTAGE RATING OF ARRESTER	IMPULSE SPARKOVER VOLTAGE			SWITCHING SURGE SPARKOVER VOLTAGE
	FRONT-OF-WAVE		1.2 × 50-μs	
	RATE OF RISE OF TEST VOLTAGE (kV/μs)	kV CREST (RANGE OF MAXIMA)	kV CREST (RANGE OF MAXIMA)	kV CREST (RANGE OF MAXIMA)
3	25	10–18	10–14	—
6	50	19–28	16–23	—
9	75	28.5–38	24–32	—
12	100	36–48	32–41	—
15	125	45–57	40–51	—
21	175	63–76	54–68	—
24	200	71–86	62–77	—
30	250	89–103	77–93	—
36	300	107–118	92–108	—
39	325	115–125	100–114	—
48	400	143–148	122–132	—
60	500	170–190	141–165	136–153
72	600	204–226	169–190	163–178
90	750	254–275	210–235	203–215
96	800	270–295	218–245	218–225
108	900	304–325	245–270	245–250
120	1000	338–360	272–300	272–275
144	1200	400–430	326–346	325–326
168	1400	460–525	380–404	380–381
180	1500	490–565	400–430	400–410
192	1600	520–600	427–460	426–435
240	2000	620–735	535–577	533–545
258	2000	766–790	575–620	573–585
276	2000	820–840	615–664	612–630
294	2000	875–885	653–675	653–675
312	2000	924–935	690–750	693–710
372	2000	1078–1100	870–890	790–830
396	2000	1140–1176	925–950	840–885
420	2000	1200–1250	980–1005	890–940
444	2000	1265–1320	1035–1055	940–990
468	2000	1326–1390	1090–1110	992–1045
492	2000	1385–1425	1160–1165	1045–1090
540	2000	1515–1555	1274–1280	1145–1200
576	2000	1616–1665	1359–1380	1225–1285
612	2000	1700–1765	1440–1480	1300–1370
648	2000	1790–1865	1525–1570	1380–1445
684	2000	1880–1960	1610–1680	1455–1525

Table 11.2 Characteristics of station-type arresters [20]

DISCHARGE VOLTAGE FOR 8 × 20-μs DISCHARGE CURRENT WAVE					
kV CREST FOR 1500 A (RANGE OF MAXIMA)	kV CREST FOR 3000 A (RANGE OF MAXIMA)	kV CREST FOR 5000 A (RANGE OF MAXIMA)	kV CREST FOR 10 000 A (RANGE OF MAXIMA)	kV CREST FOR 20 000 A (RANGE OF MAXIMA)	kV CREST FOR 40 000 A (RANGE OF MAXIMA)
4.7–6	5.3–6.5	6–7	6.7–7.5	7.7–8.3	—
9.3–11	10–12	12–13	13.4–14.3	15.3–16.3	—
13.9–17	16–18	18–19	20–21.5	22.9–24.3	—
18.5–22	21.3–24	23.5–25.5	26.7–28.5	30.1–32.1	—
23.1–27.5	26.6–30	29.5–32	33.4–36	38.2–40	—
32.3–38.5	37.2–42	41–45	46.8–50	53.4–55.5	—
36.9–44	42.5–48	47–51	53.4–57	61–63.5	—
46.1–55	53.1–60	59–64	66.9–72	76.3–79	—
55.3–66	63.7–72	70.5–76	80–85	91.5–94.5	—
60–71.5	69–78	76.5–82.5	86.5–92	99.1–102	—
73.8–88	84.9–96	94–102	106–114	122–126	—
95–109	110–120	118–130	132–143	150–158	—
114–131	130–144	141–155	159–170	180–189	—
142–163	162–180	176–194	199–213	225–237	—
151–174	173–192	188–218	212–227	240–253	—
170–196	194–216	212–245	238–256	270–284	—
188–218	216–240	235–272	265–285	300–316	—
226–262	260–288	282–311	318–342	360–379	—
263–305	303–336	329–362	371–399	420–442	—
281–327	324–360	353–388	397–455	450–505	—
300–348	346–384	376–414	424–427	480–495	—
374–436	432–480	470–518	530–570	605–630	—
402–438	465–474	505–515	569–575	650–666	—
429–468	496–507	540–570	609–615	690–714	—
458–472	528–532	576–595	653–653	735–758	—
485–530	562–574	611–620	688–693	780–805	874–961
562–610	655–680	726–738	809–826	932–955	1136–1145
599–672	697–726	772–785	861–880	990–1015	1109–1226
634–713	739–770	819–830	913–930	1050–1070	1176–1294
670–753	781–814	866–875	965–977	1110–1130	1243–1358
707–794	823–860	913–930	1018–1040	1170–1200	1310–1441
742–830	865–925	958–1000	1070–1115	1232–1290	1500–1515
814–890	949–990	1052–1070	1173–1195	1350–1390	1646–1663
868–950	1012–1060	1122–1150	1251–1285	1440–1480	1755–1780
924–1010	1076–1130	1193–1220	1330–1360	1531–1580	1865–1885
977–1070	1138–1190	1261–1290	1407–1440	1619–1670	1974–1996
1031–1130	1153–1260	1331–1360	1489–1520	1709–1765	2063–2107

Protective devices should satisfy the following four criteria:

1. Provide a high or infinite impedance during normal system voltages, to minimize steady-state losses
2. Provide a low impedance during surges, to limit voltage
3. Dissipate or store the energy in the surge without damage to itself
4. Return to open-circuit conditions after the passage of a surge

One of the simplest protective devices is the rod gap, two metal rods with a fixed air gap, which is designed to spark over at specified overvoltages. Although it satisfies the first two protective device criteria, it dissipates very little energy and it cannot clear itself after arcing over.

A surge arrester, consisting of an air gap in series with a nonlinear silicon carbide resistor, satisfies all four criteria. The gap eliminates losses at normal voltages and arcs over during overvoltages. The resistor has the property that its resistance decreases sharply as the current through it increases, thereby limiting the voltage across the resistor to a specified ceiling. The resistor also dissipates the energy in the surge. Finally, following the passage of a surge, various forms of arc control quench the arc within the gap and restore the surge arrester to normal open-circuit conditions.

The "gapless" surge arrester, consisting of a nonlinear metal oxide resistor with no air gap, also satisfies all four criteria. At normal voltages the resistance is extremely high, limiting steady-state currents to microamperes and steady-state losses to a few watts. During surges, the resistance sharply decreases, thereby limiting overvoltage while dissipating surge energy. After the surge passes, the resistance naturally returns to its original high value. One advantage of the gapless arrester is that its ceiling voltage is closer to its normal operating voltage than is the conventional arrester, thus permitting reduced BILs and potential savings in the capital cost of equipment insulation.

There are four classes of surge arresters: station, intermediate, distribution, and secondary. Station arresters, which have the heaviest construction, are designed for the greatest range of ratings and have the best protective characteristics. Intermediate arresters, which have moderate construction, are designed for systems with nominal voltages 138 kV and below. Distribution arresters are employed with lower voltage transformers and lines, where there is a need for economy. Secondary arresters are used for nominal system voltages below 1000 V. A summary of the protective characteristics of station arresters is given in Table 11.2 [20]. This summary is based on manufacturers' catalog information.

Note that arrester currents due to lightning surges are generally less than the lightning currents shown in Figure 11.23. In the case of direct strokes to transmission-line phase conductors, traveling waves are set up in two directions from the point of the stroke. Flashover of line insulation diverts part of the lightning current from the arrester. Only in the case of a direct stroke to a phase conductor very near an arrester, where no line flashover occurs, does the arrester discharge the full lightning current. The probability of this occurrence can be significantly reduced by using overhead shield wires to shield transmission lines and substations. Recommended practice for substations with unshielded lines is to select an arrester discharge current of at least 20 kA (even higher if the isokeraunic level is above 40 thunderstorm days per year). For substations with shielded lines, lower arrester discharge currents, from 5 to 20 kA, have been found satisfactory in most situations [20].

Example 11.8 Determine the protective margin for a 120-kV station arrester employed in a 115-kV three-phase system with a 450-kV BIL, based on a 1.2 × 50 standard impulse voltage wave. Also determine the maximum discharge voltage across the arrester for a 20-kA arrester discharge current.

Solution From Table 11.2, for a 120-kV surge arrester, the maximum discharge voltage for a 1.2 × 50 μs impulse voltage wave ranges from 272 to 300 kV, depending on arrester manufacturer. Therefore, the protective margin ranges from (450 − 300) = 150 kV to (450 − 272) = 178 kV. Also from Table 11.2, for a 20-kA discharge current, the maximum discharge voltage across the 120-kV arrester ranges from 300 to 316 kV. ◁

Section 11.9

Personal Computer Program: TRANSMISSION-LINE TRANSIENTS

It is evident from the discussion in this chapter that transmission-line transients problems are complicated and that closed-form solutions are impossible except for simple idealized cases. Accordingly, both analog and digital computation techniques have been developed.

Analog simulators for solving transmission-line transients are known in the industry by trade names such as the "Transient Network Analyzer" (TNA) and "ANACOM." The TNA is composed of electric circuit components including lumped RLC elements and voltage sources connected to represent the various components of the power system under study [21]. Three-phase transmission lines are represented by T or π circuits. For sufficient accuracy, each line is divided into sections—for example, into 15- to 20-km lengths—and a π circuit for each section is connected in cascade. Also, electric circuit models of transformers, shunt reactors, surge arresters, and other components have been developed to match the characteristics of the actual devices. The TNA is energized by a three-phase generator with impedances added to represent source impedance. The TNA is scaled in both voltage and current; for example, 1 A in the model represents 1 kA in the system and 1 V represents 1 kV. Electrical transients caused by line-switching operations or faults are initiated in the TNA via solid-state switches. Switching operations can be performed many times per second, with the responses observed on oscilloscopes that are synchronously triggered with switching events. Many factors influence transients, such as the instant during a 60-Hz cycle at which a switching event occurs. Accordingly, one of the great advantages of the TNA is that it can perform many simulations in a short time and collect data quickly and economically.

Digital computer programs are also available for solving transmission-line transients [9, 10]. Discrete-time models of three-phase transmission lines, transformers, shunt reactors, surge arresters, and other components can be interconnected to represent a power system under study. The program sets up and solves a set of nodal equations that determine the bus voltages at each discrete-time instant. Digital techniques for modeling nonlinear and time-varying RLC elements [11] as well as lossy lines with frequency-dependent line parameters including earth resistance [6, 7, 12] are also available. One advantage of the digital techniques is solution accuracy, especially for high-frequency transient effects above 5 kHz.

The personal computer software package that accompanies this text includes the program entitled "TRANSMISSION-LINE TRANSIENTS," which computes voltage transients for the single-phase circuit shown in Figure 11.30. This circuit consists of 3 single-phase lossless line sections, 10 buses, 4 independent current sources, and 10 lumped parallel RLC elements. The line sections and RLC elements are represented by the discrete-time equivalent circuits given in Figures 11.14–11.17. A lossy line section can be represented

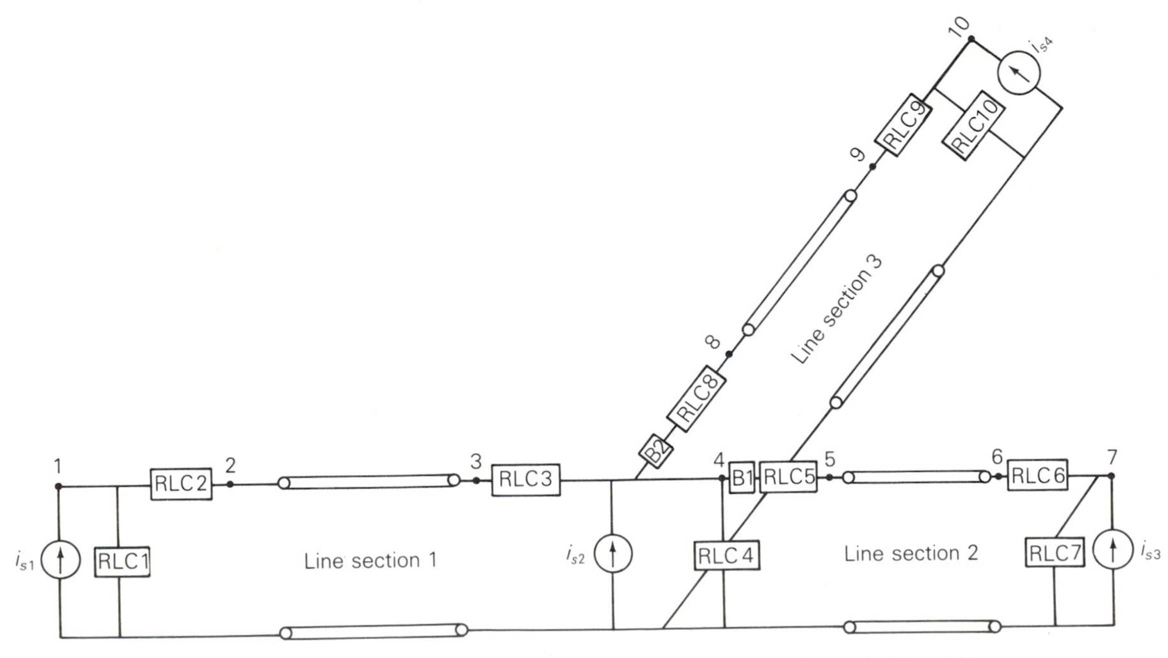

Figure 11.30 Circuit diagram for the program TRANSMISSION-LINE TRANSIENTS

by lumping half the line section series resistance at each end of the line section. BASIC functions can be used to represent the independent current sources as square waves, ramps, exponentials, or other wave shapes. The circuit in Figure 11.30 is assumed to be initially unenergized.

Input data for the computer program consist of: (1) surge impedance Z_c and transit time τ for each line section; (2) RLC values for each of the lumped elements; (3) function representation for each independent current source; (4) calculation time interval Δt and final time $t_F = N\,\Delta t$. The program user can also change the input data in order to simulate piecewise constant RLC elements.

The program replaces the circuit elements in Figure 11.30 by their discrete-time equivalent circuits and solves the resulting nodal equations at each discrete-time instant $\Delta t, 2\,\Delta t, \ldots, N\,\Delta t$. Dependent current sources arising from the models of line sections and of lumped L and C elements are rewritten in terms of bus voltages and current sources at prior times.

The program output data consist of the ten bus voltages at each discrete-time instant $0, \Delta t, 2\,\Delta t, \ldots, N\,\Delta t$.

Example 11.9 Lightning strikes the center of a single-phase 20-kV, 10-km line that is terminated at one end by a 100-MVA, 20-kV/8-kV transformer and at the other end by an open circuit. The line is initially unenergized. Run the TRANSMISSION-LINE TRANSIENTS program to compute the voltage at each end of the line and at the line center versus time. Use the following data for computations:

line:	20 kV, $l = 10$ km, $Z_c = 300\,\Omega$, $v = 3.0 \times 10^8$ m/s, $R = 0.05\,\Omega$/km
transformer:	10 MVA, 20 kV/8 kV, 200-kV BIL, winding capacitance-to-ground $C_{TR} = 6 \times 10^{-9}$ F
lightning surge:	An ideal square wave current source with magnitude 20 kA and duration 20 μs
calculation time interval:	$\Delta t = 0.1$ μs
final time:	$t_F = 150$ μs

Solution The discrete-time equivalent circuit is shown in Figure 11.31. The circuit for each 5-km line section is taken from Figure 11.14, with $Z_c = 300\,\Omega$ and $\tau = l/(2v) = 10 \times 10^3/(2 \times 3 \times 10^8) = 16.67 \times 10^{-6}\,\text{s} = 16.67\,\mu\text{s}$. The total line resistance of $(0.05\,\Omega/\text{km})(10\,\text{km}) = 0.50\,\Omega$ is divided by 4 to give $0.125\,\Omega$ lumped at each end of both line sections. The discrete-time equivalent circuit of the transformer capacitance is taken from Figure 11.16.

The program input and output data at 3 microsecond printout intervals are shown in Figure 11.32 on pages 421–423. Note that the transformer voltage exceeds its BIL rating. Also, the high transient overvoltage across the center and ends of the line indicate flashover of the line insulation.

Figure 11.31

Discrete-time equivalent circuit for Examples 11.9 and 11.10

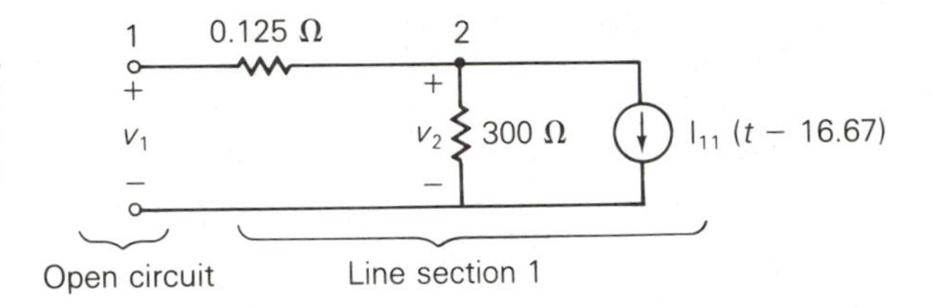

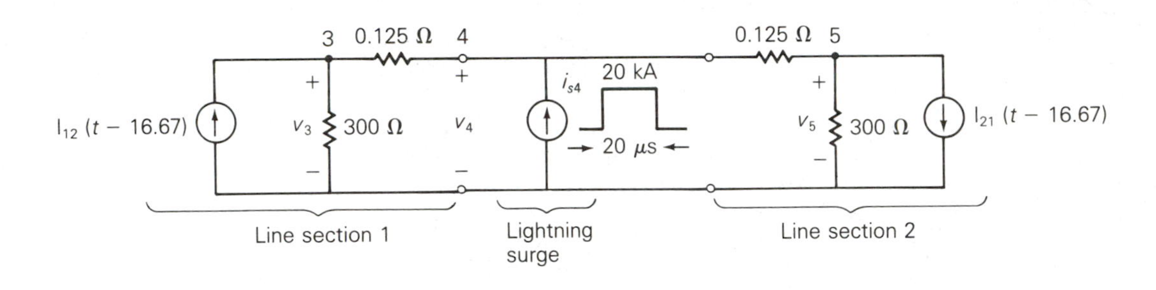

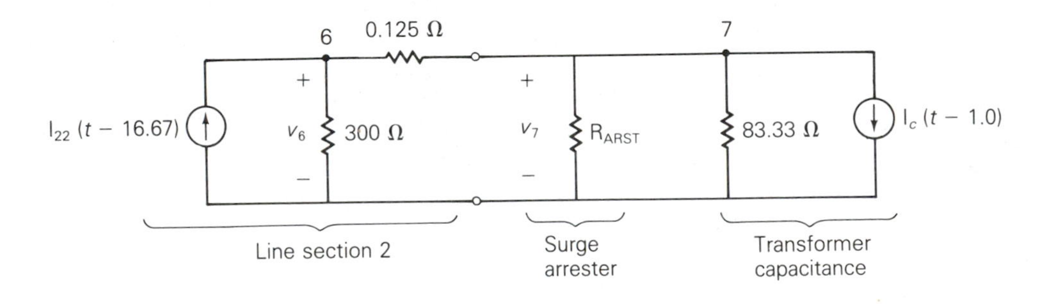

◁

Example 11.10 Repeat Example 11.9 with a 21-kV surge arrester installed adjacent to the transformer. The arrester is represented by a piecewise constant resistance given by $R_{ARST} = 2 \times 10^6$ Ω for $V_{ARST} < 55$ kV and $R_{ARST} = 4.5$ Ω for $V_{ARST} \geqq 55$ kV.

Solution The surge arrester resistor is shown in Figure 11.31. Logic in the computer program is used to change the arrester resistance when its voltage exceeds 55 kV. The output is shown in Figure 11.33. Note that the transformer voltage is held by the arrester to 88.6 kV, well below the transformer BIL. However, flashover of the line insulation is still expected to occur. ◁

The transformer in Examples 11.9 and 11.10 is represented by its winding capacitance-to-ground. Lightning or switching surges are effectively blocked by transformer leakage inductance, which acts as an open circuit at high frequencies. As such, generators are protected against transmission-line transient overvoltages by generator step-up transformers.

Figure 11.32(a)

Input data for Example 11.9

LINE DATA FOR EXAMPLE 11.9

LINE SECTION	CHARACTERISTIC IMPEDANCE	TRANSIT TIME
	OHMS	MICRO-SECONDS
1	300	16.67
2	300	16.67

CIRCUIT BREAKER ONE IS CLOSED
CIRCUIT BREAKER TWO IS OPEN

RLC DATA FOR EXAMPLE 11.9

RLC ELEMENT	1/R			1/L			C		
	FIRST VALUE	SECOND VALUE	CHANGE VOLTAGE	FIRST VALUE	SECOND VALUE	CHANGE VOLTAGE	FIRST VALUE	SECOND VALUE	CHANGE VOLTAGE
	1/OHMS	1/OHMS	kV	1/HENRIES	1/HENRIES	kV	FARADS	FARADS	kV
1	0	0	0	0	0	0	0	0	0
2	8	8	0	0	0	0	0	0	0
3	8	8	0	0	0	0	0	0	0
4	0	0	0	0	0	0	0	0	0
5	8	8	0	0	0	0	0	0	0
6	8	8	0	0	0	0	0	0	0
7	0	0	55	0	0	0	6.0D–09	6.0D–09	0

INDEPENDENT CURRENT SOURCE DATA FOR EXAMPLE 11.9

SOURCE	TYPE	I	T1	T2
		kA or kA/S	MICRO-SECONDS	MICRO-SECONDS
1	S	0	0	0
2	S	20	0	20
3	S	0	0	0

Figure 11.32(b) Output data for example 11.9

TIME	BUS VOLTAGES		
	V1	V4	V7
MICRO-SECONDS	kV	kV	kV
0	0.0	3001.3	0.0
3	0.0	3001.3	0.0
6	0.0	3001.3	0.0
9	0.0	3001.3	0.0
12	0.0	3001.3	0.0
15	0.0	3001.3	0.0
18	6000.0	3001.3	3317.6
21	6000.0	0.0	5493.2
24	6000.0	0.0	5904.3
27	6000.0	0.0	5981.9
30	6000.0	0.0	5996.6
33	6000.0	0.0	5999.4
36	6000.0	4768.0	5999.9
39	0.0	5767.2	1627.1
42	0.0	5956.0	307.4
45	0.0	5991.7	58.1
48	0.0	5998.4	11.0
51	9.8	5999.7	3002.7
54	4868.3	3955.5	5433.0
57	5786.2	747.3	5892.7
60	5959.6	141.2	5979.7
63	5992.4	26.7	5996.2
66	5998.6	5.0	5999.3
69	5999.7	3246.9	5999.9
72	3633.5	5479.5	1820.1
75	686.5	5901.6	344.3
78	129.7	5981.4	65.1
81	24.5	5996.5	12.3
84	−692.7	5999.3	−1248.6
87	4733.8	8838.0	2522.0
90	5760.5	1670.9	4944.6
93	5954.7	315.9	5725.3
96	5991.4	59.7	5933.9
99	5998.4	11.3	5984.8
102	5999.7	276.7	5996.6
105	4064.1	3949.9	6428.5
108	768.9	5429.6	2493.5
111	145.5	5857.7	712.7
114	27.5	5966.6	180.3
117	−1057.3	5992.5	−1228.6
120	−42.0	6855.5	2275.5
123	3968.9	4405.8	4849.5
126	5448.2	1420.0	5698.0

Figure 11.33 Output data for example 11.10

TIME	BUS VOLTAGES		
	V1	V4	V7
MICRO-SECONDS	kV	kV	kV
0	0.0	3001.3	0.0
3	0.0	3001.3	0.0
6	0.0	3001.3	0.0
9	0.0	3001.3	0.0
12	0.0	3001.3	0.0
15	0.0	3001.3	0.0
18	6000.0	3001.3	88.6
21	6000.0	0.0	88.6
24	6000.0	0.0	88.6
27	6000.0	0.0	88.6
30	6000.0	0.0	88.6
33	6000.0	0.0	88.6
36	6000.0	91.1	88.6
39	0.0	91.1	−7.8
42	0.0	91.1	−1.5
45	0.0	91.1	−0.3
48	0.0	91.1	−0.1
51	−5812.9	91.1	88.6
54	−5812.9	−18.9	88.6
57	−5812.9	−3.6	88.6
60	−5812.9	−0.7	88.6
63	−5812.9	−0.1	88.6
66	−5812.9	−0.0	88.6
69	−5812.9	−5813.0	88.6
72	−17.3	−5813.0	−8.7
75	−3.3	−5813.0	−1.6
78	−0.6	−5813.0	−0.3
81	−0.1	−5813.0	−0.1
84	−5813.0	−5813.0	−85.9
87	−5813.0	−42.1	−85.9
90	−5813.0	−8.0	−85.9
93	−5813.0	−1.5	−85.9
96	−5813.0	−0.3	−85.9
99	−5813.0	−0.1	−85.9
102	−5813.0	−88.3	−85.9
105	−19.4	−88.3	−10.1
108	−3.7	−88.3	−8.0
111	−0.7	−88.3	−2.7
114	−0.1	−88.3	−0.7
117	5631.7	−88.3	−85.8
120	5631.6	17.4	−85.8
123	5631.6	−11.6	−85.8
126	5631.6	−5.0	−85.8

TIME	BUS VOLTAGES			TIME	BUS VOLTAGES		
	V1	V4	V7		V1	V4	V7
MICRO-SECONDS	kV	kV	kV	MICRO-SECONDS	kV	kV	kV
129	5864.0	379.3	5926.9	129	5631.6	−1.5	−85.8
132	5968.3	92.6	5983.2	132	5631.6	−0.4	−85.8
135	5992.9	−2568.8	5996.2	135	5631.6	5631.8	−85.8
138	8817.8	2215.5	6681.2	138	3.9	5631.8	−8.8
141	4522.1	4875.3	2695.2	141	−12.9	5631.8	−8.5
144	1393.9	5710.0	780.7	144	−5.0	5631.8	−2.9
147	365.3	5930.5	198.9	147	−1.4	5631.8	−0.8
150	55.9	5984.1	771.9	150	5632.8	5631.8	82.9

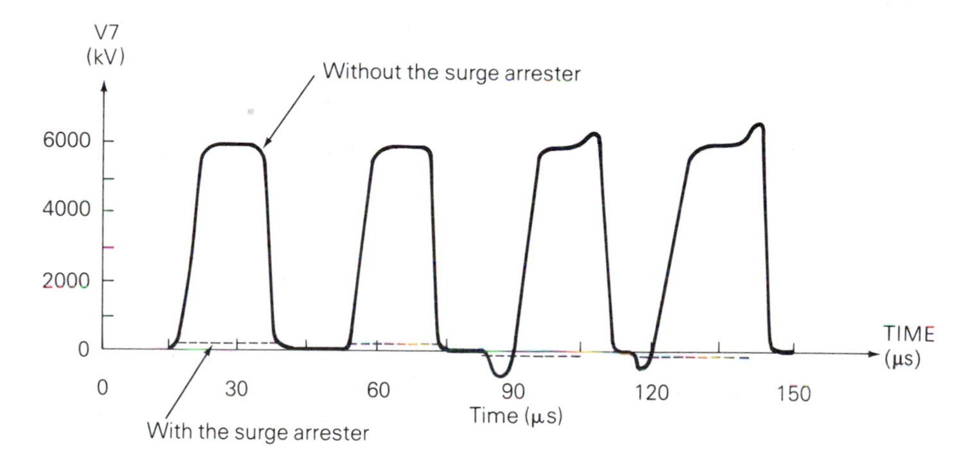

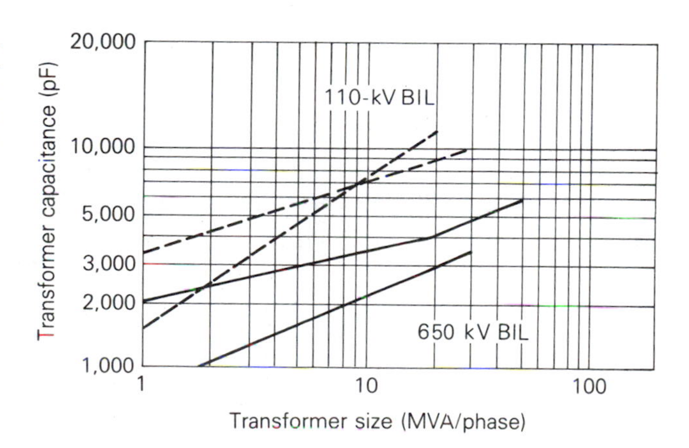

Figure 11.34 Transformer winding capacitance-to-ground versus transformer size [1]

Figure 11.34 shows typical ranges of transformer winding capacitance-to-ground versus transformer size, with BIL of the highest voltage winding as a parameter [1]. This figure gives the total capacitance-to-ground of the highest voltage winding on a per-phase basis. For a Y-connected transformer bank, 0.30 to 0.41 of the capacitance values from Figure 11.34 are the approximate values for each winding. For a Δ-connected bank, the full value from the figure is used for each winding.

Problems

Section 11.2

11.1 From the results of Example 11.2, plot the voltage and current profiles along the line at times $\tau/2$, $3\tau/2$, and 2τ. That is, plot $v(x, \tau/2)$ and $i(x, \tau/2)$ versus x for $0 \le x \le l$; then plot $v(x, 3\tau/2)$, $i(x, 3\tau/2)$, $v(x, 2\tau)$, and $i(x, 2\tau)$ versus x.

11.2 Rework Example 11.2 if the source voltage at the sending end is a ramp, $e_G(t) = Eu_{-2}(t) = Etu_{-1}(t)$, with $Z_G = Z_c$.

11.3 Referring to the single-phase two-wire lossless line shown in Figure 11.3, the receiving end is terminated by an inductor with L_R henries. The source voltage at the sending end is a step, $e_G(t) = Eu_{-1}(t)$ with $Z_G = Z_c$. Both the line and inductor are initially unenergized. Determine and plot the voltage at the center of the line $v(l/2, t)$ versus time t.

11.4 Rework Problem 11.3 if $Z_R = Z_c$ at the receiving end and the source voltage at the sending end is $e_G(t) = Eu_{-1}(t)$, with an inductive source impedance $Z_G(s) = sL_G$. Both the line and source inductor are initially unenergized.

11.5 Rework Example 11.4 with $Z_R = 3Z_c$ and $Z_G = Z_c/2$.

Section 11.3

11.6 Draw the Bewley lattice diagram for Problem 11.5, and plot $v(l/2, t)$ versus time t for $0 \le t \le 5\tau$. Also plot $v(x, 4\tau)$ versus x for $0 \le x \le l$.

11.7 Rework Problem 11.6 if the source voltage is a pulse of magnitude E and duration $\tau/10$; that is, $e_G(t) = E[u_{-1}(t) - u_{-1}(t - \tau/10)]$. $Z_R = 3Z_c$ and $Z_G = Z_c/2$ are the same as in Problem 11.6.

11.8 Rework Example 11.6 if the source impedance at the sending end of line A is $Z_G = Z_A/2 = 200\,\Omega$, and the receiving end of line B is short-circuited, $Z_R = 0$.

11.9 Rework Example 11.6 if the overhead line and cable are interchanged. That is, $Z_A = 100\,\Omega$, $v_A = 2 \times 10^8$ m/s, $l_A = 20$ km, $Z_B = 400\,\Omega$, $v_B = 3 \times 10^8$ m/s, and $l_B = 30$ km. The step voltage source $e_G(t) = Eu_{-1}(t)$ is applied to the sending end of line A with $Z_G = Z_A = 100\,\Omega$, and $Z_R = 2Z_B = 800\,\Omega$ at the receiving end. Draw the lattice diagram for $0 \le t \le 0.6$ ms and plot the junction voltage versus time t.

11.10 As shown in Figure 11.35, a single-phase two-wire lossless line with $Z_c = 400\,\Omega$, $v = 3 \times 10^8$ m/s, and $l = 100$ km has a 400-Ω resistor, denoted R_J, installed across the center of the line, thereby dividing the line into two sections, A and B. The source voltage at the sending end is a pulse of magnitude 100 V and duration 0.1 ms. The source impedance is $Z_G = Z_c = 400\,\Omega$, and the receiving end of the line is short-circuited. (a) Show that for an incident voltage wave arriving at the center of the line from either line section, the voltage reflection and refraction coefficients are given by

$$\Gamma_{BB} = \Gamma_{AA} = \frac{\left(\frac{Z_{eq}}{Z_c}\right) - 1}{\left(\frac{Z_{eq}}{Z_c}\right) + 1} \qquad \Gamma_{AB} = \Gamma_{BA} = \frac{2\left(\frac{Z_{eq}}{Z_c}\right)}{\left(\frac{Z_{eq}}{Z_c}\right) + 1}$$

where

$$Z_{eq} = \frac{R_J Z_c}{R_J + Z_c}$$

(b) Draw the Bewley lattice diagram for $0 \le t \le 6\tau$. (c) Plot $v(l/2, t)$ versus time t for $0 \le t \le 6\tau$ and plot $v(x, 6\tau)$ versus x for $0 \le x \le l$.

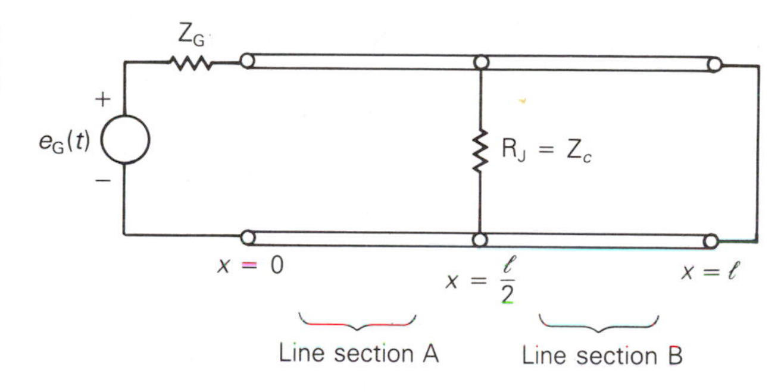

Figure 11.35 Circuit for Problem 11.10

11.11 The junction of four single-phase two-wire lossless lines, denoted A, B, C, and D, is shown in Figure 11.13. Consider a voltage wave v_A^+ arriving at the junction from line A. Using (11.3.8) and (11.3.9), determine the voltage reflection coefficient Γ_{AA} and the voltage refraction coefficients Γ_{BA}, Γ_{CA}, and Γ_{DA}.

11.12 Referring to Figure 11.3, the source voltage at the sending end is a step $e_G(t) = Eu_{-1}(t)$ with an inductive source impedance $Z_G(s) = sL_G$, where $L_G/Z_c = \tau/2$. At the receiving end, $Z_R = Z_c/3$. The line and source inductance are initially unenergized. (a) Draw the Bewley lattice diagram for $0 \le t \le 5\tau$. (b) Plot $v(l, t)$ versus time t for $0 \le t \le 5\tau$.

Section 11.4

11.13 For the circuit given in Problem 11.3, replace the circuit elements by their discrete-time equivalent circuits and write nodal equations in a form suitable for computer solution of the sending-end and receiving-end voltages. Give equations for all dependent sources. Assume $E = 1000$ V, $L_R = 10$ mH, $Z_c = 100\,\Omega$, $v = 2 \times 10^8$ m/s, $l = 40$ km, and $\Delta t = 0.02$ ms.

11.14 Repeat Problem 11.13 for the circuit given in Problem 11.10. Assume $\Delta t = 0.03333$ ms.

Section 11.5

11.15 Rework Problem 11.13 for a lossy line with a constant series resistance $R = 0.3\,\Omega$/km. Lump half of the total resistance at each end of the line.

Section 11.8

11.16 Repeat Example 11.8 for a 372-kV station arrester employed in a 345-kV system with a 1550-kV BIL.

11.17 Select a surge arrester from Table 11.2 for the high-voltage side of a three-phase 600-MVA, 500-kV Y/13.8-kV Δ transformer. The high-voltage windings of the transformer have a 1550-kV BIL and a solidly grounded neutral. A minimum protective margin of 1.4 per unit based on a $1.2 \times 50\,\mu s$ voltage waveform is required. Lightning discharge currents are limited to 40 kA.

Section 11.9

11.18 Run the TRANSMISSION-LINE TRANSIENTS program for Examples 11.9 and 11.10 if the lightning surge strikes the line 1 km from the transformer.

11.19 Investigate the effect of different transformer winding capacitances-to-ground on the transformer voltage in Example 11.9. Run two additional cases of the TRANSMISSION-LINE TRANSIENTS program for Example 11.9: case (1), $C_{TR} = 3 \times 10^{-9}$ F; case (2), $C_{TR} = 9 \times 10^{-9}$ F. Which case gives the largest transformer voltage?

11.20 Recommended practice is to install surge arresters as close as possible to the equipment they protect. As shown in Figure 11.36, a single-phase unenergized line is terminated by a transformer at the receiving end and is matched at the sending end. A surge arrester is located 50 m from the transformer. Lightning strikes the surge arrester. Run the TRANSMISSION-LINE TRANSIENTS program to compute the transformer voltage and arrester voltage. Does the arrester protect the transformer? Use the following data:

line:	20 kV, $l = 10$ km, $Z_c = 300\,\Omega$, $v = 3 \times 10^8$ m/s, $R = 0.05\,\Omega$/km
transformer:	100 MVA, 20 kV/8 kV, high-voltage winding BIL = 200 kV; high-voltage winding capacitance-to-ground, three cases: $C_{TR} = 1.5 \times 10^{-10}$, 1.5×10^{-9}, and 1.5×10^{-8} F
arrester:	$R_{ARST} = 2 \times 10^6\,\Omega$ for $V_{ARST} < 55$ kV $R_{ARST} = 4.5\,\Omega$ for $V_{ARST} \geqq 55$ kV
lightning surge:	a 30 kA, $1.5 \times 50\,\mu s$ current source which is approximated by a double exponential
calculation interval:	$\Delta t = 0.1\,\mu s$
final time:	$t_F = 5\,\mu s$

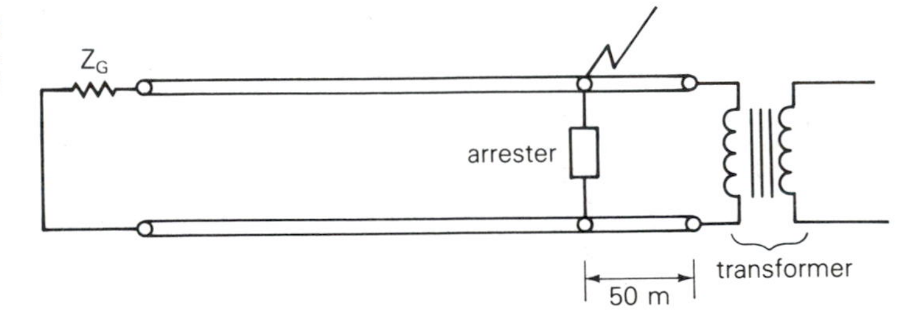

Figure 11.36

Circuit for Problem 11.20

References

1. Allan Greenwood, *Electrical Transients in Power Systems* (New York: Wiley Interscience, 1971).
2. L. V. Bewley, *Travelling Waves on Transmission Systems,* 2d ed. (New York: John Wiley and Sons, 1951).
3. L. Bergeron, *Water Hammer in Hydraulics and Wave Surges in Electricity* (New York: John Wiley and Sons, 1961).
4. H. M. Lacey, "The Lightning Protection of High-Voltage Overhead Transmission and Distribution Systems," *Proc. IEE*, Vol. 96 (1949): p. 287.
5. John R. Carson, "Wave Propagation in Overhead Wires with Ground Return," *Bell System Technical Journal* 5 (1926): p. 539–554.
6. W. S. Meyer, and H. W. Dommel, "Numerical Modelling of Frequency-Dependent Transmission Line Parameters in an Electromagnetic Transients Program," *IEEE Transactions PAS*, Vol. PAS–99 (September/October 1974): pp. 1401–1409.
7. A. Budner, "Introduction of Frequency-Dependent Line Parameters into an Electromagnetics Transients Program," *IEEE Transactions PAS*, Vol. PAS–89 (January 1970): pp. 88–97.
8. D. E. Hedman, "Propagation on Overhead Transmission Lines: I—Theory of Modal Analysis and II—Earth Conduction Effects and Practical Results," *IEEE Transactions PAS* (March 1965): pp. 200–211.
9. H. W. Dommel, "A Method for Solving Transient Phenomena in Multiphase Systems," *Proceedings 2nd Power Systems Computation Conference*, Stockholm, 1966.
10. H. W. Dommel, "Digital Computer Solution of Electromagnetic Transients in Single- and Multiphase Networks," *IEEE Transactions PAS*, Vol. PAS–88 (1969): pp. 388–399.
11. H. W. Dommel, "Nonlinear and Time-Varying Elements in Digital Simulation of Electromagnetic Transients," *IEEE Transactions PAS*, Vol. PAS 90 (November/December 1971): pp. 2561–2567.
12. S. R. Naidu, *Transitorios Eletromagnéticos em Sistemas de Potência*, Eletrobras/UFPb, Brazil, 1985.
13. "Report of Joint IEEE-EEI Committee on EHV Line Outages," *IEEE Transactions PAS*, Vol. 86 (1967): p. 547.
14. R. A. W. Connor, and R. A. Parkins, "Operations Statistics in the Management of Large Distribution System," *Proc. IEE*, Vol. 113 (1966): p. 1823.
15. C. T. R. Wilson, "Investigations on Lightning Discharges and on the Electrical Field of Thunderstorms," *Phil. Trans. Royal Soc.*, Series A, Vol. 221 (1920): p. 73.
16. G. B. Simpson, and F. J. Scrase, "The Distribution of Electricity in Thunderclouds," *Proc. Royal Soc.*, Series A, Vol. 161 (1937): p. 309.
17. B. F. J. Schonland, and H. Collens, "Progressive Lightning," *Proc. Royal Soc.*, Series A, Vol. 143 (1934): p. 654.
18. Westinghouse Electric Corporation, *Electrical Transmission and Distribution Reference Book*, 4th ed. (East Pittsburgh, Pa., 1964).
19. E. W. Kimbark, and A. C. Legate, "Fault Surge versus Switching Surge, A Study of Transient Voltages Caused by Line to Ground Faults," *IEEE Transactions PAS*, Vol. 87 (1968): p. 1762.
20. *Guide for the Application of Valve-Type Surge Arresters for Alternating-Current Systems*, ANSI C62.2–1981 (New York: American National Standards Institute, 1981).
21. C. Concordia, "The Transient Network Analyzer for Electric Power System Problems," Supplement to *CIGRE Committee No. 13 Report*, 1956.

CHAPTER 12

Transient Stability

Power-system stability refers to the ability of synchronous machines to move from one steady-state operating point following a disturbance to another steady-state operating point, without losing synchronism [1]. There are three types of power-system stability: steady-state, transient, and dynamic.

Steady-state stability, discussed in Chapter 6, involves slow or gradual changes in operating points. Steady-state stability studies, which are usually performed with a power-flow computer program (Chapter 7), ensure that phase angles across transmission lines are not too large, that bus voltages are close to nominal values, and that generators, transmission lines, transformers, and other equipment are not overloaded.

Transient stability, the main focus of this chapter, involves major disturbances such as loss of generation, line-switching operations, faults, and sudden load changes. Following a disturbance, synchronous machine frequencies undergo transient deviations from synchronous frequency (60 Hz), and machine power angles change. The objective of a transient stability study is to determine whether or not the machines will return to synchronous frequency with new steady-state power angles. Changes in power flows and bus voltages are also of concern.

Elgerd [2] gives an interesting mechanical analogy to the power-system transient stability program. As shown in Figure 12.1, a number of masses representing synchronous machines are interconnected by a network of elastic strings representing transmission lines. Assume that this network is initially at rest in steady-state, with the net force on each string below its break point, when one of the strings is cut, representing the loss of a transmission line. As a result, the masses undergo transient oscillations and the forces on the strings fluctuate. The system will then either settle down to a new steady-state operating point with a new set of string forces, or additional strings will break, resulting in an even weaker network and eventual system collapse. That is, for a given disturbance, the system is either transiently stable or unstable.

In today's large-scale power systems with many synchronous machines interconnected by complicated transmission networks, transient stability studies are best performed with a digital computer program. For a specified disturbance, the program alternately solves, step by step, algebraic power-flow equations representing a network and nonlinear differential equations representing synchronous machines. Both predisturbance, disturbance, and postdisturbance computations are performed. The program output includes power angles and frequencies of synchronous machines, bus voltages, and power flows versus time.

In many cases, transient stability is determined during the first swing of machine power angles following a disturbance. During the first swing, which typically lasts about 1 second, the mechanical output power and the internal voltage of a generating unit are often assumed constant. However, where multiswings lasting several seconds are of concern, models of turbine-governors and excitation systems (for example, see Figures 10.3 and 10.5) as well as more detailed machine models can be employed to obtain accurate transient stability results over the longer time period.

Dynamic stability involves an even longer time period, typically several minutes. It is possible for controls to affect dynamic stability even though transient stability is maintained. The action of turbine-governors, excitation systems, tap-changing transformers, and controls from a power-system dispatch center can interact to stabilize or destabilize a power system several minutes after a disturbance has occurred.

In order to simplify transient stability studies, the following assumptions are made:

1. Only balanced three-phase systems and balanced disturbances are considered. Therefore, only positive-sequence networks are employed.
2. Deviations of machine frequencies from synchronous frequency (60 Hz) are small, and dc offset currents and harmonics are neglected. Therefore, the network of transmission lines, transformers, and impedance loads is essentially in steady-state, and voltages, currents, and powers can be computed from algebraic power-flow equations.

In Section 12.1 we introduce the swing equation, which determines synchronous machine rotor dynamics. In Section 12.2 a simplified model of a synchronous machine and a Thévenin equivalent of a system consisting of lines, transformers, loads, and other machines are given. Then in Section 12.3 the equal-area criterion is presented; this gives a direct method for determining the transient stability of one machine connected to a system equivalent. Numerical integration techniques for solving swing equations step by step are discussed in Section 12.4 and used in Section 12.5 to determine multimachine stability. The personal computer program TRANSIENT STABILITY, which can be used for the design and analysis of power systems with regard to transient stability, is described in Section 12.6. Finally, Section 12.7 discusses design methods for improving power-system transient stability.

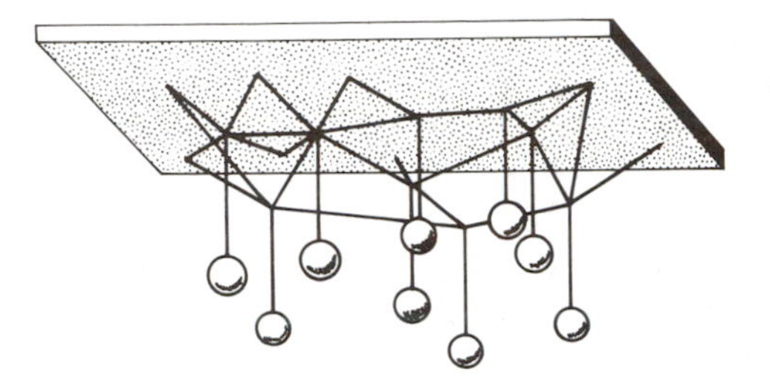

Figure 12.1 Mechanical analog of power-system transient stability [2]

Section 12.1

The Swing Equation

Consider a generating unit consisting of a three-phase synchronous generator and its prime mover. The rotor motion is determined by Newton's second law, given by

$$J\alpha_m(t) = T_m(t) - T_e(t) = T_a(t) \tag{12.1.1}$$

where J = total moment of inertia of the rotating masses, kgm^2

α_m = rotor angular acceleration, rad/s^2

T_m = mechanical torque supplied by the prime mover minus the retarding torque due to mechanical losses, Nm

T_e = electrical torque that accounts for the total three-phase electrical power output of the generator, plus electrical losses, Nm

T_a = net accelerating torque, Nm

Also, the rotor angular acceleration is given by

$$\alpha_m(t) = \frac{d\omega_m(t)}{dt} = \frac{d^2\theta_m(t)}{dt^2} \tag{12.1.2}$$

$$\omega_m(t) = \frac{d\theta_m(t)}{dt} \tag{12.1.3}$$

where ω_m = rotor angular velocity, rad/s

θ_m = rotor angular position with respect to a stationary axis, rad

T_m and T_e are positive for generator operation. In steady-state T_m equals T_e, the accelerating torque T_a is zero, and, from (12.1.1), the rotor acceleration α_m is zero, resulting in a constant rotor velocity called *synchronous speed.* When

T_m is greater than T_e, T_a is positive and α_m is therefore positive, resulting in increasing rotor speed. Similarly, when T_m is less than T_e, the rotor speed is decreasing.

It is convenient to measure the rotor angular position with respect to a synchronously rotating reference axis instead of a stationary axis. Accordingly, we define

$$\theta_m(t) = \omega_{msyn} t + \delta_m(t) \tag{12.1.4}$$

where ω_{msyn} = synchronous angular velocity of the rotor, rad/s

δ_m = rotor angular position with respect to a synchronously rotating reference, rad

Using (12.1.2) and (12.1.4), (12.1.1) becomes

$$J\frac{d^2\theta_m(t)}{dt^2} = J\frac{d^2\delta_m(t)}{dt^2} = T_m(t) - T_e(t) = T_a(t) \tag{12.1.5}$$

It is also convenient to work with power rather than torque, and to work in per-unit rather than in actual units. Accordingly, we multiply (12.1.5) by $\omega_m(t)$ and divide by S_{rated}, the three-phase voltampere rating of the generator:

$$\frac{J\omega_m(t)}{S_{rated}}\frac{d^2\delta_m(t)}{dt^2} = \frac{\omega_m(t)T_m(t) - \omega_m(t)T_e(t)}{S_{rated}}$$

$$= \frac{p_m(t) - p_e(t)}{S_{rated}} = p_{mp.u.}(t) - p_{ep.u.}(t) = p_{ap.u.}(t) \tag{12.1.6}$$

where $p_{mp.u.}$ = mechanical power supplied by the prime mover minus mechanical losses, per unit

$p_{ep.u.}$ = electrical power output of the generator plus electrical losses, per unit

Finally, it is convenient to work with a normalized inertia constant, called the H constant, which is defined as

$$H = \frac{\text{stored kinetic energy at synchronous speed}}{\text{generator voltampere rating}}$$

$$= \frac{\frac{1}{2}J\omega_{msyn}^2}{S_{rated}} \quad \text{joules/VA or per unit-seconds} \tag{12.1.7}$$

The H constant has the advantage that it falls within a fairly narrow range, normally between 1 and 10 p.u.-s, whereas J varies widely, depending on generating unit size and type. Solving (12.1.7) for J and using in (12.1.6),

$$2H\frac{\omega_m(t)}{\omega_{msyn}^2}\frac{d^2\delta_m(t)}{dt^2} = p_{mp.u.}(t) - p_{ep.u.}(t) = p_{ap.u.}(t) \tag{12.1.8}$$

Defining per-unit rotor angular velocity,

$$\omega_{p.u.}(t) = \frac{\omega_m(t)}{\omega_{msyn}} \tag{12.1.9}$$

Equation (12.1.8) becomes

$$\frac{2H}{\omega_{msyn}}\omega_{p.u.}(t)\frac{d^2\delta_m(t)}{dt^2} = p_{mp.u.}(t) - p_{ep.u.}(t) = p_{ap.u.}(t) \tag{12.1.10}$$

For a synchronous generator with P poles, the electrical angular acceleration α, electrical radian frequency ω, and power angle δ are

$$\alpha(t) = \frac{P}{2}\alpha_m(t) \tag{12.1.11}$$

$$\omega(t) = \frac{P}{2}\omega_m(t) \tag{12.1.12}$$

$$\delta(t) = \frac{P}{2}\delta_m(t) \tag{12.1.13}$$

Similarly, the synchronous electrical radian frequency is

$$\omega_{syn} = \frac{P}{2}\omega_{msyn} \tag{12.1.14}$$

The per-unit electrical frequency is

$$\omega_{p.u.}(t) = \frac{\omega(t)}{\omega_{syn}} = \frac{\frac{P}{2}\omega(t)}{\frac{P}{2}\omega_{syn}} = \frac{\omega_m(t)}{\omega_{msyn}} \tag{12.1.15}$$

Therefore, using (12.1.13–12.1.15), (12.1.10) can be written as

$$\frac{2H}{\omega_{syn}}\omega_{p.u.}(t)\frac{d^2\delta(t)}{dt^2} = p_{mp.u.}(t) - p_{ep.u.}(t) = p_{ap.u.}(t) \tag{12.1.16}$$

Equation (12.1.16), called the per-unit *swing equation*, is the fundamental equation that determines rotor dynamics in transient stability studies. Note that it is nonlinear due to $p_{ep.u.}(t)$, which is shown in Section 12.2 to be a nonlinear function of δ. Equation (12.1.16) is also nonlinear due to the $\omega_{p.u.}(t)$ term. However, in practice the rotor speed does not vary significantly from synchronous speed during transients. That is, $\omega_{p.u.}(t) \simeq 1.0$, which is often assumed in (12.1.16) for hand calculations.

Equation (12.1.16) is a second-order differential equation that can be rewritten as two first-order differential equations. Differentiating (12.1.4), and then using (12.1.3) and (12.1.12)–(12.1.14), we obtain

$$\frac{d\delta(t)}{dt} = \omega(t) - \omega_{syn} \tag{12.1.17}$$

Using (12.1.17) in (12.1.16),

$$\frac{2H}{\omega_{syn}}\omega_{p.u.}(t)\frac{d\omega(t)}{dt} = p_{mp.u.}(t) - p_{ep.u.}(t) = p_{ap.u.}(t) \qquad (12.1.18)$$

Equations (12.1.17) and (12.1.18) are two first-order differential equations.

Example 12.1 A three-phase, 60 Hz, 500 MVA, 15 kV, 32-pole hydroelectric generating unit has an H constant of 2.0 p.u.-s. (a) Determine ω_{syn} and ω_{msyn}. (b) Give the per-unit swing equation for this unit. (c) The unit is initially operating at $p_{mp.u.} = p_{ep.u.} = 1.0$, $\omega = \omega_{syn}$, and $\delta = 10°$ when a three-phase-to-ground bolted short circuit at the generator terminals causes $p_{ep.u.}$ to drop to zero for $t \geq 0$. Determine the power angle 3 cycles after the short circuit commences. Assume $p_{mp.u.}$ remains constant at 1.0 per unit. Also assume $\omega_{p.u.}(t) = 1.0$ in the swing equation.

Solution **a.** For a 60-Hz generator,

$$\omega_{syn} = 2\pi 60 = 377 \quad \text{rad/s}$$

and, from (12.1.14), with P = 32 poles,

$$\omega_{msyn} = \frac{2}{P}\omega_{syn} = \left(\frac{2}{32}\right)377 = 23.56 \quad \text{rad/s}$$

b. From (12.1.16), with H = 2.0 p.u.-s,

$$\frac{4}{2\pi 60}\omega_{p.u.}(t)\frac{d^2\delta(t)}{dt^2} = p_{mp.u.}(t) - p_{ep.u.}(t)$$

c. The initial power angle is

$$\delta(0) = 10° = 0.1745 \quad \text{radian}$$

Also, from (12.1.17), at $t = 0$,

$$\frac{d\delta(0)}{dt} = 0$$

Using $p_{mp.u.}(t) = 1.0$, $p_{ep.u.} = 0$, and $\omega_{p.u.}(t) = 1.0$, the swing equation from **(b)** is

$$\left(\frac{4}{2\pi 60}\right)\frac{d^2\delta(t)}{dt^2} = 1.0 \qquad t \geq 0$$

Integrating twice and using the above initial conditions,

$$\frac{d\delta(t)}{dt} = \left(\frac{2\pi 60}{4}\right)t + 0$$

$$\delta(t) = \left(\frac{2\pi 60}{8}\right)t^2 + 0.1745$$

$$\text{At } t = 3 \text{ cycles} = \frac{3 \text{ cycles}}{60 \text{ cycles/second}} = 0.05 \text{ second},$$

$$\delta(0.05) = \left(\frac{2\pi 60}{8}\right)(0.05)^2 + 0.1745$$

$$= 0.2923 \text{ radian} = 16.75°$$

◁

Example 12.2 A power plant has two three-phase, 60-Hz generating units with the following ratings:

Unit 1 500 MVA, 15 kV, 0.85 power factor, 32 poles, $H_1 = 2.0$ p.u.-s

Unit 2 300 MVA, 15 kV, 0.90 power factor, 16 poles, $H_2 = 2.5$ p.u.-s

(a) Give the per-unit swing equation of each unit on a 100-MVA system base. (b) If the units are assumed to "swing together," that is, $\delta_1(t) = \delta_2(t)$, combine the two swing equations into one equivalent swing equation.

Solution **a.** If the per-unit powers on the right-hand side of the swing equation are converted to the system base, then the H constant on the left-hand side must also be converted. That is,

$$H_{new} = H_{old}\frac{S_{old}}{S_{new}} \quad \text{per unit}$$

Converting H_1 from its 500-MVA rating to the 100-MVA system base,

$$H_{1new} = H_{1old}\frac{S_{old}}{S_{new}} = (2.0)\left(\frac{500}{100}\right) = 10 \quad \text{p.u.-s}$$

Similarly, converting H_2,

$$H_{2new} = (2.5)\left(\frac{300}{100}\right) = 7.5 \quad \text{p.u.-s}$$

The per-unit swing equations on the system base are then

$$\frac{2H_{1new}}{\omega_{syn}}\omega_{1p.u.}(t)\frac{d^2\delta_1(t)}{dt^2} = \frac{20.0}{2\pi 60}\omega_{1p.u.}(t)\frac{d^2\delta_1(t)}{dt^2} = p_{m1p.u.}(t) - p_{e1p.u.}(t)$$

$$\frac{2H_{2new}}{\omega_{syn}}\omega_{2p.u.}(t)\frac{d^2\delta_2(t)}{dt^2} = \frac{15.0}{2\pi 60}\omega_{2p.u.}(t)\frac{d^2\delta_2(t)}{dt^2} = p_{m2p.u.}(t) - p_{e2p.u.}(t)$$

b. Letting:

$$\delta(t) = \delta_1(t) = \delta_2(t)$$

$$\omega_{p.u.}(t) = \omega_{1p.u.}(t) = \omega_{2p.u.}(t)$$

$$p_{mp.u.}(t) = p_{m1p.u.}(t) + p_{m2p.u.}(t)$$

$$p_{ep.u.}(t) = p_{e1p.u.}(t) + p_{e2p.u.}(t)$$

and adding the above swing equations

$$\frac{2(H_{1\text{new}} + H_{2\text{new}})}{\omega_{\text{syn}}}\omega_{\text{p.u.}}(t)\frac{d^2\delta(t)}{dt^2} = \frac{35.0}{2\pi 60}\omega_{\text{p.u.}}(t)\frac{d^2\delta(t)}{dt^2} = p_{m\text{p.u.}}(t) - p_{e\text{p.u.}}(t)$$

When transient stability studies involving large-scale power systems with many generating units are performed with a digital computer, computation time can be reduced by combining the swing equations of those units that swing together. Such units, which are called *coherent machines*, usually are connected to the same bus or are electrically close, and they are usually remote from network disturbances under study. ◁

Section 12.2

Simplified Synchronous Machine Model and System Equivalents

Figure 12.2 shows a simplified model of a synchronous machine that can be used in transient stability studies. As shown, the synchronous machine is represented by a constant internal voltage E' behind its direct axis transient reactance X'_d. This model is based on the following assumptions:

1. The machine is operating under balanced three-phase positive-sequence conditions.
2. Machine excitation is constant.
3. Machine losses, saturation, and saliency are neglected.

Figure 12.2 Simplified synchronous machine model for transient stability studies

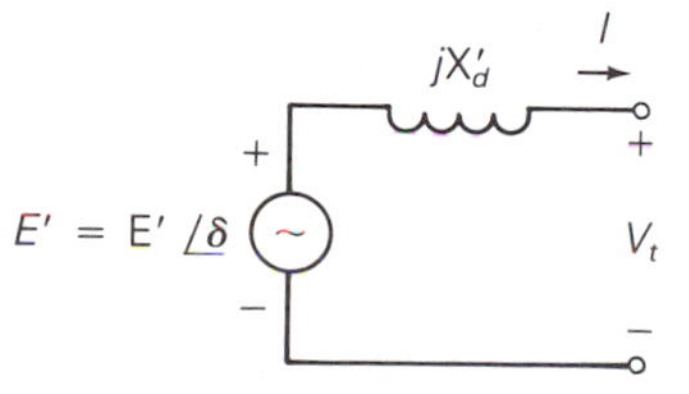

(a) Circuit diagram

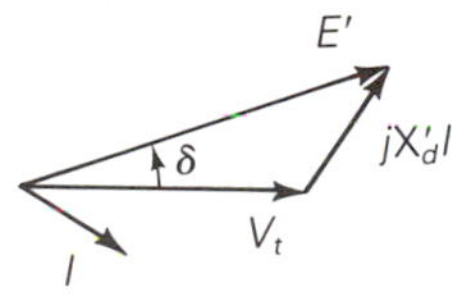

(b) Phasor diagram

In transient stability programs, more detailed models can be used to represent exciters, losses, saturation, and saliency. However, the simplified model reduces model complexity while maintaining reasonable accuracy in stability calculations.

Each generator in the model is connected to a system consisting of transmission lines, transformers, loads, and other machines. To a first approximation the system can be represented by an "infinite bus" behind a system reactance. An infinite bus is an ideal voltage source that maintains constant voltage magnitude, constant phase, and constant frequency.

Figure 12.3 shows a synchronous generator connected to a system equivalent. The voltage magnitude V_{bus} and 0° phase of the infinite bus are constant. The phase angle δ of the internal machine voltage is the machine power angle with respect to the infinite bus.

Figure 12.3

Synchronous generator connected to a system equivalent

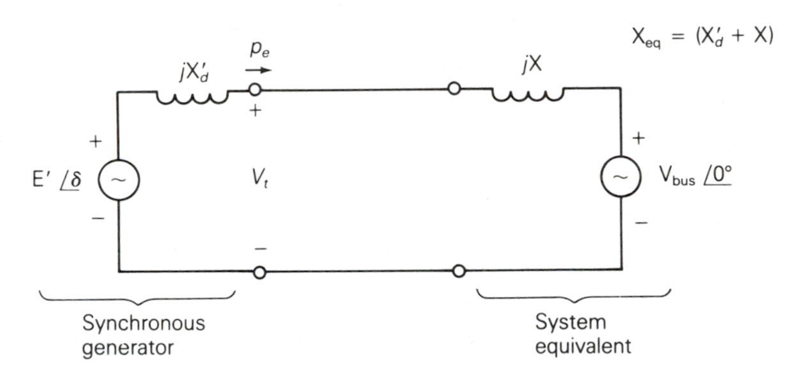

The equivalent reactance between the machine internal voltage and the infinite bus is $X_{eq} = (X'_d + X)$. From (7.10.3), the real power delivered by the synchronous generator to the infinite bus is

$$p_e = \frac{E'V_{bus}}{X_{eq}} \sin \delta \tag{12.2.1}$$

During transient disturbances both E' and V_{bus} are considered constant in (12.2.1). Thus p_e is a sinusoidal function of the machine power angle δ.

Example 12.3

Figure 12.4 shows a single-line diagram of a three-phase 60-Hz synchronous generator, connected through a transformer and parallel transmission lines to an infinite bus. All reactances are given in per-unit on a common system base.

Figure 12.4

Single-line diagram for Example 12.3

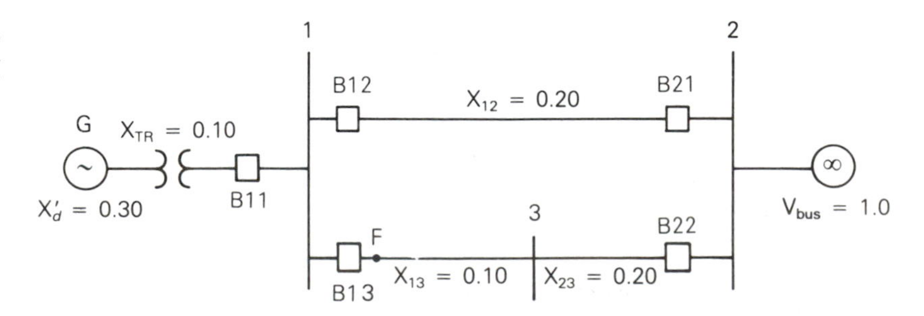

If the infinite bus receives 1.0 per unit real power at 0.95 p.f. lagging, determine (a) the internal voltage of the generator and (b) the equation for the electrical power delivered by the generator versus its power angle δ.

Solution

a. The equivalent circuit is shown in Figure 12.5, from which the equivalent reactance between the machine internal voltage and infinite bus is

$$\begin{aligned} X_{eq} &= X_d' + X_{TR} + X_{12} \| (X_{13} + X_{23}) \\ &= 0.30 + 0.10 + 0.20 \| (0.10 + 0.20) \\ &= 0.520 \quad \text{per unit} \end{aligned}$$

The current into the infinite bus is

$$\begin{aligned} I &= \frac{P}{V_{bus}(\text{p.f.})} \underline{/-\cos^{-1}(\text{p.f.})} = \frac{(1.0)}{(1.0)(0.95)} \underline{/-\cos^{-1} 0.95} \\ &= 1.05263 \underline{/-18.195^\circ} \end{aligned}$$

and the machine internal voltage is

$$\begin{aligned} E' = E' \underline{/\delta} &= V_{bus} + jX_{eq} I \\ &= 1.0\underline{/0^\circ} + (j0.520)(1.05263\underline{/-18.195^\circ}) \\ &= 1.0\underline{/0^\circ} + 0.54737\underline{/71.805^\circ} \\ &= 1.1709 + j0.5200 \\ &= 1.2812\underline{/23.946^\circ} \quad \text{per unit} \end{aligned}$$

b. From (12.2.1),

$$p_e = \frac{(1.2812)(1.0)}{0.520} \sin\delta = 2.4638 \sin\delta \quad \text{per unit}$$

◁

Figure 12.5

Equivalent circuit for Example 12.3

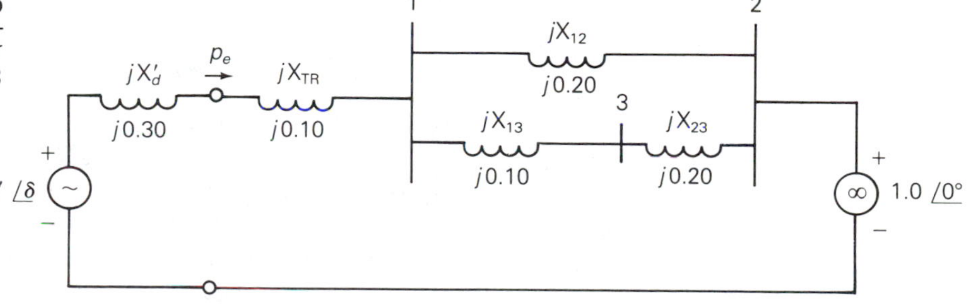

Section 12.3

The Equal-Area Criterion

Consider a synchronous generating unit connected through a reactance to an infinite bus. Plots of electrical power p_e and mechanical power p_m versus power angle δ are shown in Figure 12.6. p_e is a sinusoidal function of δ, as given by (12.2.1).

Figure 12.6

p_e and p_m versus δ

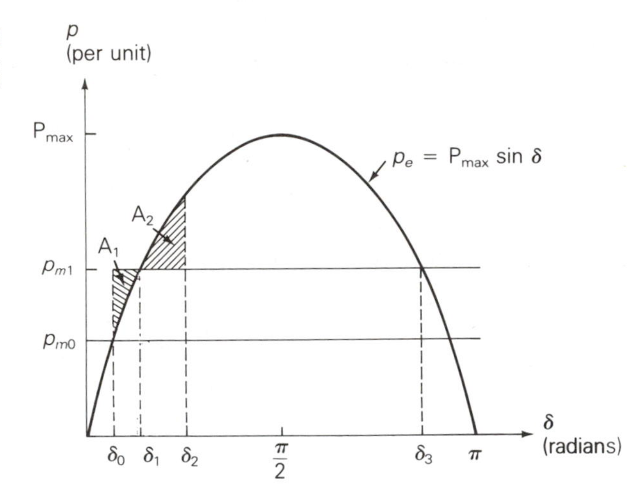

Suppose the unit is initially operating in steady-state at $p_e = p_m = p_{m0}$ and $\delta = \delta_0$, when a step change in p_m from p_{m0} to p_{m1} occurs at $t = 0$. Due to rotor inertia, the rotor position cannot change instantaneously. That is, $\delta_m(0^+) = \delta_m(0^-)$; therefore, $\delta(0^+) = \delta(0^-) = \delta_0$ and $p_e(0^+) = p_e(0^-)$. Since $p_m(0^+) = p_{m1}$ is greater than $p_e(0^+)$, the acceleration power $p_a(0^+)$ is positive and, from (12.1.16), $(d^2\delta)/(dt^2)(0^+)$ is positive. The rotor accelerates and δ increases. When δ reaches δ_1, $p_e = p_{m1}$ and $(d^2\delta)/(dt^2)$ becomes zero. However, $d\delta/dt$ is still positive and δ continues to increase, overshooting its final steady-state operating point. When δ is greater than δ_1, p_m is less than p_e, p_a is negative, and the rotor decelerates. Eventually, δ reaches a maximum value δ_2 and then swings back toward δ_1. Using (12.1.16), which has no damping, δ would continually oscillate around δ_1. However, damping due to mechanical and electrical losses causes δ to stabilize at its final steady-state operating point δ_1. Note that if the power angle exceeded δ_3, then p_m would exceed p_e and the rotor would accelerate again, causing a further increase in δ and loss of stability.

One method for determining stability and maximum power angle is to solve the nonlinear swing equation via numerical integration techniques using a digital computer. This method, which is applicable to multimachine systems, is described in Section 12.4. However, there is also a direct method for determining stability that does not involve solving the swing equation; this method is applicable for one machine connected to an infinite bus or for two machines.

The method, called the *equal-area criterion*, is described in this section.

In Figure 12.6 p_m is greater than p_e during the interval $\delta_0 < \delta < \delta_1$, and the rotor is accelerating. The shaded area A_1 between the p_m and p_e curves is called the accelerating area. During the interval $\delta_1 < \delta < \delta_2$, p_m is less than p_e, the rotor is decelerating, and the shaded area A_2 is the decelerating area. At both the initial value $\delta = \delta_0$ and the maximum value $\delta = \delta_2$, $d\delta/dt = 0$. The equal-area criterion states that $A_1 = A_2$.

In order to derive the equal-area criterion for one machine connected to an infinite bus, assume $\omega_{\text{p.u.}}(t) = 1$ in (12.1.16), giving

$$\frac{2H}{\omega_{\text{syn}}} \frac{d^2\delta}{dt^2} = p_{\text{mp.u.}} - p_{\text{ep.u.}} \tag{12.3.1}$$

Multiplying by $d\delta/dt$ and using

$$\frac{d}{dt}\left[\frac{d\delta}{dt}\right]^2 = 2\left(\frac{d\delta}{dt}\right)\left(\frac{d^2\delta}{dt^2}\right)$$

(12.3.1) becomes

$$\frac{2H}{\omega_{\text{syn}}}\left(\frac{d^2\delta}{dt^2}\right)\left(\frac{d\delta}{dt}\right) = \frac{H}{\omega_{\text{syn}}}\frac{d}{dt}\left[\frac{d\delta}{dt}\right]^2 = (p_{\text{mp.u.}} - p_{\text{ep.u.}})\frac{d\delta}{dt} \tag{12.3.2}$$

Multiplying (12.3.2) by dt and integrating from δ_0 to δ,

$$\frac{H}{\omega_{\text{syn}}}\int_{\delta_0}^{\delta} d\left[\frac{d\delta}{dt}\right]^2 = \int_{\delta_0}^{\delta} (p_{\text{mp.u.}} - p_{\text{ep.u.}})\, d\delta$$

or

$$\frac{H}{\omega_{\text{syn}}}\left[\frac{d\delta}{dt}\right]^2\Bigg|_{\delta_0}^{\delta} = \int_{\delta_0}^{\delta} (p_{\text{mp.u.}} - p_{\text{ep.u.}})\, d\delta \tag{12.3.3}$$

The above integration begins at δ_0 where $d\delta/dt = 0$, and continues to an arbitrary δ. When δ reaches its maximum value, denoted δ_2, $d\delta/dt$ again equals zero. Therefore, the left-hand side of (12.3.3) equals zero for $\delta = \delta_2$ and

$$\int_{\delta_0}^{\delta_2} (p_{\text{mp.u.}} - p_{\text{ep.u.}})\, d\delta = 0 \tag{12.3.4}$$

Separating this integral into positive (accelerating) and negative (decelerating) areas, we arrive at the equal-area criterion

$$\int_{\delta_0}^{\delta_1} (p_{\text{mp.u.}} - p_{\text{ep.u.}})\, d\delta + \int_{\delta_1}^{\delta_2} (p_{\text{mp.u.}} - p_{\text{ep.u.}})\, d\delta = 0$$

or

$$\underbrace{\int_{\delta_0}^{\delta_1} (p_{\text{mp.u.}} - p_{\text{ep.u.}})\, d\delta}_{A_1} = \underbrace{\int_{\delta_1}^{\delta_2} (p_{\text{ep.u.}} - p_{\text{mp.u.}})\, d\delta}_{A_2} \tag{12.3.5}$$

In practice, sudden changes in mechanical power usually do not occur, since the time constants associated with prime mover dynamics are on the order of seconds. However, stability phenomena similar to that described above can also occur from sudden changes in electrical power, due to system faults and line switching. The following three examples are illustrative.

Example 12.4 The synchronous generator shown in Figure 12.4 is initially operating in the steady-state condition given in Example 12.3, when a temporary three-phase-to-ground bolted short circuit occurs on line 1–3 at bus 1, shown as point F in Figure 12.4. Three cycles later the fault extinguishes by itself. Due to a relay misoperation, all circuit breakers remain closed. Determine whether stability is or is not maintained and determine the maximum power angle. The inertia constant of the generating unit is 3.0 per unit-seconds on the system base. Assume p_m remains constant throughout the disturbance. Also assume $\omega_{p.u.}(t) = 1.0$ in the swing equation.

Solution Plots of p_e and p_m versus δ are shown in Figure 12.7. From Example 12.3 the initial operating point is $p_e(0^-) = p_m = 1.0$ per unit and $\delta(0^+) = \delta(0^-) = \delta_0 = 23.95° = 0.4179$ radian. At $t = 0$, when the short circuit occurs, p_e instantaneously drops to zero and remains at zero during the fault since power cannot be transferred past faulted bus 1. From (12.1.16), with $\omega_{p.u.}(t) = 1.0$,

$$\frac{2H}{\omega_{syn}} \frac{d^2\delta(t)}{dt^2} = p_{mp.u.} \qquad 0 \le t \le 0.05 \quad \text{s}$$

Integrating twice with initial conditions $\delta(0) = \delta_0$ and $\frac{d\delta(0)}{dt} = 0$,

$$\frac{d\delta(t)}{dt} = \frac{\omega_{syn} p_{mp.u.}}{2H} t + 0$$

$$\delta(t) = \frac{\omega_{syn} p_{mp.u.}}{4H} t^2 + \delta_0$$

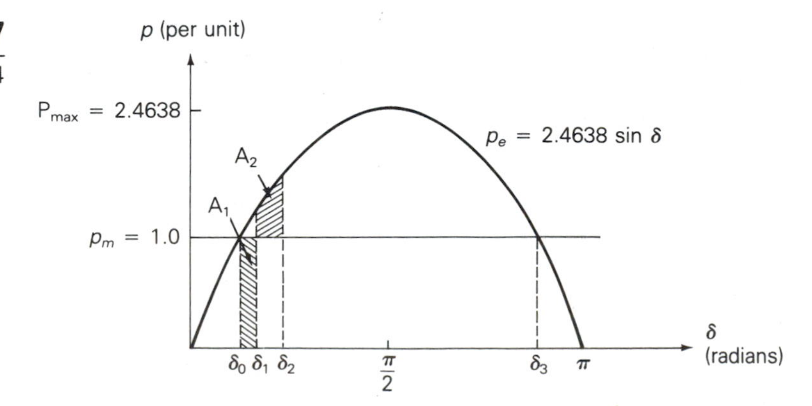

Figure 12.7

p–δ plot for Example 12.4

At $t = 3$ cycles $= 0.05$ second,

$$\delta_1 = \delta(0.05\text{ s}) = \frac{2\pi 60}{12}(0.05)^2 + 0.4179$$

$$= 0.4964 \text{ radian} = 28.44°$$

The accelerating area A_1, shaded in Figure 12.7, is

$$A_1 = \int_{\delta_0}^{\delta_1} p_m d\delta = \int_{\delta_0}^{\delta_1} 1.0 d\delta = (\delta_1 - \delta_2) = 0.4964 - 0.4179 = 0.0785$$

At $t = 0.05$ s the fault extinguishes and p_e instantaneously increases from zero to the sinusoidal curve in Figure 12.7. δ continues to increase until the decelerating area A_2 equals A_1. That is,

$$A_2 = \int_{\delta_1}^{\delta_2} (p_{\max} \sin\delta - p_m)\, d\delta$$

$$= \int_{0.4964}^{\delta_2} (2.4638 \sin\delta - 1.0)\, d\delta = A_1 = 0.0785$$

Integrating,

$$2.4638\,[\cos(0.4964) - \cos\delta_2] - (\delta_2 - 0.4964) = 0.0785$$

$$2.4638 \cos\delta_2 + \delta_2 = 2.5843$$

The above nonlinear algebraic equation can be solved iteratively to obtain

$$\delta_2 = 0.7003 \text{ radian} = 40.12°$$

Since the maximum angle δ_2 does not exceed $\delta_3 = (180° - \delta_0) = 156.05°$, stability is maintained. In steady-state, the generator returns to its initial operating point $p_{ess} = p_m = 1.0$ per unit and $\delta_{ss} = \delta_0 = 23.95°$.

Note that as the fault duration increases, the risk of instability also increases. The *critical clearing time*, denoted t_{cr}, is the longest fault duration allowable for stability. ◁

Example 12.5 Assuming the temporary short circuit in Example 12.4 lasts longer than 3 cycles, calculate the critical clearing time.

Solution The p–δ plot is shown in Figure 12.8. At the critical clearing angle, denoted δ_{cr}, the fault is extinguished. The power angle then increases to a maximum value $\delta_3 = 180° - \delta_0 = 156.05° = 2.7236$ radians, which gives the maximum decelerating area. Equating the accelerating and decelerating areas,

$$A_1 = \int_{\delta_0}^{\delta_{cr}} p_m d\delta = A_2 = \int_{\delta_{cr}}^{\delta_3} (P_{\max} \sin\delta - p_m)\, d\delta$$

$$\int_{0.4179}^{\delta_{cr}} 1.0\, d\delta = \int_{\delta_{cr}}^{2.7236} (2.4638 \sin\delta - 1.0)\, d\delta$$

Solving for δ_{cr},

$$(\delta_{cr} - 0.4179) = 2.4638\,[\cos\delta_{cr} - \cos(2.7236)] - (2.7236 - \delta_{cr})$$

$$2.4638\cos\delta_{cr} = +0.05402$$

$$\delta_{cr} = 1.5489 \text{ radians} = 88.74°$$

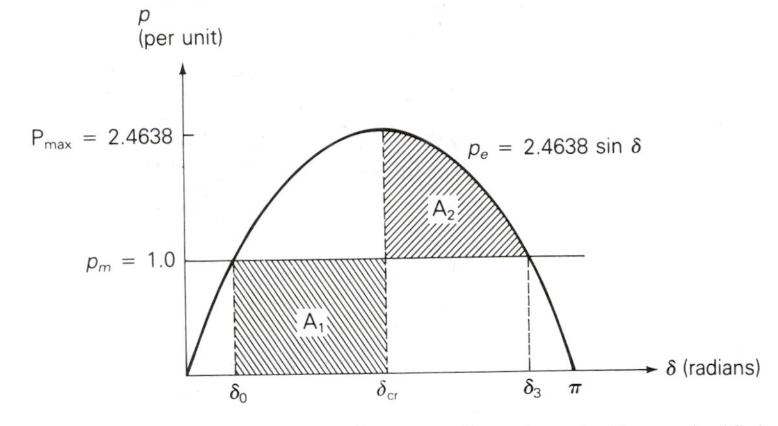

Figure 12.8

p–δ plot for Example 12.5

From the solution to the swing equation given in Example 12.4,

$$\delta(t) = \frac{\omega_{syn} p_{m\text{p.u.}}}{4H} t^2 + \delta_0$$

Solving

$$t = \sqrt{\frac{4H}{\omega_{syn} p_{m\text{p.u.}}}(\delta(t) - \delta_0)}$$

Using $\delta(t_{cr}) = \delta_{cr} = 1.5489$ and $\delta_0 = 0.4179$ radian,

$$t_{cr} = \sqrt{\frac{12}{(2\pi 60)(1.0)}(1.5489 - 0.4179)}$$

$$= 0.1897 \text{ s} = 11.38 \text{ cycles}$$

If the fault is cleared before $t = t_{cr} = 11.38$ cycles, stability is maintained. Otherwise, the generator goes out of synchronism with the infinite bus. That is, stability is lost. ◁

Example 12.6 The synchronous generator in Figure 12.4 is initially operating in the steady-state condition given in Example 12.3 when a permanent three-phase-to-ground bolted short circuit occurs on line 1–3 at bus 3. The fault is cleared by opening the circuit breakers at the ends of line 1–3 and line 2–3. These circuit breakers then remain open. Calculate the critical clearing angle. As in previous examples, H = 3.0 p.u.-s, $p_m = 1.0$ per unit and $\omega_{\text{p.u.}} = 1.0$ in the swing equation.

Solution From Example 12.3, the equation for the prefault electrical power, denoted p_{e1}

here, is $p_{e1} = 2.4638 \sin \delta$ per unit. The faulted network is shown in Figure 12.9(a), and the Thévenin equivalent of the faulted network, as viewed from the generator internal voltage source, is shown in Figure 12.9(b). The Thévenin reactance is

$$X_{Th} = 0.40 + 0.20 \| 0.10 = 0.46666 \quad \text{per unit}$$

and the Thévenin voltage source is

$$V_{Th} = 1.0\underline{/0^\circ}\left[\frac{X_{13}}{X_{13} + X_{12}}\right] = 1.0\underline{/0^\circ}\frac{0.10}{0.30}$$

$$= 0.33333\underline{/0^\circ} \quad \text{per unit}$$

Figure 12.9

Example 12.6

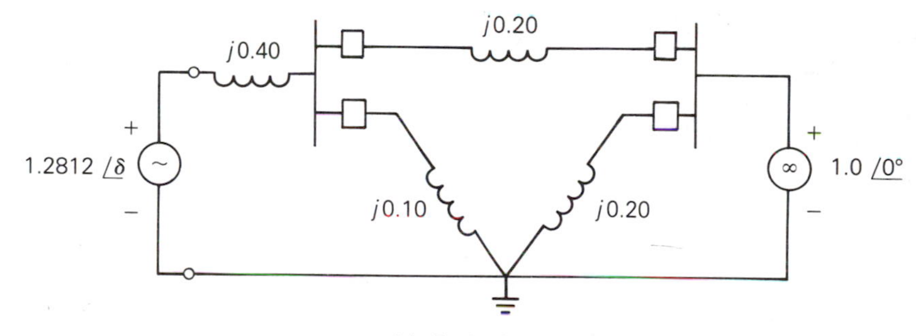

(a) Faulted network

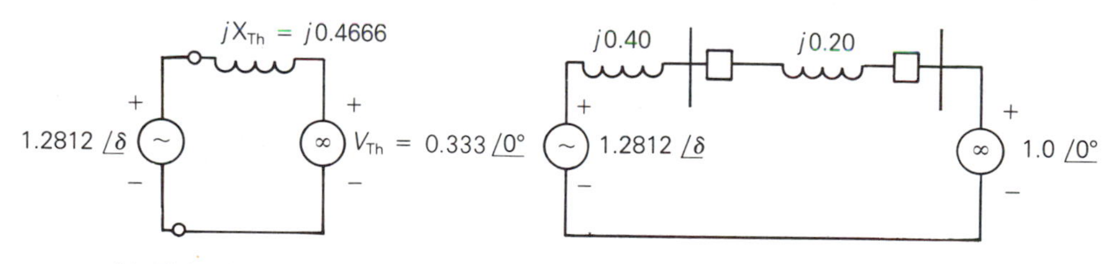

(b) Thévenin equivalent of the faulted network

(c) Postfault conditions

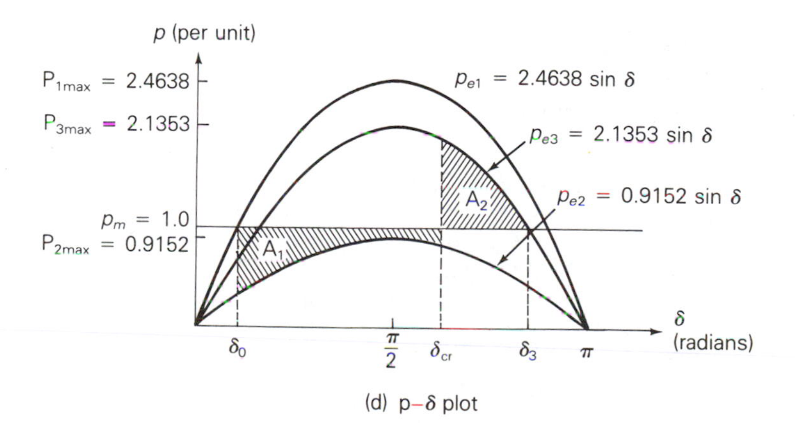

(d) p–δ plot

From Figure 12.9(b), the equation for the electrical power delivered by the generator to the infinite bus during the fault, denoted p_{e2}, is

$$p_{e2} = \frac{E'V_{Th}}{X_{Th}} \sin\delta = \frac{(1.2812)(0.3333)}{0.46666} \sin\delta = 0.9152 \sin\delta \quad \text{per unit}$$

The postfault network is shown in Figure 12.9(c), where circuit breakers have opened and removed lines 1–3 and 2–3. From this figure, the postfault electrical power delivered, denoted p_{e3}, is

$$p_{e3} = \frac{(1.2812)(1.0)}{0.60} \sin\delta = 2.1353 \sin\delta \quad \text{per unit}$$

The p–δ curves as well as the accelerating area A_1 and decelerating area A_2 corresponding to critical clearing are shown in Figure 12.9(d). Equating A_1 and A_2,

$$A_1 = \int_{\delta_0}^{\delta_{cr}} (p_m - P_{2max} \sin\delta)\, d\delta = A_2 = \int_{\delta_{cr}}^{\delta_3} (P_{3max} \sin\delta - p_m)\, d\delta$$

$$\int_{0.4179}^{\delta_{cr}} (1.0 - 0.9152 \sin\delta)\, d\delta = \int_{\delta_{cr}}^{2.6542} (2.1353 \sin\delta - 1.0)\, d\delta$$

Solving for δ_{cr},

$$(\delta_{cr} - 0.4179) + 0.9152(\cos\delta_{cr} - \cos 0.4179)$$
$$= 2.1353(\cos\delta_{cr} - \cos 2.6542) - (2.6542 - \delta_{cr})$$

$$-1.2201 \cos\delta_{cr} = 0.4868$$
$$\delta_{cr} = 1.9812 \text{ radians} = 113.5°$$

If the fault is cleared before $\delta = \delta_{cr} = 113.5°$, stability is maintained. Otherwise, stability is lost. ◁

Section 12.4

Numerical Integration of the Swing Equation

The equal-area criterion is applicable to one machine and an infinite bus or to two machines. For multimachine stability problems, however, numerical integration techniques can be employed to solve the swing equation for each machine.

Given a first-order differential equation

$$\frac{dx}{dt} = f(x) \tag{12.4.1}$$

one relatively simple integration technique is Euler's method [1], illustrated in Figure 12.10. The integration step size is denoted Δt. Calculating the slope at the beginning of the integration interval, from (12.4.1),

$$\frac{dx_t}{dt} = f(x_t) \tag{12.4.2}$$

The new value $x_{t+\Delta t}$ is calculated from the old value x_t by adding the increment Δx,

$$x_{t+\Delta t} = x_t + \Delta x = x_t + \left(\frac{dx_t}{dt}\right)\Delta t \tag{12.4.3}$$

Figure 12.10

Euler's method

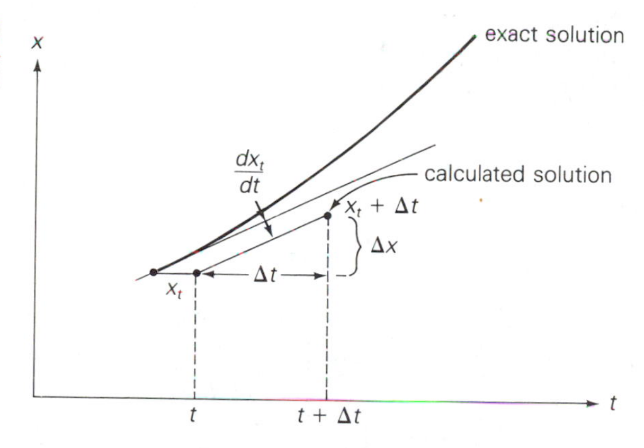

As shown in the figure, Euler's method assumes that the slope is constant over the entire interval Δt. An improvement can be obtained by calculating the slope at both the beginning and end of the interval, and then averaging these slopes. The modified Euler's method is illustrated in Figure 12.11. First, the slope at the beginning of the interval is calculated from (12.4.1) and used to calculate a preliminary value $\tilde{x}$ given by

$$\tilde{x} = x_t + \left(\frac{dx_t}{dt}\right)\Delta t \tag{12.4.4}$$

Figure 12.11

Modified Euler's method

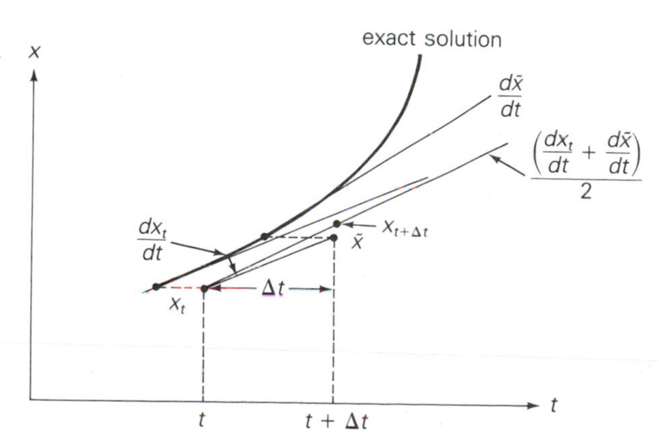

Next the slope at $\tilde{x}$ is calculated:

$$\frac{d\tilde{x}}{dt} = f(\tilde{x}) \tag{12.4.5}$$

Then the new value is calculated using the average slope:

$$x_{t+\Delta t} = x_t + \frac{\left(\dfrac{dx_t}{dt} + \dfrac{d\tilde{x}}{dt}\right)}{2} \Delta t \tag{12.4.6}$$

We now apply the modified Euler's method to calculate machine frequency ω and power angle δ. Letting x be either δ or ω, the old values at the beginning of the interval are denoted δ_t and ω_t. From (12.1.17) and (12.1.18), the slopes at the beginning of the interval are

$$\frac{d\delta_t}{dt} = \omega_t - \omega_{\text{syn}} \tag{12.4.7}$$

$$\frac{d\omega_t}{dt} = \frac{p_{a\text{p.u.}t}\,\omega_{\text{syn}}}{2\text{H}\omega_{\text{p.u.}t}} \tag{12.4.8}$$

where $p_{a\text{p.u.}t}$ is the per-unit accelerating power calculated at $\delta = \delta_t$, and $\omega_{\text{p.u.}t} = \omega_t/\omega_{\text{syn}}$. Applying (12.4.4), preliminary values are

$$\tilde{\delta} = \delta_t + \left(\frac{d\delta_t}{dt}\right)\Delta t \tag{12.4.9}$$

$$\tilde{\omega} = \omega_t + \left(\frac{d\omega_t}{dt}\right)\Delta t \tag{12.4.10}$$

Next, the slopes at $\tilde{\delta}$ and $\tilde{\omega}$ are calculated, again using (12.1.17) and (12.1.18):

$$\frac{d\tilde{\delta}}{dt} = \tilde{\omega} - \omega_{\text{syn}} \tag{12.4.11}$$

$$\frac{d\tilde{\omega}}{dt} = \frac{\tilde{p}_{a.\text{p.u.}}\,\omega_{\text{syn}}}{2\text{H}\tilde{\omega}_{\text{p.u.}}} \tag{12.4.12}$$

where $\tilde{p}_{a\text{p.u.}}$ is the per-unit accelerating power calculated at $\delta = \tilde{\delta}$, and $\tilde{\omega}_{\text{p.u.}} = \tilde{\omega}/\omega_{\text{syn}}$. Applying (12.4.6), the new values at the end of the interval are

$$\delta_{t+\Delta t} = \delta_t + \frac{\left(\dfrac{d\delta_t}{dt} + \dfrac{d\tilde{\delta}}{dt}\right)}{2} \Delta t \tag{12.4.13}$$

$$\omega_{t+\Delta t} = \omega_t + \frac{\left(\dfrac{d\omega_t}{dt} + \dfrac{d\tilde{\omega}}{dt}\right)}{2} \Delta t \tag{12.4.14}$$

This procedure, given by (12.4.7)–(12.4.13), begins at $t = 0$ with specified initial values δ_0 and ω_0, and continues iteratively until $t = T$, a specified final time. Calculations are best performed using a digital computer.

Example 12.7 Verify the critical clearing angle determined in Example 12.6, and calculate the critical clearing time by applying the modified Euler's method to solve the swing equation for the following two cases:

Case 1 The fault is cleared at $\delta = 1.95$ radians $= 112°$ (which is less than δ_{cr})

Case 2 The fault is cleared at $\delta = 2.09$ radians $= 120°$ (which is greater than δ_{cr})

For calculations, use a step size $\Delta t = 0.01$ s, and solve the swing equation from $t = 0$ to $t = T = 0.85$ s.

Solution Equations (12.4.7)–(12.4.14) are solved by a digital computer program written in BASIC. From Example 12.6, the initial conditions at $t = 0$ are

$$\delta_0 = 0.4179 \quad \text{rad}$$

$$\omega_0 = \omega_{syn} = 2\pi 60 \quad \text{rad/s}$$

Also, the H constant is 3.0 p.u.-s, and the faulted accelerating power is

$$p_{ap.u.} = 1.0 - 0.9152 \sin \delta$$

The postfault accelerating power is

$$p_{ap.u.} = 1.0 - 2.1353 \sin \delta \quad \text{per unit}$$

The computer program and results at 0.02 s printout intervals are listed in Table 12.1. As shown, these results agree with Example 12.6, since the system is stable for Case 1 and unstable for Case 2. Also from Table 12.1, the critical clearing time is between 0.34 and 0.36 s. ◁

In addition to Euler's method, there are many other numerical integration techniques, such as Runge–Kutta, Picard's method, and Milne's predictor-corrector method [1]. Comparison of the methods shows a trade-off of accuracy versus computation complexity. The Euler method is a relatively simple method to compute, but requires a small step size Δt for accuracy. Some of the other methods can use a larger step size for comparable accuracy, but the computations are more complex.

Table 12.1 Computer calculation of swing curves for Example 12.7

CASE 1 STABLE

TIME s	DELTA rad	OMEGA rad/s
0.000	0.418	376.991
0.020	0.426	377.778
0.040	0.449	378.547
0.060	0.488	379.283
0.080	0.541	379.970
0.100	0.607	380.599
0.120	0.685	381.159
0.140	0.773	381.646
0.160	0.870	382.056
0.180	0.975	382.392
0.200	1.086	382.660
0.220	1.202	382.868
0.240	1.321	383.027
0.260	1.443	383.153
0.280	1.567	383.262
0.300	1.694	383.370
0.320	1.823	383.495
0.340	1.954	383.658
FAULT CLEARED		
0.360	2.076	382.516
0.380	2.176	381.510
0.400	2.257	380.638
0.420	2.322	379.886
0.440	2.373	379.237
0.460	2.413	378.674
0.480	2.441	378.176
0.500	2.460	377.726
0.520	2.471	377.307
0.540	2.473	376.900
0.560	2.467	376.488
0.580	2.453	376.056
0.600	2.429	375.583
0.620	2.396	375.053
0.640	2.351	374.446
0.660	2.294	373.740
0.680	2.221	372.917
0.700	2.130	371.960
0.720	2.019	370.855
0.740	1.884	369.604
0.760	1.723	368.226
0.780	1.533	366.773
0.800	1.314	365.341
0.820	1.068	364.070
0.840	0.799	363.143

CASE 2 UNSTABLE

TIME s	DELTA rad	OMEGA rad/s
0.000	0.418	376.991
0.020	0.426	377.778
0.040	0.449	378.547
0.060	0.488	379.283
0.080	0.541	379.970
0.100	0.607	380.599
0.120	0.685	381.159
0.140	0.773	381.646
0.160	0.870	382.056
0.180	0.975	382.392
0.200	1.086	382.660
0.220	1.202	382.868
0.240	1.321	383.027
0.260	1.443	383.153
0.280	1.567	383.262
0.300	1.694	383.370
0.320	1.823	383.495
0.340	1.954	383.658
0.360	2.090	383.876
FAULT CLEARED		
0.380	2.217	382.915
0.400	2.327	382.138
0.420	2.424	381.546
0.440	2.511	381.135
0.460	2.591	380.902
0.480	2.668	380.844
0.500	2.746	380.969
0.520	2.828	381.288
0.540	2.919	381.824
0.560	3.022	382.609
0.580	3.145	383.686
0.600	3.292	385.111
0.620	3.472	386.949
0.640	3.693	389.265
0.660	3.965	392.099
0.680	4.300	395.426
0.700	4.704	399.079
0.720	5.183	402.689
0.740	5.729	405.683
0.760	6.325	407.477
0.780	6.941	407.812
0.800	7.551	406.981
0.820	8.139	405.711
0.840	8.702	404.819

PROGRAM LISTING

```
10 REM EXAMPLE 12.7
20 REM SOLUTION TO SWING EQUATION
30 REM THE STEP SIZE IS DELTA
40 REM THE CLEARING ANGLE IS DLTCLR
50 DELTA = .01
60 DLTCLR = 1.95
70 J = 1
80 PMAX = .9152
90 PI = 3.1415927#
100 T = 0
110 X1 = .4179
120 X2 = 2 * PI * 60
130 LPRINT "  TIME   DELTA  OMEGA"
140 LPRINT "    s       rad       rad/s "
150 LPRINT USING "#####.###"; T; X1; X2
160 FOR K = 1 TO 86
170 REM LINE 180 IS EQ(12.4.7)
180 X3 = X2 – (2 * PI * 60)
190 IF J = 2 THEN GOTO 240
200 IF X1 > DLTCLR OR X1 = DLTCLR THEN
       PMAX = 2.1353
210 IF X1 > DLTCLR OR X1 = DLTCLR THEN
       LPRINT "  FAULT CLEARED"
220 IF X1 > DLTCLR OR X1 = DLTCLR THEN
       J = 2
230 REM LINES 240 AND 250 ARE EQ(12.4.8)
240 X4 = 1 – PMAX * SIN(X1)
250 X5 = X4 * (2 * PI * 60) * (2 * PI * 60)/(6 * X2)
260 REM LINE 270 IS EQ(12.4.9)
270 X6 = X1 + X3 * DELTA
280 REM LINE 290 IS EQ(12.4.10)
290 X7 = X2 + X5 * DELTA
300 REM LINE 310 IS EQ(12.4.11)
310 X8 = X7 – 2 * PI * 60
320 REM LINES 330 AND 340 ARE EQ(12.4.12)
330 X9 = 1 – PMAX * SIN(X6)
340 X10 = X9 * (2 * PI * 60) * (2 * PI * 60)/(6 * X7)
350 REM LINE 360 IS EQ(12.4.13)
360 X1 = X1 + (X3 + X8) * (DELTA/2)
370 REM LINE 380 IS EQ(12.4.14)
380 X2 = X2 + (X5 + X10) * (DELTA/2)
390 T = K * DELTA
400 Z = K/2
```

CASE 1 STABLE			CASE 2 UNSTABLE			PROGRAM LISTING
TIME s	DELTA rad	OMEGA rad/s	TIME s	DELTA rad	OMEGA rad/s	
0.860	0.516	362.750	0.860	9.257	404.934	410 M = INT(Z) 420 IF M = Z THEN LPRINT USING "#####.###"; T; X1; X2 430 NEXT K 440 END

Section 12.5

Multimachine Stability

The numerical integration methods discussed in Section 12.4 can be used to solve the swing equations for a multimachine stability problem. However, a method is required for computing machine output powers for a general network. Figure 12.12 shows a general N-bus power system with M synchronous machines. Each machine is the same as that represented by the simplified model of Figure 12.2, and the internal machine voltages are denoted E_1', E_2', . . ., E_M'.

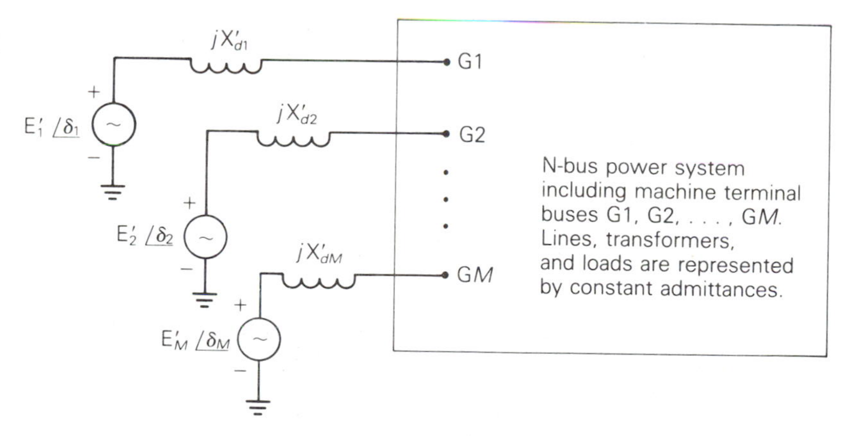

Figure 12.12 *N*-bus power-system representation for transient stability studies

The M machine terminals are connected to system buses denoted G1, G2, . . ., GM in Figure 12.12. All loads are modeled here as constant admittances. Writing nodal equations for this network,

$$\begin{bmatrix} \boldsymbol{Y}_{11} & \boldsymbol{Y}_{12} \\ \boldsymbol{Y}_{12}^{T} & \boldsymbol{Y}_{22} \end{bmatrix} \begin{bmatrix} \boldsymbol{V} \\ \boldsymbol{E} \end{bmatrix} = \begin{bmatrix} \boldsymbol{0} \\ \boldsymbol{I} \end{bmatrix} \tag{12.5.1}$$

where

$$V = \begin{bmatrix} V_1 \\ V_2 \\ \vdots \\ V_N \end{bmatrix} \quad \text{is the } N \text{ vector of bus voltages} \tag{12.5.2}$$

$$\boldsymbol{E} = \begin{bmatrix} E'_1 \\ E'_2 \\ \vdots \\ E'_M \end{bmatrix} \quad \text{is the } M \text{ vector of internal machine voltages} \tag{12.5.3}$$

$$\boldsymbol{I} = \begin{bmatrix} I_1 \\ I_2 \\ \vdots \\ I_M \end{bmatrix} \quad \begin{matrix}\text{is the } M \text{ vector of machine currents} \\ \text{(these are current sources)}\end{matrix} \tag{12.5.4}$$

$$\left[\begin{array}{c|c} \boldsymbol{Y}_{11} & \boldsymbol{Y}_{12} \\ \hline \boldsymbol{Y}_{12}^{\mathrm{T}} & \boldsymbol{Y}_{22} \end{array}\right] \quad \text{is an } (N + M) \times (N + M) \text{ admittance matrix} \tag{12.5.5}$$

The admittance matrix in (12.5.5) is partitioned in accordance with the N system buses and M internal machine buses, as follows:

$\boldsymbol{Y}_{11}$ is $N \times N$
$\boldsymbol{Y}_{12}$ is $N \times M$
$\boldsymbol{Y}_{22}$ is $M \times M$

$\boldsymbol{Y}_{11}$ is similar to the bus admittance matrix used for power flows in Chapter 7, except that load admittances and inverted generator impedances are included. That is, if a load is connected to bus n, then that load admittance is added to the diagonal element Y_{11nn}. Also, $(1/jX'_{dn})$ is added to the diagonal element Y_{11GnGn}.

Y_{22} is a diagonal matrix of inverted generator impedances. That is,

$$Y_{22} = \begin{bmatrix} \dfrac{1}{jX'_{d1}} & & & 0 \\ & \dfrac{1}{jX'_{d2}} & & \\ & & \ddots & \\ 0 & & & \dfrac{1}{jX'_{dM}} \end{bmatrix} \tag{12.5.6}$$

Also, the kmth element of Y_{12} is

$$Y_{12km} = \begin{cases} \dfrac{-1}{jX'_{dn}} & \text{if } k = \text{G}n \text{ and } m = n \\ 0 & \text{otherwise} \end{cases} \tag{12.5.7}$$

Writing (12.5.1) as two separate equations,

$$\boldsymbol{Y}_{11}\boldsymbol{V} + \boldsymbol{Y}_{12}\boldsymbol{E} = \boldsymbol{0} \tag{12.5.8}$$

$$\boldsymbol{Y}_{12}^{\mathrm{T}}\boldsymbol{V} + \boldsymbol{Y}_{22}\boldsymbol{E} = \boldsymbol{I} \tag{12.5.9}$$

Assuming $\boldsymbol{E}$ is known, (12.5.8) is a linear equation in $\boldsymbol{V}$ that can be solved either iteratively or by Gauss elimination. Using the Gauss–Seidel iterative method given by (7.2.9), the kth component of $\boldsymbol{V}$ is

$$V_k(i+1) = \frac{1}{Y_{11kk}}\left[-\sum_{n=1}^{M} Y_{12kn}E_n - \sum_{n=1}^{k-1} Y_{11kn}V_n(i+1) - \sum_{n=k+1}^{N} Y_{11kn}V_n(i)\right] \tag{12.5.10}$$

After $\boldsymbol{V}$ is computed, the machine currents can be obtained from (12.5.9). That is,

$$\boldsymbol{I} = \begin{bmatrix} I_1 \\ I_2 \\ \vdots \\ I_M \end{bmatrix} = \boldsymbol{Y}_{12}^{\mathrm{T}}\boldsymbol{V} + \boldsymbol{Y}_{22}\boldsymbol{E} \tag{12.5.11}$$

The (real) electrical power output of machine n is then

$$p_{en} = \mathrm{Re}\,[E_n I_n^*] \qquad n = 1, 2, \ldots, M \tag{12.5.12}$$

We are now ready to outline a computation procedure for solving a transient stability problem. The procedure alternately solves the swing equations representing the machines and the above algebraic power-flow equations representing the network. We use the modified Euler method of Section 12.4

to solve the swing equations and the Gauss–Seidel iterative method to solve the power-flow equations. The procedure is now given in the following 11 steps.

Transient stability computation procedure

Step 1 Run a prefault power-flow program to compute initial bus voltages V_k, $k = 1, 2, \ldots, N$, initial machine currents I_n, and initial machine electrical power outputs p_{en}, $n = 1, 2, \ldots, M$. Set machine mechanical power outputs, $p_{mn} = p_{en}$. Set initial machine frequencies, $\omega_n = \omega_{syn}$. Compute the load admittances.

Step 2 Compute the internal machine voltages:

$$E_n = \mathrm{E}_n \angle \delta_n = V_{Gn} + (jX'_{dn})I_n \qquad n = 1, 2, \ldots, M$$

where V_{Gn} and I_n are computed in Step 1. The magnitudes E_n will remain constant throughout the study. The angles δ_n are the initial power angles.

Step 3 Compute Y_{11}. Modify the $(N \times N)$ power-flow bus admittance matrix by including the load admittances and inverted generator impedances.

Step 4 Compute Y_{22} from (12.5.6) and Y_{12} from (12.5.7).

Step 5 Set time $t = 0$.

Step 6 Is there a switching operation, change in load, short circuit, or change in data? For a switching operation or change in load, modify the bus admittance matrix. For a short circuit, set the faulted bus voltage [in (12.5.10)] to zero.

Step 7 Using the internal machine voltages $E_n = \mathrm{E}_n \angle \delta_n$, $n = 1, 2, \ldots, M$, with the values of δ_n at time t, compute the machine electrical powers p_{en} at time t from (12.5.10)–(12.5.12).

Step 8 Using p_{en} computed in Step 7 and the values of δ_n and ω_n at time t, compute the preliminary estimates of power angles $\tilde{\delta}_n$ and machine speeds $\tilde{\omega}_n$ at time $(t + \Delta t)$ from (12.4.7)–(12.4.10).

Step 9 Using $E_n = \mathrm{E}_n \angle \tilde{\delta}_n$, $n = 1, 2, \ldots, M$, compute the preliminary estimates of the machine electrical powers $\tilde{p}_{en}$ at time $(t + \Delta t)$ from (12.5.10)–(12.5.12).

Step 10 Using $\tilde{p}_{en}$ computed in Step 9, as well as $\tilde{\delta}_n$ and $\tilde{\omega}_n$ computed in Step 8, compute the final estimates of power angles δ_n and machine speeds ω_n at time $(t + \Delta t)$ from (12.4.11)–(12.4.14).

Step 11 Set time $t = t + \Delta t$. Stop if $t \geq T$. Otherwise, return to Step 6.

Example 12.8 Consider a transient stability study for the power system given in Example 7.9, with the 200-Mvar shunt capacitor of Example 7.13 installed at bus 2. Machine transient reactances are $X'_{d1} = 0.20$ and $X'_{d2} = 0.10$ per unit on the system base. Determine the admittance matrices $\boldsymbol{Y}_{11}$, $\boldsymbol{Y}_{22}$, and $\boldsymbol{Y}_{12}$.

Solution From Example 7.9, the power system has $N = 5$ buses and $M = 2$ machines. The second row of the 5×5 bus admittance matrix used for power flows is calculated in Example 7.9. Calculating the other rows in the same manner, we obtain

$$\boldsymbol{Y}_{\text{bus}} = \begin{bmatrix} (0.932 - j12.43) & 0 & 0 & 0 & (-0.932 + j12.43) \\ 0 & (0.670 - j7.115) & 0 & (-0.223 + j2.480) & (-0.446 + j4.96) \\ 0 & 0 & (1.865 - j24.86) & (-1.865 + j24.86) & 0 \\ 0 & (-0.223 + j2.480) & (-1.865 + j24.86) & (2.980 - j37.0) & (-0.893 + j9.92) \\ (-0.932 + j12.43) & (-0.446 + j4.960) & 0 & (-0.893 + j9.920) & (2.271 - j27.15) \end{bmatrix} \text{ per unit}$$

To obtain $\boldsymbol{Y}_{11}$, $\boldsymbol{Y}_{\text{bus}}$ is modified by including load admittances and inverted generator impedances. From Table 7.1, the load at bus 3 is $P_{L3} + jQ_{L3} = 0.2 + j0.1$ per unit and the voltage at bus 3 is $V_3 = 1.05$ per unit. Representing this load as a constant admittance,

$$Y_{\text{load}3} = \frac{P_{L3} - jQ_{L3}}{V_3^2} = \frac{0.2 - j0.1}{(1.05)^2} = 0.1814 - j0.0907 \quad \text{per unit}$$

Similarly, the load admittance at bus 2 is

$$Y_{\text{load}2} = \frac{P_{L2} - jQ_{L2}}{V_2^2} = \frac{2 - j0.7 + j0.5}{(0.968)^2} = 2.134 - j0.213$$

where V_2 is obtained from Example 7.13 and the 200-Mvar (0.5 per unit) shunt capacitor bank is included in the bus 2 load.

The inverted generator impedances are: for machine 1 connected to bus 1,

$$\frac{1}{jX'_{d1}} = \frac{1}{j0.20} = -j5.0 \text{ per unit}$$

and for machine 2 connected to bus 3,

$$\frac{1}{jX'_{d2}} = \frac{1}{j0.10} = -j10.0 \text{ per unit}$$

To obtain $\boldsymbol{Y}_{11}$, add $(1/jX'_{d1})$ to the first diagonal element of $\boldsymbol{Y}_{\text{bus}}$, add $Y_{\text{load}2}$ to the second diagonal element, and add $Y_{\text{load}3} + (1/jX'_{d2})$ to the third diagonal element. The 5×5 matrix $\boldsymbol{Y}_{11}$ is then

$$
\boldsymbol{Y}_{11} = \begin{bmatrix}
(0.932-j17.43) & 0 & 0 & 0 & (-0.932+j12.43) \\
0 & (2.804-j7.328) & 0 & (-0.223+j2.480) & (-0.446+j4.96) \\
0 & 0 & (2.0465-j34.951) & (-1.865+j24.86) & 0 \\
0 & (-0.223+j2.480) & (-1.865+j24.86) & (2.980-j37.0) & (-0.893+j9.92) \\
(-0.932+j12.43) & (-0.446+j4.960) & 0 & (-0.893+j9.920) & (2.271-j27.15)
\end{bmatrix} \text{per unit}
$$

From (12.5.6), the 2 × 2 matrix $\boldsymbol{Y}_{22}$ is

$$
\boldsymbol{Y}_{22} = \begin{bmatrix} \dfrac{1}{jX'_{d1}} & 0 \\ 0 & \dfrac{1}{jX'_{d2}} \end{bmatrix} = \begin{bmatrix} -j5.0 & 0 \\ 0 & -j10.0 \end{bmatrix} \quad \text{per unit}
$$

From Figure 7.2, generator 1 is connected to bus 1 (therefore, bus G1 = 1) and generator 2 is connected to bus 3 (therefore G2 = 3). From (12.5.7), the 5 × 2 matrix $\boldsymbol{Y}_{12}$ is

$$
\boldsymbol{Y}_{12} = \begin{bmatrix} j5.0 & 0 \\ 0 & 0 \\ 0 & j10.0 \\ 0 & 0 \\ 0 & 0 \end{bmatrix} \quad \text{per unit}
$$

◁

Section 12.6

Personal Computer Program: TRANSIENT STABILITY

The personal computer software package that accompanies this text includes the program entitled "TRANSIENT STABILITY," which computes machine power angles and frequencies in a balanced three-phase power system subjected to disturbances. This program also computes machine angular accelerations, machine electrical power outputs, and bus voltage magnitudes.

Input data for the program include: (1) the bus admittance matrix, initial bus voltages, initial machine currents, and initial machine electrical power outputs, all obtained from the program POWER FLOW described in Chapter 7; and (2) the per-unit inertia constant and direct axis transient reactance of each synchronous machine.

The program executes the transient stability computation procedure given in Section 12.5. The program user selects the type of each disturbance and the time at which each disturbance begins. Disturbance types include: switching operations (opening or closing circuit breakers selected by the program user), three-phase short circuits, changes in loads, and changes in input data. The program user also selects the integration step size Δt and the final time T.

Output data consist of the power angle, frequency, and electrical power output of each machine versus time, as well as bus voltages versus time. The program user selects the outputs to be printed and the print time interval.

Example 12.9 Use the program TRANSIENT STABILITY to study a temporary three-phase short circuit at bus 5 of the power system given in Example 7.9. For prefault conditions, use the power-flow output given in Example 7.13, where a 200-Mvar shunt capacitor bank is installed at bus 2. Machine transient reactances are $X'_{d1} = 0.20$ and $X'_{d2} = 0.10$ per unit, as given in Example 12.8, and machine inertia constants are $H_1 = 5.0$ and $H_2 = 50$ p.u.-s (where machine 2 represents a large system). The short circuit is cleared by opening circuit breakers B1, B51, and B52 at $t = 0.05$ s (3 cycles), followed by reclosing these circuit breakers. Assume that the temporary fault has already self-extinguished when reclosure occurs.

Run the following two cases:

Case 1 Reclosure at $t = 0.27$ s (13 cycles after fault clearing)

Case 2 Reclosure at $t = 0.30$ s (15 cycles after fault clearing)

For computation purposes, select an integration step size $\Delta t = 0.01$ s and final time $T = 0.75$ s.

Solution The power-flow output given in Example 7.13 is selected as the input to the TRANSIENT STABILITY program, along with the machine reactances and inertia constants given above. The first disturbance selected is a short circuit at bus 5 (that is, $V_5 = 0$) at time $t = 0$. The second disturbance selected is the

CASE (1)				
TIME	DELTA1	OMEGA1	DELTA2	OMEGA2
SECONDS	DEGREES	RAD/SEC	DEGREES	RAD/SEC
FAULT AT BUS 5				
0.00	11.16	376.99	6.10	376.99
0.03	12.02	377.99	6.19	377.10
FAULT CLEARED				
0.06	14.61	379.02	6.46	377.17
0.09	19.03	380.10	6.75	377.15
0.12	25.31	381.18	7.02	377.14
0.15	33.43	382.26	7.27	377.13
0.18	43.41	383.33	7.50	377.12
0.21	55.22	384.40	7.72	377.11
0.24	68.87	385.47	7.91	377.10
RECLOSURE				
0.27	84.36	386.53	8.09	377.09
0.30	99.39	384.95	8.47	377.34
0.33	111.78	383.48	9.29	377.60
0.36	121.78	382.17	10.56	377.86
0.39	129.66	381.01	12.27	378.11
0.42	135.66	379.98	14.41	378.36
0.45	139.97	379.02	16.99	378.61
0.48	142.66	378.09	19.98	378.86
0.51	143.72	377.11	23.39	379.10
0.54	143.02	376.04	27.24	379.35
0.57	140.37	374.82	31.52	379.61
0.60	135.47	373.43	36.24	379.87
0.63	128.04	371.87	41.41	380.13
0.66	117.82	370.22	47.00	380.37
0.69	104.82	368.67	52.99	380.57
0.72	89.39	367.48	59.27	380.71
0.75	72.47	366.98	65.73	380.76

CASE (2)				
TIME	DELTA1	OMEGA1	DELTA2	OMEGA2
SECONDS	DEGREES	RAD/SEC	DEGREES	RAD/SEC
FAULT AT BUS 5				
0.00	11.16	376.99	6.10	376.99
0.03	12.02	377.99	6.19	377.10
FAULT CLEARED				
0.06	14.61	379.02	6.46	377.17
0.09	19.03	380.10	6.75	377.15
0.12	25.31	381.18	7.02	377.14
0.15	33.43	382.26	7.27	377.13
0.18	43.41	383.33	7.50	377.12
0.21	55.22	384.40	7.72	377.11
0.24	68.87	385.47	7.91	377.10
0.27	84.36	386.53	8.09	377.09
RECLOSURE				
0.30	101.66	387.59	8.25	377.08
0.33	118.65	386.21	8.63	377.34
0.36	133.51	385.13	9.44	377.59
0.39	146.83	384.43	10.68	377.83
0.42	159.27	384.11	12.30	378.03
0.45	171.49	384.17	14.25	378.21
0.48	184.17	384.64	16.48	378.36
0.51	198.01	385.54	18.93	378.46
0.54	213.78	386.90	21.51	378.52
0.57	232.34	388.78	24.15	378.52
0.60	254.56	391.15	26.73	378.45
0.63	281.23	393.92	29.14	378.31
0.66	312.80	396.75	31.25	378.12
0.69	348.90	399.05	33.02	377.93
0.72	388.05	400.20	34.52	377.82
0.75	427.95	399.94	35.95	377.85

Table 12.2 Transient stability output for Example 12.9

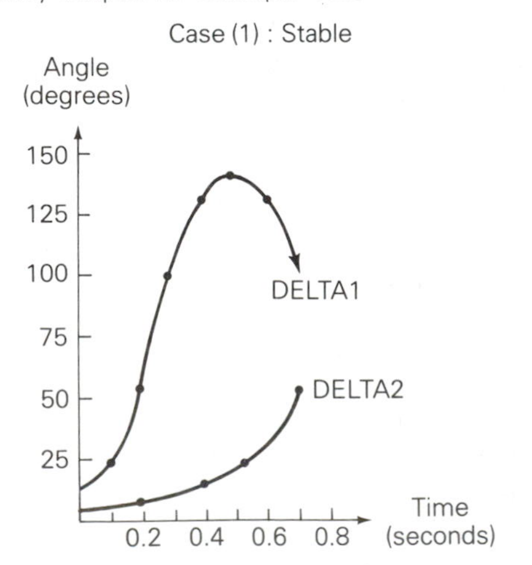

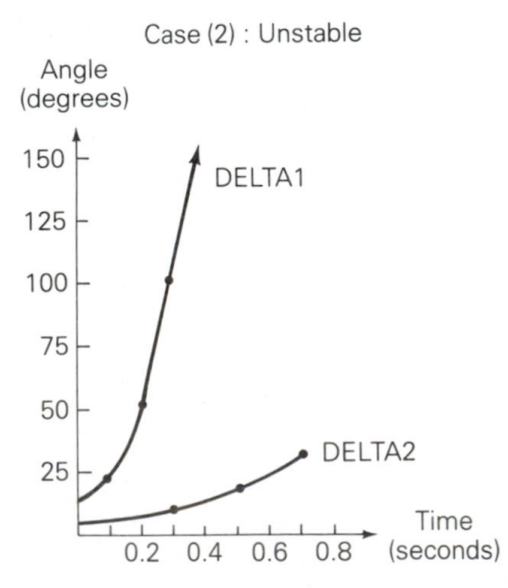

opening of circuit breakers B1, B51, and B52 to clear the fault at $t = 0.05$ s. The third disturbance selected is the reclosure of these breakers at $t = 0.27$ s for Case 1 and at $t = 0.30$ s for Case 2. The condition $V_5 = 0$ is also removed.

Machine power angle outputs are listed in Table 12.2 versus time for both cases. The printout interval is 0.03 s. As shown for Case 1 power angle δ_1 reaches a maximum value and then swings back. With damping included, δ_1 and δ_2 would settle down to new steady-state values.

Case 2 is unstable. The power angle δ_1 of machine 1 exceeds 180 degrees and diverges away from δ_2. That is, machine 1 pulls out of synchronism with machine 2 on the first swing. ◁

Section 12.7

Design Methods for Improving Transient Stability

Design methods for improving power-system transient stability include the following:

1. Improved steady-state stability
   a. Higher system voltage levels
   b. Additional transmission lines
   c. Smaller transmission-line series reactances
   d. Smaller transformer leakage reactances
   e. Series capacitive transmission-line compensation
2. High-speed fault clearing
3. High-speed reclosure of circuit breakers
4. Single-pole switching
5. Larger machine inertia, lower transient reactance
6. Fast responding, high-gain exciters
7. Fast valving
8. Braking resistors

We discuss these design methods in the following paragraphs.

1. Increasing the maximum power transfer in steady-state can also improve transient stability, allowing for increased power transfer through the unfaulted portion of a network during disturbances. Upgrading voltage on existing transmission or opting for higher voltages on new transmission increases line loadability (6.5.6). Additional parallel lines increase power-transfer capability. Reducing system reactances also increases power-transfer capability. Lines with bundled phase conductors have lower series reactances than lines that are not bundled. Oversized transformers with lower leakage reactances also help. Series capacitors reduce the total series reactances of a line by compensating for the series line inductance.

2. High-speed fault clearing is fundamental to transient stability. Standard practice for EHV systems is 1-cycle relaying and 2-cycle circuit breakers, allowing for fault clearing within 3 cycles (0.05 s). Ongoing research is presently aimed at reducing these to one-half cycle relaying and 1-cycle circuit breakers.

3. The majority of transmission-line short circuits are temporary, with the fault arc self-extinguishing within 5 to 40 cycles (depending on system voltage) after the line is deenergized. High-speed reclosure of circuit breakers can increase postfault transfer power, thereby improving transient stability. Conservative practice for EHV systems is to employ high-speed reclosure only if stability is maintained when reclosing into a permanent fault with subsequent reopening and lockout of breakers.

4. Since the majority of short circuits are single line-to-ground, relaying schemes and independent-pole circuit breakers can be used to clear a faulted phase while keeping the unfaulted phases of a line operating, thereby maintaining some power transfer across the faulted line. Studies have shown that single line-to-ground faults are self-clearing even when only the faulted phase is deenergized. Capacitive coupling between the energized unfaulted phases and the deenergized faulted phase is, in most cases, not strong enough to maintain an arcing short circuit [5].

5. Inspection of the swing equation, (12.1.16), shows that increasing the per-unit inertia constant H of a synchronous machine reduces angular acceleration, thereby slowing down angular swings and increasing critical clearing times. Stability is also improved by reducing machine transient reactances, which increases power-transfer capability during fault or postfault periods [*see* (12.2.1)]. Unfortunately, present-day generator manufacturing trends are toward lower H constants and higher machine reactances, which are a detriment to stability.

6. Modern machine excitation systems with fast thyristor controls and high amplifier gains (to overcome generator saturation) can rapidly increase generator field excitation after sensing low terminal voltage during faults. The effect is to rapidly increase internal machine voltages during faults, thereby increasing generator output power during fault and postfault periods. Critical clearing times are also increased [6].

7. Some steam turbines are equipped with fast valving to divert steam flows and rapidly reduce turbine mechanical power outputs. During faults near the generator, when electrical power output is reduced, fast valving action acts to balance mechanical and electrical power, providing reduced acceleration and longer critical clearing times. The turbines are designed to withstand thermal stresses due to fast valving [7].

8. In power systems with generation areas that can be temporarily separated from load areas, braking resistors can improve stability. When separation occurs, the braking resistor is inserted into the generation area for a second or two, preventing or slowing acceleration in the generation area. Shelton et al. [8] describe a 3-GW-s braking resistor.

Problems

Section 12.1

12.1 A three-phase, 60-Hz, 400-MVA, 13.8-kV, 4-pole steam turbine-generating unit has an H constant of 6.0 p.u.-s. Determine: (a) ω_{syn} and ω_{msyn}; (b) the kinetic energy in joules stored in the rotating masses at synchronous speed; (c) the mechanical angular acceleration α_m and electrical angular acceleration α if the unit is operating at synchronous speed with an accelerating power of 300 MW.

12.2 Calculate J in kg m^2 for the generating unit given in Problem 12.1.

12.3 Generator manufacturers often use the term WR^2, which is the weight in pounds of all the rotating parts of a generating unit (including the prime mover) multiplied by the square of the radius of gyration in feet. $WR^2/32.2$ is then the total moment of inertia of the rotating parts in slug-ft^2. (a) Determine a formula for the stored kinetic energy in ft-lb of a generating unit in terms of WR^2 and rotor angular velocity ω_m. (b) Show that

$$H = \frac{2.31 \times 10^{-4}\, WR^2 (\text{rpm})^2}{S_{rated}} \quad \text{per unit-seconds}$$

where S_{rated} is the voltampere rating of the generator and rpm is the synchronous speed in r/min. Note that 1 ft-lb = 746/550 = 1.356 joules. (c) Evaluate H for a three-phase generating unit rated 800 MVA, 3600 r/min, with WR^2 = 4,000,000 lb-ft^2.

12.4 The generating unit in Problem 12.1 is initially operating at $p_{mp.u.} = p_{ep.u.} = 0.8$ per unit, $\omega = \omega_{syn}$, and $\delta = 15°$ when a fault reduces the generator electrical power output by 70%. Determine the power angle δ five cycles after the fault commences. Assume that the accelerating power remains constant during the fault. Also assume that $\omega_{p.u.}(t) = 1.0$ in the swing equation.

12.5 A third generating unit rated 400 MVA, 15 kV, 0.90 power factor, 16 poles, with $H_3 = 3.0$ p.u.-s is added to the power plant in Example 12.2. Assuming all three units swing together, determine an equivalent swing equation for the three units.

Section 12.2

12.6 The synchronous generator in Figure 12.4 delivers 1.0 per-unit real power at 1.05 per-unit terminal voltage. Determine: (a) the reactive power output of the generator; (b) the generator internal voltage; and (c) an equation for the electrical power delivered by the generator versus power angle δ.

12.7 The generator in Figure 12.4 is initially operating in the steady-state condition given in Problem 12.6 when a three-phase-to-ground bolted short circuit occurs at bus 3. Determine an equation for the electrical power delivered by the generator versus power angle δ during the fault.

Section 12.3

12.8 The generator in Figure 12.4 is initially operating in the steady-state condition given in Example 12.3 when circuit breaker B12 inadvertently opens. Use the equal-area criterion to calculate the maximum value of the generator power angle δ. Assume $\omega_{\text{p.u.}}(t) = 1.0$ in the swing equation.

12.9 The generator in Figure 12.4 is initially operating in the steady-state condition given in Example 12.3 when a temporary three-phase-to-ground short circuit occurs at point F. Three cycles later, circuit breakers B13 and B22 permanently open to clear the fault. Use the equal-area criterion to determine the maximum value of the power angle δ.

12.10 If breakers B13 and B22 in Problem 12.9 open later than 3 cycles after the fault commences, determine the critical clearing time.

12.11 Rework Problem 12.9 if circuit breakers B13 and B22 open after 3 cycles and then reclose when the power angle reaches 35°. Assume that the temporary fault has already self-extinguished when the breakers reclose.

Section 12.4

12.12 Verify the maximum power angle determined in Problem 12.8 by applying the modified Euler's method to numerically integrate the swing equation. Write and run a computer program.

12.13 Investigate the effect of generating-unit damping torque on the maximum power angle in Problem 12.12. Damping, which is caused by friction and windage, can be represented by subtracting from the per-unit accelerating power $p_{a\text{p.u.}}(t)$ [used in (12.4.8) and (12.4.12)] the term $\text{B}\omega_{\text{p.u.}}(t)$, where B is a per-unit damping coefficient. Compare the maximum power angle using $\text{B} = 0.01$ per unit with that computed in Problem 12.12. Discuss the effect of generating-unit damping torques on stability.

12.14 Verify the critical clearing time determined in Problem 12.10 by applying the modified Euler's method. Write and run a computer program.

12.15 In Problem 12.11, assume that the circuit breakers open at $t = 3$ cycles and then reclose at $t = 24$ cycles (instead of when δ reaches 90°). Determine the maximum power angle by applying the modified Euler method. Write and run a computer program.

Section 12.5

12.16 Consider the six-bus power system shown in Figure 10b which appears in the user's manual that accompanies this text, where all data is given in per-unit on a common system base. All resistances as well as transmission-line capacitances are neglected. (a) Determine the 6×6 per-unit bus admittance matrix $\boldsymbol{Y}_{\text{bus}}$ suitable for a power-flow computer program. (b) Determine the per-unit admittance matrices $\boldsymbol{Y}_{11}$, $\boldsymbol{Y}_{12}$, and $\boldsymbol{Y}_{22}$ given in (12.5.5), which are suitable for a transient stability study.

12.17 Modify the matrices Y_{11}, Y_{12}, and Y_{22} determined in Problem 12.16 for (a) the case when circuit breakers B12 and B22 open to remove line 1–2; and (b) the case when the load $P_{L4} + jQ_{L4}$ is removed.

Sections 12.6 and 12.7

12.18 Run the program TRANSIENT STABILITY for Example 12.9. Verify that Case 1 is stable and Case 2 is unstable.

12.19 Investigate the effect of varying the transient reactance X'_{d1} of machine 1 in Example 12.9. Run the program TRANSIENT STABILITY for (a) $X'_{d1} = 0.1$ and (b) $X'_{d1} = 0.4$ per unit. Discuss the effect of X'_{d1} on stability.

12.20 Investigate the effect of varying the inertia constant H in Example 12.9. Run the program TRANSIENT STABILITY for (a) $H_1 = 3.0$ and (b) $H_1 = 8.0$ p.u.-s. Discuss the effect of H_1 on stability.

12.21 Investigate the effect of varying the fault-clearing time in Example 12.9. Run the program TRANSIENT STABILITY, with the circuit breakers opening at (a) $t = 0.025$ s (1.5 cycles) and (b) $t = 0.075$ s (4.5 cycles), instead of at $t = 0.05$ s. In both cases, assume that reclosure occurs 13 cycles after fault clearing. Discuss the effect of fault-clearing time on stability.

12.22 Rework Example 12.9 with an additional line installed between buses 4 and 5. The parameters of the added line equal those of existing line 4–5. First run the program POWER FLOW with the additional line to determine prefault conditions. Then, using the POWER FLOW output as prefault conditions, run the program TRANSIENT STABILITY for both cases given in Example 12.9. Discuss the effect of the additional line on stability.

12.23 Use the program TRANSIENT STABILITY to study a three-phase short circuit at bus 2 of the power system given in Figure 10 b . First run the program POWER FLOW with bus 1 as the swing bus, where $V_1 = 1.0$ per unit and $\delta_1 = 0°$, and with buses 2 and 6 as constant voltage magnitude buses, where $V_6 = V_2 = 1.0$ per unit, $P_{G2} = 2.0$ per unit, and $P_{G3} = 3.0$ per unit. Using the POWER FLOW output as prefault conditions, run the program TRANSIENT STABILITY. Assume that the short circuit at bus 2 is cleared by permanently opening breakers B21, B22, and B23. Determine the critical clearing time. Figure 10b appears in the software manual.

References

1. G. W. Stagg, and A. H. El-Abiad, *Computer Methods in Power Systems* (New York: McGraw-Hill, 1968).
2. O. I. Elgerd, *Electric Energy Systems Theory*, 2d ed. (New York: McGraw-Hill, 1982).
3. C. A. Gross, *Power System Analysis* (New York: John Wiley and Sons, 1979).
4. W. D. Stevenson, Jr., *Elements of Power System Analysis*, 4th ed. (New York: McGraw-Hill, 1982).
5. E. W. Kimbark, "Suppression of Ground-Fault Arcs on Single-Pole Switched EHV Lines by Shunt Reactors," *IEEE TRANS PAS*, Vol. 83 (March 1964): pp. 285–290.
6. K. R. McClymont, et. al., "Experience with High-Speed Rectifier Excitation Systems," *IEEE TRANS PAS*, Vol. PAS-87 (June 1986): pp. 1464–1470.

7. E. W. Cushing, et. al., "Fast Valving as an Aid to Power System Transient Stability and Prompt Resynchronization and Rapid Reload after Full Load Rejection," *IEEE TRANS PAS*, Vol. PAS-90 (November/December 1971): pp. 2517–2527.

8. M. L. Shelton, et. al., "Bonneville Power Administration 1400 MW Braking Resistor," *IEEE TRANS PAS*, Vol. PAS-94 (March/April 1975): pp. 602–611.

Appendix

CONSTANT (UNITS)	TYPE	SYMBOL	TURBO-GENERATOR (SOLID ROTOR)	WATER-WHEEL GENERATOR (WITH DAMPERS)	SYNCHRON-OUS CONDENSER	SYNCHRON-OUS MOTOR
Reactances (per unit)	Synchronous	X_d	1.1	1.15	1.80	1.20
		X_q	1.08	0.75	1.15	0.90
	Transient	X'_d	0.23	0.37	0.40	0.35
		X'_q	0.23	0.75	1.15	0.90
	Subtransient	X''_d	0.12	0.24	0.25	0.30
		X''_q	0.15	0.34	0.30	0.40
	Negative-sequence	X_2	0.13	0.29	0.27	0.35
	Zero-sequence	X_0	0.05	0.11	0.09	0.16
Resistances (per unit)	Positive-sequence	R (dc)	0.003	0.012	0.008	0.01
		R (ac)	0.005	0.012	0.008	0.01
	Negative-sequence	R_2	0.035	0.10	0.05	0.06
Time constants (seconds)	Transient	T'_{do}	5.6	5.6	9.0	6.0
		T'_d	1.1	1.8	2.0	1.4
	Subtransient	$T''_d = T''_q$	0.035	0.035	0.035	0.036
	Armature	T_a	0.16	0.15	0.17	0.15

Adapted from E. W. Kimbark, *Power-System Stability: Synchronous Machines* (New York: Dover Publications, 1956/1968), Chap. 12.

Table A.1 Typical average values of synchronous-machine constants

Table A.2

Typical transformer leakage reactances

RATING OF HIGHEST VOLTAGE WINDING	BIL OF HIGHEST VOLTAGE WINDING	LEAKAGE REACTANCE	
kV	kV	per unit*	
DISTRIBUTION TRANSFORMERS			
2.4	30	0.023–0.049	
4.8	60	0.023–0.049	
7.2	75	0.026–0.051	
12	95	0.026–0.051	
23	150	0.052–0.055	
34.5	200	0.052–0.055	
46	250	0.057–0.063	
69	350	0.065–0.067	
POWER TRANSFORMERS 10 MVA AND BELOW			
8.7	110	0.050–0.058	
25	150	0.055–0.058	
34.5	200	0.060–0.065	
46	250	0.060–0.070	
69	350	0.070–0.075	
92	450	0.070–0.085	
115	550	0.075–0.100	
138	650	0.080–0.105	
161	750	0.085–0.011	
POWER TRANSFORMERS ABOVE 10 MVA			
		SELF-COOLED OR FORCED-AIR-COOLED	FORCED-OIL-COOLED
8.7	110	0.050–0.063	0.082–0.105
34.5	200	0.055–0.075	0.090–0.128
46	250	0.057–0.085	0.095–0.143
69	350	0.063–0.095	0.103–0.158
92	450	0.060–0.118	0.105–0.180
115	550	0.065–0.135	0.107–0.195
138	650	0.070–0.140	0.117–0.245
161	750	0.075–0.150	0.125–0.250
230	900	0.070–0.160	0.120–0.270
345	1300	0.080–0.170	0.130–0.280
500	1550	0.100–0.200	0.160–0.340
765		0.110–0.210	0.190–0.350

*Per-unit reactances are based on the transformer rating

Table A.3 Characteristics of copper conductors, hard drawn, 97.3% conductivity

Size of Conductor		Number of Strands	Diameter of Individual Strands (Inches)	Outside Diameter (Inches)	Breaking Strength (Pounds)	Weight (Pounds per Mile)	Approx. Current Carrying Capacity* (Amps)	Geometric Mean Radius at 60 Hz (Feet)	r_a Resistance (Ohms per Conductor per Mile)								x_a Inductive Reactance (Ohms per Conductor Per Mile At 1 Ft Spacing)			x'_a Shunt Capacitive Reactance (Megohms per Conductor Per Mile At 1 Ft Spacing)		
									25°C (77°F)				50°C (122°F)									
Circular Mils	A.W.G. or B. & S.								dc	25 Hz	50 Hz	60 Hz	dc	25 Hz	50 Hz	60 Hz	25 Hz	50 Hz	60 Hz	25 Hz	50 Hz	60 Hz
1 000 000	...	37	0.1644	1.151	43 830	16 300	1 300	0.0368	0.0585	0.0594	0.0620	0.0634	0.0640	0.0648	0.0672	0.0685	0.1666	0.333	0.400	0.216	0.1081	0.0901
900 000	...	37	0.1560	1.092	39 510	14 670	1 220	0.0349	0.0650	0.0658	0.0682	0.0695	0.0711	0.0718	0.0740	0.0752	0.1693	0.339	0.406	0.220	0.1100	0.0916
800 000	...	37	0.1470	1.029	35 120	13 040	1 130	0.0329	0.0731	0.0739	0.0760	0.0772	0.0800	0.0806	0.0826	0.0837	0.1722	0.344	0.413	0.224	0.1121	0.0934
750 000	...	37	0.1424	0.997	33 400	12 230	1 090	0.0319	0.0780	0.0787	0.0807	0.0818	0.0853	0.0859	0.0878	0.0888	0.1739	0.348	0.417	0.226	0.1132	0.0943
700 000	...	37	0.1375	0.963	31 170	11 410	1 040	0.0308	0.0836	0.0842	0.0861	0.0871	0.0914	0.0920	0.0937	0.0947	0.1759	0.352	0.422	0.229	0.1145	0.0954
600 000	...	37	0.1273	0.891	27 020	9 781	940	0.0285	0.0975	0.0981	0.0997	0.1006	0.1066	0.1071	0.1086	0.1095	0.1799	0.360	0.432	0.235	0.1173	0.0977
500 000	...	37	0.1162	0.814	22 510	8 151	840	0.0260	0.1170	0.1175	0.1188	0.1196	0.1280	0.1283	0.1296	0.1303	0.1845	0.369	0.443	0.241	0.1205	0.1004
500 000	...	19	0.1622	0.811	21 590	8 151	840	0.0256	0.1170	0.1175	0.1188	0.1196	0.1280	0.1283	0.1296	0.1303	0.1853	0.371	0.445	0.241	0.1206	0.1005
450 000	...	19	0.1539	0.770	19 750	7 336	780	0.0243	0.1300	0.1304	0.1316	0.1323	0.1422	0.1426	0.1437	0.1443	0.1879	0.376	0.451	0.245	0.1224	0.1020
400 000	...	19	0.1451	0.726	17 560	6 521	730	0.0229	0.1462	0.1466	0.1477	0.1484	0.1600	0.1603	0.1613	0.1619	0.1909	0.382	0.458	0.249	0.1245	0.1038
350 000	...	19	0.1357	0.679	15 590	5 706	670	0.0214	0.1671	0.1675	0.1684	0.1690	0.1828	0.1831	0.1840	0.1845	0.1943	0.389	0.466	0.254	0.1269	0.1058
350 000	...	12	0.1708	0.710	15 140	5 706	670	0.0225	0.1671	0.1675	0.1684	0.1690	0.1828	0.1831	0.1840	0.1845	0.1918	0.384	0.460	0.251	0.1253	0.1044
300 000	...	19	0.1257	0.629	13 510	4 891	610	0.01987	0.1950	0.1953	0.1961	0.1966	0.213	0.214	0.214	0.215	0.1982	0.396	0.476	0.259	0.1296	0.1080
300 000	...	12	0.1581	0.657	13 170	4 891	610	0.0208	0.1950	0.1953	0.1961	0.1966	0.213	0.214	0.214	0.215	0.1957	0.392	0.470	0.256	0.1281	0.1068
250 000	...	19	0.1147	0.574	11 360	4 076	540	0.01813	0.234	0.234	0.235	0.235	0.256	0.256	0.257	0.257	0.203	0.406	0.487	0.266	0.1329	0.1108
250 000	...	12	0.1443	0.600	11 130	4 076	540	0.01902	0.234	0.234	0.235	0.235	0.256	0.256	0.257	0.257	0.200	0.401	0.481	0.263	0.1313	0.1094
211 600	4/0	19	0.1055	0.528	9 617	3 450	480	0.01668	0.276	0.277	0.277	0.278	0.302	0.303	0.303	0.303	0.207	0.414	0.497	0.272	0.1359	0.1132
211 600	4/0	12	0.1328	0.552	9 483	3 450	490	0.01750	0.276	0.277	0.277	0.278	0.302	0.303	0.303	0.303	0.205	0.409	0.491	0.269	0.1343	0.1119
211 600	4/0	7	0.1739	0.522	9 154	3 450	480	0.01579	0.276	0.277	0.277	0.278	0.302	0.303	0.303	0.303	0.210	0.420	0.503	0.273	0.1363	0.1136
167 800	3/0	12	0.1183	0.492	7 556	2 736	420	0.01559	0.349	0.349	0.349	0.350	0.381	0.381	0.382	0.382	0.210	0.421	0.505	0.277	0.1384	0.1153
167 800	3/0	7	0.1548	0.464	7 366	2 736	420	0.01404	0.349	0.349	0.349	0.350	0.381	0.381	0.382	0.382	0.216	0.431	0.518	0.281	0.1405	0.1171
133 100	2/0	7	0.1379	0.414	5 926	2 170	360	0.01252	0.440	0.440	0.440	0.440	0.481	0.481	0.481	0.481	0.222	0.443	0.532	0.289	0.1445	0.1205
105 500	1/0	7	0.1228	0.368	4 752	1 720	310	0.01113	0.555	0.555	0.555	0.555	0.606	0.607	0.607	0.607	0.227	0.455	0.546	0.298	0.1488	0.1240
83 690	1	7	0.1093	0.328	3 804	1 364	270	0.00992	0.699	0.699	0.699	0.699	0.765				0.233	0.467	0.560	0.306	0.1528	0.1274
83 690	1	3	0.1670	0.360	3 620	1 351	270	0.01016	0.692	0.692	0.692	0.692	0.757				0.232	0.464	0.557	0.299	0.1495	0.1246
66 370	2	7	0.0974	0.292	3 045	1 082	230	0.00883	0.881	0.882	0.882	0.882	0.964				0.239	0.478	0.574	0.314	0.1570	0.1308
66 370	2	3	0.1487	0.320	2 913	1 071	240	0.00903	0.873				0.955				0.238	0.476	0.571	0.307	0.1537	0.1281
66 370	2	1	....	0.258	3 003	1 061	220	0.00836	0.864				0.945				0.242	0.484	0.581	0.323	0.1614	0.1345
52 630	3	7	0.0867	0.260	2 433	858	200	0.00787	1.112				1.216				0.245	0.490	0.588	0.322	0.1611	0.1343
52 630	3	3	0.1325	0.285	2 359	850	200	0.00805	1.101		Same as dc		1.204		Same as dc		0.244	0.488	0.585	0.316	0.1578	0.1315
52 630	3	1	....	0.229	2 439	841	190	0.00745	1.090				1.192				0.248	0.496	0.595	0.331	0.1656	0.1380
41 740	4	3	0.1180	0.254	1 879	674	180	0.00717	1.388				1.518				0.250	0.499	0.599	0.324	0.1619	0.1349
41 740	4	1	....	0.204	1 970	667	170	0.00663	1.374				1.503				0.254	0.507	0.609	0.339	0.1697	0.1415
33 100	5	3	0.1050	0.226	1 505	534	150	0.00638	1.750				1.914				0.256	0.511	0.613	0.332	0.1661	0.1384
33 100	5	1	....	0.1819	1 591	529	140	0.00590	1.733				1.895				0.260	0.519	0.623	0.348	0.1738	0.1449
26 250	6	3	0.0935	0.201	1 205	424	130	0.00568	2.21				2.41				0.262	0.523	0.628	0.341	0.1703	0.1419
26 250	6	1	....	0.1620	1 280	420	120	0.00526	2.18				2.39				0.265	0.531	0.637	0.356	0.1779	0.1483
20 820	7	1	....	0.1443	1 030	333	110	0.00468	2.75				3.01				0.271	0.542	0.651	0.364	0.1821	0.1517
16 510	8	1	....	0.1285	826	264	90	0.00417	3.47				3.80				0.277	0.554	0.665	0.372	0.1862	0.1552

*For conductor at 75°C, air at 25°C, wind 1.4 miles per hour (2 ft/sec), frequency = 60 Hz.

Table A.4 Characteristics of aluminum cable, steel, reinforced (Aluminum Company of America)

Code Word	Circular Mils Aluminum	Aluminum		Aluminum Strand Diameter (Inches)	Steel	Steel Strand Diameter (Inches)	Outside Diameter (Inches)	Copper Equivalent* Circular Mils or A.W.G.	Ultimate Strength (Pounds)	Weight (Pounds per Mile)	Geometric Mean Radius at 60 Hz (Feet)	Approx. Current Carrying Capacity† (Amps)	r_a Resistance (Ohms per Conductor per Mile) 25°C (77°F) Small Currents				r_a Resistance 50°C (122°F) Current Approx. 75% Capacity‡				x_a Inductive Reactance (Ohms per Conductor per Mile at 1 Ft Spacing All Currents)	x'_a Shunt Capacitive Reactance (Megohms per Conductor per Mile at 1 Ft Spacing)
													dc	25 Hz	50 Hz	60 Hz	dc	25 Hz	50 Hz	60 Hz	60 Hz	60 Hz
Joree	2 515 000	76	...	0.1819	19	0.0849	1.880		61 700		0.0621									0.0450	0.337	0.0755
Thrasher	2 312 000	76	...	0.1744	19	0.0814	1.802		57 300		0.0595									0.0482	0.342	0.0767
Kiwi	2 167 000	72	4	0.1735	7	0.1157	1.735		49 800		0.0570									0.0511	0.348	0.0778
Bluebird	2 156 000	84	4	0.1602	19	0.0961	1.762		60 300		0.0588									0.0505	0.344	0.0774
Chukar	1 781 000	84	4	0.1456	19	0.0874	1.602		51 000		0.0534									0.0598	0.355	0.0802
Falcon	1 590 000	54	3	0.1716	19	0.1030	1.545	1 000 000	56 000	10 777	0.0520	1 380	0.0587	0.0588	0.0590	0.0591	0.0646	0.0656	0.0675	0.0684	0.359	0.0814
Parrot	1 510 500	54	3	0.1673	19	0.1004	1.506	950 000	53 200	10 237	0.0507	1 340	0.0618	0.0619	0.0621	0.0622	0.0680	0.0690	0.0710	0.0720	0.362	0.0821
Plover	1 431 000	54	3	0.1628	19	0.0977	1.465	900 000	50 400	9 699	0.0493	1 300	0.0652	0.0653	0.0655	0.0656	0.0718	0.0729	0.0749	0.0760	0.365	0.0830
Martin	1 351 000	54	3	0.1582	19	0.0949	1.424	850 000	47 600	9 160	0.0479	1 250	0.0691	0.0692	0.0694	0.0695	0.0761	0.0771	0.0792	0.0803	0.369	0.0833
Pheasant	1 272 000	54	3	0.1535	19	0.0921	1.382	800 000	44 800	8 621	0.0465	1 200	0.0734	0.0735	0.0737	0.0738	0.0808	0.0819	0.0840	0.0851	0.372	0.0847
Grackle	1 192 500	54	3	0.1486	19	0.0892	1.338	750 000	43 100	8 082	0.0450	1 160	0.0783	0.0784	0.0786	0.0788	0.0862	0.0872	0.0894	0.0906	0.376	0.0857
Finch	1 113 000	54	3	0.1436	19	0.0862	1.293	700 000	40 200	7 544	0.0435	1 110	0.0839	0.0840	0.0842	0.0844	0.0924	0.0935	0.0957	0.0969	0.380	0.0867
Curlew	1 033 500	54	3	0.1384	7	0.1384	1.246	650 000	37 100	7 019	0.0420	1 060	0.0903	0.0905	0.0907	0.0909	0.0994	0.1005	0.1025	0.1035	0.385	0.0878
Cardinal	954 000	54	3	0.1329	7	0.1329	1.196	600 000	34 200	6 479	0.0403	1 010	0.0979	0.0980	0.0981	0.0982	0.1078	0.1088	0.1118	0.1128	0.390	0.0890
Canary	900 000	54	3	0.1291	7	0.1291	1.162	566 000	32 300	6 112	0.0391	970	0.104	0.104	0.104	0.104	0.1145	0.1155	0.1175	0.1185	0.393	0.0898
Crane	874 500	54	3	0.1273	7	0.1273	1.146	550 000	31 400	5 940	0.0386	950	0.107	0.107	0.107	0.108	0.1178	0.1188	0.1218	0.1228	0.395	0.0903
Condor	795 000	54	3	0.1214	7	0.1214	1.093	500 000	28 500	5 399	0.0368	900	0.117	0.118	0.118	0.119	0.1288	0.1308	0.1358	0.1378	0.401	0.0917
Drake	795 000	26	2	0.1749	7	0.1360	1.108	500 000	31 200	5 770	0.0375	900	0.117	0.117	0.117	0.117	0.1288	0.1288	0.1288	0.1288	0.399	0.0912
Mallard	795 000	30	2	0.1628	19	0.0977	1.140	500 000	38 400	6 517	0.0393	910	0.117	0.117	0.117	0.117	0.1288	0.1288	0.1288	0.1288	0.393	0.0904
Crow	715 500	54	3	0.1151	7	0.1151	1.036	450 000	26 300	4 859	0.0349	830	0.131	0.131	0.131	0.132	0.1442	0.1452	0.1472	0.1482	0.407	0.0932
Starling	715 500	26	2	0.1659	7	0.1290	1.051	450 000	28 100	5 193	0.0355	840	0.131	0.131	0.131	0.131	0.1442	0.1442	0.1442	0.1442	0.405	0.0928
Redwing	715 500	30	2	0.1544	19	0.0926	1.081	450 000	34 600	5 865	0.0372	840	0.131	0.131	0.131	0.131	0.1442	0.1442	0.1442	0.1442	0.399	0.0920
Flamingo	666 600	54	3	0.1111	7	0.1111	1.000	419 000	24 500	4 527	0.0337	800	0.140	0.140	0.141	0.141	0.1541	0.1571	0.1591	0.1601	0.412	0.0943
Rook	636 000	54	3	0.1085	7	0.1085	0.977	400 000	23 600	4 319	0.0329	770	0.147	0.147	0.148	0.148	0.1618	0.1638	0.1678	0.1688	0.414	0.0950
Grosbeak	636 000	26	2	0.1564	7	0.1216	0.990	400 000	25 000	4 616	0.0335	780	0.147	0.147	0.147	0.147	0.1618	0.1618	0.1618	0.1618	0.412	0.0946
Egret	636 000	30	2	0.1456	19	0.0874	1.019	400 000	31 500	5 213	0.0351	780	0.147	0.147	0.147	0.147	0.1618	0.1618	0.1618	0.1618	0.406	0.0937
Peacock	605 000	54	3	0.1059	7	0.1059	0.953	380 500	22 500	4 109	0.0321	750	0.154	0.155	0.155	0.155	0.1695	0.1715	0.1755	0.1775	0.417	0.0957
Squab	605 000	26	2	0.1525	7	0.1186	0.966	380 500	24 100	4 391	0.0327	760	0.154	0.154	0.154	0.154	0.1700	0.1720	0.1720	0.1720	0.415	0.0953
Dove	556 500	26	2	0.1463	7	0.1138	0.927	350 000	22 400	4 039	0.0313	730	0.168	0.168	0.168	0.168	0.1849	0.1859	0.1859	0.1859	0.420	0.0965
Eagle	556 500	30	2	0.1362	7	0.1362	0.953	350 000	27 200	4 588	0.0328	730	0.168	0.168	0.168	0.168	0.1849	0.1859	0.1859	0.1859	0.415	0.0957
Hawk	477 000	26	2	0.1355	7	0.1054	0.858	300 000	19 430	3 462	0.0290	670	0.196	0.196	0.196	0.196	0.216				0.430	0.0988
Hen	477 000	30	2	0.1261	7	0.1261	0.883	300 000	23 300	3 933	0.0304	670	0.196	0.196	0.196	0.196	0.216				0.424	0.0980
Ibis	397 500	26	2	0.1236	7	0.0961	0.783	250 000	16 190	2 885	0.0265	590	0.235				0.259				0.441	0.1015
Lark	397 500	30	2	0.1151	7	0.1151	0.806	250 000	19 980	3 277	0.0278	600	0.235		Same as dc		0.259		Same as dc		0.435	0.1006
Linnet	336 400	26	2	0.1138	7	0.0855	0.721	4/0	14 050	2 442	0.0244	530	0.278				0.306				0.451	0.1039
Oriole	336 400	30	2	0.1059	7	0.1059	0.741	4/0	17 040	2 774	0.0255	530	0.278				0.306				0.445	0.1032
Ostrich	300 000	26	2	0.1074	7	0.0835	0.680	188 700	12 650	2 178	0.0230	490	0.311				0.342				0.458	0.1057
Piper	300 000	30	2	0.1000	7	0.1000	0.700	188 700	15 430	2 473	0.0241	500	0.311				0.342				0.462	0.1049
Partridge	266 800	26	2	0.1013	7	0.0788	0.642	3/0	11 250	1 936	0.0217	460	0.350				0.385				0.465	0.1074

*Based on copper 97%, aluminum 61% conductivity.

†For conductor at 75°C, air at 25°C, wind 1.4 miles per hour (2 ft/sec), frequency = 60 Hz.

‡"Current Approx. 75% Capacity" is 75% of the "Approx. Current Carrying Capacity in Amps" and is approximately the current which will produce 50°C conductor temp. (25°C rise) with 25°C air temp., wind 1.4 miles per hour.

CHAPTER OPENING PHOTO CAPTIONS AND CREDITS

Chapter One: 1300 MW coal-fired power plant. (Courtesy of American Electric Power.)

Chapter Two: Fossil fuel (oil/gas) power plant with two 850-MVA generating units. (Courtesy of Florida Power and Light Company.)

Chapter Three: Generator stator showing completed windings for a 757-MVA, 3600-RPM, 60-Hz synchronous generator. (Courtesy of General Electric.)

Chapter Four: Core and coil assemblies of a three-phase 20.3 kV $\triangle$/345 kVY step-up transformer. This oil-immersed transformer is rated 325 MVA self-cooled (OA)/542 MVA forced oil, forced air-cooled (FOA)/ 607 MVA forced oil, forced air-cooled (FOA). (Courtesy of General Electric.)

Chapter Five: 765 kV transmission line with aluminum guyed-V towers. (Courtesy of American Electric Power.)

Chapter Six: 345 kV substation with SF6 gas-insulated circuit-breakers. (Courtesy of New England Electric.)

Chapter Seven: Supervisory control and data acquisition (SCADA) center. (Courtesy of Boston Edison.)

Chapter Eight: Three-phase 500 kV puffer-type circuit breaker. (Courtesy of Florida Power and Light Company.)

Chapter Nine: 500/220 kV substation with SF6 gas-insulated switchgear and conventional oil-filled transformers. (Courtesy of Southern California Edison.)

Chapter Ten: New England Power Exchange (NEPEX) control room. (Courtesy of New England Electric.)

Chapter Eleven: Three-phase 1500-MVA, 525-kV/241-kV substation transformer bank. (Courtesy of Florida Power and Light Company.)

Chapter Twelve: 1300 MW generating unit consisting of a cross-compound steam turbine and two 722-MVA synchronous generators. (Courtesy of American Electric Power.)

INDEX